Redox Homeostasis and Oxidative Stress in Human Metabolism and Disease

Guest Editors

Alina Woźniak
Jaroslaw Nuszkiewicz

MDPI

Basel • Beijing • Wuhan • Barcelona • Belgrade • Novi Sad • Cluj • Manchester

Guest Editors

Alina Woźniak
Department of Medical
Biology and Biochemistry
Nicolaus Copernicus
University in Toruń
Bydgoszcz
Poland

Jaroslaw Nuszkiewicz
Department of Medical
Biology and Biochemistry
Nicolaus Copernicus
University in Toruń
Bydgoszcz
Poland

Editorial Office

MDPI AG
Grosspeteranlage 5
4052 Basel, Switzerland

This is a reprint of the Special Issue, published open access by the journal *International Journal of Molecular Sciences* (ISSN 1422-0067), freely accessible at: https://www.mdpi.com/journal/ijms/special_issues/0X70M0058H.

For citation purposes, cite each article independently as indicated on the article page online and as indicated below:

Lastname, A.A.; Lastname, B.B. Article Title. *Journal Name* **Year**, *Volume Number*, Page Range.

ISBN 978-3-7258-4921-5 (Hbk)
ISBN 978-3-7258-4922-2 (PDF)
https://doi.org/10.3390/books978-3-7258-4922-2

Contents

About the Editors

Alina Woźniak

Alina Woźniak is the head of the Department of Medical Biology and Biochemistry, Faculty of Medicine, Ludwik Rydygier Collegium Medicum in Bydgoszcz, Nicolaus Copernicus University in Toruń, Poland. She obtained her PhD in medical biology in 1996 and has been pursuing an academic career since 1989, progressing through the ranks of assistant, assistant professor, and associate professor. In 2014, she was awarded the title of Professor of Medical Sciences by the President of the Republic of Poland. Her research focuses on redox balance, oxidative stress, and biomarkers of inflammation in relation to human health and disease. A substantial part of her work investigates the impact of environmental stressors, such as whole-body cryostimulation, sauna exposure, hyperbaric oxygen therapy, and cold-water immersion, on oxidative stress markers and inflammatory responses. She has contributed extensively to advancing knowledge in molecular medicine and continues to play an active role in both research and academic leadership.

Jaroslaw Nuszkiewicz

Jaroslaw Nuszkiewicz is an Assistant Professor in the Department of Medical Biology and Biochemistry, Faculty of Medicine, Ludwik Rydygier Collegium Medicum in Bydgoszcz, Nicolaus Copernicus University in Toruń, Poland. He obtained his PhD in medical and health sciences in 2023. His research focuses on oxidative stress, redox imbalance, inflammation, and metabolic dysregulation, particularly in the context of cancer, neurodegenerative diseases, and cardiometabolic disorders. He also investigates the endocrine activity of adipose tissue, the role of adipokines, and the biological effects of radiation exposure. He has been involved in national and international interdisciplinary research projects that integrate molecular biochemistry with medical biology to uncover mechanisms linking oxidative processes with disease risk and progression. In addition to his research, he is actively engaged in academic teaching, mentoring, and participation in university committees and professional organizations related to laboratory medicine and diagnostics. His scientific contributions aim to bridge fundamental research with translational and diagnostic applications in human health.

Preface

The Reprint *Redox Homeostasis and Oxidative Stress in Human Metabolism and Disease* brings together a comprehensive body of research focused on the dualistic nature of redox processes in physiological and pathological states. Redox signaling and oxidative stress have emerged as central themes in biomedical science, with far-reaching implications for understanding cellular metabolism, aging, inflammation, neurodegeneration, cancer, fertility, and cardiovascular and metabolic diseases.

This Reprint includes both original research articles and critical reviews that explore molecular mechanisms of redox balance, the generation and detoxification of reactive oxygen species, and the consequences of oxidative damage at the cellular and systemic level. A particular emphasis is placed on mitochondrial dysfunction, antioxidant defense systems, redox-sensitive pathways, and the development of biomarkers and therapeutic interventions.

The aim of this Reprint is to provide readers with a multidisciplinary view of current advancements in redox biology, integrating experimental models and clinical insights. It is intended for a broad scientific audience including molecular biologists, biochemists, clinicians, pharmacologists, and graduate students engaged in medical and life sciences.

Our motivation in compiling this volume stems from the increasing relevance of redox imbalance in the pathophysiology of chronic diseases, and the need to foster dialogue between basic science and clinical application. We hope this Reprint will serve as a useful resource for researchers seeking to deepen their understanding of redox mechanisms and translate this knowledge into novel diagnostic and therapeutic approaches.

Alina Woźniak and Jaroslaw Nuszkiewicz
Guest Editors

International Journal of *Molecular Sciences*

MDPI

Article

Melatonin Improves Lipid Homeostasis, Mitochondrial Biogenesis, and Antioxidant Defenses in the Liver of Prediabetic Rats

Milena Cremer de Souza [1,2], Maria Luisa Gonçalves Agneis [2], Karoliny Alves das Neves [2], Matheus Ribas de Almeida [2], Geórgia da Silva Feltran [2], Ellen Mayara Souza Cruz [1], João Paulo Ferreira Schoffen [3], Luiz Gustavo de Almeida Chuffa [2] and Fábio Rodrigues Ferreira Seiva [2,*]

1 Department of Parasitology, Immunology and General Pathology, State University of Londrina (UEL), Londrina 86057-970, Paraná, Brazil
2 Department of Chemistry and Biochemistry, São Paulo State University (UNESP), Botucatu 18618-693, São Paulo, Brazil
3 Center of Biological Sciences, State University of Northern Paraná (UENP), Bandeirantes 86360-000, Paraná, Brazil
* Correspondence: fabio.seiva@unesp.br; Tel.: +55-14-3880-0601

Abstract: Type 2 diabetes mellitus represents a major global health burden and is often preceded by a prediabetic state characterized by insulin resistance and metabolic dysfunction. Mitochondrial alterations, oxidative stress, and disturbances in lipid metabolism are central to the prediabetes pathophysiology. Melatonin, a pleiotropic indolamine, is known to regulate metabolic and mitochondrial processes; however, its therapeutic potential in prediabetes remains poorly understood. This study investigated the effects of melatonin on energy metabolism, oxidative stress, and mitochondrial function in a rat model of prediabetes induced by chronic sucrose intake and low-dose streptozotocin administration. Following prediabetes induction, animals were treated with melatonin (20 mg/kg) for four weeks. Biochemical analyses were conducted to evaluate glucose and lipid metabolism, and mitochondrial function was assessed via gene expression, enzymatic activity, and oxidative stress markers. Additionally, hepatic mitochondrial dynamics were examined by quantifying key regulators genes associated with biogenesis, fusion, and fission. Prediabetic animals exhibited dyslipidemia, hepatic lipid accumulation, increased fat depots, and impaired glucose metabolism. Melatonin significantly reduced serum glucose, triglycerides, and total cholesterol levels, while enhancing the hepatic high-density lipoprotein content. It also stimulated β-oxidation by upregulating hydroxyacyl-CoA dehydrogenase and citrate synthase activity. Mitochondrial dysfunction in prediabetic animals was evidenced by the reduced expression of peroxisome proliferator-activated receptor gamma coactivator-1 alpha and mitochondrial transcription factor A, both of which were markedly upregulated by melatonin. The indolamine also modulated mithocondrial dynamics by regulating fusion and fission markers, including mitosuin 1 and 2, optic atrophy protein, and dynamin-related protein. Additionally, melatonin mitigated oxidative stress by enhancing the activity of superoxide dismutase and catalase while reducing lipid peroxidation. These findings highlight melatonin's protective role in prediabetes by improving lipid and energy metabolism, alleviating oxidative stress, and restoring mitochondrial homeostasis. This study provides novel insights into the therapeutic potential of melatonin in addressing metabolic disorders, particularly in mitigating mitochondrial dysfunction associated with prediabetes.

Keywords: melatonin; prediabetes; mitochondrial dynamics; energy metabolism; oxidative stress

1. Introduction

Diabetes mellitus (DM) is a chronic non-communicable disease (CNCD) characterized by persistent hyperglycemia and represents one of the most significant global health challenges of the 21st century. Epidemiological data indicate that approximately 537 million adults aged 20–79 years were living with DM in 2021, with an estimated 6.7 million deaths attributed to the disease [1,2]. Among its subtypes, type 2 diabetes mellitus (T2DM) accounts for approximately 90% of all cases and is primarily associated with disturbances in carbohydrate and lipid metabolism. These metabolic disruptions result in insulin resistance (IR) and impaired insulin secretion due to pancreatic β-cell dysfunction [3].

The progression of T2DM is often preceded by an intermediate stage known as prediabetes, characterized by elevated blood glucose levels that remain below the diagnostic threshold for diabetes. According to the World Health Organization [4] and the American Diabetes Association [5], prediabetes is defined by fasting blood glucose levels between 6.1–6.9 mmol/L (110–125 mg/dL), impaired glucose tolerance in the two-hour oral glucose tolerance test (OGTT) with values between 7.8 and 11 mol/L (140–199 mg/dL), or glycated hemoglobin (HbA1c) levels ranging from 5.7 to 6.4%.

The transition from prediabetes to T2DM is strongly influenced by lifestyle factors, particularly dietary habits. The excessive consumption of sugar-sweetened beverages, often associated with to obesity, is a major contributor. Studies indicate that the daily intake of 250 mL of these beverages can result in an annual weight gain of 0.12 kg in adults and a body mass index (BMI) increase of 0.05 kg/m^2 in children [6]. Over the past four decades, obesity—driven by an imbalance between energy intake and expenditure—has reached epidemic proportions worldwide [7].

Altered energy metabolism is a hallmark of both prediabetes and T2DM. In prediabetes, glucose intolerance primarily stems from IR in skeletal muscle and adipose tissue, leading to reduced glucose uptake in the postprandial state. Elevated fasting glucose levels are often associated with hepatic IR and increased gluconeogenesis [8]. As IR progresses, the dysregulation of glucose and fatty acid metabolism becomes more pronounced, characterized by impaired glucose uptake, enhanced glycolysis, and glycogenolysis, increased gluconeogenesis, and sustained lipogenesis—all of which are closely linked to alterations in key enzymatic pathways [9].

Mitochondria, the central organelles in cellular metabolism, play essential roles in oxidative phosphorylation, fatty acid metabolism, and the generation of reactive oxygen species (ROS). Mitochondrial function is regulated by three fundamental processes: biogenesis, dynamics, and bioenergetics. Mitochondrial biogenesis involves the coordinated synthesis of new mitochondria, including mitochondrial DNA (mtDNA) replication and the expression of nuclear-encoded proteins critical to mitochondrial function [10–12]. Mitochondrial dynamics refers to the continuous processes of fusion and fission that maintain organelle integrity, distribution, and adaptability. Mitochondrial bioenergetics relies on the electron transport chain (ETC) to generate ATP via oxidative phosphorylation through the transfer of electrons across complexes embedded in the inner mitochondrial membrane [10–12].

Mitochondrial dysfunction is a key driver of T2DM pathogenesis, contributing to energy imbalance by reducing ATP production, increasing ROS accumulation, and impairing ETC function [13]. Abnormalities in mitochondrial biogenesis and dynamics are strongly linked to the progression of IR [14]. Impaired mitochondrial function promotes metabolic inflexibility, reducing the cellular ability to switch between glucose and fatty acid substrates. Excessive mitochondrial fission and reduced fusion, commonly observed in diabetic conditions, decrease ATP generation and shift cells toward glycolysis. This metabolic shift leads to the accumulation of intermediates that disrupt insulin signaling pathways. Moreover, increased ROS generation further aggravates IR by inducing oxidative damage and promoting inflammation [15–17].

Melatonin (N-acetyl-5-methoxytryptamine) is a lipophilic indolamine synthesized from tryptophan, primarily by the pineal gland, though also produced in peripheral tissues [18]. While melatonin is well known for its role in regulating circadian rhythm, it exhibits pleiotropic biological effects, including anti-carcinogenic, immunomodulatory, and antioxidant properties, depending on the physiological context. Emerging evidence suggests that melatonin plays a significant role in metabolic modulation, influencing multiple pathways involved with the metabolic syndrome and DM [19–22]. In T2DM, melatonin modulates the hepatic energy metabolism by enhancing glycogenesis and suppressing gluconeogenesis, contributing to glycemic control. It also influences lipid metabolism by reducing total cholesterol, triglycerides, and LDL levels while elevating HDL concentrations [23–28]. Additionally, melatonin's antioxidant properties counteract oxidative stress by scavenging ROS and enhancing the activity of endogenous antioxidant enzymes, thus protecting mitochondrial function in IR and T2DM [29–31]. Melatonin also regulates mitochondrial dynamics by inhibiting excessive fission and promoting fusion, stabilizing the ETC, and enhancing ATP production—mechanisms that support cellular energy homeostasis and mitigate metabolic disturbances linked to IR and T2DM [21].

Despite growing evidence supporting the beneficial effects of melatonin in DM, its role in the metabolic alterations associated with IR and prediabetes remains poorly understood. Further research is needed to elucidate its impact on energy metabolism, lipid homeostasis, OS, and mitochondria-associated events in the context of prediabetes. To this end, we investigated the effects of melatonin treatment in a Wistar rat model of prediabetes induced by a high-sugar diet combined with low-dose streptozotocin (STZ). This study aims to provide novel insights into the therapeutic potential of melatonin in alleviating metabolic dysfunction and mitochondrial impairments associated with prediabetes.

2. Results

2.1. Sucrose Solution Consumption Alters Daily Feed Intake in Prediabetic Wistar Rats

Weight gain remained comparable across all experimental groups throughout the study, with no significant differences observed in body mass index (BMI), Lee index, abdominal circumference, or body length. However, daily food intake (g/day) was significantly reduced in the pre-diabetes (PD) group compared to the control (C) group ($p < 0.0001$). Consequently, energy intake from chow was 43.96% lower in the PD group relative to control ($p < 0.0001$), but 16.63% higher in the PD-Mel group compared to PD ($p = 0.0437$). Energy intake from liquid consumption was elevated in both PD and PD-Mel groups, as the C group did not receive sucrose solution. Overall, total energy intake (kcal/day) in the PD group was 73.25% higher than in the control group ($p < 0.0001$), and melatonin treatment did not significantly alter this parameter. No differences in daily liquid intake were observed between groups (Table 1).

Table 1. Morphometric and nutritional parameters of rats from control, prediabetic, and prediabetic treated with melatonin groups.

Parameters	Groups		
	C	PD	PD-Mel
Body weight gain (g)	212.3 ± 18.27	244.4 ± 20.39	197.7 ± 76.36
BMI (g/cm^2)	0.63 ± 0.02	0.63 ± 0.01	0.55 ± 0.04
Lee index (g/cm)	11.71 ± 0.22	11.85 ± 0.37	11.08 ± 0.76
Abdominal circumference (cm)	16.1 ± 1.37	16.00 ± 0.81	15.22 ± 1.3
Body length (cm)	24.5 ± 0.53	24.38 ± 0.74	25.13 ± 1.8
Food consumption (g/day)	27.87 ± 0.47	15.51 ± 0.5 [a]	16.85 ± 1.18
Liquid consumption (mL/day)	52.7 ± 1.47	50.14 ± 2.4	65 ± 7.57
Energy intake from food (kcal/day)	83.98 ± 1.45	47.06 ± 2.27 [a]	50.89 ± 2.55
Energy intake from liquid (kcal/day)	0 ± 0.00	87.53 ± 4.23 [a]	95.6 ± 3.77
Total energy intake (kcal/day)	83.98 ± 1.45	145.5 ± 8.73 [a]	149.1 ± 6.64

Data expressed as mean ± SEM and analyzed by one-way ANOVA followed by Tukey's test for multiple comparisons. [a] Differs statistically from C group. C—control group (n = 10); PD—prediabetic group (n = 8); PD-Mel—prediabetic-melatonin-treated group (n = 9).

2.2. Melatonin Reduces Serum Glucose but Does Not Improve Glucose Intolerance

Fasting serum glucose was measured via cardiac puncture, while glucose tolerance was evaluated using glucose tolerance test (GTT), insulin tolerance test (ITT), and pyruvate tolerance test (PTT), via tail puncture. The combination of sucrose intake and STZ injection led to a 25.51% increase in serum glucose in the PD group compared to control ($p = 0.0219$). The melatonin treatment reduced serum glucose by 22.33% in the PD-Mel group ($p = 0.0240$) (Figure 1A). During the GTT, baseline glucose levels were similar across all groups. However, in PD groups, blood glucose remained elevated at 30 ($p = 0.0052$) and 60 min ($p = 0.004$). In the PD-Mel group, glucose levels increased significantly at 60 min ($p < 0.0001$) and remained high until 90 min ($p = 0.0048$) (Figure 1B). The area under the curve (AUC) analysis confirmed higher glucose intolerance in the PD group vs. control ($p < 0.0001$) and in the PD-Mel vs. PD ($p = 0.0168$) (Figure 1E). No significant differences were observed in ITT or PTT among groups (Figure 1C,D). The TyG index, an insulin resistance marker, was 6.03% higher in the PD group vs. control ($p = 0.002$), and melatonin reduced this index by 4.57% in the PD-Mel group ($p = 0.0067$) (Figure 1F).

2.3. Melatonin Reduces Fat Deposits and Influences Serum Lipid Profiles

Total fat content in the PD group was 109.79% higher than in the Control ($p = 0.0002$). Melatonin treatment the reduced fat content by 61.07% in the PD-Mel group ($p < 0.0001$), restoring levels to those of Control (Table 2). Retroperitoneal and epididymal fat increased by 191% ($p < 0.0001$) and 75.06% ($p = 0.0068$) in the PD group, respectively, and were both reduced by ~60% with melatonin ($p < 0.0001$).

The PD group exhibited dyslipidemia, with elevated triglyceride (TG) and total cholesterol (TC) levels compared to Control ($p = 0.0008$ and $p = 0.0002$, respectively). Melatonin normalized TG levels and reduced TC by 12.8% ($p = 0.0065$). (Table 2). Aspartate transaminase (AST) levels were approximately 30% lower in PD-Mel compared to PD group ($p = 0.0226$), with no changes in alanine transaminase (ALT) or lipoprotein levels.

Hepatic TG increased by 18.63% in PD vs. Control ($p = 0.0049$), and melatonin did not affect this parameter. Hepatic HDL was 4.95% lower in PD vs. Control ($p < 0.0001$), with no changes in hepatic TC or LDL levels (Table 2).

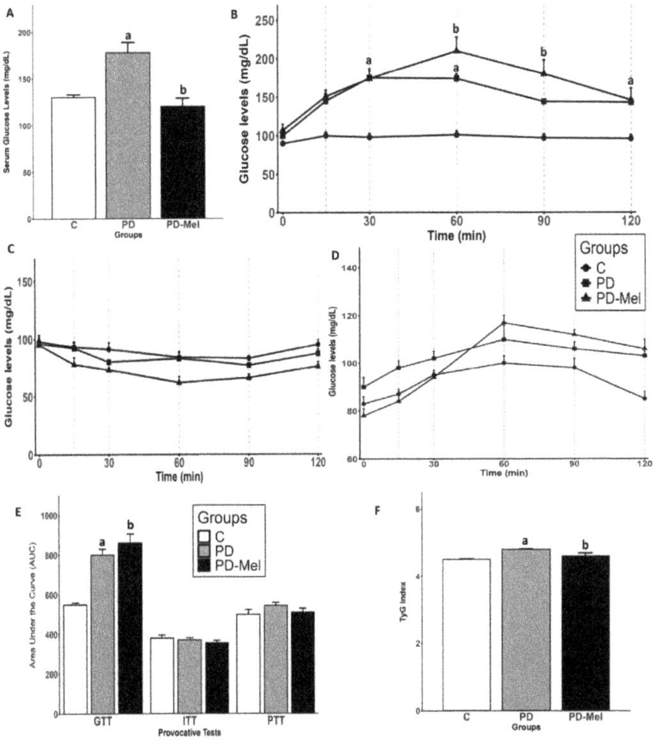

Figure 1. Glycemic parameters of rats from control, prediabetic, and prediabetic treated with melatonin groups. (**A**) Fasting serum glucose levels; (**B**) Intraperitoneal glucose tolerance test (GTT); (**C**) Insulin tolerance test (ITT); (**D**) Pyruvate tolerance test (PTT); (**E**) Area under curve (AUC) of previous provocative tests; and (**F**) TyG Index. Data are expressed as mean ± SEM and analyzed by one-way ANOVA followed by Tukey's test for multiple comparisons with p value ≤ 0.05. Letters a: Differs statistically from C group; b: Differs statistically from PD group. C—control group (n = 10); PD—prediabetic group (n = 8); PD-Mel—prediabetic-melatonin-treated group (n = 9).

Table 2. Fat deposits, serum and liver biochemical parameters of rats from control, prediabetic, and prediabetic treated with melatonin groups.

Parameters	Groups		
	C	**PD**	**PD-Mel**
Epididymal fat deposit (mg)	3.28 ± 0.22	5.75 ± 0.38 [a]	2.08 ± 0.49 [b]
Visceral fat deposit (mg)	2.33 ± 0.18	4.14 ± 0.34	2.31 ± 0.57
Retroperitoneal fat deposit (mg)	2.24 ± 0.31	6.54 ± 0.57 [a]	2.48 ± 0.81 [b]
Sum of fat depots (mg)	7.84 ± 1.85	16.45 ± 3 [a]	6.4 ± 4.89 [b]
Serum triglycerides levels (mg/dL)	101 ± 9.50	143 ± 29.31 [a]	118 ± 34.11
Serum cholesterol levels (mg/dL)	105 ± 2.8	128 ± 4.51 [a]	112 ± 2.41 [b]
Serum HDL levels (mg/dL)	55 ± 2.81	61 ± 3.49	50 ± 5.51
Serum LDL levels (mg/dL)	28 ± 4.17	36 ± 3.35	27 ± 3.00
Serum ALT levels (U/L)	29 ± 2.07	20 ± 2.44	24 ± 4.35
Serum AST levels (U/L	43 ± 2.42	44 ± 2.74	30 ± 4.81 [b]
Liver triglycerides levels (mg/dL)	87 ± 2.28	103 ± 4.02 [a]	100 ± 3.10

Table 2. *Cont.*

Parameters	Groups		
	C	PD	PD-Mel
Liver cholesterol levels (mg/dL)	94 ± 0.84	97 ± 1.17	95 ± 0.76
Liver HDL levels (mg/dL)	21 ± 0.17	20 ± 0.12 [a]	20 ± 0.11
Liver LDL levels (mg/dL)	57 ± 7.43	55 ± 5.34	53 ± 2.22

Data expressed as mean ± SEM and analyzed by one-way ANOVA followed by Tukey's test for multiple comparisons with p value ≤ 0.05. Letters [a]: Differs statistically from C group; [b]: Differs statistically from PD group. C—control group (n = 10); PD—prediabetic group (n = 8); PD-Mel—prediabetic-melatonin-treated group (n = 9).

2.4. Melatonin Modulates the Expression of Genes Related to Energy Metabolism

Gene expression analysis revealed significant changes in hepatic energy metabolism and ETC components (Figure 2). The ATP synthase subunit 6 (*Atp6*) was 139.02% higher in PD vs. Control (p = 0.0003), and 30.39% higher in PD-Mel vs. PD (p = 0.0408). Cytochrome c oxidase 1 (*Cox1;* ETC—complex IV) was reduced by 57.41% in the PD group (p = 0.0077), but increased by 607% in PD-Mel compared to PD (p < 0.0001). Citrate synthase (*Cs*) was downregulated in PD-Mel compared to PD (~39%, p < 0.0001). The expression of cytochrome b (*Cytb*), a subunit of ETC-complex III, was 73.84% higher in PD-Mel vs. PD (p < 0.0001).

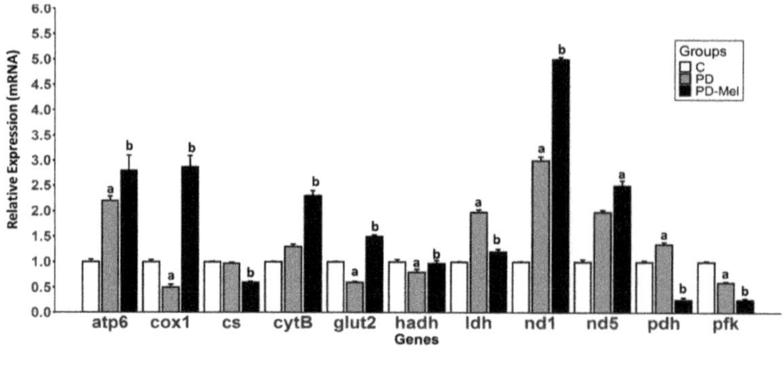

Figure 2. Gene expression panel of key enzymes related to hepatic energy metabolism of rats from Control, prediabetic, and prediabetic treated with melatonin groups. Differences were calculated using the ΔΔCt method, with β-actin used as the housekeeping gene. For comparative analysis, the control group was set as a calibrator (value = 1), and the relative expression levels in the other groups were calculated accordingly. Data expressed as mean ± SEM and analyzed by one-way ANOVA followed by Tukey's test for multiple comparisons with p value ≤ 0.05. Letters a: Differs statistically from C group; b: Differs statistically from PD group. C—Control group. *Atp6*: ATP synthase subunit 6; *cox1*: cytochrome c oxidase 1; *cs*: citrate synthase, cytB: cytochrome b; *glut2*: glucose transporter type 2; *hadh*: hydroxyacyl-coA dehydrogenase; *ldh*: lactate dehydrogenase; nd1: NADH dehydrogenase subunit 1; *nd5*: NADH dehydrogenase subunit 5; *pdh*: pyruvate dehydrogenase subunit; *pfk*: phosphofrutokinase-1; C—control group (n = 10); PD—prediabetic group (n = 8); PD-Mel—prediabetic-melatonin-treated group (n = 9).

In the PD group, glucose transporter type 2 (*Glut2*) expression was 45.1% lower in PD (p < 0.0001) and increased by 187.27% with melatonin (p < 0.0001). (Figure 2). Hydroxyacyl-CoA dehydrogenase (*Hadh*) expression was 24.15% higher in PD-Mel vs. PD (p = 0.0031), and the expression of lactate dehydrogenase (*Ldha*) increased by 99.7% in PD (p < 0.0001) and decreased by 45.98% after melatonin treatment (p < 0.0001).

Both the prediabetic state and melatonin treatment altered the mRNA expression of the components of the respiratory complex 1 (Figure 2). Expression of NADH dehydrogenase subunit 1 (*Nd1*) was 180% higher in PD vs. Control (p = 0.0462), and 110% higher in PD-Mel vs. PD (p = 0.0011). Pyruvate dehydrogenase subunit 1 (*Pdh1*) was 33.39% higher in PD group ($p < 0.0001$), but decreased by 80% with melatonin ($p < 0.0001$). Expression of phosphofrutokinase-1 liver-type (*Pfkl*) was lower in PD compared to Control ($p < 0.0001$), and lower in the PD-Mel compared to PD ($p < 0.0001$).

2.5. Melatonin Modulates the Activity of Enzymes Related to Energy Metabolism

Activity analysis revealed significant alterations in key metabolic enzymes in prediabetic rats (Figure 3). Lactate dehydrogenase (LDH) activity was 62.67% higher in PD vs. control (p = 0.0380) (Figure 3C). β-hydroxyacyl-coA dehydrogenase (β-OHADH) activity increased by 79.45% in PD-Mel vs. PD (p = 0.0004). Citrate synthase (CS) activity was 22.21% higher in PD-Mel vs. Control (p = 0.0137), and 41.15% vs. PD (p = 0.0002) (Figure 3D,E). Both PD and PD-Mel groups exhibited an ~18% reduction in the pyruvate dehydrogenase complex (PiDH) activity ($p < 0.0001$ for both) (Figure 3B).

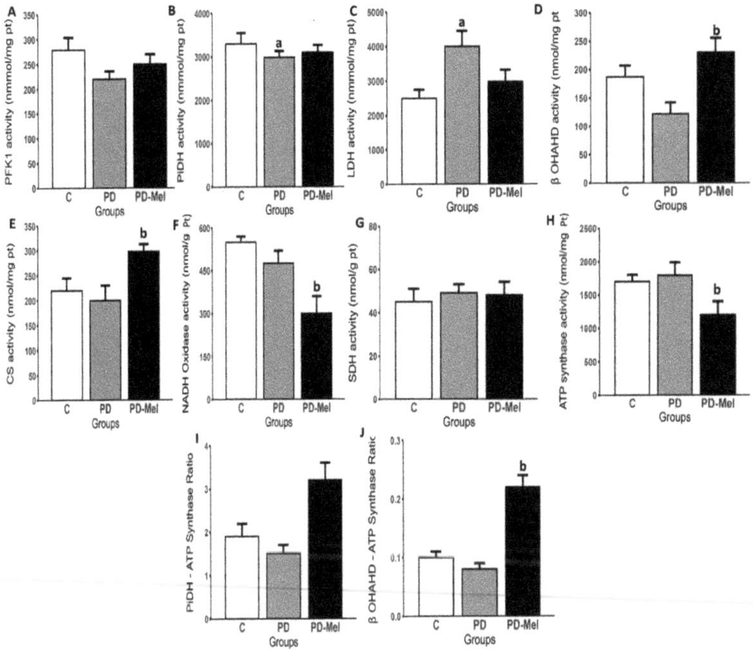

Figure 3. Activity of key enzymes of the hepatic energy metabolism of rats from Control, prediabetic, and prediabetic treated with melatonin groups. (**A**) PFK1 (Phosphofructokinase1); (**B**) PiDH (pyruvate dehydrogenase) complex; (**C**) LDH (lactate dehydrogenase); (**D**) β-OHADH (β-hydroxyacyl-coA dehydrogenase;) (**E**) CS (citrate synthase); (**F**) NADH oxidase; (**G**) SDH (succinate dehydrogenase); (**H**) ATP synthase; (**I**) Ratio between PiDH/ATP synthase activities; (**J**) Ratio between β-OHADH/ATP synthase activities. Data are expressed as the mean ± SEM and analyzed by one-way ANOVA followed by Tukey's test for multiple comparisons with p value ≤ 0.05. Letters a: Differs statistically from group C; b: Differs statistically from PD group. C—control group (n = 10); PD—prediabetic group (n = 8); PD-Mel—prediabetic-melatonin-treated group (n = 9).

NADH oxidase activity was 36.44% lower in PD-Mel vs. PD ($p = 0.0357$), and ATP synthase activity decreased by ~35% ($p = 0.0002$) (Figure 3F,H). No significant changes were observed in phosphofructokinase (PFK1) or succinate dehydrogenase (SDH) activity (Figure 3A,G). To further explore the functional implications of these enzymatic changes, the ratios between PiDH/ATP synthase and β-OHADH/ATP synthase activities were analyzed. The β-OHADH/ATP synthase ratio was significantly higher in the PD-Mel group compared to the PD group ($p < 0.0001$) (Figure 3I,J).

2.6. Melatonin Alters the Expression of Mitochondrial Biogenesis and Dynamics Markers

Cyclophilin D (*Cypd*) expression was 31% lower in PD ($p = 0.0023$), and increased by 192% in PD-Mel ($p < 0.0001$) (Figure 4). Dynamin-related protein 1 (*Drp1*), a marker of mitochondrial fission, rose by ~40% in PD vs. Control ($p < 0.0001$). Mitofusin 1 (*Mfn1*) was 49% lower in PD vs. control ($p < 0.0001$), and increased by 209% with melatonin ($p <$ 0.0001). The *Mfn2* expression also increased by 200% in PD-Mel ($p < 0.0001$).

Optic atrophy protein 1 (*Opa1*), a marker of mitochondrial fusion, had its expression increased by 267% in PD ($p < 0.0001$). Peroxisome proliferator-activated receptor gamma coactivator-1 alpha (*Pgc1-α*) was 57% lower in PD ($p = 0.004$) and increased by 502% in PD-Mel ($p < 0.0001$). Similarly, the mitochondrial transcription factor A (*Tfam*) expression was 73% lower in PD ($p = 0.0024$), and increased by 827% in PD-Mel ($p < 0.0001$) (Figure 4).

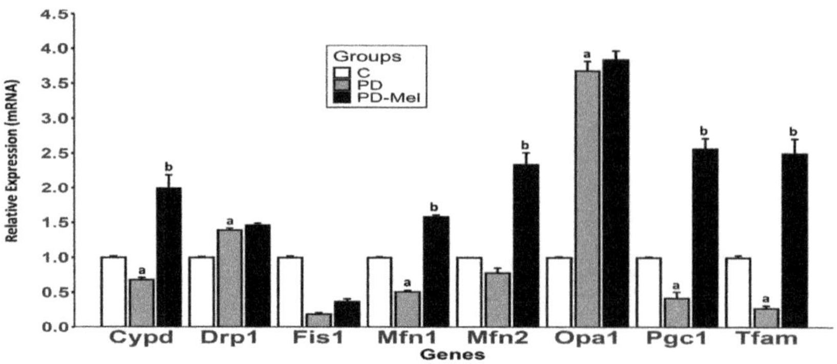

Figure 4. Gene expression panel of proteins related to hepatic mitochondrial events of rats from Control, prediabetic, and prediabetic treated with melatonin groups. Gene expression differences were calculated using the ΔΔCt method, with β-actin used as the housekeeping gene. For comparative analysis, the control group was set as a calibrator (value = 1), and the relative expression levels in the other groups were calculated accordingly. Data expressed as mean ± SEM and analyzed by one-way ANOVA followed by Tukey's test with p value ≤ 0.05. Letters a: Differs statistically from C group; b: Differs statistically from PD group. Cypd: cyclophilin D; Drp1: dynamin-related protein; Fis1: mitochondrial fission protein 1; Mfn1: mitofusin-1; Mfn2: mitofusin-2; Opa: optic atrophy 1; Pgc1 peroxisome proliferator-activated receptor gamma coactivator; Tfam: mitochondrial transcription factor A. C—control group (n = 10); PD—prediabetic group (n = 8); PD-Mel—prediabetic-melatonin-treated group (n = 9).

2.7. Melatonin Reduces Hepatic Oxidative Stress

Malondialdehyde (MDA) levels increased by 135.44% in PD ($p = 0.0002$), and were reduced by 22.64% in PD-Mel ($p = 0.0275$) (Figure 5A). Superoxide dismutase (SOD) activity was reduced by 11% in PD ($p < 0.0001$) and increased by 16.42% in PD-Mel ($p = 0.0018$) (Figure 5B). Similarly, catalase (CAT) activity decreased by 29.39% in PD ($p = 0.0246$) and increased by 65.96% in PD-Mel ($p = 0.0006$) (Figure 5C).

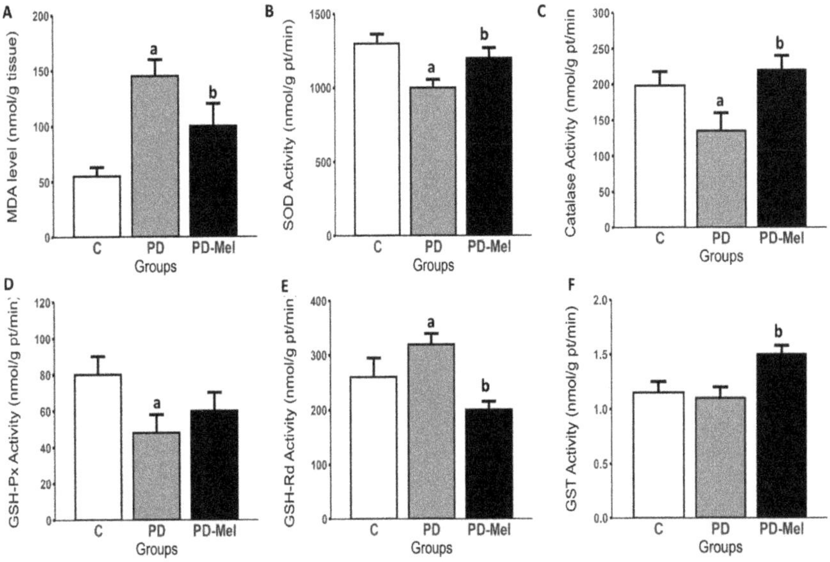

Figure 5. Hepatic oxidative stress parameters of rats from Control, prediabetic, and prediabetic treated with melatonin groups. (**A**) Malondialdehyde levels (MDA), and activity of (**B**) Superoxide dismutase (SOD); (**C**) Catalase; (**D**) Glutathione peroxidase (GSH-PX); (**E**) Glutathione reductase (GSH-Rd); (**F**) Glutathione transferase (GST). Data are expressed as mean ± SEM and analyzed by one-way ANOVA followed by Tukey's test for multiple comparisons with p value ≤ 0.05. Letters a: Differs statistically from group C; b: Differs statistically from PD group. (C) control group (n = 10); PD—prediabetic group (n = 8); PD-Mel—prediabetic-melatonin-treated group (n = 9).

Glutathione S-transferase (GST) activity increased significantly in PD-Mel ($p = 0.0017$) (Figure 5F). Glutathione reductase (GSH-Rd) activity rose by 25.10% in PD ($p = 0.0001$), and decreased by 35.67% in PD-Mel ($p < 0.0001$) (Figure 5E). Glutathione peroxidase (GSH-Px) activity decreased by 39.68% in PD ($p = 0.0183$) (Figure 5D). Reduced glutathione (GSH) levels dropped by ~37% in PD ($p < 0.0001$), and oxidized glutathione (GSSG) levels were 33.33% lower in PD vs. PD-Mel. OSI was elevated in PD and reduced with melatonin (Table 3).

Table 3. Liver glutathione pool and OSI of rats from the Control, prediabetic and prediabetic treated with melatonin groups.

Parameters	Groups		
	C	PD	PD-Mel
Total Glutathione (nmol/mg tissue)	2.97 ± 0.10	2.21 ± 0.15	1.93 ± 0.29
GSH (nmol/mg tissue)	2.11 ± 0.08	1.31 ± 0.07 [a]	1.33 ± 0.15
GSSG (nmol/mg tissue)	0.52 ± 0.04	0.45 ± 0.02	0.3 ± 0.04 [b]
OSI (%)	10.55 ± 0.72	14.47 ± 1.75 [a]	10.34 ± 1.56 [b]

Data expressed as mean ± SEM and analyzed by one-way ANOVA followed by Tukey's test for multiple comparisons. Letters [a]: Differs statistically from C group; [b]: Differs statistically from PD group. GSH—reduced glutathione; GSSG—oxidized glutathione; OSI—oxidative stress index. C—control group (n = 10); PD—prediabetic group (n = 8); PD-Mel—prediabetic-melatonin-treated group (n = 9).

2.8. Prediabetic Animals Exhibit Increased Collagen Deposition and Enlarged Hepatocyte Diameter

Hepatocyte diameter 122.97% higher in PD vs. C group (p = 0.0009). Melatonin treatment partially reversed this effect, reducing hepatocyte diameter by 22.82% (p = 0.0001). Collagen content increased by 351.61% in PD vs. control. Although melatonin reduced collagen by 22.85%, the difference was not statistically significant. Hepatocyte counts were unchanged among groups (Table 4).

Table 4. Histopathological parameters of rats from the Control, prediabetic, and prediabetic treated with melatonin groups.

Parameters	Groups		
	C	PD	PD-Mel
Hepatocyte number	237.3 ± 6.86	221.6 ± 8.99	232.6 ± 10.85
Hepatocyte morphometry (μm)	2.96 ± 0.07	3.64 ± 0.09 [a]	2.82 ± 0.10 [b]
Collagen (%)	0.31 ± 0.02	1.40 ± 0.19 [a]	1.08 ± 0.14

Data expressed as mean ± SEM and analyzed by one-way ANOVA followed by Tukey's test for multiple comparisons. Percentage of collagen refers to the proportion of the stained area in liver histological sections. Letters [a]: Differs statistically from C group; [b]: Differs statistically from PD group. C—control group (n = 10); PD—prediabetic group (n = 8); PD-Mel—prediabetic-melatonin-treated group (n = 9).

3. Discussion

The findings of this study provide novel insights into the therapeutic potential of melatonin in mitigating metabolic dysfunction and mitochondrial abnormalities associated with prediabetes. Our results demonstrate that melatonin treatment significantly improves lipid and energy metabolism, reduces oxidative stress, and restores mitochondrial homeostasis in a rat model of prediabetes induced by chronic sucrose consumption and low-dose STZ administration. These effects were evident at both biochemical and molecular levels, highlighting melatonin's pleiotropic actions in modulating key pathways involved in the pathophysiology of prediabetes.

High-sugar diets, particularly among overweight or obese individuals, are strongly associated with the development of T2DM. Sugar-sweetened beverages contribute significantly to daily caloric intake, leading to weight gain and metabolic disturbances [29]. The impact of liquid carbohydrate consumption on energy metabolism is well documented and contributes to a positive energy balance [30,31]. However, in our study, chronic sucrose consumption did not significantly alter body weight or BMI. Furthermore, the dietary pattern, i.e., sucrose and food consumption, remained unchanged throughout the entire experiment. Neither STZ administration nor melatonin treatment affected these parameters. These finding may be related to the study duration, as morphometric changes often require prolonged exposure to high-sucrose diets [19]. STZ may also have influenced weight regulation, given its known β-cell toxicity and associated catabolic effects [32].

Although sucrose solution may be more palatable than water, no differences in liquid consumption were observed—a recurring finding in our previous studies [20]. However, the presence of sucrose led to a compensatory decrease in food intake. This shift is critical as sucrose-supplemented water is energy dense, contributing to increased total caloric intake, even without increased liquid volume. This increased energy intake likely contributed to the fat accumulation, dyslipidemia, and insulin resistance observed in the PD group.

Despite the absence of weight gain, prediabetic animals exhibited increased fat deposition, evidencing the metabolic consequences of high-sucrose intake. Chronic sucrose consumption thus promoted adipose tissue expansion and dyslipidemia—both hallmark features of metabolic syndrome [33]. Importantly, melatonin treatment significantly attenuated these alterations, reducing fat deposits, lowering TC and TG levels, and increasing hepatic HDL. These results are consistent with previous reports using melatonin doses ranging from 2 to 50 mg/kg [27,28], including a study by Hadjzadeh et al. [26] employing 20 mg/kg, the same dose used herein. A limitation in this context is that we did not evaluate skeletal muscle mass or subcutaneous fat depots, which are important components of body composition. While adiposity was estimated using internal fat pads, more comprehensive and precise assessments—such as dual-energy X-ray absorptiometry (DEXA) in live animals—would provide better insights into total fat distribution and lean mass, thereby strengthening the interpretation of our findings.

The role of melatonin in regulating glycolytic homeostasis has been previously described [23,34,35]. Although melatonin did not improve glucose intolerance, it effectively normalized fasting serum glucose, which is typically disrupted due to hepatic IR [8]. The reduction in the TyG index, a robust predictor of IR, following melatonin treatment further supports an improvement in insulin sensitivity. Elevated ALT and AST levels are commonly recognized markers of hepatic damage and have been reported to increase following chronic sucrose consumption [19]. Interestingly, ALT and AST levels remained unchanged, suggesting no overt hepatic injury. These results suggest that metabolic dysfunction preceded hepatocellular damage, which is consistent with early-stage prediabetes.

The increased hepatic *Glut2* expression in melatonin-treated animals suggests improved glucose uptake. The observed downregulation of *Pfk1* expression, a gene regulated by insulin [36], may be associated with insulin deficiency. However, PFK1 enzymatic activity remained unchanged, possibly due to post-transcriptional regulatory mechanisms. Additionally, melatonin has been shown to modulate PXR, which downregulates *Glut2* and contributes to glucose intolerance [37,38], raising the possibility that melatonin's effects are mediated in part through PXR inhibition.

Hepatic metabolic alterations are commonly reported in diabetic models under hypercaloric conditions [39]. Prediabetic animals showed reduced PiDH activity and increased *Ldha* expression, indicating a shift in pyruvate metabolism toward lactate production. Melatonin normalized *ldh* expression and LDH activity, a finding consistent with previous reports from our group [40]. This reduction in LDH activity also supports the safety of melatonin relative to biguanide-class drugs, e.g., metformin, which may increase lactate levels and pose a risk of lactic acidosis [41–43].

Melatonin also promoted a shift toward lipid oxidation. The upregulation of *Hadh* expression and increased β-OHADH activity in the PD-Mel group, alongside a higher β-OHADH/ATP synthase ratio, suggest that β-oxidation became the predominant ATP source. Under diabetic conditions, hepatic lipogenesis is often sustained, contributing to metabolic imbalances [39]. The increased CS activity may reflect a compensatory response to elevated lipid metabolism, supporting energy production while limiting lipotoxicity. Together with the reduced fat depots and improved lipid profiles, these results emphasize melatonin's role in regulating lipid metabolism. Nevertheless, it is important to acknowledge certain methodological limitations. We did not assess state 3, mitochondrial respiration (i.e., ADP phosphorylation capacity), which could be evaluated using high-resolution respirometry to provide a more direct assessment of mitochondrial bioenergetic function. Additionally, ketone body levels were not measured, which would further clarify whether enhanced β-oxidation was accompanied by increased ketogenesis. These parameters should be considered in future studies to fully elucidate the metabolic consequences of melatonin treatment.

Mitochondria function is regulated by biogenesis, fusion, and fission processes, all of which are essential for cellular homeostasis [10]. To our knowledge, few prediabetes models assessed these parameters after melatonin's treatment. Mitochondrial biogenesis genes have already been directly associated with the development of T2DM; decreased expression of *Pgc1-α* leads directly to decreased oxidative phosphorylation, lipid oxidation and thus contributes to increasing IR [44]. In this study, the downregulation of *Pgc1-α* and *Tfam* in the PD group suggests impaired mitochondrial biogenesis, consistent with metabolic disease models [15]. Thiazolidinediones (e.g., pioglitazone), which are known for upregulating PGC1-α, have shown similar metabolic benefits [45]. Melatonin treatment reversed these changes, upregulating *Pgc1-α* and *Tfam*, thereby promoting mitochondrial biogenesis and stability. In an in vitro model of sepsis-induced hepatocyte injury, melatonin treatment increased the protein levels of PGC-1α, NRF1, and TFAM, as well as reduced MDA levels and increased SOD activity [46].

In the T2DM scenario, mitochondrial dynamics are also affected by the imbalance in fission and fusion processes. Elevated *Drp1* expression in both PD and PD-Mel groups indicates persistent fission activity, a common feature in IR [16]. The persistent increase in *Drp1* expression even after melatonin treatment may indicate an incomplete restoration of mitochondrial homeostasis. However, melatonin markedly increased the expression of *Mfn1* and *Mfn2*, suggesting a partial recovery of fusion capacity. The significant downregulation of *Mfn1* in the PD group points to impaired fusion, contributing to mitochondrial fragmentation and dysfunction.

With high serum glucose levels, the Drp1/Mfn2 ratio increases, leading to mitochondrial fission and contributing to increased IR [47]. In our study, even with greater expression of Drp1 in the PD-Mel group, levels of *Mfn1 and Mfn2* were also elevated. *CypD*, a key regulator of the mitochondrial permeability transition pore (mPTP), plays a critical role in cell survival and apoptosis. The reduced *CypD* expression in PD animals suggests a shift toward a more closed mPTP state, possibly as an adaptive response to prevent excessive cell death under metabolic stress. *CypD* upregulation in the PD-Mel group may have restored mPTP homeostasis, potentially improving mitochondrial function by modulating permeability transition events. *Opa1*, another key fusion protein, was markedly upregulated in both PD and PD-Mel groups. *Opa1* is essential for maintaining mitochondrial cristae structure and function and alleviates ROS accumulation [48]. While excessive expression in the PD group may reflect a compensatory mechanism for impaired mitochondrial dynamics, the further increase in *Opa1* following melatonin treatment indicates that the indolamine enhances mitochondrial inner membrane remodeling, which could improve the ATP production efficiency and overall mitochondrial function.

Melatonin's influence on mitochondrial metabolism was further reflected in the increased expression of ETC components and CS activity. However, these transcriptional changes did not translate into a corresponding increase in enzymatic activity, as evidenced by the reduction in complex I and V activities in the PD-Mel group. This may be attributed to excessive ROS production during enhanced β-oxidation, as the elevated NADH/NAD^+ ratio can disrupt proton gradients and oxidative phosphorylation efficiency [49–51]. Complex I is particularly susceptible to ROS-mediated damage due to its limited electron-processing capacity, which can contribute to decreased ATP synthesis [52].

These previous findings are further supported by melatonin's antioxidant effects, which occur through direct ROS scavenging or the indirect modulation of antioxidant enzymes [53]. The increased activity of SOD and CAT, coupled with reduced lipid peroxidation, supports this hypothesis. Elevated GST and reduced GSH-Rd activity may indicate a shift toward phase II detoxification and enhanced melatonin metabolism [53]. Despite a decline in GSH, GSH-Px activity remained low, possibly reflecting an overreliance on GSH for direct ROS neutralization. These findings are consistent with prior studies showing melatonin's capacity to modulate oxidative stress in diabetic models [26,35,54,55]. Interestingly, while melatonin improved SOD and CAT activity, MDA and GSH may reflect more chronic oxidative damage or other non-enzymatic pathways.

Liver damage in metabolic disorders can progress from steatosis to fibrosis and, eventually, cirrhosis. Studies have demonstrated melatonin's protective role in diabetic conditions by mitigating hydropic degeneration, pro-necrotic lesions [53], and fibrosis, including a reduction in collagen accumulation [56]. Our histological analysis revealed increased collagen deposition and hepatocyte hypertrophy in prediabetic animals, consistent with early fibrotic remodeling [20]. While melatonin treatment did not significantly reduce collagen accumulation, it attenuated hepatocyte hypertrophy, a marker of hepatic dysfunction. These data suggest that melatonin exerts hepatoprotective effects primarily through modulation of cellular hypertrophy rather than fibrotic resolution. This aligns with previous studies showing melatonin's ability to modulate lipid and energy metabolism in the liver [23,24,27,28].

In this study, melatonin treatment improved key biochemical markers, including TC and TG levels, decreased fat deposits, and increased hepatic HDL levels. Additionally, melatonin enhanced β-oxidation, a key process in lipid metabolism likely contributing to the observed modulation of lipid parameters. While melatonin did not improve ETC function directly, possibly due to the persistent OS from fatty acid oxidation byproducts, it significantly reduced anaerobic glucose oxidation, an important finding given that elevated lactate levels are a common side effect of antidiabetic drugs such as metformin. For the first time, we demonstrated that melatonin treatment impacts mitochondrial events which are involved in the prediabetes stage (Figure 6).

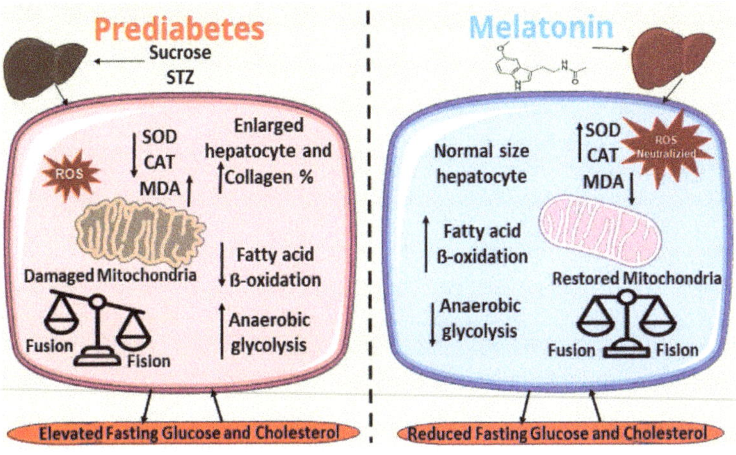

Figure 6. Serum and hepatic alterations caused by sucrose and streptozotocin (STZ) exposition and possible mechanisms by which melatonin treatment improves prediabetic condition.

Study Limitations

Despite the promising results, this study has additional limitations. While we assessed hepatic metabolic flux through enzyme expression and activity assays, advanced techniques such as isolated hepatocyte studies or ex vivo liver perfusion could provide a more comprehensive evaluation of metabolic pathways. Western blot analysis was also not conducted to validate protein expression levels, limiting the depth of our molecular insights.

4. Materials and Methods

4.1. Animals and Experimental Protocol

This study was conducted in compliance with the guidelines of the National Council for the Control of Animal Experimentation (CONCEA), Brazil, and was approved by the Ethics Committee for the Use of Animals (CEUA) of the State University of Northern Paraná (UENP), Brazil (protocol number: 001/2023). The experimental model was initially designed to induce T2DM in Wistar rats, adapting the protocol described by Salido and collaborators [57], by combining chronic sucrose consumption with a single injection of STZ. According to established criteria, T2DM in rats is confirmed when blood glucose levels exceed 200 mg/dL [58]. While this threshold was not reached in our study, the observed hyperglycemia and glucose intolerance were consistent with a prediabetic state, as characterized by Barrière et al. [59].

Thirty male Wistar rats (30 days old) were obtained from the central animal facility of the State University of Londrina, Brazil and housed individually under controlled conditions (temperature: 22 ± 3 °C, light/dark cycle: 12 h, humidity: 60 ± 5%). Following an acclimatization period, the animals were randomly assigned into two experimental groups. Both groups received ad libitum access to a standard rodent chow (Nuvilab-CR1, Nuvital, Colombo, Brazil); however, the control group (C; n = 10) was provided with autoclaved, filtered water, while the prediabetic group (PD; n = 20) received a 40% sucrose solution as the sole drinking source.

After four weeks, the PD group was further subdivided into two groups: untreated prediabetic rats (PD; n = 10) and prediabetic rats receiving melatonin (PD-Mel; n = 10). To induce prediabetes, both subgroups received a single intraperitoneal injection of STZ (50 mg/kg) dissolved in ice-cold 100 mM citrate buffer (pH 4.5). Seven days after STZ administration, blood glucose levels were assessed to confirm the prediabetic condition. Subsequently, the PD-Mel group was treated with melatonin (25 mg/kg), administered intraperitoneally three times per week for four weeks. Melatonin was prepared in a vehicle solution containing 99.5% ethanol and diluted in 0.9% NaCl. The absence of a melatonin-only control group is justified on ethical, scientific, and practical grounds. To stay in few rationalities: by not including this group, the study remains focused on the primary objective of understanding the effects of prediabetes in the presence of melatonin treatment, and by following the 3Rs, treating a control group with melatonin when melatonin's effect on healthy animals is already well established could expose the animals to unnecessary treatments.

The dose of melatonin was selected based on previous studies demonstrating therapeutic properties in T2DM models using dose ranging from 0.2 to 50 mg/kg [54]. We adopted an intermediate dose of 25 mg/kg to reflect a balance between safety and efficacy. The chosen dose is consistent with studies using similar models and dosing regimens [26,27], although different routes of administration were used. Notably, during the experimental period, three animals died following STZ administration.

4.2. Nutritional and Morphometric Parameters

Throughout the study, body weight, food consumption, and water intake were recorded weekly in the morning. Food and liquid consumption were determined by calculating the difference between the total amount provided and the residual quantity in each cage. These data, along with their respective caloric values (3.09 kcal/g and 1.6 kcal/mL) [19], were used to calculate the energy derived from chow, from the sucrose solution, and the total energy intake. At the end of the experimental period, animals were anesthetized, and body morphometric parameters—including body weight and total body length—were measured to calculate the Body Mass Index (BMI: g/cm^2 = body weight/length^2) and the Lee index ((g/cm) = cube root of body weight/naso-anal length). The Lee index, similarly to BMI, is a quick, and non-invasive method to estimate adiposity in rats, which correlates with total body fat mass and is often reflective of metabolic alterations [60,61].

4.3. Intraperitoneal Glucose, Insulin, and Pyruvate Tolerance Tests

In the final week of the experiment, metabolic tolerance tests were conducted to evaluate glucose metabolism. The Glucose Tolerance Test (GTT), Insulin Tolerance Test (ITT), and Pyruvate Tolerance Test (PTT) were performed eight, five, and two days before euthanasia, respectively. For the GTT and ITT, animals underwent a 8 h fasting period. The GTT was conducted by administering a 20% sucrose solution (2 g/kg, i.p.), followed by blood glucose measurements at 0, 15, 30, 60, 90, and 120 min. For the ITT, animals received an intraperitoneal injection of human insulin (1 U/kg, HumulinTM Eli Lilly, São Paulo, Brazil), with blood glucose levels monitored at the same time intervals. The PTT was performed after 16 h of fasting to evaluate gluconeogenesis. Rats were administered an intraperitoneal injection of a 25% pyruvate solution (2 g/kg), and blood glucose levels were measured at 0, 15, 30, 60, 90, and 120 min. All blood glucose readings were obtained using a digital glucometer (On Call® Plus II, ACON Biotech, Hangzhou, China).

4.4. Collection of Biological Material

At the end of the experimental period, animals fasted for 6 h were anesthetized with ketamine (90 mg/kg) and xylazine (10 mg/kg) and euthanized via cardiac puncture. Blood samples were centrifuged (RCF 85.75× g, 10 min), and serum aliquots were stored at −80 °C for biochemical analyses, including alanine transaminase (ALT), aspartate aminotransferase (AST), glucose, triacylglycerol (TG), urea, creatinine, total cholesterol (TC), and high-density lipoprotein (HDL) levels. All biochemical measurements were performed using commercial kits (Gold Analisa®) and read in a microplate reader (MultiSkan Skyhigh, Thermo Fisher Scientific Inc., MA, USA). The Friedewald equation was used to estimate low-density lipoprotein (LDL) levels: [LDL] = (TC − HDL) − (TG/5). The Triglyceride–Glucose (TyG) Index was calculated as an indirect marker of insulin resistance using the equation: Tyg = ln (fasting TG × fasting glucose/2).

Following euthanasia, white adipose tissue deposits, epididymal, visceral, and retroperitoneal, were excised and weighed. Liver samples were collected, weighed, and immediately stored at −80 °C for subsequent oxidative stress assessments and metabolic analyses. The same biochemical determinations performed on serum were also conducted on hepatic tissue.

4.5. Total RNA Extraction and cDNA Synthesis

Liver samples were homogenized in TRIzol® reagent and stored at −80 °C until RNA extraction. Total RNA was isolated using the TRIzol®/Chloroform/Isopropanol method, following the manufacturer's instructions. RNA purity and concentration were assessed using a NanoDrop 2000 spectrophotometer (Thermo Fisher Scientific, Waltham, MA, USA), with quality thresholds set at OD 260/280 ≥ 1.8 and OD 260/230 ≥ 1.0. Complementary DNA (cDNA) synthesis was performed using the SuperScript First Strand Supermix kit (Invitrogen, Carlsbad, CA, USA), following the manufacturer's protocol. Random primers were used to ensure the comprehensive reverse transcription of RNA into cDNA.

4.6. Gene Expression Analysis

Quantitative PCR (qPCR) was employed to assess gene expression profiles using the SYBR Green Master Mix (qPCR®, Promega, Madison, WI, USA). Each reaction was performed in triplicate, with gene-specific primers (Table 5). $2^{-\Delta\Delta Ct}$ method was employed to quantify relative gene expression. β-ACTIN was selected as the internal reference gene for normalization. Fold changes and standard deviations were determined from three independent experiments, each conducted in duplicate.

Table 5. Sequence of primers used for gene expressions.

Gene (ID)	Primer	5′–3′ Sequence	Cycle Condition	Product Size (bp)
ACTB (81822)	Forward	CCTCTATGCCAACACAGTGC	95 °C—8 s; 61.5 °C—8 s; 72 °C—8 s	206
	Reverse	CCTGCTTGCTGATCCACATC		
Slc2a2 (25351)	Forward	TGTGAAAGTCATGCATGTGGC	95 °C—8 s; 61.5 °C—8 s; 72 °C—8 s	104
	Reverse	AGGCCAGGGATTGGTGTTAAA		
Hadh (113965)	Forward	AGTCTGGACTTGACCTTCTTGG	95 °C—8 s; 61.5 °C—8 s; 72 °C—8 s	138
	Reverse	TCTGAGGGCCCATTTTGATGT		
Pdh1 (29554)	Forward	TCCCCAGCTTGGCTTATTGT	95 °C—8 s; 61.5 °C—8 s; 72 °C—8 s	132
	Reverse	GGGTGGCTTCAAGTTTGCTTT		
Pfkl (25741)	Forward	AACAATGTCCCTGGCACTGA	95 °C—8 s; 61.5 °C—8 s; 72 °C—8 s	189
	Reverse	ACTCTCCATTGCAGCGTTGA		
Cs (170587)	Forward	ACCATGACGGTGGCAATGTA	95 °C—8 s; 61.5 °C—8 s; 72 °C—8 s	210
	Reverse	TGGTTTGCTAGTCCATGCAGA		
Ldha (24533)	Forward	TCCTCAGCGTCCCATGTATC	95 °C—8 s; 61.5 °C—8 s; 72 °C—8 s	193
	Reverse	TCCATAGAAACCCTGCTGCA		
Tfam (21780)	Forward	TCCACAGAACAGCTACCCAA	95 °C—8 s; 66 °C—8 s; 72 °C—8 s	84
	Reverse	CCACAGGGCTGCAATTTTCC		
Cypd (105675)	Forward	CCCTCCTGGTCACTGTGAAT	95 °C—8 s; 66 °C—8 s; 72 °C—8 s	186
	Reverse	GCCAGAAGACACTTCCCTCT		
Nd1 (17716)	Forward	ATTACTTCTGCCAGCCTGACC	95 °C—8 s; 66 °C—8 s; 72 °C—8 s	70
	Reverse	GGCCCGGTTTGTTTCTGCTA		
Nd5 (4540)	Forward	CCTGGCACTGAGTCACCATA	95 °C—8 s; 66 °C—8 s; 72 °C—8 s	214
	Reverse	TTGTTGGCTGAGGTGAGGAT		
Atp6 (100503946)	Forward	TCCCAATCGTTGTAGCCATCA	95 °C—8 s; 66 °C—8 s; 72 °C—8 s	76
	Reverse	AGACGGTTGTTGATTAGGCGT		
Cox1 (17708)	Forward	ATCACTACCAGTGCTAGCCG	95 °C—8 s; 66 °C—8 s; 72 °C—8 s	84
	Reverse	CCTCCAGCGGGATCAAAGAA		
Drp1 (74006)	Forward	GTCGAGTCCCCATTCATTGC	95 °C—8 s; 66 °C—8 s; 72 °C—8 s	151
	Reverse	ACTACGACGATCTGAGGCAG		
Mfn1 (67414)	Forward	GCAGACAGCACATGGAGAGA	95 °C—8 s; 61 °C—8 s; 72 °C—8 s	83
	Reverse	GATCCGATTCCGAGCTTCCG		
Mfn2 (170731)	Forward	TGCACCGCCATATAGAGGAAG	95 °C—8 s; 61 °C—8 s; 72 °C—8 s	78
	Reverse	TCTGCAGTGAACTGGCAATG		
Opa1 (74143)	Forward	ACCTTGCCAGTTTAGCTCCC	95 °C—8 s; 61 °C—8 s; 72 °C—8 s	82
	Reverse	TTGGGACCTGCAGTGAAGAA		

Table 5. *Cont.*

Gene (ID)	Primer	5′-3′ Sequence	Cycle Condition	Product Size (bp)
Fis1 (66437)	Forward	CAAAGAGGAACAGCGGGACT	95 °C—8 s; 61 °C—8 s; 72 °C—8 s	95
	Reverse	ACAGCCCTCGCACATACTTT		
Cytb (17711)	Forward	GGCTACGTCCTTCCATGAGG	95 °C—8 s; 61 °C—8 s; 72 °C—8 s	75
	Reverse	TGGGATGGCTGATAGGAGGT		

ACTB: β-actin; Slc2a2: coding gene for GLUT2; Hadh: coding gene for hydroxyacyl-CoA dehydrogenase; Pdhl: coding gene for alpha 1 subunit of pyruvate dehydrogenase complex; Pfkl: coding gene for phosphofrutokinase-1 liver-type; Cs: coding gene for citrate synthase; Ldha: coding gene for lactate dehydrogenase A; Tfam: mitochondrial transcription factor A; Cypd: cyclophilin D; Nd1: NADH dehydrogenase subunit 1; Nd5: NADH dehydrogenase subunit 5; Atp6: ATP synthase subunit 6; Cox1: cytochrome c oxydase 1; Drp1: dynamin-related protein 1; Mfn1: mitofusin-1; Mfn2; mitofusin-2; Opa1: optic atrophy 1; Fis1: mitochondrial fission protein 1; Cytb: cytochrome b.

4.7. Hepatic Energy Metabolism

To evaluate the hepatic energy metabolism, 200 mg of liver tissue was homogenized and subjected to centrifugation (RCF 10,360× *g*, 10 min). The resulting supernatant was collected, while the pellet was resuspended in 0.1 mol/L sodium phosphate buffer containing 250 mmol/L sucrose and 2 mmol/L EDTA for further enzymatic analyses. The activity of the enzymes NADH-oxidase and ATP synthase was analyzed using the cell precipitate and the supernatant was used for the other analyses. The enzymatic activity of key metabolic markers was assessed, including phosphofructokinase (PFK; E.C.2.7.1.11.), citrate synthase (CS; E.C.4.1.3.7.), lactate dehydrogenase (LDH; E.C.1.1.1.27.), β-hydroxyacyl-coenzyme A dehydrogenase (β-OHADH; E.C.1.1.1.35.), pyruvate dehydrogenase (PiDH; E.C.1.2.1.51.), succinate dehydrogenase (SDH; E.C.1.3.99.1.), NADH-oxidase (E.C.1.6.99.5.), and ATP synthase (E.C.7.1.2.2).

The reaction medium for each enzyme was prepared as follows:

- PFK: 50 mmol/L Tris-HCl buffer (pH 8.0), 10 mmol/L $MgCl_2$, 1U glyceraldehyde-3-phosphate dehydrogenase, 1 U aldolase, 2.5 U triose phosphate isomerase, 0.12 mmol/L NADH, 0.75 mmol/L ATP, and 6 mmol/L fructose-6-phosphate.
- CS: 50 mmol/L Tris-HCl buffer (pH 8.0), 0.3 mmol/L acetyl-CoA, 0.1 mmol/L 5,5'-dithiobis-(2-nitrobenzoic acid) (DTNB), and 0.5 mmol/L oxaloacetate.
- LDH: 50 mmol/L Tris-HCl buffer (pH 7.4), 0.14 mmol/L NADH, and 1 mmol/L sodium pyruvate.
- β-OHADH: 0.1 mol/L Tris-HCl buffer (pH 7.0), 5 mmol/L EDTA, 0.05 mmol/L acetoacetyl-coenzyme A, and 0.1 mmol/L NADH.
- PiDH: 2.5 mmol/L NAD, 0.1 mmol/L coenzyme A, 0.2 mmol/L thiamine pyrophosphate, 0.3 mmol/L dithiothreitol (DTT), 1 mmol/L $MgCl_2$, 5 mmol/L pyruvate, 0.08 mmol/L nitrotetrazolium blue (NBT), and 1 mg bovine serum albumin (BSA), with 3 mmol/L sodium pyruvate and 0.05 mmol/L phenazine metasulfate in 50 mmol/L potassium phosphate buffer (pH 7.4).
- SDH: 50 mmol/L potassium phosphate buffer (pH 7.4), 10 mmol/L sodium succinate, 0.36 mmol/L phenazine metasulfate, and 0.12 mmol/L dichlorophenolindophenol (DPIP).
- NADH-oxidase: 80 mmol/L sodium phosphate buffer (pH 7.4), 50 mmol/L EDTA, and 0.2 mmol/L NADH.
- ATP synthase: 0.11 mmol/L $MgCl_2$, 0.02 mmol/L NADH, 0.25 mmol/L phosphoenolpyruvate (PEP), 0.2 mmol/L ATP, 1686 U LDH, and 4001.18 U pyruvate kinase.

Enzymatic activity was determined at 25 °C using a microplate reader (μQuant-MQX 200) controlled via Kcjunior software (Bio-Tec Instruments, Winooski, VT, USA).

4.8. Hepatic Oxidative Stress

To assess oxidative stress markers, 200 mg of liver tissue was homogenized in 5 mL of ice-cold 0.1 mol/L sodium phosphate buffer (pH 7.4) using a Potter–Elvehjem homogenizer on ice. The homogenate was centrifuged (RCF 10,360× *g*, 10 min), and the supernatant was used for biochemical analyses. Markers of oxidative stress included total protein quantification [62], lipid peroxidation (malondialdehyde levels), total glutathione, reduced glutathione (GSH), and oxidized glutathione [63], and the activities of key antioxidant enzymes, such as glutathione peroxidase (GSH-Px; E.C.1.11.1.9.) [64], glutathione-S-transferase (GST; E.C.2.5.1.18.) [65], superoxide dismutase (SOD; E.C.1.15.1.1.) [66], and catalase (CAT; E.C.1.11.1.6.) [67]. All enzymatic assays were conducted at 25 °C using a microplate reader (μQuant-MQX 200) with Kcjunior software (Bio-Tec Instruments, Winooski, VT, USA).

4.9. Histopathological Analysis

For histopathological evaluation, liver samples were fixed in Bouin's solution for 24 h, dehydrated, cleared, and embedded in paraffin. Semi-serial sections (5 μm thick) were obtained and stained for microscopic analysis.

Liver architecture was examined using a Zeiss AxioLab 5 light microscope (Zeiss, Oberkochen, Germany) equipped with a high-resolution digital camera. Thirty images per animal were captured at 200× magnification, focusing on the peri-central vein region. Hematoxylin and eosin (H&E) staining was used to determine hepatocyte density and morphometry (100 hepatocytes/animal) within a fixed field area (0.216 mm^2). Image analysis was performed using ImageProPlus 4.5 software (Media Cybernetics, Rockville, MD, USA). To assess fibrosis, additional sections were stained with PicroSirius Red to detect type I and III collagen deposition. Quantitative collagen analysis was performed using ImageJ FIJI 1.54i software, applying RGB stack segmentation with the same parameters used for H&E-stained sections.

4.10. Statistical Analysis

Data were analyzed using one-way analysis of variance (ANOVA), followed by Tukey's post-hoc test. Results are expressed as mean ± standard error of mean (SEM). Statistical significance was considered at $p < 0.05$. All analyses were conducted using GraphPad Prism 8.0.2 (GraphPad Software, San Diego, CA, USA).

5. Conclusions

In this study, melatonin treatment partially restored glycolytic homeostasis in prediabetic rats by lowering fasting glucose levels and reducing reliance on anaerobic glycolysis, as evidenced by decreased LDH expression and activity. The treatment also improved lipid metabolism, as demonstrated by reduced fat accumulation, lower serum triglyceride and total cholesterol levels, and increased hepatic HDL content. These findings indicate that melatonin promotes a favorable shift in both glucose and lipid handling in the liver.

In parallel, melatonin enhanced the activity of key antioxidant enzymes, including SOD and CAT, contributing to reduced hepatic oxidative stress. Importantly, our data provide compelling evidence of mitochondrial dysfunction in the prediabetic state and demonstrate that melatonin modulates the expression of genes involved in mitochondrial biogenesis, integrity, fusion, and fission. Taken together, these results highlight the potential of melatonin as a therapeutic agent to counteract metabolic and mitochondrial disturbances associated with prediabetes.

Author Contributions: Conceptualization: M.C.d.S. and F.R.F.S.; Methodology: M.C.d.S., M.L.G.A., M.R.d.A., G.d.S.F., K.A.d.N., J.P.F.S. and E.M.S.C.; Formal analysis and research: M.C.d.S. and F.R.F.S. Writing of the original draft: M.C.d.S. Writing—proofreading and editing: F.R.F.S. and L.G.d.A.C.; Acquisition of funding: F.R.F.S.; Supervision: F.R.F.S. All authors have read and agreed to the published version of the manuscript.

Funding: This work was supported in part by the Coordenação de Aperfeiçoamento de Pessoal de Nível Superior—Brasil (CAPES)—Finance Code 001, by CAPES-PROEX (process number: 1287/2023/88881.844932/2023-1), and by PROPG/UNESP/Brazil.

Institutional Review Board Statement: Not applicable.

Informed Consent Statement: Not applicable.

Data Availability Statement: Data is contained within the article.

Conflicts of Interest: The authors declare no conflicts of interest.

References

1. Alam, S.; Hasan, M.K.; Neaz, S.; Hussain, N.; Hossain, M.F.; Rahman, T. Diabetes Mellitus: Insights from Epidemiology, Biochemistry, Risk Factors, Diagnosis, Complications and Comprehensive Management. *Diabetology* **2021**, *2*, 36–50. [CrossRef]
2. Sun, H.; Saeedi, P.; Karuranga, S.; Pinkepank, M.; Ogurtsova, K.; Duncan, B.B.; Stein, C.; Basit, A.; Chan, J.C.N.; Mbanya, J.C.; et al. IDF Diabetes Atlas: Global, Regional and Country-Level Diabetes Prevalence Estimates for 2021 and Projections for 2045. *Diabetes Res. Clin. Pract.* **2022**, *183*, 109119. [CrossRef]
3. DeFronzo, R.A.; Ferrannini, E.; Groop, L.; Henry, R.R.; Herman, W.H.; Holst, J.J.; Hu, F.B.; Kahn, C.R.; Raz, I.; Shulman, G.I.; et al. Type 2 Diabetes Mellitus. *Nat. Rev. Dis. Primers* **2015**, *1*, 15019. [CrossRef]
4. World Health Organization. Definition and Diagnosis of Diabetes Mellitus and Intermediate Hyperglycemia. 2006. Available online: https://www.who.int/publications/i/item/definition-and-diagnosis-of-diabetes-mellitus-and-intermediate-hyperglycaemia (accessed on 1 March 2024).
5. American Diabetes Association Diagnosis and Classification of Diabetes Mellitus. *Diabetes Care* **2014**, *37*, 81–90. [CrossRef] [PubMed]
6. Rosinger, A.; Herrick, K.; Gahche, J.; Park, S. Sugar-Sweetened Beverage Consumption Among U.S. Adults, 2011–2014; NCHS Data Brief No. 270. 2017; pp. 1–8. Available online: https://www.cdc.gov/nchs/products/databriefs/db270.htm (accessed on 1 March 2024).
7. Blüher, M. Obesity: Global Epidemiology and Pathogenesis. *Nat. Rev. Endocrinol.* **2019**, *15*, 288–298. [CrossRef] [PubMed]
8. Echouffo-Tcheugui, J.B.; Perreault, L.; Ji, L.; Dagogo-Jack, S. Diagnosis and Management of Prediabetes: A Review. *JAMA* **2023**, *329*, 1206–1216. [CrossRef] [PubMed]
9. Jiang, S.; Young, J.L.; Wang, K.; Qian, Y.; Cai, L. Diabetic-Induced Alterations in Hepatic Glucose and Lipid Metabolism: The Role of Type 1 and Type 2 Diabetes Mellitus (Review). *Mol. Med. Rep.* **2020**, *22*, 603–611. [CrossRef]
10. Pfanner, N.; Warscheid, B.; Wiedemann, N. Mitochondrial Proteins: From Biogenesis to Functional Networks. *Nat. Rev. Mol. Cell Biol.* **2019**, *20*, 267–284. [CrossRef]
11. Popov, L.D. Mitochondrial Biogenesis: An Update. *J. Cell Mol. Med.* **2020**, *24*, 4892–4899. [CrossRef]
12. Liu, L.; Li, Y.; Chen, G.; Chen, Q. Crosstalk between Mitochondrial Biogenesis and Mitophagy to Maintain Mitochondrial Homeostasis. *J. Biomed. Sci.* **2023**, *30*, 86. [CrossRef]
13. Chen, W.; Zhao, H.; Li, Y. Mitochondrial Dynamics in Health and Disease: Mechanisms and Potential Targets. *Signal Transduct. Target. Ther.* **2023**, *8*, 333. [CrossRef] [PubMed]
14. Rovira-Llopis, S.; Bañuls, C.; Diaz-Morales, N.; Hernandez-Mijares, A.; Rocha, M.; Victor, V.M.; Region, V. Mitochondrial Dynamics in Type 2 Diabetes: Pathophysiological Implications. *Redox Biol.* **2017**, *11*, 637–645. [CrossRef]
15. Veluthakal, R.; Esparza, D.; Hoolachan, J.M.; Balakrishnan, R.; Ahn, M.; Oh, E.; Jayasena, C.S.; Thurmond, D.C. Mitochondrial Dysfunction, Oxidative Stress, and Inter-Organ Miscommunications in T2D Progression. *Int. J. Mol. Sci.* **2024**, *25*, 1504. [CrossRef] [PubMed]

16. Maneechote, C.; Kerdphoo, S.; Jaiwongkam, T.; Chattipakorn, S.C.; Chattipakorn, N. Chronic Pharmacological Modulation of Mitochondrial Dynamics Alleviates Prediabetes-Induced Myocardial Ischemia–Reperfusion Injury by Preventing Mitochondrial Dysfunction and Programmed Apoptosis. *Cardiovasc. Drugs Ther.* **2023**, *37*, 89–105. [CrossRef]
17. Park, I.R.; Chung, Y.G.; Won, K.C. Overcoming β-Cell Dysfunction in Type 2 Diabetes Mellitus: CD36 Inhibition and Antioxidant System. *Diabetes Metab. J.* **2025**, *49*, 1–12. [CrossRef]
18. Tan, D.X.; Manchester, L.C.; Esteban-Zubero, E.; Zhou, Z.; Reiter, R.J. Melatonin as a Potent and Inducible Endogenous Antioxidant: Synthesis and Metabolism. *Molecules* **2015**, *20*, 18886–18906. [CrossRef] [PubMed]
19. Souza Cruz, E.M.; Bitencourt de Morais, J.M.; Dalto da Rosa, C.V.; da Silva Simões, M.; Comar, J.F.; de Almeida Chuffa, L.G.; Seiva, F.R.F. Long-Term Sucrose Solution Consumption Causes Metabolic Alterations and Affects Hepatic Oxidative Stress in Wistar Rats. *Biol. Open* **2020**, *9*, bio047282. [CrossRef]
20. de Morais, J.M.B.; Cruz, E.M.S.; da Rosa, C.V.D.; Cesário, R.C.; Comar, J.F.; Moreira, C.C.L.; de Almeida Chuffa, L.G.; Seiva, F.R.F. Pterostilbene Influences Glycemia and Lipidemia and Enhances Antioxidant Status in the Liver of Rats That Consumed Sucrose Solution. *Life Sci.* **2021**, *269*, 119048. [CrossRef]
21. Patel, R.; Parmar, N.; Pramanik Palit, S.; Rathwa, N.; Ramachandran, A.V.; Begum, R. Diabetes Mellitus and Melatonin: Where Are We? *Biochimie* **2022**, *202*, 2–14. [CrossRef]
22. Reiter, R.J.; Mayo, J.C.; Tan, D.X.; Sainz, R.M.; Alatorre-Jimenez, M.; Qin, L. Melatonin as an Antioxidant: Under Promises but over Delivers. *J. Pineal Res.* **2016**, *61*, 253–278. [CrossRef]
23. Shieh, J.M.; Wu, H.T.; Cheng, K.C.; Cheng, J.T. Melatonin Ameliorates High Fat Diet-Induced Diabetes and Stimulates Glycogen Synthesis via a PKCzeta-Akt-GSK3beta Pathway in Hepatic Cells. *J. Pineal Res.* **2009**, *47*, 339–344. [CrossRef] [PubMed]
24. Fernández Vázquez, G.; Reiter, R.J.; Agil, A. Melatonin Increases Brown Adipose Tissue Mass and Function in Zücker Diabetic Fatty Rats: Implications for Obesity Control. *J. Pineal Res.* **2018**, *64*, e12472. [CrossRef] [PubMed]
25. Hidayat, M.; Mahar, Y.; Wasim, B. Effect of Melatonin on Serum Glucose and Body Weights in Strptozotocin Induced Diabetes in Albino Rats. *J. Ayub Med. Coll. Abbottabad* **2015**, *27*, 274–276.
26. Hadjzadeh, M.A.R.; Alikhani, V.; Hosseinian, S.; Zarei, B.; Keshavarzi, Z. The Effect of Melatonin against Gastric Oxidative Stress and Dyslipidemia in Streptozotocin-Induced Diabetic Rats. *Acta Endocrinol.* **2018**, *14*, 453. [CrossRef]
27. Uluışık, D.; Keskin, E.; Hatipoğlu, D. A Study on Effect of Melatonin in Dyslipidemia Caused by Experimentally Diabetes. *Turk. J. Sport Exerc.* **2019**, *21*, 117–121. [CrossRef]
28. Mirunalini, S.; Jayakumar, P.; Gomathy, T. Modulation of Antioxidant Status, Carbohydrate and Lipid Metabolism by Melatonin on Streptozotocin Induced Diabetic Rats. *J. Biochem. Technol.* **2012**, *4*, 524–530.
29. Malik, V.S.; Hu, F.B. The Role of Sugar-Sweetened Beverages in the Global Epidemics of Obesity and Chronic Diseases. *Nat. Rev. Endocrinol.* **2022**, *18*, 205–218. [CrossRef]
30. Munsters, M.J.M.; Saris, W.H.M. The Effect of Sugar-Sweetened Beverage Intake on Energy Intake in an Ad Libitum 6-Month Low-Fat High-Carbohydrate Diet. *Ann. Nutr. Metab.* **2010**, *57*, 116–123. [CrossRef]
31. Dimeglio, D.P.; Mattes, R.D. Liquid versus Solid Carbohydrate: Effects on Food Intake and Body Weight. *Int. J. Obes.* **2000**, *24*, 794–800. [CrossRef]
32. De Magalhães, D.A.; Kume, W.T.; Correia, F.S.; Queiroz, T.S.; Allebrandt Neto, E.W.; Dos Santos, M.P.; Kawashita, N.H.; De França, S.A. High-Fat Diet and Streptozotocin in the Induction of Type 2 Diabetes Mellitus: A New Proposal. *An. Acad. Bras. Cienc.* **2019**, *91*, e20180314. [CrossRef]
33. von Krüchten, R.; Lorbeer, R.; Müller-Peltzer, K.; Rospleszcz, S.; Storz, C.; Askani, E.; Kulka, C.; Schuppert, C.; Rathmann, W.; Peters, A.; et al. Association between Adipose Tissue Depots and Dyslipidemia: The KORA-MRI Population-Based Study. *Nutrients* **2022**, *14*, 797. [CrossRef] [PubMed]
34. Abdulwahab, D.A.; El-Missiry, M.A.; Shabana, S.; Othman, A.I.; Amer, M.E. Melatonin Protects the Heart and Pancreas by Improving Glucose Homeostasis, Oxidative Stress, Inflammation and Apoptosis in T2DM-Induced Rats. *Heliyon* **2021**, *7*, e06474. [CrossRef]
35. Wan, X.D.; Li, S.Q.; Xi, S.M.; Wang, J.F.; Guo, Y.C.; Wang, X.M. Long-Term Melatonin Administration Improves Glucose Homeostasis and Insulin Resistance State in High-Fat-Diet Fed Rats. *Cent. Eur. J. Biol.* **2013**, *8*, 958–967. [CrossRef]
36. Ausina, P.; Da Silva, D.; Majerowicz, D.; Zancan, P.; Sola-Penna, M. Insulin Specifically Regulates Expression of Liver and Muscle Phosphofructokinase Isoforms. *Biomed. Pharmacother.* **2018**, *103*, 228–233. [CrossRef] [PubMed]
37. Jang, M.; Manz, A.; Volk, T.; Kleber, A. Study of Melatonin-Mediated Effects on Various Hepatic Inflammatory Responses Stimulated by IL-6 in a New HepG2-on-a-Chip Platform. *Biomed. Microdevices* **2018**, *20*, 54. [CrossRef]
38. Hassani-Nezhad-Gashti, F.; Rysä, J.; Kummu, O.; Näpänkangas, J.; Buler, M.; Karpale, M.; Hukkanen, J.; Hakkola, J. Activation of Nuclear Receptor PXR Impairs Glucose Tolerance and Dysregulates GLUT2 Expression and Subcellular Localization in Liver. *Biochem. Pharmacol.* **2018**, *148*, 253–264. [CrossRef] [PubMed]

39. Czech, M.P. Insulin Action and Resistance in Obesity and Type 2 Diabetes. *Nat. Med.* **2017**, *23*, 804–814. [CrossRef]
40. Cruz, E.M.S.; Concato, V.M.; de Morais, J.M.B.; Silva, T.F.; Inoue, F.S.R.; de Souza Cremer, M.; Bidóia, D.L.; Machado, R.R.B.; de Almeida Chuffa, L.G.; Mantovani, M.S.; et al. Melatonin Modulates the Warburg Effect and Alters the Morphology of Hepatocellular Carcinoma Cell Line Resulting in Reduced Viability and Migratory Potential. *Life Sci.* **2023**, *319*, 121530. [CrossRef]
41. Wang, G.S.; Hoyte, C. Review of Biguanide (Metformin) Toxicity. *J. Intensive Care Med.* **2019**, *34*, 863–876. [CrossRef]
42. Defronzo, R.; Fleming, G.A.; Chen, K.; Bicsak, T.A. Metformin-Associated Lactic Acidosis: Current Perspectives on Causes and Risk. *Metabolism* **2016**, *65*, 20–29. [CrossRef]
43. Akhter, M.S.; Uppal, P. Toxicity of Metformin and Hypoglycemic Therapies. *Adv. Chronic Kidney Dis.* **2020**, *27*, 18–30. [CrossRef] [PubMed]
44. Liang, H.; Ward, W.F. Staying Current PGC-1: A Key Regulator of Energy Metabolism. *Adv. Physiol. Educ.* **2006**, *30*, 145–151. [CrossRef]
45. Wilson-Fritch, L.; Nicoloro, S.; Chouinard, M.; Lazar, M.A.; Chui, P.C.; Leszyk, J.; Straubhaar, J.; Czech, M.P.; Corvera, S. Mitochondrial Remodeling in Adipose Tissue Associated with Obesity and Treatment with Rosiglitazone. *J. Clin. Investig.* **2004**, *114*, 1281–1289. [CrossRef]
46. Hu, B.; Chen, Z.; Liang, L.; Zheng, M.; Chen, X.; Zeng, Q. Melatonin Promotes Mitochondrial Biogenesis and Mitochondrial Degradation in Hepatocytes During Sepsis. *Altern. Ther. Health Med.* **2023**, *29*, 284–289.
47. Cooper, I.D.; Brookler, K.H.; Kyriakidou, Y.; Elliott, B.T.; Crofts, C.A.P. Metabolic Phenotypes and Step by Step Evolution of Type 2 Diabetes: A New Paradigm. *Biomedicines* **2021**, *9*, 800. [CrossRef]
48. Quintana-Cabrera, R.; Manjarrés-Raza, I.; Vicente-Gutiérrez, C.; Corrado, M.; Bolaños, J.P.; Scorrano, L. Opa1 Relies on Cristae Preservation and ATP Synthase to Curtail Reactive Oxygen Species Accumulation in Mitochondria. *Redox Biol.* **2021**, *41*, 101944. [CrossRef] [PubMed]
49. Maechler, P.; Jornot, L.; Wollheim, C.B. Hydrogen Peroxide Alters Mitochondrial Activation and Insulin Secretion in Pancreatic Beta Cells. *J. Biol. Chem.* **1999**, *274*, 27905–27913. [CrossRef] [PubMed]
50. Hajam, Y.A.; Rani, R.; Ganie, S.Y.; Sheikh, T.A.; Javaid, D.; Qadri, S.S.; Pramodh, S.; Alsulimani, A.; Alkhanani, M.F.; Harakeh, S.; et al. Oxidative Stress in Human Pathology and Aging: Molecular Mechanisms and Perspectives. *Cells* **2022**, *11*, 552. [CrossRef]
51. Brand, M.D. The Sites and Topology of Mitochondrial Superoxide Production. *Exp. Gerontol.* **2010**, *45*, 466. [CrossRef]
52. Hardeland, R. Melatonin and the Electron Transport Chain. *Cell. Mol. Life Sci.* **2017**, *74*, 3883–3896. [CrossRef]
53. Grigorov, I.; Bogojević, D.; Jovanović, S.; Petrović, A.; Ivanović-Matić, S.; Zolotarevski, L.; Poznanović, G.; Martinović, V. Hepatoprotective Effects of Melatonin against Pronecrotic Cellular Events in Streptozotocin-Induced Diabetic Rats. *J. Physiol. Biochem.* **2014**, *70*, 441–450. [CrossRef] [PubMed]
54. Kahya, M.C.; Nazıroğlu, M. Melatonin Reduces Lens Oxidative Stress Level in STZ-Induced Diabetic Rats through Supporting Glutathione Peroxidase and Reduced Glutathione Values. *J. Cell. Neurosci. Oxidative Stress* **2016**, *8*, 588–594. [CrossRef]
55. Agarkov, A.A.; Popova, T.N.; Matasova, L.V. Effects of Melatonin on the Activity of the Glutathione Antioxidant System and Various NADPH-Generating Enzymes in the Liver and Blood of Rats with Type 2 Diabetes Mellitus. *Pharm. Chem. J.* **2011**, *45*, 385–388. [CrossRef]
56. Zhu, L.; Zhang, Q.; Hua, C.; Ci, X. Melatonin Alleviates Particulate Matter-Induced Liver Fibrosis by Inhibiting ROS-Mediated Mitophagy and Inflammation via Nrf2 Activation. *Ecotoxicol. Environ. Saf.* **2023**, *268*, 115717. [CrossRef]
57. Salido, E.M.; Bordone, M.; De Laurentiis, A.; Chianelli, M.; Keller Sarmiento, M.I.; Dorfman, D.; Rosenstein, R.E. Therapeutic Efficacy of Melatonin in Reducing Retinal Damage in an Experimental Model of Early Type 2 Diabetes in Rats. *J. Pineal Res.* **2013**, *54*, 179–189. [CrossRef] [PubMed]
58. Furman, B.L. Streptozotocin-Induced Diabetic Models in Mice and Rats. *Curr. Protoc. Pharmacol.* **2015**, *70*, 5.47.1–5.47.20. [CrossRef]
59. Barrière, D.A.; Noll, C.; Roussy, G.; Lizotte, F.; Kessai, A.; Kirby, K.; Belleville, K.; Beaudet, N.; Longpré, J.M.; Carpentier, A.C.; et al. Combination of High-Fat/High-Fructose Diet and Low-Dose Streptozotocin to Model Long-Term Type-2 Diabetes Complications. *Sci. Rep.* **2018**, *8*, 424. [CrossRef]
60. Macêdo, A.P.A.; Cordeiro, G.S.; Santos, L.S.; Santo, D.A.E.; Perez, G.S.; Couto, R.D.; Machado, M.E.P.C.; Medeiros, J.M.B. Murinometric Measurements and Retroperitoneal Adipose Tissue in Young Rats Exposed to the High-Fat Diet: Is There Correlation? *Braz. J. Biol.* **2021**, *81*, 246–250. [CrossRef]
61. Seiva, F.R.F.; Berbert, C.M.S.; Souza, G.A.; Rocha, K.K.H.; Ebaid, G.M.X.; Burneiko, R.C.M.; Novelli, E.L.B. Energy Expenditure, Lipid Profile, Oxidative Stress, and Cardiac Energy Metabolism after Growth Hormone Treatment in Obese Young Rats. *Horm. Metab. Res.* **2010**, *42*, 496–501. [CrossRef]
62. Zheng, K.; Wu, L.; He, Z.; Yang, B.; Yang, Y. Measurement of the Total Protein in Serum by Biuret Method with Uncertainty Evaluation. *Measurement* **2017**, *112*, 16–21. [CrossRef]

63. Alisik, M.; Neselioglu, S.; Erel, O. A Colorimetric Method to Measure Oxidized, Reduced and Total Glutathione Levels in Erythrocytes. *J. Lab. Med.* **2019**, *43*, 269–277. [CrossRef]
64. Hopkins, J.; Tudhope, G.R. Glutathione Peroxidase in Human Redcells in Health and Disease. *J. Haematol.* **1973**, *25*, 563–575. [CrossRef] [PubMed]
65. Keen, J.; Habig, W.H.; Jakoby, W.B. Mechanism for Several Activities of the Glutathione-S-Transferase. *J. Biol. Chem.* **1976**, *20*, 6183–6188. [CrossRef]
66. Crouch, R.K.; Gandy, S.E.; Kimsey, G.; Galbraith, R.A.; Galbraith, G.M.; Buse, M.G. The Inhibition of Islet Superoxide Dismutase by Diabetogenic Drugs. *Diabetes* **1981**, *30*, 235–241. [CrossRef]
67. Aebi, H. Catalase in Vitro. *Methods Enzymol.* **1984**, *105*, 121–126. [CrossRef]

International Journal of
Molecular Sciences

MDPI

Article

Biomarkers of Oxidative Stress in COVID-19 Patients

Elitsa Pavlova [1,*], Petar Atanasov [2], Ivaylo Ivanov [2] and Georgi Dyankov [3]

1 Faculty of Physics, Sofia University "St. Kliment Ohridski", 1164 Sofia, Bulgaria
2 University Multiprofile Hospital for Active Treatment and Emergency Medicine "N. I. Pirogov", 1606 Sofia, Bulgaria; gemel@mail.bg (P.A.); ivo0816@gmail.com (I.I.)
3 Institute of Optical Materials and Technology, Bulgarian Academy of Sciences, 1113 Sofia, Bulgaria; ge.dyankov@gmail.com
* Correspondence: elli_pavlova@abv.bg; Tel.: +359-887982172

Abstract: We focused on evaluating oxidative stress as a major mechanism of cell damage in patients with COVID-19 infection by simultaneously assessing standard oxidative stress biomarkers in vivo—for the very first time in this specific combination—alongside typical clinical biomarkers of inflammation. Standard biomarkers were used to evaluate the oxidative stress status and antioxidant activity in the blood plasma of COVID-19 patients and healthy controls. These included TBARSs (Thiobarbituric Acid-Reactive Substances), SOD (Super Oxide Dismutase), CAT (catalase), GRA (glutathione reductase) activities, and AOC (antioxidant capacity). All clinical inflammation data confirmed a highly activated immune response in the tested COVID-19 patients: WBCs (white blood cells) were increased by nearly 100%, LYMs (lymphocytes) increased by ~30%, CRP (C-reactive protein) rose by over 2200%, and the ESR (erythrocyte sedimentation rate) increased by ~320% compared to established maximum control levels. The results confirmed that the infection involved a free-radical-mediated damage mechanism: TBARS levels increased almost 3-fold, the AOC decreased more than 4-fold, SOD was increased nearly 5-fold, CAT was increased by 1.4 times, and GRA was suppressed by 2.5 times. COVID-19 was associated with oxidative stress and suppressed antioxidant activity. All these changes contribute to the severity of the disease, complications, and mortality in COVID-19 patients.

Keywords: COVID-19; inflammation; biomarkers; oxidative stress; correlation

1. Introduction

Severe acute respiratory syndrome-related coronavirus 2 (SARS-CoV-2) causes severe systemic symptoms [1–4]. Its pathogenesis also involves injury to cells that are not infected by the virus. This is due to the attack of ROS (Reactive Oxygen Species) that are generated from activated white blood cells, which are infiltrated into the virus-infected organs and spread in the body via the bloodstream [5,6]. When oxidative stress causes tissue damage in viral infections, the ROS pathways are a typical component of the immune response mechanisms [7,8].

The exact mechanisms by which the SARS-CoV-2 virus induces oxidative stress are not yet fully understood. However, it is known that the virus infects and harms respiratory tract cells, causing cytokines and chemokines to be released. These can activate immune cells like macrophages and neutrophils. These immune cells produce ROS as a part of their defense mechanisms against pathogens. However, excessive ROS production can cause tissue damage and oxidative stress. Furthermore, the virus can directly cause oxidative stress by interfering with the mitochondria's function. Inflammatory pathways are triggered as a result, adding to the burden of oxidative stress. Oxidative stress, including reductive stress, can have many detrimental effects on the body, including inflammation, tissue damage, and cell death. Oxidative stress in COVID-19 can lead to the development of acute respiratory distress syndrome, a potentially fatal illness marked by lung inflammation and fluid buildup. It also plays a role in complications involving other systems, particularly the cardiovascular system, leading to conditions such as myocarditis, thrombosis, and arrhythmias.

Various diseases can complicate and burden the symptoms of COVID-19 infection. Hypertension and cardiovascular diseases are highlighted as critical comorbidities that elevate the risk of severe COVID-19 outcomes. The endothelial dysfunction that is commonly observed in hypertensive individuals may exacerbate the pro-thrombotic state that is induced by COVID-19, increasing the likelihood of thromboembolic complications [9].

Chronic kidney disease and chronic obstructive pulmonary disease are also associated with increased COVID-19 severity and mortality. Patients with chronic kidney disease often have an impaired immune response and are at higher risk of fluid overload and electrolyte imbalances, which can complicate the clinical management of COVID-19. Similarly, individuals with chronic obstructive pulmonary disease have pre-existing pulmonary damage and reduced lung function, making them more susceptible to severe respiratory complications when infected with SARS-CoV-2 [9].

Among the comorbidities, obesity stands out as a significant risk factor. Individuals with obesity often experience compromised respiratory function due to decreased diaphragmatic excursion and reduced lung volumes, which can exacerbate the respiratory challenges posed by COVID-19. Additionally, obesity is associated with chronic low-grade inflammation, characterized by elevated levels of pro-inflammatory cytokines such as interleukin-6 and tumor necrosis factor-α. This inflammatory medium may amplify the cytokine storm that is observed in severe COVID-19 cases, leading to acute respiratory distress syndrome and multi-organ failure. Furthermore, adipose tissue expresses the angiotensin-converting enzyme 2 receptor, which the virus utilizes for cellular entry [9].

Diabetes mellitus is another prevalent comorbidity that significantly impacts COVID-19 severity. The underlying mechanisms include chronic inflammation, impaired immune response, and potential direct pancreatic damage by SARS-CoV-2, leading to dysregulated glucose metabolism. Hyperglycemia itself can impair neutrophil function and reduce complement activity, further compromising the host's ability to combat the viral infection [9].

The early identification of high-risk individuals through detailed medical history and appropriate diagnostic evaluations is crucial for implementing targeted therapeutic strategies.

In particular, ROS, such as hydrogen peroxide and hydroxyl radicals, superoxide anion radicals, singlet oxygen, and other free oxygen radicals that are produced in all free radical conditions and diseases are considered reagents that attack the cell membranes and give rise to lipid peroxidation [1,10]. Numerous studies have identified lipid peroxidation as a major mechanism underlying cell membrane degradation and damage.

Cellular injury or organ dysfunction caused by oxidative stress occurs when ROS accumulate in excess and damage the host defense mechanisms. These effects include the oxidation of proteins and lipids, which leads to the loss of cell membrane integrity and barrier dysfunction; cytoskeleton fragility and disrupted cell signaling; damage to the nuclear membrane and genetic material; immunological overload, which causes both acute and chronic diseases; and imbalance in the concentrations of various metal–protein complexes and the cell redox status [1,11–13].

The body naturally uses the free radical processes to metabolize oxygen and the harmful byproducts of metabolism, immunological response, and regeneration. The antioxidant systems typically maintain a balance between the synthesis of free radicals and ROS and their neutralization. Oxidative stress is defined as the disturbance of this equilibrium [1,14,15]. It is generally accepted that free radical damage processes are the cause of more than 60 distinct diseases [16–18].

Enzymes like superoxide dismutase (SOD), glutathione peroxidase (GPx), glutathione reductase (GR), and catalase (CAT) are key components of the body's enzymatic antioxidant defense system. Non-enzymatic antioxidants include vitamin E, vitamin C, ß-carotene, uric acid, different proteins like albumin and bilirubin, and many other compounds [7,10]. In general, antioxidants are compounds that can delay, inhibit, or prevent oxidation by trapping free radicals and reducing the level of oxidative stress [17–21]. The body's antioxidant capacity is continually challenged by free radical oxidation processes, particularly during the course of viral infections.

We focused on evaluating oxidative stress as a major mechanism of cell damage in patients with COVID-19 infection by simultaneously assessing standard oxidative stress biomarkers in vivo—for the very first time in this specific combination—alongside typical clinical biomarkers of inflammation.

The selected oxidative biomarkers (Thiobarbituric Acid-Reactive Substances—TBARSs), as well as the antioxidant ones (the activities of the following enzymes: superoxide dismutase, catalase, glutathione reductase, and antioxidant capacity), provided a comprehensive profile of the oxidative stress status in the patients studied. Such a complete study is very informative. All data obtained were compared to those of healthy subjects, who were not infected with SARS-CoV-2 or another infection recently and did not suffer from any chronic condition.

TBARSs are formed as a byproduct of the lipid peroxidation caused by the induced oxidative burst resulting from the infection. They are accepted as a standard biomarker. Because ROS have extremely short half-lives, their direct measurement is difficult. Instead, several products of the damage can be measured, such as TBARSs. Malondialdehyde (MDA) is one of the several low-molecular-weight end products that are formed via the decomposition of certain primary and secondary lipid peroxidation products, measurable as TBARSs [22]. That test displays the total amount of free radical peroxidation products, as well as the oxidation process trend in the infection's target organs.

The SOD amounts in cellular and extracellular media are essential for preventing oxidative-stress-related diseases, including viral infections [23,24]. The reaction catalyzed by SOD is extremely fast, with a turnover of 2×10^6 $mol^{-1}s^{-1}$, and the presence of sufficient amounts of the enzyme in cells and tissues typically keeps the concentration of superoxide radicals very low [25]. Therefore, quantifying SOD activity is crucial to accurately characterizing a biological system's antioxidant capacity. Notably, SOD treatment has been shown to reduce the lethal or toxic effects of influenza virus infection in mouse models, although it does not inhibit the viral replication itself [26–29].

Catalase is an enzyme that is found in practically all living creatures who are exposed to oxygen. It catalyzes the conversion of hydrogen peroxide into oxygen and water [30]. It is a crucial enzyme for shielding the cell from ROS-induced oxidative damage. One catalase molecule may convert millions of hydrogen peroxide molecules into oxygen and water per second, making it one of the enzymes with the fastest turnover rates [31]. CAT is a tetramer made from four polypeptide chains, each over 500 amino acids long [32]. It contains four Fe-containing heme groups that allow the enzyme to react with H_2O_2.

Glutathione reductase is a flavoprotein that catalyzes the NADP.H-dependent reduction of oxidized glutathione (GSSG) to glutathione (GSH) [33]. This enzyme is a part of the GSH redox cycle. It is responsible for the levels of reduced cellular GSH. To defend against oxidative stress, a high GSH/GSSG ratio is necessary. GPx1 is predominantly present in erythrocytes, kidneys, and the liver, and GPx4 is highly expressed in renal epithelial cells and testes. The same substrate, H_2O_2, is used by both GPx and CAT. However, GPx is the main source of defense against oxidative stress, because it can react with lipids and other organic hydroperoxides in an efficient manner [34,35]. Therefore, oxidative stress or other circumstances that reduce GSH in the oral, nasal, and upper airway epithelium may increase the vulnerability to viral infections.

The total antioxidant capacity reflects the cumulative action of both water-soluble and lipid-soluble antioxidants. Since various antioxidants and antioxidant enzymes function synergistically in vivo, assessing the total antioxidant capacity provides a more comprehensive overview of the body's antioxidant status than evaluating individual antioxidants or enzymes in isolation.

By concurrently evaluating common oxidative stress biomarkers in vivo—for the first time in this particular combination—alongside common clinical biomarkers of inflammation, we aimed to assess oxidative stress as a primary mechanism of cell damage in patients with COVID-19 infection.

2. Results

2.1. Clinical Data (Figure 1)

The standard clinical data obtained for the inflammation biomarkers showed that all tested COVID-19 patients exhibited markedly elevated inflammatory activity:

- The number of WBCs in the healthy controls was approximately 70% compared to the established control levels, whereas in the infected individuals, it was 30% higher.
- The number of LYMs was decreased more than two times in the infected patients in comparison to the healthy individuals.

The white blood cells are a key component of the immune system that helps fight infections. A much higher number of WBCs is generally associated with inflammation. This was observed in the chosen patients too. In some cases, COVID-19 can cause a decrease in the number of white blood cells, especially lymphocytes, a condition called leukopenia, especially in the early stages of the disease or in severe cases. That was the case in the tested patients as well. The mechanism by which COVID-19 causes lymphopenia is not fully understood, but it may be due to a direct viral attack on the lymphocytes or to the immune response against the virus as a whole. In other cases of this infection, the opposite condition may be observed—an increase in the number of white blood cells, especially lymphocytes (leukocytosis). This increase may be due to the infection-stimulated immune response fighting the virus.

- The most informative and sensitive parameter for inflammation processes in the body is CRP. This was raised to more than 2200% in the COVID-19 patients. Its level in the healthy control was about 4% compared to the highest control levels.

The CRP levels in the blood usually rise rapidly in response to inflammation, especially in patients with severe COVID-19 infection. They are usually high in cases of respiratory failure or other complications. CRP levels can also be used to monitor the course of COVID-19 infection and the patient's response to the chosen treatment.

- The ESR was also much higher in the tested COVID-19 patients, by more than 320%; it was calculated to be about 35% in the healthy subjects, in comparison to the established highest control levels.

The ESR measures the rate of red blood cell sedimentation. Elevated levels are generally indicative of the presence of inflammation. It has been found that the ESR is elevated in many COVID-19 patients, particularly those with severe disease. Similar to CRP, the ESR is a useful predictor of the severity of the disease in these patients.

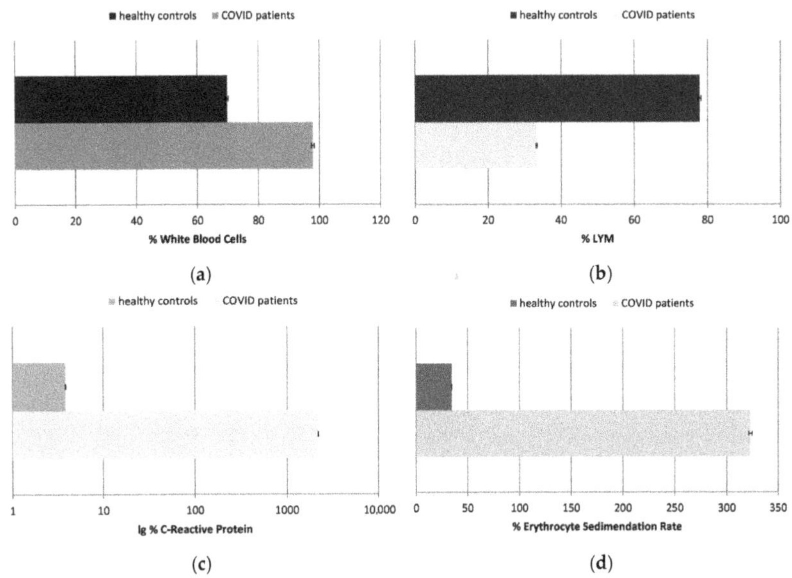

Figure 1. The number of white blood cells (WBCs) (**a**) and lymphocytes (LYMs) (**b**); the concentration of C-reactive protein (CRP) (**c**); and the erythrocyte sedimentation rate (ESR) (**d**), presented as a percentage in healthy control subjects and COVID-19 patients ($p \leq 0.05$). The data are not presented in standard units due to different, but standard, methodologies and reference ranges being applied in the laboratories where the samples of the patients and healthy controls were tested; data are presented in percentages, with 100% being the maximum acceptable control value.

The inflammation markers in the COVID-19 patients included in our study were significantly altered, reflecting a heightened systemic inflammatory response. However, it is important to note that inflammation in COVID-19 can vary widely depending on the severity of the infection and the individual immune responses. Our findings confirmed a strong, direct correlation between the inflammation status and oxidative stress levels in these patients, reinforcing the role of both mechanisms in the pathophysiology of the COVID-19 disease.

2.2. *Thiobarbituric Acid-Reactive Substances (TBARSs) (Figure 2)*

Our data showed that TBARSs increased by almost three times in the COVID-19 patients in comparison to the healthy controls. Elevated levels of TBARSs indicate oxidative stress, and our studies confirmed that those products accumulated in the bloodstream as a result of the infection. Thus, further exacerbated cellular damage may easily contribute to the severity of the disease.

High levels of TBARSs have been found in COVID-19 patients and have been associated with disease severity and thrombotic events. These findings suggest that oxidative stress may play a role in the pathogenesis of COVID-19 and that antioxidant therapies may be beneficial in the management of COVID-19 [36]. Some results of other authors have also shown that the TBARS levels in erythrocytes are higher in COVID-19 patients than in control groups, showing the existence of an oxidative imbalance in these patients and the overproduction of free radicals [37].

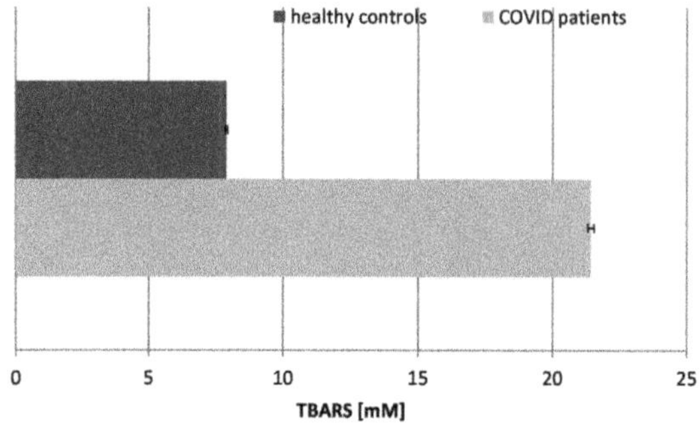

Figure 2. Concentration of TBARSs in the blood plasma of COVID-19 patients and healthy control subjects ($p \leq 0.05$).

2.3. *Antioxidant Capacity (AOC) (Figure 3)*

The antioxidant capacity (AOC) and activity refer to the ability of the body to neutralize harmful free radicals and ROS and to protect against oxidative stress. Some authors have reported that the total antioxidant capacity levels are considerably lower in patients compared to healthy individuals ($p < 0.05$), including patients with mild and severe disease ($p < 0.05$). These findings suggest that COVID-19 patients may be susceptible to a depleted total antioxidant capacity. Moreover, observing such variations in the blood samples of infected individuals could be considered a predictive marker of COVID-19 severity [38]. The antioxidant capacity reflects the overall antioxidant content and activity of the body in fighting oxidative stress of various etiologies. Our results showed that the AOC was greatly reduced in the blood plasma of COVID-19 patients by more than 4-fold, which proves the overall exhaustion of antioxidants and antioxidant enzymes in the organism. The drastic reduction in AOC indicates that the overall antioxidant capacity is overwhelmed, failing to neutralize the excess free radicals.

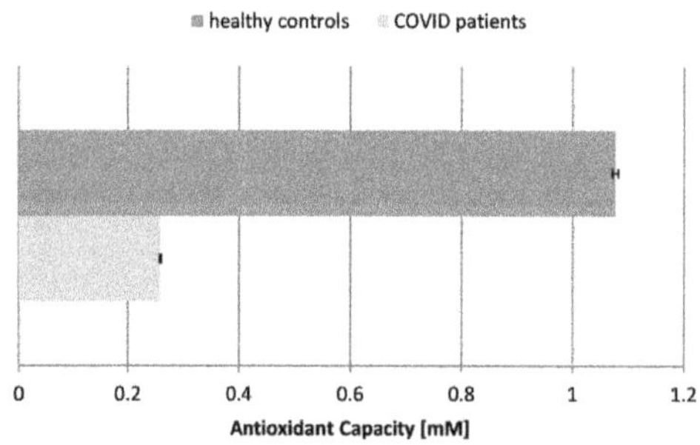

Figure 3. Antioxidant capacity (AOC) in the blood plasma of COVID-19 patients and healthy control subjects ($p \leq 0.05$).

A major part of the antioxidant defense system of the organism consists of antioxidant enzymes, which are formed and kept on an ongoing basis to regulate everyday levels of oxidation in the body. We only tested the major ones—superoxide dismutase (SOD), catalase (CAT), and glutathione reductase (GR). They work together and neutralize specific ROS, maintaining the cellular redox balance.

2.4. Activity of Super Oxide Dismutase (SOD) (Figure 4)

Our results demonstrated that the SOD activity in COVID-19 patients was significantly elevated, being nearly 5-fold higher than in healthy individuals. This is indicative of stimulation of the enzyme in the course of the disease and its potential to fight superoxide radicals ($O_2^{\cdot -}$), generated during the inflammatory response to SARS-CoV-2 infection. The significant increase in SOD activity suggests a reactive upregulation to counteract the high levels of superoxide radicals. Decreased SOD activity may be associated with serious complications and lethal outcomes in COVID-19 patients.

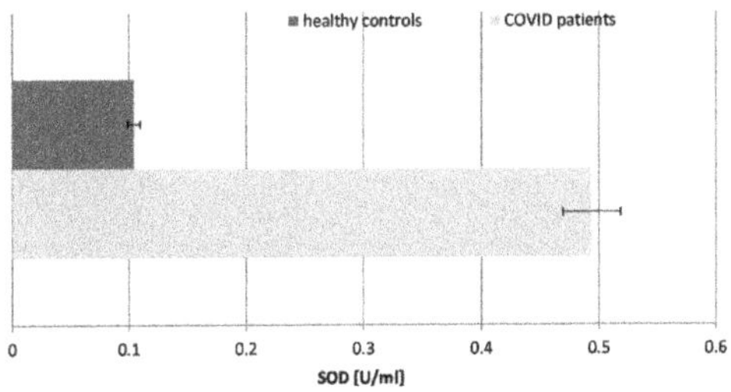

Figure 4. The activity of superoxide dismutase (SOD) in the blood plasma of COVID-19 patients and healthy control subjects ($p \leq 0.05$).

SOD converts superoxide radicals ($O_2^{\cdot -}$) into hydrogen peroxide (H_2O_2), necessitating enhanced CAT activity.

2.5. Catalase (CAT) Activity (Figure 5)

CAT is a key antioxidant enzyme that plays a crucial role in protecting cells from oxidative damage by catalyzing the decomposition of hydrogen peroxide (H_2O_2), which accumulates rapidly during disease states. Yan et al., 2020 found altered, decreased catalase activity in COVID-19 patients, suggesting that oxidative stress may contribute to the pathogenesis of the disease. Furthermore, the authors observed an inverse correlation between catalase levels and inflammatory markers, implicating oxidative stress in the cytokine storm that is commonly seen in severe COVID-19 cases [39].

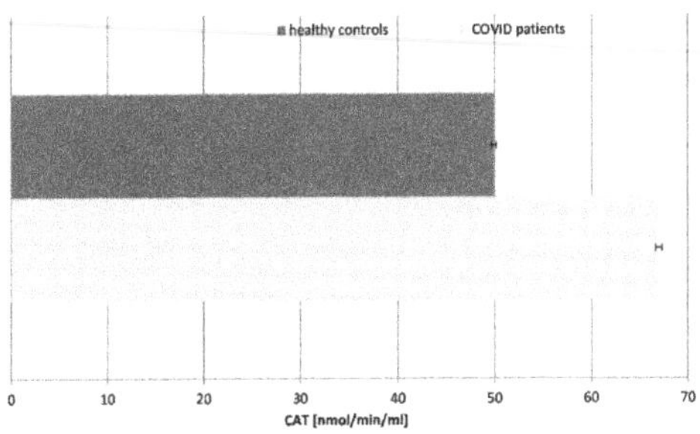

Figure 5. Catalase (CAT) activity in blood plasma of COVID-19 patients and healthy control subjects ($p \leq 0.05$).

Our results revealed that the CAT activity was significantly increased (36%) in the COVID-19 patients in comparison to the healthy controls, following the pattern of the SOD enzyme. The turnover of CAT is the highest among all enzymes. The increase in CAT activity suggests an adaptive response to elevated hydrogen peroxide levels. It controls the concentration of H_2O_2, not only as an inflammatory but also as a messenger molecule, controlling the activity of other enzymes and genes. The activity of CAT and all antioxidant enzymes is exhausted over time, especially in COVID-19 cases with additional complications, as has been confirmed by many authors. It should be noted that all patients tested in that research were chosen during active COVID-19 infection, without any additional acute or chronic conditions or diseases.

The elevated SOD and CAT activity may provide temporary protection against oxidative damage during the acute phase of infection, limiting lipid peroxidation and cellular injury. This is supported by studies showing enhanced antioxidant enzyme activity in the early disease stages as part of systemic defense mechanisms. However, prolonged stimulation may lead to enzyme exhaustion, particularly in severe cases or patients with complications [40,41].

2.6. Glutathione Reductase Activity (GRA) (Figure 6)

Glutathione is a tripeptide composed of glutamate, cysteine, and glycine and appears to be a crucial antioxidant molecule in the body. It acts as a substrate for glutathione reductase and peroxidase and is also involved in detoxifying ROS and other toxins. Glutathione reductase activity refers to the rate at which this enzyme catalyzes the reduction of oxidized glutathione (GSSG) to its reduced form (GSH), using the cofactor NADPH. Glutathione reductase activity is important for maintaining the levels of reduced glutathione within cells, which can scavenge ROS and protect cells from oxidative damage. Deficiencies in glutathione reductase activity have been associated with various pathological conditions, including neurodegenerative diseases, liver dysfunction, cancer, etc. Therefore, measuring glutathione reductase activity is an important tool in assessing the oxidative stress status of cells and tissues. The suppression of GRA indicates a compromised detoxification pathway for hydrogen peroxide and lipid peroxides.

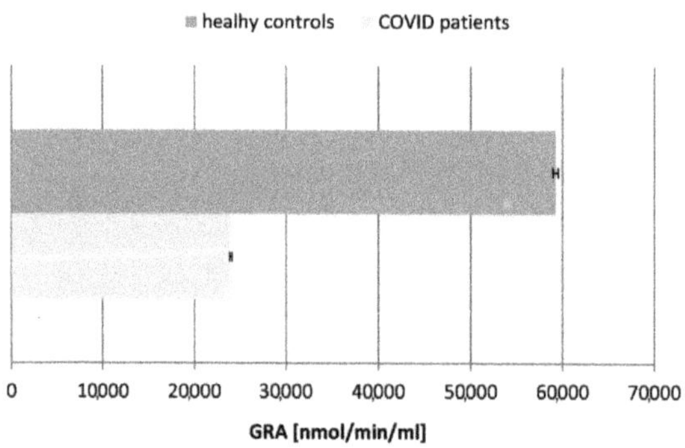

Figure 6. Glutathione reductase activity (GRA) in blood plasma of COVID-19 patients and healthy control subjects ($p \leq 0.05$).

Our results confirmed a ~2.5-fold decrease in the activity of glutathione reductase in the tested COVID-19 patients in comparison to the controls (healthy subjects), proving the exhaustion of glutathione as a major antioxidant molecule in the organism during and after the acute phase of the infection.

The contrasting trends, the upregulation of SOD and CAT and downregulation of GRA, reflect a dynamic interplay between compensatory mechanisms and exhaustion phases:

- Early adaptive response: During the initial infection stages, increased SOD and CAT activities mitigate ROS accumulation, protecting cellular components from oxidative damage
- Progressive exhaustion: As the infection persists, antioxidant systems like GRA become depleted due to sustained ROS production and inflammation. This exhaustion phase reduces the cellular resilience against oxidative stress, potentially contributing to complications such as organ failure or severe systemic inflammation.

Monitoring SOD, CAT, and GRA activities could serve as biomarkers for disease progression or therapeutic efficacy. Elevated SOD and CAT may indicate early compensatory responses, while reduced GRA could signal impending oxidative-stress-related complications [40,41].

2.7. Correlation of Clinical and Experimental Biomarkers Between Healthy and Infected Subjects

The analysis of the clinical parameters (number of white blood cells, % lymphocytes, C-reactive protein, and erythrocyte sedimentation rate) as standard clinical inflammatory markers and the measured basic parameters of oxidative status in the organism (TBARSs, AOC, SOD, CAT, GRA) showed complete correlation between the data of the tested healthy subjects and the COVID-19 patients ($r = +0.995$), (Figure S1). The used function computes the Pearson correlation coefficient, which measures the strength and direction of the linear relationship between two variables. This is another confirmation that oxidative stress and the free radical and oxidative mechanism, on the one hand, and the cytokine storm, on the other, appear to be a damaging pathway in patients with "pure" COVID-19 infection.

A major strength of our study lies in the simultaneous assessment of multiple standardized biomarkers with strong clinical relevance, allowing for a comprehensive overview of both oxidative and inflammatory processes. However, a key limitation is that although our findings establish a robust correlation, they do not provide direct evidence of causation. The exact mechanisms by which oxidative stress may drive disease severity in COVID-19 remain to be elucidated through targeted experimental studies.

A cytokine storm, a severe immune response marked by excessive cytokine production, can arise not only from viral evasion strategies but also from genetic abnormalities in the host. Cytokines contribute to endothelial permeability by damaging junctional proteins like VE-cadherin or suppressing their synthesis. SARS-CoV-2 has been shown to activate Toll-like receptor (TLR) 9 and NF-κB, which increases the expression of inflammatory genes while reducing nitric oxide, a key vasodilator and antithrombotic agent. Additionally, SARS-CoV-2 can hyperactivate neutrophils via the NLR family pyrin domain containing 3 (NLRP3), causing leukocytes to release ROS, extracellular traps (NETs), and proteolytic enzymes that perpetuate inflammation [42].

SARS-CoV-2 also drives chronic inflammation through mitochondrial dysfunction and the activation of the cyclic GMP-AMP synthase (cGAS)-stimulator of the interferon genes (STING) pathway. Elevated circulating inflammatory agents, endothelial dysfunction, and micro-thrombosis increase the risk of cardiovascular complications, including the progression of atherosclerotic plaques, arterial or venous thrombosis, and heightened arterial stiffness [42].

SARS-CoV-2 enters the host cells primarily through the ACE2 receptor. The spike (S) protein plays a crucial role in this process. The host proteases, such as TMPRSS2 and cathepsins B and L, facilitate the viral entry by activating the S protein.

Based on these factors, the possible therapeutic strategies for COVID-19 may include the following:

- Protease inhibitors that target host proteases like TMPRSS2 and cathepsins, blocking the virus's entry;
- Antiviral agents—drugs inhibiting the viral RNA replication by various mechanisms;
- Immunomodulators, such as JAK inhibitors, reducing the inflammatory response;
- Plasma therapy—the application of antibodies from recovered patients to stimulate the immune response;
- Monoclonal antibodies—engineered antibodies targeting viral proteins, e.g., the spike protein;
- Vaccines, which are more effective against SARS-CoV-2 and its variants;
- Antioxidants—a supportive therapy against the inflammation and oxidative stress in the organism [43].

The free radical mechanism is a basic damaging pathway. It has been proven to have great clinical significance. Cell damage and oxidative stress contribute to the disease severity, complications, and mortality in COVID-19 patients [44]. Therefore, antioxidant agents targeting oxidative stress are suitable for reducing its severity [45]. The combinations of antioxidants and antiviral drugs could synergistically reduce the complications and lethal effects of the infection. Further research is needed to explore targeted antioxidant therapies as potential adjunct treatments for COVID-19 to mitigate oxidative damage and improve the clinical outcomes.

The elucidation of the mechanisms of prophylaxis, damage, and therapy in this type of viral infection is of great importance and social significance. Despite the number of articles available on the topic, the information reported by other authors on the damage mechanisms is still insufficient or missing for many specific reactions, products, and activities. The simultaneous evaluation of the chosen inflammatory and oxidative major biomarkers investigated in our research clarified, to some extent, the role of oxidative stress in COVID-19.

3. Materials and Methods

3.1. Human Subjects

Blood and blood plasma were obtained and tested on the same day from individuals who were eligible for participation: 27 hospitalized male adults (n = 27), aged 23–62, with a confirmed COVID-19 diagnosis and active, severe infection, and 25 healthy male and 5 healthy female individuals (same age range, n = 30), assigned to the tested healthy group. All COVID-19 patients had recovered for shorter or longer periods in the hospital. No antioxidant supplements or therapy were applied to any of the tested people. The exclusion criteria included all other acute or chronic diseases or conditions and accompanying medications. All these restrictions significantly reduced the number of chosen patients and control individuals.

The gender imbalance (100% male patients vs. 83% male controls) between the two tested groups may affect the antioxidant responses—estrogen enhances glutathione synthesis, meaning a better antioxidant defense and more stable control levels, while testosterone may suppress immune responses, meaning lower oxidative stress levels. Nevertheless, the differences registered between both groups are significant and credible.

The tested blood plasma was stored frozen at −30 °C. The procedures were performed in compliance with the relevant laws and institutional guidelines and according to the ethical standards of the Declaration of Helsinki and preliminarily approved by the Ethics

Committees of the involved institutions (ДК1/10 29 March 2021). Informed consent was obtained from every individual taking part in the research.

3.2. Clinical Data

The standard clinical data used for the number of white blood cells (WBCs), C-reactive protein (CRP) content, erythrocyte sedimentation rate (ESR), and number/percentage of lymphocytes (LYMs) were obtained from the University Multiprofile Hospital for Active Treatment and Emergency Medicine "N. I. Pirogov", (COVID-19 patients), the National Center for Transfusion Hematology and the Diagen Lab (healthy controls), Sofia, Bulgaria (Table 1).

Table 1. Reference ranges of the tested clinical inflammation biomarkers.

Reference Ranges	UMHATEM "N. I. Pirogov" Lab	Diagen Lab
WBCs [$\times 10^9$]	4.1–11	3.5–10.5
LYMs	0.6–4.1 [$\times 10^9$]	20–40%
CRP	0.0–0.5 [mg/dL]	0.00–10.00 [mg/L]
ESR [mm/h]	0–15	2–20

3.3. Reagents

TBARSs, SOD, CAT, GR activities, and AOC were measured using kits from Cayman Chemicals, Ann Arbor, MI, USA, following the manufacturer's protocols. Detailed information on each method, including reagents and its protocol, can be found at https://www.caymanchem.com/products/categories and the following links that were first visited on 18 May 2020:

TBARS Assay Kit, Cat. # 10009055
https://cdn.caymanchem.com/cdn/seawolf/insert/10009055.pdf
SuperOxide Dismutase Assay Kit, Cat. # 706002
https://cdn.caymanchem.com/cdn/seawolf/insert/706002.pdf
Catalase Assay Kit Cat. # 707002
https://cdn.caymanchem.com/cdn/seawolf/insert/707002.pdf
Glutathione Reductase Assay Kit, Cat. # 703202,
https://cdn.caymanchem.com/cdn/seawolf/insert/703202.pdf
Antioxidant Assay Kit, Cat. # 709001
https://cdn.caymanchem.com/cdn/seawolf/insert/709001.pdf

Spectrophotometrically, all samples were measured by SPECTROstar® Nano, BMG LABTECH, Ortenberg, Germany.

3.4. Statistical Analysis

The limited statistical power in this study poses risks of Type II errors, inflated effect sizes, and reduced generalizability, while introducing confounding factors like a gender imbalance. We employed strategies to improve the power and reduce bias during the measurement, data analysis, and interpretation stages. All results were obtained as triple reproducible measurements. All random and systematic errors were excluded. Statistical analysis (Student's t-test) was performed using Origin 8.5 (OriginLab, Northampton, MA, USA) and Microsoft Office Excel 2010 (Microsoft Corp., Redmond, WA, USA) software. Data were accepted as statistically significant at $p \leq 0.05$.

4. Conclusions

All clinical biomarkers of inflammation confirmed the highly activated status of the immune system in the tested COVID-19 patients: WBCs were increased by almost 100%, LYMs were increased by ~30%, CRP was raised by more than 2200%, and ESR was increased by ~320% in comparison to the established maximum control levels.

The biomarkers of oxidative stress confirmed that the infection possesses a free radical damage mechanism: TBARSs were increased almost 3-fold, AOC was reduced more than 4-fold, SOD was increased almost 5-fold, CAT was increased 1.4-fold, and GRA was suppressed 2.5-fold in the patients with "pure" COVID-19 infection. None of the tested COVID-19 patients suffered from any other acute or chronic conditions.

In conclusion, COVID-19 is associated with oxidative stress and suppressed antioxidant activity. COVID-19 infection can be described as a free radical disease.

Supplementary Materials: The following supporting information can be downloaded at https://www.mdpi.com/article/10.3390/ijms26083869/s1.

Author Contributions: Conceptualization, E.P.; Methodology, E.P., P.A., and I.I.; Formal Analysis and Investigation, E.P. and I.I.; Resources, G.D.; Data Curation, E.P. and I.I.; Writing—Original draft preparation—E.P.; Writing—Review and Editing—E.P., P.A., I.I., and G.D.; Visualization, E.P.; Supervision, Project Administration, and Funding Acquisition—G.D. All authors have read and agreed to the published version of the manuscript.

Funding: This research was funded by the National Science Fund of Bulgaria, Project Study of the interaction of specific structural proteins of SARS-CoV-2 with biologically active molecules and their application for the creation of rapid antigen tests for early diagnosis of COVID-19, Contract Number: ДК1/10 29 March 2021 and the Bulgarian Ministry of Education and Science, Contract Number: D01-352 13 December 2023.

Institutional Review Board Statement: The study was conducted in accordance with the Declaration of Helsinki and approved by the Institutional Review Boards. It was also preliminarily approved by the Ethics Committees of the involved institutions—Sofia University "St. Kliment Ohridski", University Multiprofile Hospital for Active Treatment and Emergency Medicine "N. I. Pirogov" and Bulgarian Academy of Sciences, Bulgaria under Contract Number: ДК1/10 29 March 2021.

Informed Consent Statement: Informed consent was obtained from all subjects involved in the study.

Data Availability Statement: All research data are available upon request.

Acknowledgments: We would like to give our special thanks to our colleagues Julia Veleva and Zdravka Remeni from the National Center for Transfusion Hematology, 112 Bratya Miladinovi Str, 1202, Sofia, Bulgaria, who provided all the samples from the healthy individuals, along with the data on the standard clinical parameters of those individuals.

Conflicts of Interest: The authors declare no conflicts of interest.

Abbreviations

The following abbreviations are used in this manuscript:

COVID-19	Coronavirus Disease of 2019
TBARSs	Thiobarbituric Acid-Reactive Substances
SOD	Super Oxide Dismutase
CAT	Catalase
GRA	Glutathione Reductase Activity

AOC	Antioxidant Capacity
WBCs	White Blood Cells
LYMs	Lymphocytes
CRP	C-Reactive Protein
ESR	Erythrocyte Sedimentation Rate
SARS-CoV-2	Severe Acute Respiratory Syndrome-Related Coronavirus 2
ROS	Reactive Oxygen Species
GPx	Glutathione-Peroxidase
GR	Glutathione-Reductase
MDA	Malondialdehyde
GSSG	Oxidized Glutathione
GSH	Glutathione

References

1. Sies, H. *Oxidative Stress*; Elsevier: Amsterdam, The Netherlands, 2013.
2. Tumpey, T.M.; García-Sastre, A.; Taubenberger, J.K.; Palese, P.; Swayne, D.E.; Pantin-Jackwood, M.J.; Schultz-Cherry, S.; Solórzano, A.; Van Rooijen, N.; Katz, J.M.; et al. Pathogenicity of influenza viruses with genes from the 1918 pandemic virus: Functional roles of alveolar macrophages and neutrophils in limiting virus replication and mortality in mice. *J. Virol.* **2005**, *79*, 14933–14944. [CrossRef]
3. Yilmaz, N.; Eren, E.; Kalayci, Z.; Saribek, F.; Oz, C. COVID-19 may be acting like a trojan horse: Novel target hemoglobin and free iron. In Proceedings of the GETAT 2020 3rd International Traditional and Complementary Medicine Congress: GETAT Congress 2020, Istanbul, Turkey, 1–3 April 2020. Available online: https://www.researchgate.net/publication/340540096_Covid-19_may_be_acting_like_a_Trojan_Horse_Novel_target_hemoglobin_and_free_iron_Trojan_Horse_and_Covid_-19 (accessed on 9 August 2023).
4. Delgado-Roche, L.; Mesta, F. Oxidative stress as key player in severe acute respiratory syndrome Coronavirus (SARS-CoV) infection. *Arch. Med. Res.* **2020**, *51*, 384–387. [CrossRef]
5. Peterhans, E. Reactive oxygen species and nitric oxide in viral diseases. *Biol. Trace Elem. Res.* **1997**, *56*, 107–116. [CrossRef]
6. Uchide, N.; Toyoda, H. Current status of monocyte differentiation-inducing (MDI) factors derived from human fetal membrane chorion cells undergoing apoptosis after influenza virus infection. *Gene Regul. Syst. Biol.* **2007**, *1*, GRSB.S374-4. [CrossRef]
7. Akaike, T.; Suga, M.; Maeda, H. Free radicals in viral pathogenesis: Molecular mechanisms involving superoxide and NO. *Exp. Biol. Med.* **1998**, *217*, 64–73. [CrossRef]
8. Halliwell, B. Free radicals. antioxidants. and human disease: Curiosity. cause. or consequence? *Lancet* **1994**, *344*, 721–724. [CrossRef]
9. Fitero, A.; Bungau, S.G.; Tit, D.M.; Endres, L.; Khan, S.A.; Bungau, A.F.; Romanul, I.; Vesa, C.M.; Radu, A.-F.; Tarce, A.G.; et al. Comorbidities, Associated Diseases, and Risk Assessment in COVID-19—A Systematic Review. *Int. J. Clin. Pr.* **2022**, *2022*, 1571826. [CrossRef]
10. Halliwell, B.; Gutteridge, J.M.C. *Free Radicals in Biology and Medicine*; Oxford University Press: Oxford, UK, 2015.
11. Ames, B.N.; Shigenaga, M.K.; Hagen, T.M. Oxidants. antioxidants. and the degenerative diseases of aging. *Proc. Natl. Acad. Sci. USA* **1993**, *90*, 7915–7922. [CrossRef]
12. Blumberg, J. Use of biomarkers of oxidative stress in research studies. *J. Nutr.* **2004**, *134*, 3188S–3189S. [CrossRef]
13. Cerne, D.; Lukac-Bajalo, J. Oxidative stress assays for disease risk stratification. *PubMed* **2006**, *56*, 1–17.
14. Halliwell, B. Oxidative stress. nutrition and health. Experimental strategies for optimization of nutritional antioxidant intake in humans. *Free. Radic. Res.* **1996**, *25*, 57–74. [CrossRef]
15. Liu, T.Z.; Stern, A. Assessment of the role oxidative stress in human diseases. *J. Biomed. Lab. Sci.* **1998**, *10*, 12–18.
16. Oda, T.; Akaike, T.; Hamamoto, T.; Suzuki, F.; Hirano, T.; Maeda, H. Oxygen radicals in influenza-induced pathogenesis and treatment with pyran polymer-conjugated SOD. *Science* **1989**, *244*, 974–976. [CrossRef]
17. Pavlova, E.L.; Lilova, M.; Savov, V. Oxidative stress in children with kidney disease. *Pediatr. Nephrol.* **2005**, *20*, 1599–1604. [CrossRef]
18. Ioannides, C. *Cytochromes P450: Metabolic and Toxicological Aspects*; CRC Press: Boca Raton, FL, USA, 1996.
19. Pavlova, E.L.; Savov, V. Chemiluminescent in vitro estimation of the inhibitory constants of antioxidants ascorbic and uric acids in Fenton's reaction in urine. *Biokhimiya* **2006**, *71*, 861–863. [CrossRef]
20. Rice-Evans, C. *Free Radicals, Oxidant Stress and Drug Action*; Richelieu Press: London, UK, 1987; ISSN 0961-0308.

21. Savov, V.M.; Galabov, A.S.; Tantcheva, L.P.; Mileva, M.M.; Pavlova, E.L.; Stoeva, E.S.; Braykova, A.A. Effects of rutin and quercetin on monooxygenase activities in experimental influenza virus infection. *Exp. Toxicol. Pathol.* **2006**, *58*, 59–64. [CrossRef]
22. Janero, D.R. Malondialdehyde and thiobarbituric acid-reactivity as diagnostic indices of lipid peroxidation and peroxidative tissue injury. *Free Radic. Biol. Med.* **1990**, *9*, 515–540. [CrossRef]
23. Maier, C.M.; Chan, P.H. Book Review: Role of superoxide dismutases in oxidative damage and neurodegenerative disorders. *Neuroscientist* **2002**, *8*, 323–334. [CrossRef]
24. Liu, D. The roles of free radicals in amyotrophic lateral sclerosis. *J. Mol. Neurosci.* **1996**, *7*, 159–167. [CrossRef] [PubMed]
25. Malstrom, B.; Andreasson, L.; Reinhammer, B. *The Enzymes*; Boyer, P., Ed.; XIIB; Academic Press: New York, NY, USA, 1975; p. 533.
26. Stýblo, M.; Walton, F.S.; Harmon, A.W.; Sheridan, P.A.; Beck, M.A. Activation of superoxide dismutase in selenium-deficient mice infected with influenza virus. *J. Trace Elem. Med. Biol.* **2007**, *21*, 52–62. [CrossRef] [PubMed]
27. Sidwell, R.W.; Huffman, J.H.; Bailey, K.W.; Wong, M.H.; Nimrod, A.; Panet, A. Inhibitory effects of recombinant manganese superoxide dismutase on influenza virus infections in mice. *Antimicrob. Agents Chemother.* **1996**, *40*, 2626–2631. [CrossRef]
28. Serkedjieva, J.; Roeva, I.; Angelova, M.; Dolashka, P.; Voelter, W.G. Combined protective effect of a fungal Cu/Zn-containing superoxide dismutase and rimantadine hydrochloride in experimental murine influenza virus infection. *PubMed* **2003**, *47*, 53–56.
29. Akaike, T.; Ando, M.; Oda, T.; Doi, T.; Ijiri, S.; Araki, S.; Maeda, H. Dependence on O_2^- generation by xanthine oxidase of pathogenesis of influenza virus infection in mice. *J. Clin. Investig.* **1990**, *85*, 739–745. [CrossRef]
30. Chelikani, P.; Fita, I.; Loewen, P.C. Diversity of structures and properties among catalases. *Cell. Mol. Life Sci.* **2004**, *61*, 192–208. [CrossRef]
31. Goodsell, D. PDB101: Molecule of the Month: Catalase [Internet]. RCSB: PDB-101. 2004. Available online: https://pdb101.rcsb.org/motm/57 (accessed on 9 August 2023).
32. Catalase Structure Tutorial Text [Internet]. earth.callutheran.edu. Available online: https://earth.callutheran.edu/Academic_Programs/Departments/BioDev/omm/catalase/frames/cattx.htm#intro (accessed on 9 August 2023).
33. Carlberg, I.; Mannervik, B. Glutathione reductase. In *Glutamate, Glutamine, Glutathione, and Related Compounds. Methods in Enzymology*; Meister, A., Ed.; Academic Press: Cambridge, MA, USA; Volume 113, pp. 484–490.
34. Matés, J.M.; Pérez-Gómez, C.; Núñez de Castro, I. Antioxidant enzymes and human diseases. *Clin. Biochem.* **1999**, *32*, 595–603. [CrossRef]
35. Cai, J.; Chen, Y.; Seth, S.; Furukawa, S.; Compans, R.W.; Jones, D.P. Inhibition of influenza infection by glutathione. *Free Radic. Biol. Med.* **2003**, *34*, 928–936. [CrossRef]
36. Golabi, S.; Ghasemi, S.; Adelipour, M.; Bagheri, R.; Suzuki, K.; Wong, A.; Seyedtabib, M.; Naghashpour, M. Oxidative stress and inflammatory status in COVID-19 outpatients: A health center-based analytical cross-sectional study. *Antioxidants* **2022**, *11*, 606. [CrossRef]
37. Weber, A.A.P.; Bulegon, J.S.; de Souza, M.D.; Vendrame, S.A.; Venturini, L.; Mombaque Dos Santos, W.; Gonçalves, T.L. Clinical, demographic and oxidative profile of patients with COVID-19 and disease severity. *Diagn. Microbiol. Infect. Dis.* **2023**, *105*, 115886. [CrossRef]
38. Yaghoubi, N.; Youssefi, M.; Jabbari Azad, F.; Farzad, F.; Yavari, Z.; Zahedi Avval, F. Total antioxidant capacity as a marker of severity of COVID-19 infection: Possible prognostic and therapeutic clinical application. *J. Med Virol.* **2021**, *94*, 1558–1565. [CrossRef]
39. Yan, L.; Zhang, H.T.; Goncalves, J.; Xiao, Y.; Wang, M.; Guo, Y.; Sun, C.; Tang, X.; Jing, L.; Zhang, M.; et al. An interpretable mortality prediction model for COVID-19 patients. *Nat. Mach. Intell.* **2020**, *2*, 283–288. [CrossRef]
40. Tsermpini, E.E.; Glamočlija, U.; Ulucan-Karnak, F.; Redenšek Trampuž, S.; Dolžan, V. Molecular mechanisms related to responses to oxidative stress and antioxidative therapies in COVID-19: A systematic review. *Antioxidants* **2022**, *11*, 1609. [CrossRef]
41. Muhammad, Y.; Kani, Y.A.; Iliya, S.; Muhammad, J.B.; Binji, A.; El-Fulaty Ahmad, A.; Kabir, M.B.; Umar Bindawa, K.; Ahmed, A. Deficiency of antioxidants and increased oxidative stress in COVID-19 patients: A cross-sectional comparative study in Jigawa, Northwestern Nigeria. *SAGE Open Med.* **2021**, *9*, 2050312121991246. [CrossRef]
42. Karakasis, P.; Nasoufidou, A.; Sagris, M.; Fragakis, N.; Tsioufis, K. Vascular alterations following COVID-19 infection: A comprehensive literature review. *Life* **2024**, *14*, 545. [CrossRef]
43. Kabir, M.T.; Uddin, M.S.; Hossain, M.F.; Abdulhakim, J.A.; Alam, M.A.; Ashraf, G.M.; Bungau, S.G.; Bin-Jumah, M.N.; Abdel-Daim, M.M.; Aleya, L. nCOVID-19 pandemic: From molecular pathogenesis to potential investigational therapeutics. *Front. Cell Dev. Biol.* **2020**, *8*, 616. [CrossRef]
44. Mehri, F.; Rahbar, A.H.; Ghane, E.T.; Souri, B.; Esfahani, M. Changes in oxidative markers in COVID-19 patients. *Arch. Med Res.* **2021**, *52*, 843–849. [CrossRef]

45. Mohiuddin, M.; Kasahara, K. The emerging role of oxidative stress in complications of COVID-19 and potential therapeutic approach to diminish oxidative stress. *Respir. Med.* **2021**, *187*, 106605. [CrossRef]

International Journal of
Molecular Sciences

MDPI

Article

Cardiac Tyrosine 97 Phosphorylation of Cytochrome *c* Regulates Respiration and Apoptosis

Paul T. Morse [1], Vignesh Pasupathi [1,2], Susanna Vuljaj [1,2], Nabil Yazdi [1], Matthew P. Zurek [1], Junmei Wan [1], Icksoo Lee [3], Asmita Vaishnav [2], Brian F.P. Edwards [2], Tasnim Arroum [1] and Maik Hüttemann [1,2,*]

1 Center for Molecular Medicine and Genetics, Wayne State University, Detroit, MI 48201, USA; morsepa@wayne.edu (P.T.M.)
2 Department of Biochemistry, Microbiology, and Immunology, Wayne State University, Detroit, MI 48201, USA
3 College of Medicine, Dankook University, Cheonan-si 31116, Chungcheongnam-do, Republic of Korea
* Correspondence: mhuttema@med.wayne.edu; Tel.: +1-313-577-9150

Abstract: It was previously reported that tyrosine 97 (Y97) of cytochrome *c* is phosphorylated in cow heart tissue under physiological conditions. Y97 phosphorylation was shown to partially inhibit respiration in vitro in the reaction with purified cytochrome *c* oxidase. Here, we use phosphomimetic Y97E Cyt*c* to further characterize the functional effects of this modification both in vitro and in cell culture models. In vitro, phosphomimetic Y97E Cyt*c* showed lower activity in the reaction with purified cow heart cytochrome *c* oxidase (COX), decreased caspase-3 activity, and reduced rate of reduction. Additionally, the phosphomimetic Y97E Cyt*c* tended to be resistant to heme degradation and showed an increased rate of oxidation. Intact mouse Cyt*c* double knockout fibroblasts were transfected with plasmids coding for phosphomimetic Y97E Cyt*c* and other variants. Compared to cells expressing wild-type Cyt*c*, the cells expressing phosphomimetic Y97E Cyt*c* showed reduced respiration, mitochondrial membrane potential, and reactive oxygen species production, and protection from apoptosis. In an oxygen–glucose deprivation/reoxygenation cell culture model of ischemia/reperfusion injury, mitochondrial membrane potential and reactive oxygen species production were decreased. These data show that Cyt*c* phosphorylation controls the overall flux through the electron transport chain by maintaining optimal intermediate $\Delta\Psi m$ potentials for efficient ATP production while minimizing reactive oxygen species production, thus protecting the cell from apoptosis.

Keywords: cytochrome *c*; mitochondrial respiration; apoptosis: reactive oxygen species; myocardial infarction; ischemia-reperfusion

1. Introduction

Cytochrome *c* (Cyt*c*) is a small globular protein with a single covalently linked heme group per peptide chain. Most mitochondria Cyt*c* species contain between 94 and 114 amino acids, depending on the species, and 104 amino acids in mammals [1]. Cyt*c* acts as a molecular switch between cellular life and death. The primary life-sustaining role of Cyt*c* is to shuttle the electrons from complex III to complex IV (cytochrome *c* oxidase; COX), also known as the terminal O_2 acceptor within the electron transport chain (ETC) [2]. The electron transfer from Cyt*c* to COX is the proposed rate-limiting step of the ETC (reviewed in [3]) and, therefore, is subject to tight regulation. Cyt*c* also acts as a detoxifying agent to dispose of reactive oxygen species (ROS) [4,5] and as part of the Erv1-Mia40 redox-coupled protein import pathway [6,7]. Cyt*c* has even been shown to adopt different conformations depending on the location of the protein within the mitochondria [8]. Under stress conditions, Cyt*c* is released from the mitochondria into the cytosol, acting as an inducer of cell death by initiating intrinsic apoptosis [9–11]. During this process, Cyt*c* binds to apoptotic protease activating factor 1 (Apaf-1) to form the apoptosome, activating procaspase-9 to caspase-9, which cleaves procaspase-3 to caspase-3 and commits the cell to apoptosis. However, this is a delicate balance, as the release of sublethal amounts of Cyt*c* has been shown to desensitize cancer cells to apoptosis [12]. Cyt*c* has other pro-apoptotic functions, such as cardiolipin peroxidase activity and ROS formation via reduction of $p66^{Shc}$ [11,13–18]. Recent work has also shown that Cyt*c* can translocate to the nucleus as part of the DNA damage response [19,20]. Altogether, these diverse functions of Cyt*c* require tight regulatory control by the cell, such as allosteric regulation via ATP binding, expression of tissue-specific isoforms (somatic and testis), and post-translational modifications (PTMs) (reviewed in [21,22]).

Many PTMs of Cyt*c* have been studied, including acetylation, carbonylation [23–25], deamidation [26], glycation [27–30], glycosylation [31,32], homocysteinylation [33–36], nitration [37–41], nitrosylation [42], phosphorylation, and sulfoxidation [43–46]. To date, six tissue-specific phosphorylations of Cyt*c* have been mapped by mass spectrometric analysis (reviewed in [21]): T28 and T58 in the kidney, S47 in the brain, Y48 in the liver, and T49 and Y97 in the heart. Additionally, there are two characterized disease-specific acetylations of Cyt*c* (reviewed in [21]) K39 in ischemic skeletal muscle and K53 in prostate cancer. We and others have characterized the functional effects of the phosphorylations directly and by using in vitro and in-cell culture models with phosphomimetics. Typically, phosphorylations of Cyt*c* and their respective phosphomimetics demonstrate reduced respiration, mitochondrial membrane potential ($\Delta\Psi_m$), ROS production, and apoptosis. Recently, Cyt*c* phosphorylation has been shown to modify the newly discovered nuclear functions of Cyt*c* [47]. Additionally, Cyt*c* phosphorylation can be lost under ischemia-reperfusion (I/R) injury, likely due to activation of Ca^+-sensitive phosphatases, such as S47 dephosphorylation in ischemic brain [48,49], which promotes cell death. We propose that Cyt*c* phosphorylation maintains optimal intermediate $\Delta\Psi_m$ levels for efficient ATP production but limited ROS generation while dephosphorylation induces pathologically high $\Delta\Psi_m$ levels [7–10,13] and excessive ROS production, leading to cell death.

We previously identified phosphorylation of Cyt*c* on Y97 in the cow heart under basal conditions using our specialized purification protocol, which preserves the physiological phosphorylation status [48]. Phosphorylated Cyt*c* Y97 demonstrated sigmoidal kinetics in the reaction with purified cow COX, while the dephosphorylated protein produced a hyperbolic response [50]. Additionally, phosphorylation at Y97 shifted the heme iron-methionine 80 absorption peak from 695 nm to 687 nm, indicating perturbations close to the catalytic heme group. Other studies using a phosphomimetic system demonstrated reduced thermal stability [51] and caspase-3 activity [52]. We also reported that neuroprotective insulin treatment induced this modification in the brain, resulting in reduced Cyt*c* release from the mitochondria and reduced overall neuronal death after an ischemic insult [53].

In the present study, we aim to expand our understanding of the effects of Cyt*c* Y97 phosphorylation on multiple Cyt*c* functions by using mutagenesis with purified Cyt*c* variants and with cell lines stably expressing Cyt*c* variants, including WT, phosphomimetic mutant Y97E, and phenylalanyl mutant Y97F as an additional control. In vitro, we found that phosphomimetic Y97E substitution causes inhibition of respiration in the reaction with COX, similar to what is observed with the purified Y97 phosphorylated Cyt*c*, and reduced caspase-3 activity. In the cell culture model, we found that by introducing plasmids coding for phosphomimetic Y97E and WT Cyt*c* into Cyt*c* double knockout cells, phosphomimetic substitution reduces intact cell respiration, $\Delta\Psi_m$, and ROS production. Additionally, in an oxygen–glucose deprivation/reoxygenation (OGD/R) model, which mimics I/R injury, Y97E is protective against cell death by maintaining an optimal $\Delta\Psi_m$ and reduced levels of ROS. These data suggest that phosphorylation of Y97 regulates the overall ETC flux under normal conditions, maintains a healthy intermediate $\Delta\Psi_m$, and controls mitochondrial respiration, whereas under stress conditions such as I/R injury, dephosphorylated Cyt*c* drives maximal ETC flux, thus causing overproduction of ROS and cell death.

2. Results

2.1. Overexpression and Purification of Functional Cytochrome c Variants in Bacterial Cells

Y97 phosphorylation was studied in vitro via mutagenesis to create phosphomimetic Cyt*c* using an *E. coli* overexpression system. We have used the approach by replacing the rodent (somatic mouse and rat Cyt*c* have the same sequence) wild-type (WT) Y97 Cyt*c* residue (Figure 1A) with the negatively charged amino acid glutamate (Y97E), which mimics the in vivo phosphorylated Y97 due to glutamate and phospho-tyrosine both possessing a negative charge. Using glutamate as a mimetic for phospho-tyrosine is an established technique for both Cyt*c* [21] and other proteins in general [54]. We also generated additional control Y97 phenylalanine (Y97F), which cannot be phosphorylated. Using an *E. coli* overexpression system and our purification protocol, we generated significant amounts of the recombinant proteins. The Coomassie blue stained gel (Figure 1B) and reduced UV–Vis spectra (Figure 1C) show that isolated Cyt*c* variants were highly pure and demonstrated correct folding due to the presence of the characteristic α, β, and γ peaks. In the previous publication where Y97 Cyt*c* phosphorylation was initially discovered, in vivo, partially Y97 phosphorylated Cyt*c* demonstrated a blue shift of the 695 nm peak to 687 nm. Here, phosphomimetic Y97E Cyt*c* also demonstrates a shift and broadening of the 695 nm peak to 687 nm (Figure 1C inset), indicating that Y97E replacement is a good mimetic for in vivo Y97 phosphorylation.

2.2. Y97E Phosphomimetic Cytc Shows Decreased COX Activity

In the previous publication where Y97 Cyt*c* phosphorylation was initially discovered, in vivo partially Y97 phosphorylated Cyt*c* purified from bovine heart demonstrated sigmoidal kinetics with a maximal turnover of 32 s^{-1} and a K_m of 5.5 μM, compared to the un-phosphorylated Cyt*c* showing a hyperbolic response with a maximal turnover of 32 s^{-1} and a K_m of 2.5 μM [50]. Here, COX activity was measured with the recombinant Cyt*c* variants in reaction with bovine heart COX (Figure 1D). Similarly to the data previously seen with Y97 phosphorylated Cyt*c*, phosphomimetic Y97E Cyt*c* showed pronounced sigmoidal kinetics and a maximal turnover number of 7.0 s^{-1}. In contrast, WT Cyt*c* and Y97F Cyt*c* show hyperbolic responses with a maximum turnover of 9.3 s^{-1} and 9.2 s^{-1}, respectively. The K_m for Y97E Cyt*c* in the reaction with COX was 9.1 μM, while the K_m for WT Cyt*c* and Y97F Cyt*c* were 4.3 μM and 4.8 μM, respectively. While the absolute values for V_{max} and K_m seen here differ from those of the previous publication, likely due to differences between using bovine Cyt*c* (previous publication) and rodent Cyt*c* (here) with bovine COX, the trends are overall similar. These data suggest that Y97E replacement is a good mimetic for in vivo Y97 phosphorylation, showing sigmoidal kinetics compared to WT.

2.3. Y97E Phosphomimetic Cytc Shows Reduced Caspase-3 Activity

Cyt*c* plays a central role in I/R injury via Apaf-1 interaction and caspase-3 activation. To test the effect of Y97 phosphorylation on apoptosis, where Cyt*c* interacts with

Apaf-1 to form the apoptosome, we analyzed Cyt*c* variants in a cell-free caspase-3 assay. Phosphomimetic Y97E Cyt*c* showed 64% decreased caspase-3 activity compared to the WT (Figure 1E).

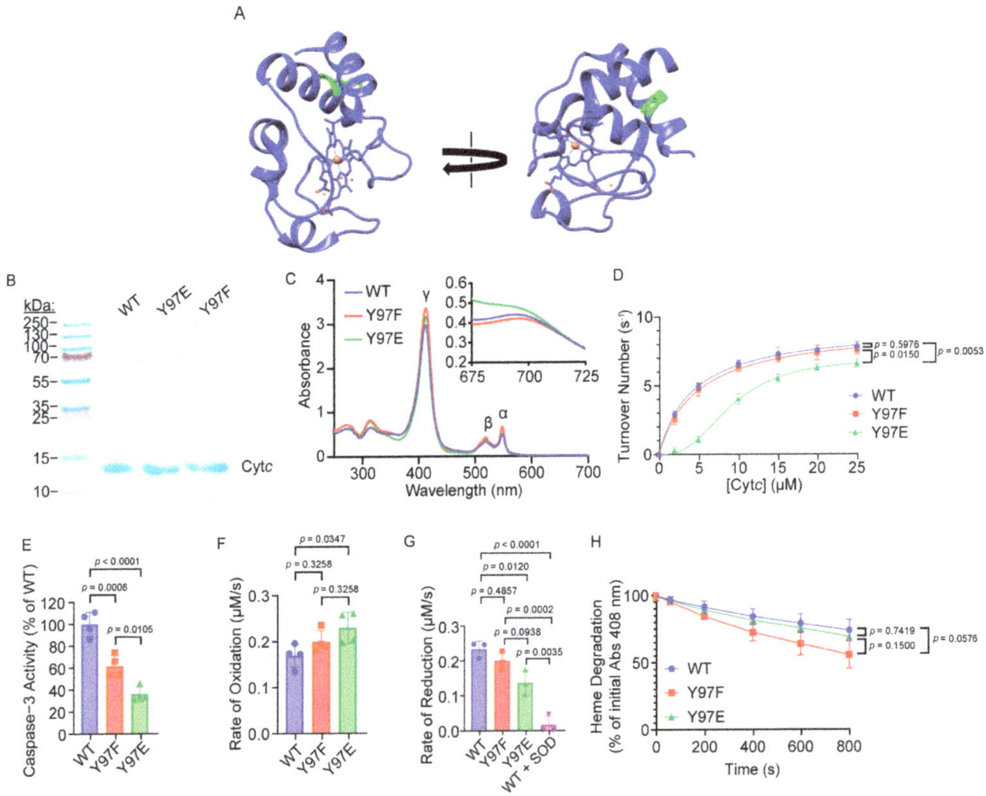

Figure 1. (**A**) Crystal structure (5C0Z.pdb) of rodent WT Cyt*c* (blue) [48] with Y97 (green) labeled. The PDB file was processed with the program Chimera (version 1.16). (**B**) Representative Coomassie Blue stained 10% tris-tricine SDS-PAGE gel showing purity of the recombinant Cyt*c* variants. (**C**) Reduced Cyt*c* UV–Vis spectra and oxidized Cyt*c* UV–Vis spectra (inset) indicating correct folding of the recombinant Cyt*c* variants with characteristic α, β, and γ peaks labeled. (**D**) The oxygen consumption rate of bovine heart COX in reaction with the recombinant Cyt*c* variants was measured using an Oxygraph+ system at 25 °C (n = 3). (**E**) Downstream caspase-3 activity in reaction with the recombinant Cyt*c* variants was measured using rhodamine fluorescence resulting from caspase-3 mediated cleavage of Z-DEVD-R110 (n = 4). (**F**) Initial rate of oxidation of reduced recombinant Cyt*c* variants by 100 μM of H_2O_2 (n = 4). (**G**) Initial rate of reduction of oxidized recombinant Cyt*c* variants by superoxide generated with a hypoxanthine/xanthine oxidase system (n = 3). (**H**) Heme degradation of recombinant Cyt*c* variants by 3 mM of H_2O_2 (n = 3).

2.4. Phosphomimetic Y97E Cytc Displays a Higher Oxidation Rate, Lower Reduction Rate, and Is Partially Resistant to Heme Degradation by Hydrogen Peroxide

Cyt*c* plays an important role as a ROS scavenger. To study this behavior, we analyzed the rate of oxidation of ferro Cyt*c* variants in the presence of 100 μM of H_2O_2 and the rate of reduction of ferri Cyt*c* variants in the presence of superoxide generated by a hypoxanthine/xanthine oxidase reaction system. Phosphomimetic Y97E Cyt*c* displayed a 36% higher oxidation rate (Figure 1F) and a 42% lower reduction rate, both compared to the

WT (Figure 1G). For the rate of reduction, an additional group was included as a negative control which contained superoxide dismutase (SOD), detoxifying superoxide generated by the hypoxanthine/xanthine oxidase system. While Cyt*c* plays an important role in ROS scavenging, Cyt*c* can lose its functionality at high ROS concentrations via the destruction of the heme group. To assess the stability of the heme group, we spectrophotometrically tracked the degradation of the heme catalytic site of Cyt*c* over time upon the addition of 3 mM H_2O_2. Loss of absorbance of the Soret peak of Cyt*c* at 408 nm is characteristic of heme degradation. Phosphomimetic Y97E Cyt*c* did not show any statistically significant difference in heme degradation at 800 s compared to the WT. Interestingly, Y97F Cyt*c* did show increased, but not statistically significant, heme degradation, implying that a nonpolar residue at the site destabilizes the heme group (Figure 1H).

2.5. Mutagenesis and Expression of Cytochrome c Variants in Stable Cell Line

To study the functional effects of a Y97E phosphomimetic substitution of Cyt*c* in an intact cell culture system, we created cell lines stably expressing WT Cyt*c*, Y97F Cyt*c*, and Y97E Cyt*c*. Plasmids encoding these Cyt*c* variants were transfected into a Cyt*c* double knockout mouse lung fibroblast cell line, where both the rodent testes and somatic isoforms of Cyt*c* were knocked out [55]. Cell lines that were equally expressing WT, Y97F, and Y97E Cyt*c*, along with a Cyt*c*-null empty vector (EV) control, were used for the functional studies. Western blot analysis showed that Cyt*c* variants were expressed at similar levels in the cell lines (Figure 2A).

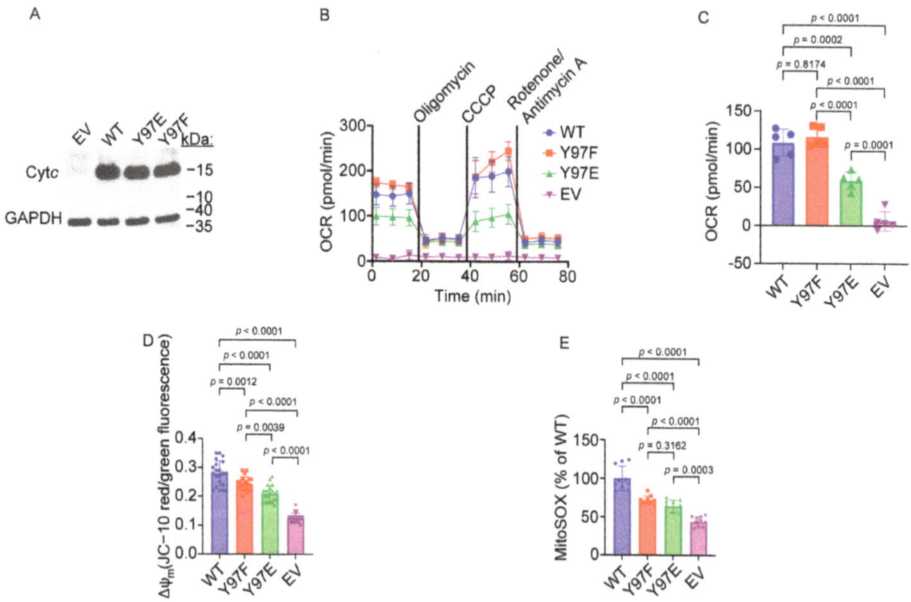

Figure 2. (**A**) Representative Western blot of Cyt*c* double knockout cells transfected with Cyt*c* variants, showing equal Cyt*c* expression and GAPDH as a loading control. (**B**) Mitochondrial stress test performed with sequential injections of 1 μM of oligomycin, 2.5 μM of carbonylcyanide-3-chlorophenylhydrazone, and 1 μM of rotenone/antimycin A (n = 5). (**C**) Basal oxygen consumption rate (OCR) from mitochondrial stress test (n = 5). (**D**) Mitochondrial membrane potential (ΔΨm) measured using a JC-10 probe (n = 21). (**E**) Mitochondrial ROS production measured using a MitoSOX probe (n = 10).

2.6. Mitochondrial Respiration Is Inhibited in Cells Expressing Phosphomimetic Y97E Cytc

Glycolysis and mitochondrial oxidative phosphorylation are two primary energy metabolism pathways that support all other cellular functions. The reaction between Cyt*c* and COX has been proposed to be the rate-limiting step of the ETC (reviewed in [3]), and therefore, alterations of Cyt*c* may impact overall cellular respiration. Intact cellular respiration of the cells expressing the Cyt*c* variants was measured via oxygen consumption rate (OCR) during a mitochondrial stress test using a Seahorse bioanalyzer (Figure 2B). The cells expressing phosphomimetic Y97E Cyt*c* displayed a reduced basal respiration rate of 46% compared to the cells expressing the WT (Figure 2C). Additionally, cells expressing phosphomimetic Y97E Cyt*c* showed decreased ATP-coupled respiration, maximal respiration, and spare respiratory capacity compared to cells expressing the WT (Supplementary Figure S1).

2.7. Mitochondrial Membrane Potential and ROS Production Are Decreased in Cells Expressing Phosphomimetic Y97E Cytc

Since the respiration rate was decreased in cells expressing phosphomimetic Y97E Cyt*c*, we hypothesized that there should be reduced mitochondrial membrane potential ($\Delta\Psi_m$) and, consequently, reduced ROS levels. $\Delta\Psi_m$ was measured with JC-10, a voltage-dependent probe. We found that the red-to-green fluorescence ratio was decreased by 25% in cells expressing phosphomimetic Y97E Cyt*c* compared to cells expressing the WT, indicating a reduction of $\Delta\Psi_m$ (Figure 2D).

Previous research has demonstrated a correlation between $\Delta\Psi_m$ and ROS production [56], which is particularly relevant in conditions of cellular stress, such as I/R injury. As high membrane potentials drive ROS production, we measured ROS levels using the MitoSOX probe. We found that MitoSOX fluorescence was 36% decreased in cells expressing phosphomimetic Y97E Cyt*c* compared to cells expressing the WT, indicating reduced mitochondrial ROS levels (Figure 2E).

2.8. Cells Expressing Phosphomimetic Y97E Cytc Are Protected from Cell Death

We analyzed apoptosis induced by H_2O_2 or staurosporine treatment for the cell lines expressing the Cyt*c* variants. Necrosis and early and late apoptosis were assessed using annexin V/propidium iodide (PI) staining followed by flow cytometry analysis. H_2O_2 and staurosporine are known to induce cell death via oxidative stress [57] or inhibition of protein kinases [58], respectively. The cells were subjected to 400 μM of H_2O_2 for 16 h and 1 μM of staurosporine for 5 h, respectively. In both stress-inducing conditions, cells expressing the phosphomimetic Y97E variant showed decreases in total cell death. Treatment with H_2O_2 resulted in 36% cell death in cells expressing the phosphomimetic Y97E Cyt*c* compared to 45% cell death in cells expressing the WT (Figure 3A). Similarly, treatment with staurosporine resulted in 16% cell death in cells expressing the phosphomimetic Y97E Cyt*c* compared to 40% cell death in cells expressing the WT (Figure 3B). Overall, the decrease in total cell death for the cells expressing the phosphomimetic Y97E Cyt*c* mimics the decreases seen with the in vitro caspase-3 activity assay, highlighting the role of Y97 phosphorylation in reducing the pro-apoptotic features of Cyt*c*.

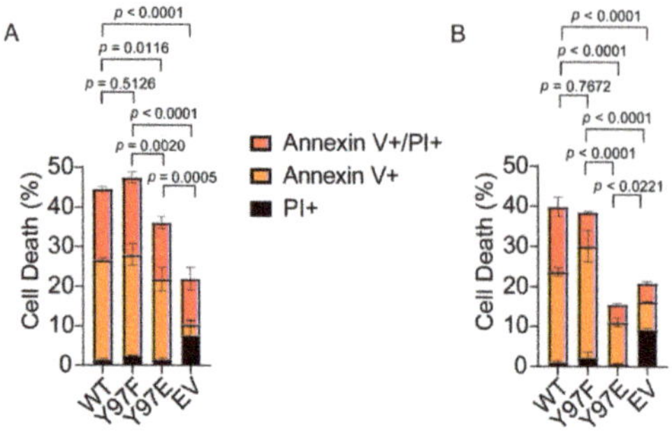

Figure 3. (**A**) Annexin V/propidium iodide flow cytometry data after exposure to 400 μM of H_2O_2 for 16 h (n = 3). (**B**) Annexin V/propidium iodide flow cytometry data after exposure to 1 μM of staurosporine for 5 h (n = 3).

2.9. Cells Expressing Phosphomimetic Y97E Cytc Suppress ROS Production upon Oxygen–Glucose Deprivation/Reoxygenation (OGD/R)

Mitochondrial dysfunction during reperfusion following ischemia has been linked to increased mitochondrial membrane potential and ROS-induced damage of mitochondrial components [59]. During reperfusion, $\Delta\Psi_m$ hyperpolarizes, which drives the production of ROS. Other phosphorylations of Cytc that are present under basal conditions are lost during ischemia due to the depletion of the kinase substrate ATP and the loss of calcium homeostasis, which results in the activation of mitochondrial phosphatases [48,49]. We employed the oxygen–glucose deprivation/reoxygenation (OGD/R) model to further study this relationship. The cells expressing the Cytc variants were exposed to 1% oxygen in glucose-depleted media for 90 min, mimicking the ischemic condition. This was followed by 30 min of simulated reperfusion in oxygen- and glucose-containing media supplemented with either JC-10 or MitoSOX probe to study $\Delta\Psi_m$ and ROS production. Excluding EV cells, which do not contain Cytc and are thus protected from Cytc-mediated apoptosis induction, cells expressing the phosphomimetic Y97E Cytc showed the lowest $\Delta\Psi_m$ levels after OGD/R compared to cells expressing other Cytc variants (Figure 4A). Similar findings were observed after OGD/R for MitoSOX, i.e., the cells expressing phosphomimetic Y97E Cytc showed the lowest mitochondrial ROS production after OGD/R, excluding the EV (Figure 4B). These results suggest a protective role of Y97E phosphomimetic replacement and, thus, Y97 phosphorylation in I/R injury.

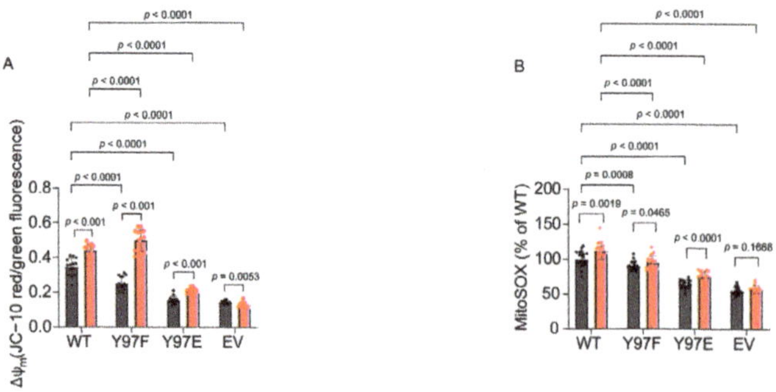

Figure 4. (**A**,**B**) $\Delta\Psi$m (n = 14) and mitochondrial ROS production (n = 21) were measured at normoxia (gray bars) or after 90 min of oxygen–glucose deprivation followed by 30 min of reoxygenation (red bars).

3. Discussion

Generally, tissue-specific phosphorylations of Cyt*c* tend to be present under physiological conditions, leading to partial inhibition of oxidative phosphorylation and significant inhibition of apoptosis (reviewed in [21]). These provide slight tweaks to mitochondrial functioning to optimize tissue metabolism and make the tissues resilient to physiological stressors. The loss of these phosphorylations during ischemia, due to rapid depletion of the kinase substrate ATP, in addition to the loss of mitochondrial Ca^{2+} homeostasis leading to activation of phosphatases, results in reperfusion injury due to $\Delta\Psi_m$ hyperpolarization driving increased ETC activity, reverse electron transfer, and ROS bursts (reviewed in [60]). We propose that the loss of the inhibitory phosphorylations during ischemia is a futile attempt of the cell trying to upregulate oxidative phosphorylation and energy production, with another deleterious effect of sensitizing the tissue to apoptosis [48,49]. Given that Y97 phosphorylation of Cyt*c* was discovered under physiological conditions in the mammalian heart, an organ with known vulnerability to I/R injury resulting from myocardial infarction, it is highly likely that loss of this phosphorylation contributes to I/R injury in a way similar to that seen of S47 phosphorylation of Cyt*c* in the brain [48].

Our lab previously showed that Y97 phosphorylation of Cyt*c* is an important modification that regulates COX enzyme activity in the mammalian heart [50]. Here, we expand on that previous work by characterizing phosphomimetic Y97E Cyt*c* using both purified protein and transfected cell culture models. Previous work on Cyt*c* has shown that glutamate is a suitable mimic for phospho-tyrosine [51]. Y97 is conserved in mammals, with only a few species of plants or microorganisms expressing a different residue at the site, most commonly F97 [1]. One of the two main functions of Cyt*c* is as a single electron carrier from complex III to COX in the ETC, which is the proposed rate-limiting step of the ETC under physiological conditions [3]. The publication that initially discovered phosphorylated Y97 Cyt*c*, which was purified from bovine heart, showed that the phosphorylation resulted in sigmoidal kinetics in reaction with COX, whereas the dephosphorylated protein showed a hyperbolic response [50]. Our data here with the phosphomimetic Y97E Cyt*c* agree with these results, with the phosphomimetic Y97E Cyt*c* showing sigmoidal kinetics and the WT, non-phosphorylated protein showing a hyperbolic response. Furthermore, the UV–Vis spectra of phosphomimetic Y97E show the characteristic blue shift of the typical 695 nm peak to 687 nm, similar to the in vivo Y97 phosphorylated Cyt*c*. These results further support the notion that Y97E replacement mimics the effects seen with Y97 phosphorylated Cyt*c*.

The second main function of Cyt*c* is as a key pro-apoptotic signal [61]. Upon its release from the mitochondria to the cytosol, Cyt*c* interacts with Apaf-1 to form the apoptosome. This results in the activation of caspase-9 and triggers the downstream caspase cascade. Our data here shows statistically significant reductions in caspase-3 activity, with a 64% reduction in caspase-3 activity when using phosphomimetic Y97E Cyt*c* compared to WT. Interestingly, when analyzed with a Swiss PDB viewer using a published docking model of Cyt*c* and Apaf-1 (3JBT.pdb), Y97 is not part of the Apaf-1 interaction site [62]. It is spatially located closest to Apaf-1 residues E659 and D1106 with an approximate distance of 14 Å (Figure 5). The binding between Cyt*c* and Apaf-1 is known to be driven by electrostatic interactions, with the positively charged Cyt*c* binding to its negatively charged binding pocket on Apaf-1 [62]. The gain of a negative charge with phosphomimetic Y97E Cyt*c* may cause globally decreased binding to Apaf-1. Additionally, when looking at purified phosphomimetic Y97E Cyt*c* protein, we identified an increased rate of oxidation with H_2O_2 and a decreased rate of reduction with superoxide.

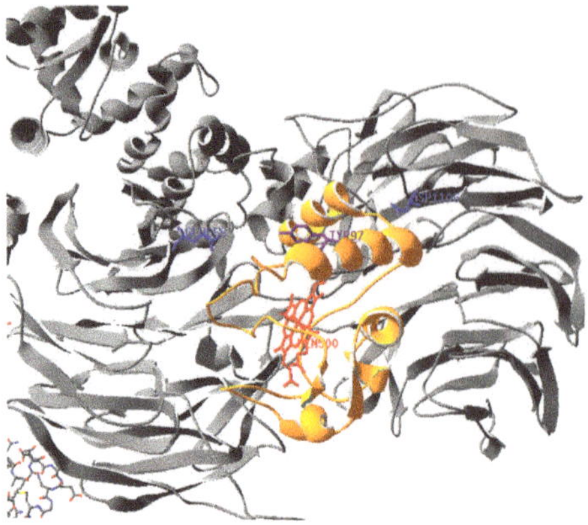

Figure 5. Docking model of Cyt*c* and Apaf-1 (3JBT.pdb) [62]. The distance between the amino acid residues E659 and D1106 (blue sticks) of Apaf-1 (gray) to Y97 (purple) of Cyt*c* (yellow with red heme group) is approximately 14 Å. The PDB file was processed with the program Swiss PDB viewer (version 4.1.0).

Furthermore, we characterized phosphomimetic Y97E Cyt*c* in a cell culture model. Rodents possess two active Cyt*c* isoforms, somatic and testes, while humans possess a single active Cyt*c* isoform that possesses features of both rodent isoforms. Knocking out the rodent somatic isoform induces expression of the testes isoform to ensure metabolic competence [55]. Therefore, we employed a double knockout model where both the somatic and testes isoforms are knocked out. These double knockout cells were transfected with the recombinant Cyt*c* variants to mimic the effects of Y97 phosphorylation. Our results show decreased respiration, decreased $\Delta\Psi_m$, and decreased ROS production in cells expressing the phosphomimetic Y97E Cyt*c*, highlighting the relationship between $\Delta\Psi_m$, respiration, and ROS production highlighted above (Figure 6). When looking at apoptosis triggered by H_2O_2 and staurosporine, the cells expressing phosphomimetic Y97E Cyt*c* showed decreased total cell death, which matches the data observed using the recombinant phosphomimetic Y97E Cyt*c* protein, which demonstrated an increased rate of oxidation with H_2O_2 and decreased caspase-3 activity.

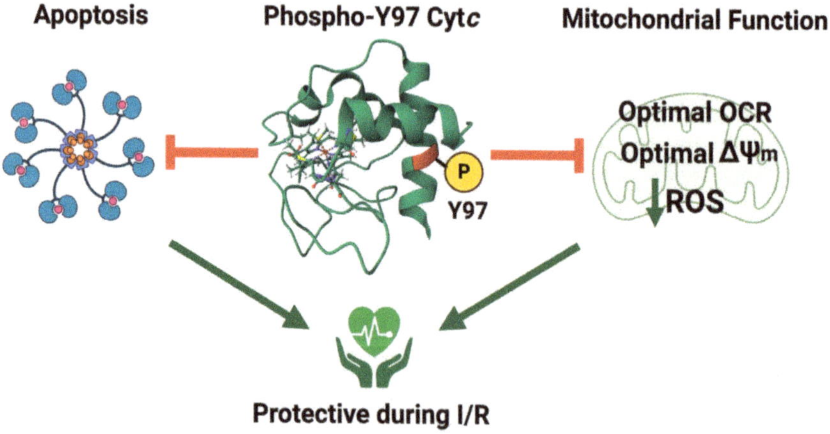

Figure 6. Model of Cyt*c* Y97 phosphorylation as protective to cardiac muscle.

The partial inhibition of respiration observed in intact cells expressing the phosphomimetic Y97E Cyt*c* is significant because of the relationship between respiration, $\Delta\Psi_m$, and ROS production. It is known that 95% of ROS are generated from electron leaks in the respiratory complexes I and III [63]. Under normal conditions, maintenance of intermediate $\Delta\Psi_m$ levels (80–140 mV) avoids ROS generation but provides the full capability to produce ATP because maximal rates of ATP synthesis by ATP synthase take place at an intermediate $\Delta\Psi_m$ of 80–140 mV [64]. At $\Delta\Psi_m$ exceeding 140 mV, ROS production at ETC complexes I and III increases exponentially, leading to cell death [65]. This relationship between $\Delta\Psi_m$ and ROS is further supported by the results presented here under basal conditions and using the OGD/R model, which simulates I/R injury. In the OGD/R model, cells expressing unphosphorylated WT Cyt*c* showed increased $\Delta\Psi_m$, which triggered a similar increase in ROS production. The cells expressing the phosphomimetic Y97E Cyt*c* demonstrated lower $\Delta\Psi_m$, which attenuated ROS production. However, on the other hand, phosphomimetic Y97E Cyt*c* protein showed a lower rate of reduction by superoxide. It is possible that this may be partially due to the repulsion between the negative charges of the phosphomimetic glutamate and superoxide. An alternate explanation provided by previous research is that point mutations at Y97 lead to subtle changes in the hydrogen bonding network surrounding the heme group, leading to perturbed redox properties under higher electrical fields [66].

The secondary control, unphosphorylated Y97F, often displayed an intermediate phenotype compared to the WT and the phosphomimetic Y97E Cyt*c*, such as the results seen for the caspase-3 activity, rate of oxidation, and rate of reduction assays. This highlights the regulatory importance of the tyrosine residue at site 97, which allows for phenotypic shifting depending on the phosphorylation state. However, phenylalanine mutation is still tolerated at site 97 and is seen in many species of plants [1]. This provides a possible evolutionary reason for the intermediate phenotype: plants lack the regulatory capability of the phosphorylation site but instead possess a mutation that shows features of both the phosphorylated and unphosphorylated Cyt*c* seen in mammals. Surprisingly, cells expressing Y97F Cyt*c* tend to behave either like cells expressing the WT or the Y97E Cyt*c*, depending on the assay. For the basal respiration and cell death assays, cells expressing Y97F Cyt*c* tend to behave like cells expressing the WT Cyt*c*. On the other hand, for the $\Delta\Psi_m$ and mitochondrial ROS production assays, cells expressing Y97F Cyt*c* tend to behave more like the cells expressing the Y97E Cyt*c*. This highlights the complex relationship between mitochondrial respiration, $\Delta\Psi_m$, and mitochondrial ROS production.

A previous publication also looked at the WT, Y97F, and Y97E Cyt*c* variants [51]. In that publication, García-Heredia et al. found that the midpoint redox potential and cardiolipin peroxidase activity of phosphomimetic Y97E Cyt*c* were not different compared to the WT. Interestingly, the authors did find that the phosphomimetic Y97E Cyt*c* showed dramatically decreased thermal stability, with the melting temperature dropping by 30 °C compared to the WT. However, the authors attributed this not to the effect of the negative charge but rather to the loss of the aromatic ring, as Y97F Cyt*c* was comparable to the WT.

Another publication studied Y97 Cyt*c* phosphorylation by substituting Y97 with the noncanonical amino acid *p*-carboxymethyl-L-phenylalanine (pCMF) [52]. This study reported increased COX activity using phosphomimetic Y97pCMF Cyt*c*, which contradicts data previously obtained with Y97 phosphorylated Cyt*c*—i.e., the gold standard—and data presented here on phosphomimetic Y97E Cyt*c*, which displays altered COX kinetics and slight enzymatic inhibition, matching those observed with the in vivo phosphorylated protein. The authors proposed that this stimulation was due to a double regulatory mechanism via the phosphorylation and increased supercomplex formation, which the authors observed in a yeast model. They also reported reduced ROS production and caspase-3 activity using phosphomimetic Y97pCMF Cyt*c*, although ROS scavenging activity and cardiolipin binding were unchanged.

Interestingly, another study identified that Y97 phosphorylation of Cyt*c* could be induced in ischemic pig and rat brains using insulin [53]. Insulin-induced Y97 phosphorylation of Cyt*c* resulted in significant neuroprotection, with a 49% increase in NeuN-positive neurons compared to the ischemia condition alone.

In conclusion, our results suggest that Cyt*c* Y97 phosphorylation results in optimal, intermediate membrane potential levels, lower ROS production, and significant protection from cell death upon OGD/R (Figure 6). Therefore, Cyt*c* Y97 phosphorylation is a beneficial modification in cardiac muscle under physiological conditions, but its loss during ischemia sensitizes cardiac tissue to reperfusion injury. Finally, recent research studying posttranslational modifications of Cyt*c* supports the concept that such small chemical modifications of this evolutionarily highly optimized protein control electron flux in the ETC, putting it in a central position of regulating $\Delta\Psi_m$, energy production and ROS in a highly fine-tuned tissue-specific manner.

4. Materials and Methods

All chemicals and reagents were purchased from MilliporeSigma (Burlington, MA, USA) unless otherwise specified.

4.1. Bacterial Overexpression and Recombinant Protein Purification

The codon corresponding to Y97 of the somatic rodent Cyt*c* (WT) cDNA cloned into the pLW01 expression vector was mutated to a glutamate residue (Y97E) as a phosphomimetic replacement or phenylalanine as an additional control (Y97F) which cannot be phosphorylated using the quick-change site-directed mutagenesis kit (Agilent Technologies, Santa Clara, CA, USA) as directed by the manufacturer. The pLW01 plasmid also contained the sequence for *S. cerevisiae* heme lyase, which bacteria lack, in order to successfully covalently attach the heme group to Cyt*c* [67]. The following primers were used to mutate Y97 of the Cyt*c* sequence: Y97E forward primer 5′-GACCTAATAGCT**GAG**CTTAAAAAGGC-3′, Y97E reverse primer 5′-GCCTTTTTAAG**CTC**AGCTATTAGGTC-3′, Y97F forward primer 5′-GGCAGACCTAATAGCT**TTC**CTTAAAAAGGC-3′, and Y97F reverse primer 5′-GCCTTTTTAAG**GAA**AGCTATTAGGTCTGCC-3′. Using the Dpn I restriction enzyme, the parental supercoiled dsDNA was digested for 1 h at 37 °C. The resulting PCR products were transformed into XL 10-Gold Ultracompetent cells. The plasmids from the overnight culture were purified from individual colonies using a QIAprep spin column miniprep kit (#27106, Qiagen; Valencia, CA, USA). Plasmid purity and quantity were assessed using a Nanodrop 1000 spectrophotometer, and the desired mutations were confirmed by sequencing.

The plasmids were transformed into chemically competent *Escherichia coli* strain, C41(DE3) cells (Biosearch technologies; Tucson, AZ, USA) according to the manufacturer's protocol in order to overexpress the Cyt*c* variants. Selected clones were inoculated into 10 mL of terrific broth containing 0.1 mg/mL of carbenicillin and grown overnight at 37 °C under 225 to 250 RPM agitation. The overnight culture was inoculated into several flasks containing 5 L of carbenicillin-containing TB medium and grown until an OD_{600} of 1 was reached. The addition of 100 μM of isopropyl-B-D-thiogalactoside induced Cyt*c* expression. Cell pellets were collected after 8 h of expression via centrifugation at 1900 RPM for 55 min at 4 °C (Shovell SS-34, Thermo Fischer Scientific, Waltham, MA, USA). Bacterial pellets were harvested and lysed, and Cyt*c* was purified using modified ion exchange chromatography; a DE52 anion-exchange column (20 mM phosphate buffer, pH 7.5, 4.0 mS/cm conductivity) followed by CM52 cation-exchange (30 mM phosphate buffer, pH 6.5, 6.0 mS/cm conductivity). Next, the protein was eluted with a high salt elution buffer (40 mM phosphate buffer, 0.5 M NaCl). Finally, the protein was desalted and concentrated by centrifugation using Amicon Ultra-15 3 kDa centrifugal filter units (Millipore; Billerica, MA, USA) and stored at −80 °C [48]. Protein concentration was determined spectrophotometrically, as detailed below. Protein purity was determined via Coomassie blue staining of a 10% tris-tricine sodium dodecyl sulfate-polyacrylamide gel electrophoresis (SDS-PAGE) gel loaded with 500 ng of each Cyt*c* variant and ran as described below.

4.2. Concentration Determination of Cyt*c*

The concentration of Cyt*c* was measured using a Jasco V-570 double beam spectrophotometer (2 nm bandwidth, 200 nm/min scanning speed) using a 0.1 mm path length quartz cuvette. The Cyt*c* variants were diluted and reduced with sodium dithionite ($Na_2S_2O_4$) or oxidized with potassium ferricyanide ($K_3Fe(CN)_6$). Cyt*c* was desalted using Sephadex G-25 NAP5 columns (GE Healthcare, Chicago, IL, USA). Cyt*c* concentration was calculated using differential spectra at 550 nm by subtracting the oxidized form from the reduced form using the formula: [Cyt*c*] = (A_{550}reduced − A_{550}oxidized)/(19.6 mM/cm × 1 cm) × dilution factor. The presented reduced UV–Vis spectra were recorded at 25 μM of Cyt*c*, while the oxidized UV–Vis spectra were recorded at 200 μM of Cyt*c*.

4.3. Cytochrome *c* Oxidase Activity Measurement

Regulatory-competent COX was previously purified from bovine heart tissue as described [50] and diluted to 3 μM in solubilization buffer (10 mM of K-HEPES, 40 mM of KCl, 1% Tween 20, 2 mM of EGTA, 10 mM of KF, pH 7.4 supplemented with a 40-fold molar excess of cardiolipin and 0.2 mM of ATP). Prior to use, COX was dialyzed overnight in 1 L of solubilization buffer using a 12–14 kD Spectra/Por 2 dialysis membrane (Spectrum Laboratories, Inc.; Rancho Dominguez, CA, USA) at 4 °C to remove cholate bound to the COX during the purification process. COX activity was measured in a Clark-type oxygen electrode (Oxygraph system, Hansatech; Pentney, UK) at a final concentration of 40 nM of bovine heart COX in the chamber at 25 °C in 220 μL of solubilization with 20 mM of ascorbate as an electron donor. Oxygen consumption of COX was measured upon titration of Cyt*c* variants injected into the airtight oxygen electrode at different concentrations ranging from 0 to 25 μM. COX activity is reported as turnover number (s^{-1}).

4.4. Measurement of Caspase-3 Activity

Caspase-3 activation was measured using a cell-free apoptosis detection system. Cyt*c*$^{-/-}$ mouse embryonic fibroblasts (CRL 2613, ATCC; Manassas, VA, USA) were cultured in 150 mm dishes, pelleted, and washed with ice-cold phosphate-buffered saline (PBS) twice, followed by one wash with cytosol extraction buffer. The cell suspension was transferred to a 2 mL Dounce homogenizer with CEB buffer (20 mM K-HEPES, 10 mM KCl, 1.5 mM $MgCl_2$, 1 mM EDTA, 1 mM EGTA, 1 mM dithiothreitol, 1 mM PMSF, pH 7.5) for 15 min at 4 °C to allow the cells to swell. The cells were ruptured using a pestle for 20–30 strokes. The lysates were centrifuged at 15,000× *g* for 15 min at 4 °C to remove cellular debris. The protein concentration of the supernatant was measured using a DC assay kit (Bio-Rad, Hercules, CA, USA). The Enzchek caspase-3 assay kit 2 (Thermo Fisher Scientific) was used which contains Rhodamine 110 linked DEVD tetrapeptide, an artificial caspase-3 substrate that fluoresces after cleavage by caspase-3. A 1 mg/mL protein concentration of cell extracts was incubated with mutant Cyt*c* variants at 20 μg/mL for 2 h at 37 °C.

Fluorescence was detected using a Fluorskan Ascent FL plate reader (Lab Systems, Thermo Scientific; Waltham, MA, USA) with an excitation filter of 485 nm with 14 nm bandwidth and an emission filter of 527 nm with 10 nm bandwidth. Readings were corrected using background measurement in the absence of Cyt*c*. The final reading is expressed in the percentage of change compared to WT.

4.5. Measurement of Cytochrome c Oxidation and Reduction Rate

The kinetics of oxidation of ferro-Cyt*c* by H_2O_2 were measured spectrophotometrically at 550 nm [68]. Cyt*c* variants were fully reduced with sodium dithionite ($Na_2S_2O_4$) and desalted using NAP5 columns (GE Healthcare, Piscataway, NJ, USA). The initial absorbance of 15 μM Cyt*c* in 0.2 M Tris-Cl, pH 7.0, was measured at 550 nm and 630 nm as a background reading using 0.2 M Tris-Cl as a blank. Then, 100 μM of H_2O_2 was added to the cuvette as an oxidizing agent, and the reduction in absorbance at 550 nm was measured every 10 s for 1 min. The rate of Cyt*c* oxidation is expressed in μM/s.

The kinetics of reduction of ferri-Cyt*c* by superoxide were measured spectrophotometrically at 550 nm. The rate of reduction was measured using a hypoxanthine/xanthine oxidase system to generate superoxide [69]. Cyt*c* variants were fully oxidized with potassium ferricyanide ($K_3Fe(CN)_6$) and desalted using NAP5 columns (GE Healthcare, Piscataway, NJ, USA). A reaction mixture consisting of 10 μM of Cyt*c*, 100 μM of hypoxanthine, and 14.2 nM of catalase in 1×PBS was prepared in a cuvette, and an initial absorbance was measured at 550 nm using 1×PBS as a blank. The reaction was initiated with the addition of 181.5 nM xanthine oxidase to the cuvette. The increase in absorbance at 550 nm was measured every 15 s for 3 min. Superoxide dismutase 2 (925 nM) was used as a negative control by detoxifying superoxide. The rate of Cyt*c* reduction is expressed in μM/s.

4.6. Heme Degradation Assay

The degradation of the covalently attached heme group of Cyt*c* using high concentration H_2O_2 was measured spectrophotometrically. Cyt*c* variants were oxidized using potassium ferricyanide ($K_3Fe(CN)_6$) and desalted using NAP5 columns (GE Healthcare, Piscataway, NJ, USA). A reaction mixture containing 5 μM of Cyt*c* in 50 mM of phosphate buffer with pH 6.1 was prepared, and the initial absorbance at 408 nm, which corresponds to the characteristic heme Soret band, was measured [68]. Heme degradation was initiated with the addition of 3 mM of H_2O_2. The loss in absorbance at 408 nm was measured at 60, 200, 400, 600, and 800 s. The heme degradation is reported as a percent change in absorbance compared to the baseline absorbance.

4.7. Cell Culture and Stable Transfection of Cytc Constructs

The codon corresponding to Y97 of the somatic rodent Cyt*c* (WT) cDNA cloned into the pBABE-puro expression plasmid (Addgene, Cambridge, MA, USA) was mutated to a glutamate residue (Y97E) as a phosphomimetic replacement or phenylalanine as an additional control (Y97F) which cannot be phosphorylated using site-directed mutagenesis (Agilent Technologies, Santa Clara, CA, USA). The same mutagenesis primers as above were used to generate the variant Cyt*c* expression constructs, which were stably transfected into Cyt*c* double knockout mouse lung fibroblasts (a kind gift from Dr. Carlos Moraes, University of Miami, Coral Gables, FL, USA) using Transfast transfection reagent (Promega, Madison, WI, USA) in a 1:1 transfection reagent to DNA ratio, as described in the manufacturer's protocol. As a negative control, cells were also transfected with a pBABE plasmid that did not contain the sequence for Cyt*c*, resulting in an empty vector (EV) cell line. The transfected cells were cultured at 37 °C with 5% CO_2 in DMEM supplemented with

10% fetal bovine serum (FBS) (Sigma-Aldrich, St. Louis, MO, USA), 100 μg/mL primocin (Invivogen, San Diego, CA, USA), 1 mM sodium pyruvate, and 50 μg/mL uridine. Initial selection was performed with the above media supplemented with 2 μg/mL puromycin.

4.8. Gel Electrophoresis and Western Blotting

Cells expressing Cyt*c* variants were lysed using 100 μL of a RIPA lysis buffer (150 mM of NaCl, 5 mM of EDTA, pH 8.0, 50 mMTris, pH 8.0, 1% NP-40, 0.5% sodium deoxycholate, 0.1% SDS) supplemented with protease inhibitor cocktail (#P8340, MilliporeSigma, Burlington, MA, USA) to prevent lysate degradation. Mixtures were then sonicated and centrifuged at 16,900× *g* for 20 min at 4 °C to remove cellular debris. SDS-PAGE of resulting lysates was carried out using 50 μg total protein of each cell line on a 10% tris-tricine gel in the presence of anode (200 mM Tris, pH 8.9) and cathode buffers (100 mM of Tris, 100 mM of Tricine, 0.1% SDS, pH 8.25). The gel was transferred onto a PVDF membrane (Bio-Rad) using a Trans-Blot SD semi-dry apparatus (#1703940; Bio-Rad) at 75 mA for 15 min and blocked in 5% milk for 1 h at room temperature. Membranes were cut between 25 and 35 kDa ladder markers. The lower and upper membranes were incubated in 1:4000 mouse anti-Cyt*c* (#556433; BD Pharmingen; San Jose, CA, USA) or 1:8000 mouse anti-GAPDH (#60004-1-Ig; Proteintech; Rosemont, IL, USA) respectively, in 5% milk overnight at 4 °C. The next day, membranes were incubated in 1:10,000 sheep anti-mouse IgG conjugated to horseradish peroxidase secondary antibody (#NA931V; GE Healthcare; Chicago, IL, USA) in 5% milk for 2 h at room temperature. The blots were visualized using Pierce ECL western blotting substrate (#32106; Thermo Fisher Scientific; Waltham, MA, USA).

4.9. Mitochondrial Stress Test

The OCR and extracellular acidification rates (ECAR) of the cells expressing the Cyt*c* variants were measured. Cells were seeded at a density of 10,000 cells/well in a 0.1% gelatin-coated XF24 plate (Agilent, #100777-004) overnight in 250 μL/well growth media. After overnight incubation, the growth medium was replaced with 675 μL of Seahorse media was prepared with 4.15 g of DMEM powder (#D5030, Millipore Sigma) dissolved in 500 mL of ddH_2O, pH 7.4, sterile filtered (#431097, Corning Incorporated, Corning, NY, USA), supplemented with 10 mM glucose and 10 mM sodium pyruvate without FBS or phenol red. The cells were incubated in a CO_2-free incubator for 1 h with seahorse media. Then, the OCR and ECAR were measured in an XFe24 Seahorse extracellular flux analyzer (Seahorse Biosciences, North Billerica, MA, USA). Following basal measurements, sequential injections of oligomycin (1 μM), carbonyl cyanide chlorophenylhydrazone (CCCP, 2.5 μM), and rotenone/antimycin A (1 μM) were performed. The mitochondrial stress test parameters (basal respiration, non-mitochondrial respiration, proton leak, ATP-coupled respiration, maximal respiration, and spare respiratory capacity) were calculated according to the manufacturer's instructions. OCR is reported as pmol O_2/min, and ECAR is reported as mpH/min.

4.10. Measurement of Membrane Potential

Mitochondrial membrane potential ($\Delta\Psi m$) was measured using the JC-10 (5,5′,6,6′-tetrachloro-1,1′,3,3′-tetraethylbenzimidazolylcarbocyanine iodide) probe (#ENZ-52305, Enzo Life Sciences; Farmingdale, NY, USA). When the membrane potential is low, the JC-10 probe occurs as monomers and emits green fluorescence. When the membrane potential is high, the JC-10 probe aggregates inside the mitochondria, producing red fluorescence. Measuring this ratio between red and green fluorescence provides a relative measure of the mitochondrial membrane potential. Cells were seeded at a density of 10,000 cells/well in a black 96-well plate (Corning, #CLS3603) coated with 0.1% gelatin. The growth medium was exchanged with FBS-free, phenol red-free DMEM cell culture medium supplemented with 3.0 μM of JC-10 and cultured for 30 min. Cells were washed with 1×PBS twice. Green fluorescence (excitation 485 nm/emission 527 nm) and red fluorescence (excitation 485 nm/emission 590 nm) of the cells were measured using the Synergy H1 microplate reader (BioTek Instruments Inc.; Winooski, VT, USA). $\Delta\Psi m$ is represented as a ratio of red/green fluorescence.

4.11. Measurement of Mitochondrial ROS Production

Mitochondrial ROS production was measured using a MitoSOX red mitochondrial superoxide indicator probe (Invitrogen, #M36008). This mitochondrial ROS indicator accumulates inside the mitochondria due to its delocalized positive charge and emits red fluorescence after superoxide oxidation. Cells were seeded at a density of 10,000 cells/well in a black 96-well plate coated with 0.1% gelatin. The growth medium was exchanged with FBS-free, phenol red-free DMEM cell culture medium supplemented with 5 μM of MitoSOX and cultured for 30 min. Cells were washed with 1×PBS. Fluorescence (excitation 510 nm/emission 580 nm) was measured using a Synergy H1 plate reader (BioTek, Winooski, VT, USA). Data is reported as a percentage change compared to WT.

4.12. Measurement of Cell Death Using Annexin v/PI Staining

Cells expressing the Cyt*c* variants were seeded at a density of 1×10^6 in 10 cm culture dishes. Cells were cultured overnight as described above in growth media. After 24 h, the cells were treated with H_2O_2 (400 μM for 16 h) or staurosporine (1 μM for 3 h). The dead cells were harvested by collecting the culture medium. Live cells were collected using trypsinization. The cell suspension was washed twice with 1×PBS, and 1×10^6 cells were resuspended in a 1 mL 1× binding buffer (BD Pharmingen, Lot: 1026022). A total of 450 μL cell suspension was incubated with 6 μL of annexin V-FTIC (BD Pharmingen, Lot: 1026022) and 6 μL of propidium iodide (BD Pharmingen, Lot: 1026022) at room temperature in the dark for 15 min. After incubation, reagents were diluted with 3 mL of 1× binding buffer. Cells were counted on a CyFlow Space flow cytometer (Sysmex America, Inc.; Lincolnshire, IL, USA) and analyzed using FCS Express 7 software (De Novo Software; Glendale, CA, USA).

4.13. Oxygen Glucose Deprivation/Reoxygenation Model

Cell lines expressing Cyt*c* variants were studied using an OGD/R model to simulate I/R injury. Cells were seeded in 96 black well plates with clear bottoms and cultured to our standard protocol, as described above. Before the experiment JC-10 or MitoSOX experiment, the media was exchanged with glucose-free, FBS-free, phenol-red-free DMEM (Gibco, Waltham, MA, USA, #A1443001) that had been equilibrated with 95% N^2 and 5% CO^2, which mimics the ischemic condition [49]. The cells were incubated with this ischemic media for 90 min at 37 °C with 1% O_2 and 5% CO_2 gas in a hypoxic chamber under the control of ProOx 110 oxygen and ProCO_2 120 carbon dioxide probes (Biospehrix, Redfield, NY, USA). The ischemic media were exchanged with glucose, FBS-free, phenol red-free DMEM culture medium (Gibco, #31053028) and reoxygenated for 30 min under normal conditions along with the respective probe (JC-10 or MitoSOX). The control plates were maintained in normoxic conditions with regular media supplemented with FBS. The JC-10 or MitoSOX experiments were then carried out as described above.

4.14. Statistical Analyses

The data shown represent the mean. Error bars represent the standard deviation. For most assays, statistical analyses for the data were analyzed for statistical significance using one-way ANOVA comparing the mean of each column with the mean of every other column followed by post-hoc Tukey test using GraphPad version 10.4.0 (GraphPad Software, San Diego, CA, USA). For COX activity, the Km and Vmax for WT and Y97F were calculated using nonlinear regression followed by Michaelis-Menten calculation, while the Km (Khalf) and Vmax for Y97E were calculated using nonlinear regression followed by allosteric sigmoidal calculation. The statistical significance was calculated using one-way ANOVA with post-hoc Tukey test as described above, specifically on the 25 uM of Cyt*c* condition. For heme degradation, statistical significance was calculated using one-way ANOVA with post-hoc Tukey test as described above, specifically on the 800 s condition. For annexin V/propidium iodide experiments, statistical significance was calculated using one-way ANOVA with post-hoc Tukey test as described above on total cell death (PI+ cells, annexin V+ cells, and annexin V+/PI+ cells combined). For OGD/R experiments, statistical significance within the hypoxia and normoxia groups was calculated using one-way ANOVA comparing the mean of each column with the mean of the control column (WT) with the post-hoc Dunnett test. However, a student's two-tailed *t*-test assuming equal variance was used to compare the hypoxia and normoxia conditions for a specific Cyt*c* variant. *p* values are indicated in the figures.

Supplementary Materials: The following supporting information can be downloaded at: https://www.mdpi.com/article/10.3390/ijms26031314/s1.

Author Contributions: P.T.M. was responsible for investigation, formal analysis, validation, visualization, and writing—original draft. V.P. was responsible for investigation and writing—original draft. S.V. was responsible for investigation. N.Y. was responsible for investigation. M.P.Z. was responsible for investigation. J.W. was responsible for investigation. I.L. was responsible for methodology. A.V. was responsible for investigation. B.F.P.E. was responsible for project administration, resources, and supervision. T.A. was responsible for visualization and writing—review and editing. M.H. was responsible for conceptualization, funding acquisition, methodology, project administration, resources, supervision, and writing—review and editing. All authors have read and agreed to the published version of the manuscript.

Funding: This work and some of the work discussed in this article was primarily supported by NIH R01 GM116807, with minor support from the Office of the Assistant Secretary of Defense for Health Affairs under Award No. HT9425-24-1-0073, NIH grant R01 NS120322, NIH grant U44 NS125160, NIH grant R01 NS134695. Opinions, interpretations, conclusions, and recommendations are those of the authors and are not necessarily endorsed by the funding agencies including the National Institutes of Health and the Department of Defense.

Institutional Review Board Statement: Not applicable.

Informed Consent Statement: Not applicable.

Data Availability Statement: The original contributions presented in this study are included in the article/Supplementary Materials. Further inquiries can be directed to the corresponding author.

Acknowledgments: We thank Carlos Moraes at the University of Miami for providing the Cyt*c* double knockout cell line. We thank Lucy Waskell at the University of Michigan for providing the pLW01 plasmid.

Conflicts of Interest: The authors declare no conflicts of interest.

Abbreviations

Apaf-1	apoptotic protease-activating factor
CCCP	carbonyl cyanide m-chlorophenyl hydrazone
COX	cytochrome *c* oxidase
Cyt*c*	cytochrome *c*
ECAR	extracellular acidification rate
ETC	electron transport chain
EV	empty vector
FBS	fetal bovine serum
I/R	ischemia-reperfusion
OCR	oxygen consumption rate
OGD/R	oxygen–glucose deprivation/reoxygenation
PBS	phosphate-buffered saline
*p*CMF	*p*-carboxymethyl-L-phenylalanine
PI	propidium iodide
PTMs	post-translational modifications
ROS	reactive oxygen species
SDS-PAGE	sodium dodecyl sulfate-polyacrylamide gel electrophoresis
SOD	superoxide dismutase
WT	wild-type
Y97E	tyrosine 97 of cytochrome *c* replaced with glutamate
Y97F	tyrosine 97 of cytochrome *c* replaced with phenylalanine
$\Delta\Psi_m$	mitochondrial membrane potential

References

1. Zaidi, S.; Hassan, M.I.; Islam, A.; Ahmad, F. The role of key residues in structure, function, and stability of cytochrome-*c*. *Cell. Mol. Life Sci.* **2014**, *71*, 229–255. [CrossRef] [PubMed]
2. Dickerson, R.E.; Timkovich, R. 7 Cytochromes c. In *The Enzymes*; Elsevier: Amsterdam, The Netherlands, 1975; pp. 397–547.
3. Kalpage, H.A.; Wan, J.; Morse, P.T.; Zurek, M.P.; Turner, A.A.; Khobeir, A.; Yazdi, N.; Hakim, L.; Liu, J.; Vaishnav, A.; et al. Cytochrome c phosphorylation: Control of mitochondrial electron transport chain flux and apoptosis. *Int. J. Biochem. Cell Biol.* **2020**, *121*, 105704. [CrossRef]
4. Min, L.; Jian-Xing, X. Detoxifying function of cytochrome c against oxygen toxicity. *Mitochondrion* **2007**, *7*, 13–16. [CrossRef]
5. Zhao, Y.; Wang, Z.B.; Xu, J.X. Effect of cytochrome c on the generation and elimination of O_2*– and H_2O_2 in mitochondria. *J. Biol. Chem.* **2003**, *278*, 2356–2360. [CrossRef] [PubMed]
6. Peker, E.; Erdogan, A.J.; Volkov, A.N.; Riemer, J. Erv1 and Cytochrome c Mediate Rapid Electron Transfer via A Colli-sion-Type Interaction. *J. Mol. Biol.* **2021**, *433*, 167045. [CrossRef]
7. Bourens, M.; Dabir, D.V.; Tienson, H.L.; Sorokina, I.; Koehler, C.M.; Barrientos, A. Role of Twin Cys-Xaa9-Cys Motif Cysteines in Mitochondrial Import of the Cytochrome c Oxidase Biogenesis Factor Cmc1. *J. Biol. Chem.* **2012**, *287*, 31258–31269. [CrossRef] [PubMed]
8. Zhan, J.; Zeng, D.; Xiao, X.; Fang, Z.; Huang, T.; Zhao, B.; Zhu, Q.; Liu, C.; Jiang, B.; Zhou, X.; et al. Real-Time Observation of Conformational Changes and Translocation of Endogenous Cytochrome c within Intact Mitochondria. *J. Am. Chem. Soc.* **2024**, *146*, 4455–4466. [CrossRef] [PubMed]
9. Zou, H.; Li, Y.; Liu, X.; Wang, X. An APAF-1·Cytochrome c Multimeric Complex Is a Functional Apoptosome That Activates Procaspase-9. *J. Biol. Chem.* **1999**, *274*, 11549–11556. [CrossRef]
10. González-Arzola, K.; Velázquez-Cruz, A.; Guerra-Castellano, A.; Casado-Combreras, M.Á.; Pérez-Mejías, G.; Díaz-Quintana, A.; Díaz-Moreno, I.; De la Rosa, M.Á. New moonlighting functions of mitochondrial cytochrome *c* in the cytoplasm and nucleus. *FEBS Lett.* **2019**, *593*, 3101–3119. [CrossRef]
11. Diaz-Quintana, A.; Perez-Mejias, G.; Guerra-Castellano, A.; De la Rosa, M.A.; Diaz-Moreno, I. Wheel and Deal in the Mitochondrial Inner Membranes: The Tale of Cytochrome c and Cardiolipin. *Oxid. Med. Cell. Longev.* **2020**, 6813405. [CrossRef] [PubMed]
12. Kalkavan, H.; Chen, M.J.; Crawford, J.C.; Quarato, G.; Fitzgerald, P.; Tait, S.W.; Goding, C.R.; Green, D.R. Sublethal cytochrome c release generates drug-tolerant persister cells. *Cell* **2022**, *185*, 3356–3374.e22. [CrossRef] [PubMed]

13. Belikova, N.A.; Vladimirov, Y.A.; Osipov, A.N.; Kapralov, A.A.; Tyurin, V.A.; Potapovich, M.V.; Basova, L.V.; Peterson, J.; Kurnikov, I.V.; Kagan, V.E. Peroxidase activity and structural transitions of cytochrome c bound to cardi-olipin-containing membranes. *Biochemistry* **2006**, *45*, 4998–5009. [CrossRef] [PubMed]
14. Kapralov, A.A.; Yanamala, N.; Tyurina, Y.Y.; Castro, L.; Samhan-Arias, A.; Vladimirov, Y.A.; Maeda, A.; Weitz, A.A.; Peterson, J.; Mylnikov, D.; et al. Topography of tyrosine residues and their involvement in peroxidation of polyunsaturated cardiolipin in cytochrome c/cardiolipin peroxidase complexes. *Biochim. Biophys. Acta (BBA)–Biomembr.* **2011**, *1808*, 2147–2155. [CrossRef] [PubMed]
15. Aluri, H.S.; Simpson, D.C.; Allegood, J.C.; Hu, Y.; Szczepanek, K.; Gronert, S.; Chen, Q.; Lesnefsky, E.J. Electron flow into cytochrome c coupled with reactive oxygen species from the electron transport chain converts cytochrome c to a cardiolipin peroxidase: Role during ischemia–reperfusion. *Biochim. Biophys. Acta (BBA)—Gen. Subj.* **2014**, *1840*, 3199–3207. [CrossRef] [PubMed]
16. O'Brien, E.S.; Nucci, N.V.; Fuglestad, B.; Tommos, C.; Wand, A.J. Defining the Apoptotic Trigger: The interaction of cytochrome c and cardiolipin. *J. Biol. Chem.* **2015**, *290*, 30879–30887. [CrossRef] [PubMed]
17. Giorgio, M.; Migliaccio, E.; Orsini, F.; Paolucci, D.; Moroni, M.; Contursi, C.; Pelliccia, G.; Luzi, L.; Minucci, S.; Marcaccio, M.; et al. Electron Transfer between Cytochrome c and p66Shc Generates Reactive Oxygen Species that Trigger Mitochondrial Apoptosis. *Cell* **2005**, *122*, 221–233. [CrossRef] [PubMed]
18. Di Lisa, F.; Giorgio, M.; Ferdinandy, P.; Schulz, R. New aspects of p66Shc in ischaemia reperfusion injury and other cardiovascular diseases. *Br. J. Pharmacol.* **2017**, *174*, 1690–1703. [CrossRef] [PubMed]
19. González-Arzola, K.; Díaz-Quintana, A.; Bernardo-García, N.; Martínez-Fábregas, J.; Rivero-Rodríguez, F.; Casado-Combreras, M.Á.; Elena-Real, C.A.; Velázquez-Cruz, A.; Gil-Caballero, S.; Velázquez-Campoy, A.; et al. Nucleus-translocated mitochondrial cytochrome c liberates nucleophosmin-sequestered ARF tumor suppressor by changing nucleolar liquid–liquid phase separation. *Nat. Struct. Mol. Biol.* **2022**, *29*, 1024–1036. [CrossRef] [PubMed]
20. Rivero-Rodríguez, F.; Díaz-Quintana, A.; Velázquez-Cruz, A.; González-Arzola, K.; Gavilan, M.P.; Velázquez-Campoy, A.; Ríos, R.M.; De la Rosa, M.A.; Díaz-Moreno, I. Inhibition of the PP2A activity by the histone chaperone ANP32B is long-range allosterically regulated by respiratory cytochrome c. *Redox Biol.* **2021**, *43*, 101967. [CrossRef] [PubMed]
21. Morse, P.T.; Arroum, T.; Wan, J.; Pham, L.; Vaishnav, A.; Bell, J.; Pavelich, L.; Malek, M.H.; Sanderson, T.H.; Edwards, B.F.P.; et al. Phosphorylations and Acetylations of Cytochrome c Control Mitochondrial Respiration, Mitochondrial Membrane Potential, Energy, ROS, and Apoptosis. *Cells* **2024**, *13*, 493. [CrossRef]
22. Guerra-Castellano, A.; Márquez, I.; Pérez-Mejías, G.; Díaz-Quintana, A.; De la Rosa, M.A.; Díaz-Moreno, I. Post-Translational Modifications of Cytochrome *c* in Cell Life and Disease. *Int. J. Mol. Sci.* **2020**, *21*, 8483. [CrossRef] [PubMed]
23. Yin, V.; Mian, S.H.; Konermann, L. Lysine carbonylation is a previously unrecognized contributor to peroxidase acti-vation of cytochrome c by chloramine-T. *Chem. Sci.* **2019**, *10*, 2349–2359. [CrossRef] [PubMed]
24. Yin, V.; Shaw, G.S.; Konermann, L. Cytochrome c as a Peroxidase: Activation of the Precatalytic Native State by $H_{(2)}O_{(2)}$-Induced Covalent Modifications. *J. Am. Chem. Soc.* **2017**, *139*, 15701–15709. [CrossRef] [PubMed]
25. Barayeu, U.; Lange, M.; Méndez, L.; Arnhold, J.; Shadyro, O.I.; Fedorova, M.; Flemmig, J. Cytochrome c autocatalyzed carbonylation in the presence of hydrogen peroxide and cardiolipins. *J. Biol. Chem.* **2019**, *294*, 1816–1830. [CrossRef] [PubMed]
26. Flatmark, T. On the heterogeneity of beef heart cytochrome c. 3. A kinetic study of the non-enzymic deamidation of the main subfractions (Cy I-Cy 3). *Acta Chem. Scand.* **1966**, *20*, 1487–1496. [CrossRef] [PubMed]
27. Sharma, G.S.; Warepam, M.; Bhattacharya, R.; Singh, L.R. Covalent Modification by Glyoxals Converts Cytochrome c Into its Apoptotically Competent State. *Sci. Rep.* **2019**, *9*, 4781. [CrossRef]
28. Hildick-Smith, G.J.; Downey, M.C.; Gretebeck, L.M.; Gersten, R.A.; Sandwick, R.K. Ribose 5-phosphate glycation reduces cytochrome c respiratory activity and membrane affinity. *Biochemistry* **2011**, *50*, 11047–11057. [CrossRef]
29. Oliveira, L.M.; Gomes, R.A.; Yang, D.; Dennison, S.R.; Família, C.; Lages, A.; Coelho, A.V.; Murphy, R.M.; Phoenix, D.A.; Quintas, A. Insights into the molecular mechanism of protein native-like aggregation upon glycation. *Biochim. Biophys. Acta (BBA)—Proteins Proteom.* **2013**, *1834*, 1010–1022. [CrossRef] [PubMed]
30. Mercado-Uribe, H.; Andrade-Medina, M.; Espinoza-Rodríguez, J.H.; Carrillo-Tripp, M.; Scheckhuber, C.Q. Analyzing structural alterations of mitochondrial intermembrane space superoxide scavengers cytochrome-c and SOD1 after methylglyoxal treatment. *PLoS ONE* **2020**, *15*, e0232408. [CrossRef]
31. Delgado, Y.; Morales-Cruz, M.; Hernández-Román, J.; Martínez, Y.; Griebenow, K. Chemical glycosylation of cytochrome c improves physical and chemical protein stability. *BMC Biochem.* **2014**, *15*, 16. [CrossRef] [PubMed]
32. Méndez, J.; Cruz, M.M.; Delgado, Y.; Figueroa, C.M.; Orellano, E.A.; Morales, M.; Monteagudo, A.; Griebenow, K. Delivery of Chemically Glycosylated Cytochrome c Immobilized in Mesoporous Silica Nanoparticles Induces Apoptosis in HeLa Cancer Cells. *Mol. Pharm.* **2014**, *11*, 102–111. [CrossRef]

33. Sharma, G.S.; Singh, L.R. Conformational status of cytochrome c upon N-homocysteinylation: Implications to cyto-chrome c release. *Arch. Biochem. Biophys.* **2017**, *614*, 23–27. [CrossRef]
34. Dopner, S.; Hildebrandt, P.; Rosell, F.I.; Mauk, A.G.; von Walter, M.; Buse, G.; Soulimane, T. The structural and func-tional role of lysine residues in the binding domain of cytochrome c in the electron transfer to cytochrome c oxidase. *Eur. J. Biochem.* **1999**, *261*, 379–391. [CrossRef] [PubMed]
35. Perła-Kaján, J.; Marczak, Ł.; Kaján, L.; Skowronek, P.; Twardowski, T.; Jakubowski, H. Modification by Homocysteine Thiolactone Affects Redox Status of Cytochrome *c*. *Biochemistry* **2007**, *46*, 6225–6231. [CrossRef] [PubMed]
36. Gates, A.T.; Moore, L.; Sylvain, M.R.; Jones, C.M.; Lowry, M.; El-Zahab, B.; Robinson, J.W.; Strongin, R.M.; Warner, I.M. Mechanistic Investigation of N-Homocysteinylation-Mediated Protein–Gold Nanoconjugate Assembly. *Langmuir* **2009**, *25*, 9346–9351. [CrossRef] [PubMed]
37. García-Heredia, J.M.; Díaz-Moreno, I.; Nieto, P.M.; Orzáez, M.; Kocanis, S.; Teixeira, M.; Pérez-Payá, E.; Díaz-Quintana, A.; De la Rosa, M.A. Nitration of tyrosine 74 prevents human cytochrome c to play a key role in apoptosis signaling by blocking caspase-9 activation. *Biochim. Biophys. Acta (BBA)—Bioenerg.* **2010**, *1797*, 981–993. [CrossRef]
38. Garcia-Heredia, J.M.; Diaz-Moreno, I.; Diaz-Quintana, A.; Orzaez, M.; Navarro, J.A.; Hervas, M.; De la Rosa, M.A. Specific nitration of tyrosines 46 and 48 makes cytochrome c assemble a non-functional apoptosome. *FEBS Lett.* **2012**, *586*, 154–158. [CrossRef] [PubMed]
39. Díaz-Moreno, I.; Nieto, P.M.; Del Conte, R.; Gairí, M.; García-Heredia, J.M.; De la Rosa, M.A.; Díaz-Quintana, A. A Non-damaging Method to Analyze the Configuration and Dynamics of Nitrotyrosines in Proteins. *Chem.–Eur. J.* **2012**, *18*, 3872–3878. [CrossRef] [PubMed]
40. Nakagawa, H.; Ohshima, Y.; Takusagawa, M.; Ikota, N.; Takahashi, Y.; Shimizu, S.; Ozawa, T. Functional Modification of Cytochrome c by Peroxynitrite in an Electron Transfer Reaction. *Chem. Pharm. Bull.* **2001**, *49*, 1547–1554. [CrossRef] [PubMed]
41. Díaz-Moreno, I.; García-Heredia, J.M.; Díaz-Quintana, A.; Teixeira, M.; De la Rosa, M.A. Nitration of tyrosines 46 and 48 induces the specific degradation of cytochrome c upon change of the heme iron state to high-spin. *Biochim. Biophys. Acta (BBA)—Bioenerg.* **2011**, *1807*, 1616–1623. [CrossRef] [PubMed]
42. Vlasova, I.I.; Tyurin, V.A.; Kapralov, A.A.; Kurnikov, I.V.; Osipov, A.N.; Potapovich, M.V.; Stoyanovsky, D.A.; Kagan, V.E. Nitric oxide inhibits peroxidase activity of cytochrome c.cardiolipin complex and blocks cardiolipin oxidation. *J. Biol. Chem.* **2006**, *281*, 14554–14562. [CrossRef]
43. Chen, Y.-R.; Deterding, L.J.; Sturgeon, B.E.; Tomer, K.B.; Mason, R.P. Protein Oxidation of Cytochrome c by Reactive Halogen Species Enhances Its Peroxidase Activity. *J. Biol. Chem.* **2002**, *277*, 29781–29791. [CrossRef]
44. Ivanetich, K.M.; Bradshaw, J.J.; Kaminsky, L.S. Methionine sulfoxide cytochrome c. *Biochemistry* **1976**, *15*, 1144–1153. [CrossRef] [PubMed]
45. Capdevila, D.A.; Rouco, S.O.; Tomasina, F.; Tortora, V.; Demicheli, V.; Radi, R.; Murgida, D.H. Active Site Structure and Peroxidase Activity of Oxidatively Modified Cytochrome c Species in Complexes with Cardiolipin. *Biochemistry* **2015**, *54*, 7491–7504. [CrossRef]
46. Capdevila, D.A.; Marmisolle, W.A.; Tomasina, F.; Demicheli, V.; Portela, M.; Radi, R.; Murgida, D.H. Specific methi-onine oxidation of cytochrome c in complexes with zwitterionic lipids by hydrogen peroxide: Potential implications for apoptosis. *Chem. Sci.* **2015**, *6*, 705–713. [CrossRef] [PubMed]
47. Tamargo-Azpilicueta, J.; Casado-Combreras, M.A.; Giner-Arroyo, R.L.; Velazquez-Campoy, A.; Marquez, I.; Olloqui-Sariego, J.L.; De la Rosa, M.A.; Diaz-Moreno, I. Phosphorylation of cytochrome c at tyrosine 48 finely regulates its binding to the histone chaperone SET/TAF-Ibeta in the nucleus. *Protein Sci.* **2024**, *33*, e5213. [CrossRef] [PubMed]
48. Kalpage, H.A.; Vaishnav, A.; Liu, J.; Varughese, A.; Wan, J.; Turner, A.A.; Ji, Q.; Zurek, M.P.; Kapralov, A.A.; Kagan, V.E.; et al. Serine-47 phosphorylation of cytochrome c in the mammalian brain regulates cytochrome c oxidase and caspase-3 activity. *FASEB J.* **2019**, *33*, 13503–13514. [CrossRef] [PubMed]
49. Kalpage, H.A.; Wan, J.; Morse, P.T.; Lee, I.; Hüttemann, M. Brain-Specific Serine-47 Modification of Cytochrome *c* Regulates Cytochrome *c* Oxidase Activity Attenuating ROS Production and Cell Death: Implications for Ischemia/Reperfusion Injury and Akt Signaling. *Cells* **2020**, *9*, 1843. [CrossRef] [PubMed]
50. Lee, I.; Salomon, A.R.; Yu, K.; Doan, J.W.; Grossman, L.I.; Hüttemann, M. New Prospects for an Old Enzyme: Mammalian Cytochrome *c* Is Tyrosine-Phosphorylated in Vivo. *Biochemistry* **2006**, *45*, 9121–9128. [CrossRef]
51. García-Heredia, J.M.; Díaz-Quintana, A.; Salzano, M.; Orzáez, M.; Pérez-Payá, E.; Teixeira, M.; De la Rosa, M.A.; Díaz-Moreno, I. Tyrosine phosphorylation turns alkaline transition into a biologically relevant process and makes human cytochrome c behave as an anti-apoptotic switch. *JBIC J. Biol. Inorg. Chem.* **2011**, *16*, 1155–1168. [CrossRef]

52. Guerra-Castellano, A.; Díaz-Quintana, A.; Pérez-Mejías, G.; Elena-Real, C.A.; González-Arzola, K.; García-Mauriño, S.M.; De la Rosa, M.A.; Díaz-Moreno, I. Oxidative stress is tightly regulated by cytochrome *c* phosphorylation and respirasome factors in mitochondria. *Proc. Natl. Acad. Sci. USA* **2018**, *115*, 7955–7960. [CrossRef]
53. Sanderson, T.H.; Mahapatra, G.; Pecina, P.; Ji, Q.; Yu, K.; Sinkler, C.; Varughese, A.; Kumar, R.; Bukowski, M.J.; Tousignant, R.N.; et al. Cytochrome c Is Tyrosine 97 Phosphorylated by Neuroprotective Insulin Treatment. *PLoS ONE* **2013**, *8*, e78627. [CrossRef] [PubMed]
54. Yang, A.; Cho, K.; Park, H.-S. Chemical biology approaches for studying posttranslational modifications. *RNA Biol.* **2018**, *15*, 427–440. [CrossRef]
55. Vempati, U.D.; Diaz, F.; Barrientos, A.; Narisawa, S.; Mian, A.M.; Millan, J.L.; Boise, L.H.; Moraes, C.T. Role of cy-tochrome c in apoptosis: Increased sensitivity to tumor necrosis factor alpha is associated with respiratory defects but not with lack of cytochrome c release. *Mol. Cell. Biol.* **2007**, *27*, 1771–1783. [CrossRef] [PubMed]
56. Starkov, A.A.; Fiskum, G. Regulation of brain mitochondrial H2O2 production by membrane potential and NAD(P)H redox state. *J. Neurochem.* **2003**, *86*, 1101–1107. [CrossRef] [PubMed]
57. Park, W.H. $H_{(2)}O_{(2)}$ inhibits the growth of human pulmonary fibroblast cells by inducing cell death, GSH depletion and G1 phase arrest. *Mol. Med. Rep.* **2013**, *7*, 1235–1240. [CrossRef] [PubMed]
58. Johansson, A.-C.; Steen, H.; Öllinger, K.; Roberg, K. Cathepsin D mediates cytochrome c release and caspase activation in human fibroblast apoptosis induced by staurosporine. *Cell. Death Differ.* **2003**, *10*, 1253–1259. [CrossRef] [PubMed]
59. Kushnareva, Y.; Murphy, A.N.; Andreyev, A. Complex I-mediated reactive oxygen species generation: Modulation by cytochrome c and NAD(P)+ oxidation–reduction state. *Biochem. J.* **2002**, *368*, 545–553. [CrossRef] [PubMed]
60. Morse, P.T.; Wan, J.; Bell, J.; Lee, I.; Goebel, D.J.; Malek, M.H.; Sanderson, T.H.; Hüttemann, M. Sometimes less is more: Inhibitory infrared light during early reperfusion calms hyperactive mitochondria and suppresses reperfusion injury. *Biochem. Soc. Trans.* **2022**, *50*, 1377–1388. [CrossRef] [PubMed]
61. Green, D.R.; Reed, J.C. Mitochondria and apoptosis. *Science* **1998**, *281*, 1309–1312. [CrossRef]
62. Cheng, T.C.; Hong, C.; Akey, I.V.; Yuan, S.; Akey, C.W. A near atomic structure of the active human apoptosome. *eLife* **2016**, *5*, e17755. [CrossRef]
63. Kalogeris, T.; Bao, Y.; Korthuis, R.J. Mitochondrial reactive oxygen species: A double edged sword in ischemia/reperfusion vs preconditioning. *Redox Biol.* **2014**, *2*, 702–714. [CrossRef] [PubMed]
64. Kaim, G.; Dimroth, P. ATP synthesis by F-type ATP synthase is obligatorily dependent on the transmembrane voltage. *EMBO J.* **1999**, *18*, 4118–4127. [CrossRef] [PubMed]
65. Hüttemann, M.; Helling, S.; Sanderson, T.H.; Sinkler, C.; Samavati, L.; Mahapatra, G.; Varughese, A.; Lu, G.; Liu, J.; Ramzan, R.; et al. Regulation of mitochondrial respiration and apoptosis through cell signaling: Cytochrome c oxidase and cytochrome c in ischemia/reperfusion injury and inflammation. *Biochim. Biophys. Acta (BBA)—Bioenerg.* **2012**, *1817*, 598–609. [CrossRef] [PubMed]
66. Olloqui-Sariego, J.L.; Pérez-Mejías, G.; Márquez, I.; Guerra-Castellano, A.; Calvente, J.J.; De la Rosa, M.A.; Andreu, R.; Díaz-Moreno, I. Electric field-induced functional changes in electrode-immobilized mutant species of human cytochrome c. *Biochim. Biophys. Acta (BBA)—Bioenerg.* **2022**, *1863*, 148570. [CrossRef]
67. Deep, S.; Im, S.C.; Zuiderweg, E.R.; Waskell, L. Characterization and calculation of a cytochrome c-cytochrome b5 complex using NMR data. *Biochemistry* **2005**, *44*, 10654–10668. [CrossRef] [PubMed]
68. Liu, Z.; Lin, H.; Ye, S.; Liu, Q.-Y.; Meng, Z.; Zhang, C.-M.; Xia, Y.; Margoliash, E.; Rao, Z.; Liu, X.-J. Remarkably high activities of testicular cytochrome *c* in destroying reactive oxygen species and in triggering apoptosis. *Proc. Natl. Acad. Sci. USA* **2006**, *103*, 8965–8970. [CrossRef]
69. Fridovich, I. Quantitative Aspects of the Production of Superoxide Anion Radical by Milk Xanthine Oxidase. *J. Biol. Chem.* **1970**, *245*, 4053–4057. [CrossRef]

International Journal of
Molecular Sciences

MDPI

Article

Plasma Redox Balance in Advanced-Maternal-Age Pregnant Women and Effects of Plasma on Umbilical Cord Mesenchymal Stem Cells

Elena Grossini [1,*], Carmen Imma Aquino [2], Sakthipriyan Venkatesan [1], Libera Troìa [2], Eleonora Tizzoni [2], Federica Fumagalli [2], Daniela Ferrante [3], Rosanna Vaschetto [4], Valentino Remorgida [2] and Daniela Surico [2]

1 Laboratory of Physiology, Department of Translational Medicine, Università del Piemonte Orientale, Via Solaroli 17, 28100 Novara, Italy; sakthipriyan.venkatesan@uniupo.it
2 Gynecology and Obstetrics, Department of Translational Medicine, Università del Piemonte Orientale, "Maggiore della Carità" Hospital, 28100 Novara, Italy; 20033548@studenti.uniupo.it (C.I.A.); 20021710@studenti.uniupo.it (E.T.); 20020579@studenti.uniupo.it (F.F.); valentino.remorgida@med.uniupo.it (V.R.); daniela.surico@med.uniupo.it (D.S.)
3 Medical Statistics, Department of Translational Medicine, Università del Piemonte Orientale, 28100 Novara, Italy; daniela.ferrante@med.uniupo.it
4 Anesthesia and Intensive Care, Department of Translational Medicine, Università del Piemonte Orientale, 28100 Novara, Italy; rosanna.vaschetto@med.uniupo.it
* Correspondence: elena.grossini@med.uniupo.it

Abstract: Pregnancy at advanced maternal age (AMA) is a condition of potential risk for the development of maternal–fetal complications with possible repercussions even in the long term. Here, we analyzed the changes in plasma redox balance and the effects of plasma on human umbilical cord mesenchymal cells (hUMSCs) in AMA pregnant women (patients) at various timings of pregnancy. One hundred patients and twenty pregnant women younger than 40 years (controls) were recruited and evaluated at various timings during pregnancy until after delivery. Plasma samples were used to measure the thiobarbituric acid reactive substances (TBARS), glutathione and nitric oxide (NO). In addition, plasma was used to stimulate the hUMSCs, which were tested for cell viability, reactive oxygen species (ROS) and NO release. The obtained results showed that, throughout pregnancy until after delivery in patients, the levels of plasma glutathione and NO were lower than those of controls, while those of TBARS were higher. Moreover, plasma of patients reduced cell viability and NO release, and increased ROS release in hUMSCs. Our results highlighted alterations in the redox balance and the presence of potentially harmful circulating factors in plasma of patients. They could have clinical relevance for the prevention of complications related to AMA pregnancy.

Keywords: aging; antioxidants; lipid peroxidation; nitric oxide; pregnancy

1. Introduction

The historical definition of the term "Advanced Maternal Age" (AMA) refers to a condition of maternal age over 35 years [1,2]. The analysis of the data collected in the last three decades evidenced an increase in the average age of conception and delivery in women belonging to developed countries [3]. These findings concern both women aged 35–39 years, in whom the increase in the birth rate was from 45.9 per 1000 women in 2010 to 52.7 per 1000 women in 2019, and women aged 40–44 years, who showed an increase in the birth rate from 10.2 to 12 per 1000 women in the same range of years [4–6]. In addition to reasons related to work and career needs, the increase in AMA pregnant women could be explained also by the development in advanced reproductive technology, which has extended the reproductive window [7,8]. Thus, on the grounds of these considerations, the above reported definition of AMA appears outdated and it would be more appropriate to use it for pregnant women older than 40 years [9,10].

It should be kept in mind that AMA pregnancy could represent a risk factor for adverse maternal complications, such as pre-eclampsia, gestational diabetes mellitus, gestational hypertension, and Cesarean delivery, as well as for fetal outcomes [7,8]. In addition to increased maternal mortality and morbidity, the above health complications also have repercussions in terms of increased costs to health systems worldwide [11]; a great deal of effort should be made to try to counteract or even prevent them.

Actions aimed at preventing AMA pregnancy-related complications could be focused on the modulation of its pathophysiological mechanisms.

In this regard, an important role is played by the loss of placental functions [12]. Indeed, changes in the placentation process may have serious consequences in pregnancy for women presenting specific characteristics, including AMA, which are also related to endothelial dysfunction [11].

In physiologic pregnancy, reactive oxygen species (ROS) and reactive nitrogen species (RNS) are generated mainly by the placenta and vascular endothelium, as well as through the action of xanthine oxidase and nitric oxide (NO) synthase [13,14].

However, when the feto-placental unit is poorly perfused, such as during AMA pregnancy, an increased oxidant release not counteracted by antioxidants like superoxide dismutase or glutathione (GSH) and glutathione peroxidase may affect normal vasodilation and become an important factor in the pathogenesis of AMA pregnancy-related complications, like those affecting the cardiovascular system, kidney and liver as well [11,13,15].

In particular, increased concentrations of lipid peroxides in the villous and decidual tissues have been reported to be common to all miscarriages and have been correlated with proteinuria, uricemia and pulsatility index of umbilical arteries in pre-eclamptic pregnant women [16].

Furthermore, changes in lipid peroxides could play an important role in AMA pregnancy since they can affect endothelial function through changes in thromboxane A2 and NO release in the utero-placental and maternal systemic vasculature as well [17].

Also, as regarding GSH, which is one of the most important antioxidants that can reduce unstable ROS and turn into oxidized GSH, it was reported to be strongly reduced in the plasma of pre-eclamptic women [11,16], in placental aging and in adverse pregnancy outcomes [18].

It is important to underline that oxidative stress resulting from unbalance between oxidants and antioxidants can be involved in the physiopathology of AMA pregnancy through changes in endothelial NO availability and endothelial dysfunction.

Another possible target of the prevention/modulation of AMA pregnancy-related complications could be represented by the human umbilical cord mesenchymal stem cells (hUMSCs), which can protect placental function by increasing its development and angiogenesis through paracrine actions and the release of immunomodulatory factors [16,19–21]. Also, those cells can be involved in the modulation of oxidative stress/inflammation, have the ability to recover ovarian function in premature ovarian insufficiency or natural aging animal models, and exert cardioprotective effects through the release of microRNA [22].

Despite existing knowledge about this issue, however, additional information regarding the plasma redox balance and the changes in hUMSCs function induced by circulating factors in AMA pregnant women at different times of pregnancy would be mandatory to better understand the pathophysiological aspects.

Considering what has been reported above regarding the role of lipid peroxides, GSH and NO in the physiopathology of AMA pregnancy and the lack of detailed information, here, our primary endpoint was to examine the plasma levels of lipid markers of peroxidation, GSH and NO at the first trimester (T0), at the second trimester during the morphologic ultrasound (T1) and at 48–72 h after the delivery (T2) in pregnant women over 40 years (median age 41) compared with pregnant women younger than 40 years (median age 30). In particular, we used the thiobarbituric acid reactive substances (TBARS) assay for measurement of plasma lipid peroxides such as malonyldialdeide (MDA) and specific assays for GSH and NO, as previously conducted [23–25]. Secondly, we wanted to analyze the effects of plasma taken from AMA pregnant women at different time points on hUMSCs in order to highlight the role of circulating factors in the modulation of their viability and oxidant release.

2. Results

2.1. Patients

All patients were subjected to careful anamnestic investigation. Risk factors were also investigated for pregnancy disorders and previous pathological pregnancies (Table 1). The anamnestic characteristics of the two groups, compared by means of Fisher's test, were found to be superimposable, with no statistically significant differences ($p > 0.05$) except for maternal age (Table 1). Therefore, the only characteristic that significantly differentiated the group of AMA patients and the group of controls was obviously the age (41 versus 30 years, $p < 0.0001$).

Table 1. Characteristics of the total population at enrollment.

	AMA Patients (n = 100)	Controls (n = 20)	p-Value
Age median (IQR)	41 (40–42)	30 (27–32)	<0.0001
BMI before pregnancy (kg/m^2) median (IQR)	23.7 (20.9–27.4)	21.5 (19.8–23.3)	0.04
Ethnicity n (%)			
Caucasic	80 (80.0)	18 (0.90)	
Other	20 (20.0)	2 (0.10)	0.47
Nulliparous n (%)	36 (36.0)	7 (35.0)	0.93
Miscarriages n (%)			
0	45 (45.0)	12 (60.0)	
1	37 (37.0)	7 (35.0)	
>=2	18 (18.0)	1 (5.0)	0.29
Before-pregnancy pathologies n (%)	18 (18.0)	1 (5.0)	0.25
Positive combined prenatal screening test and positive at second-level test	26 (26.0)	1 (5.0)	0.06

Although the body mass index (BMI) before pregnancy of AMA patients was higher than that of controls ($p < 0.05$), it was within the normal weight range.

A positive first trimester screening was found in 26 of 100 AMA pregnant patients, while it was positive in only one patient of the controls. In the calculation of the risk, the age of the patients had a substantial influence.

In Table 2, we describe the main delivery outcomes of patients and controls. The statistical analysis did not show a significant high percentage of complications in AMA patients vs. controls ($p = 0.12$). Among AMA patients, spontaneous delivery occurred in 23 patients (43.4%), Cesarean section in 22 (41.5%), induced childbirths in 5 (9.4%) and dystocic deliveries in 3 (5.7%) (Table 2). In controls, 11 (61.1%) spontaneous delivery occurred, childbirths were induced 3 times (16.7%), as Cesarean section, and dystocic delivery occurred only once (5.6%). A total of 28 patients presented no complications and 36 developed some gestational comorbidities, such as gestational diabetes, preterm birth, intrahepatic cholestasis of pregnancy, gestational hypertension, pre-eclampsia, placenta low lying, intrauterine growth restriction (IUGR) and deep vein thrombosis.

Table 2. Delivery outcomes in patients and controls who gave birth by the end of the study.

	AMA Patients (n = 53)	Controls (n = 18)	p-Value
Type of delivery n (%)			
Spontaneous vaginal	23 (43.4)	11 (61.1)	0.19
Induced	5 (9.4)	3 (16.7)	0.40

Table 2. *Cont.*

	AMA Patients (n = 53)	Controls (n = 18)	p-Value
Cesarean section	22 (41.5)	3 (16.7)	0.06
Dystocic	3 (5.7)	1 (5.6)	0.99
Gestational age at delivery median (IQR)	39 weeks (38–40)	39 weeks (38–40)	0.50
Birth weight at delivery (g) mean (SD)	3073 (575.2)	3288 (445.9)	0.15

2.2. *Plasma TBARS, GSH and NO*

The results of plasma TBARS, GSH and NO measurements performed in AMA patients and controls at various times during pregnancy up to postpartum are shown in Table 3. All models were adjusted by type of delivery (classified as in Table 2).

Table 3. Response trends over time obtained by comparing groups averaged over time and measurement time within a group in plasma measurements.

	F Value	p Value	Outcomes
Group	42.45	<0.0001	TBARS
Time	0.40	0.67	
Interaction group × time	0.43	0.65	
Type of delivery	0.61	0.61	
Group	29.20	<0.0001	NO
Time	8.56	0.0004	
Interaction group × time	1.21	0.30	
Type of delivery	0.69	0.56	
Group	101.49	<0.0001	GSH
Time	3.89	0.02	
Interaction group × time	2.31	0.11	
Type of delivery	0.30	0.82	

GSH: glutathione; NO: nitric oxide; TBARS: thiobarbituric acid reactive substances.

It is of note that the overall mean of the considered outcomes (TBARS, GSH and NO) between AMA patients and controls was significantly different (group effect).

It is also of note that, at T3, in AMA patients, the levels of TBARS remained higher than those measured in controls ($p < 0.05$): in AMA patients, 9 πM 5.3–12 and, in controls, 5.1 πM (4.6–5.7). Also, plasma GSH of AMA patients was lower than that found in controls ($p < 0.05$): in AMA patients, 5 πM (3.4–7.2) and, in controls, 8 πM (7–8.9). As regards NO, in AMA patients, the plasma values were lower than those of controls ($p < 0.05$): in AMA patients, 9.6 πM (5.6–13) and, in controls, 14.5 πM (11.9–16.7).

Also, in the plasma of the umbilical cords, there was an alteration in the redox state in AMA patients. Indeed, GSH levels were lower than those observed in the controls ($p < 0.05$; Figure 1A), while TBARS levels were higher ($p < 0.05$; Figure 1B). Instead, we did not detect any significant differences regarding NO (Figure 1C).

2.3. *Effects of Plasma on hUMSCs*

The in vitro experiments showed that the treatment of hUMSCs with plasma taken from AMA patients was able to reduce cell viability and NO release and increase ROS production (Figure 2A–C). As evidenced in Figure 2A, the cell viability was lower than that measured both in hUMSCs treated with plasma of the controls and in the untreated hUMSCs.

If we analyze the results obtained at the three different time points, it can be observed that there are no significant variations in the median values of cell viability in hUMSCs treated with plasma of AMA patients (Figure 2A). In the case of the controls, cell viability was even increased at T3 in comparison with that found in the untreated hUMSCs ($p < 0.05$).

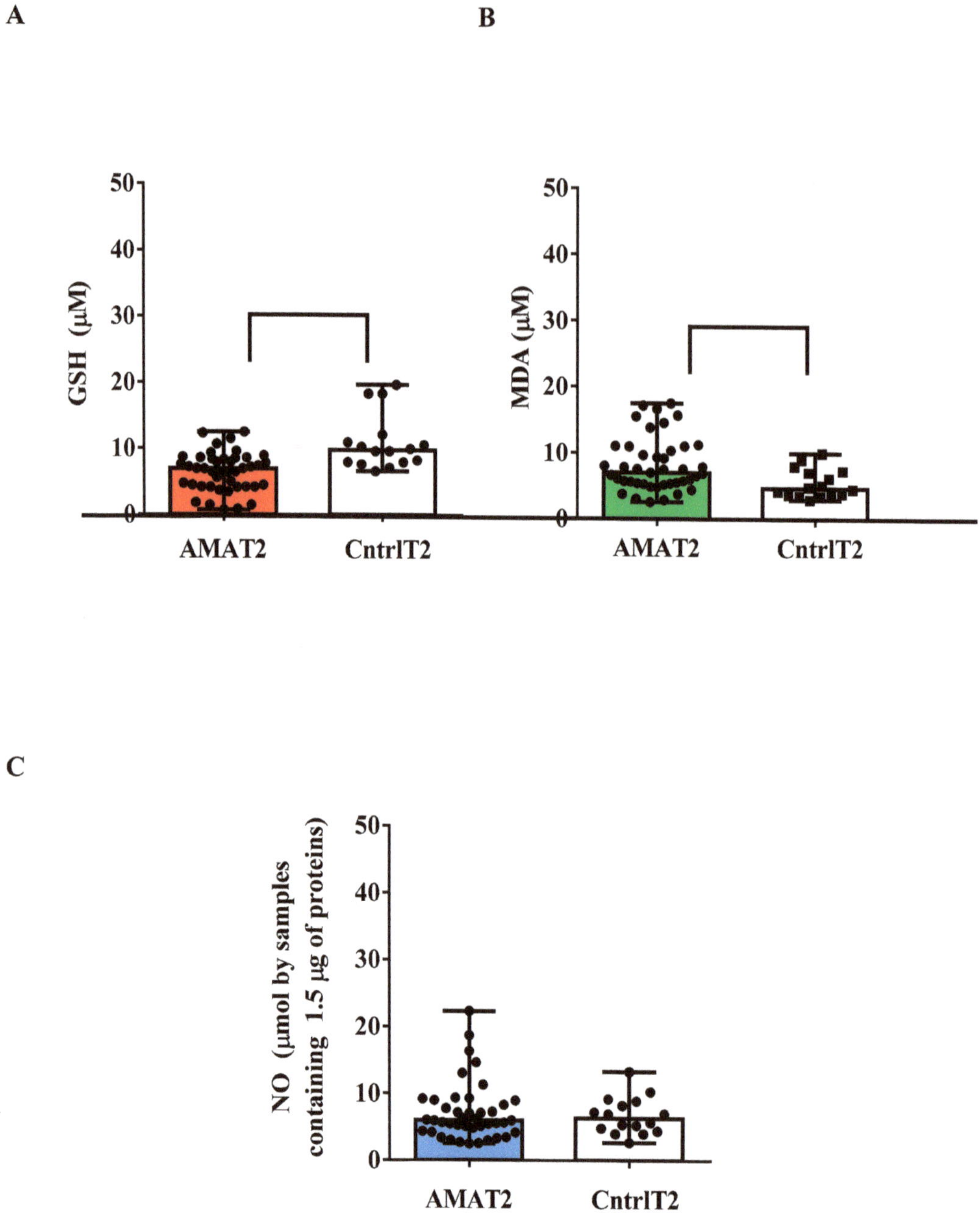

Figure 1. Glutathione (GSH, (**A**)), malonyldialdeide (MDA, (**B**)) and nitric oxide (NO, (**C**)) in umbilical cord plasma older (AMA > 40 yrs) and younger (Cntrl < 40 yrs) pregnant women collected at delivery. MDA was measured through the TBARS assay. T2: at the delivery. The results are expressed as median and range of different measurements. Square brackets indicate significance between groups ($p < 0.05$).

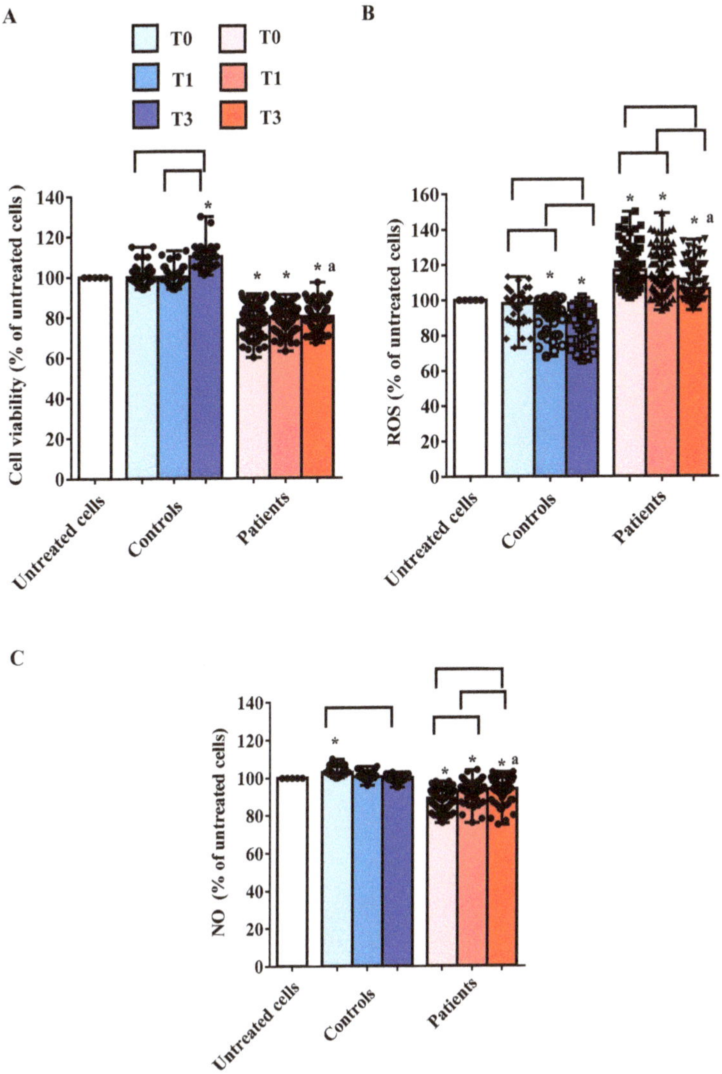

Figure 2. Effects of plasma taken from pregnant women older (patients) and younger than 40 years (controls) on cell viability (**A**), reactive oxygen species (ROS) release (**B**) and nitric oxide (NO) release (**C**) in hUMSCs. The bars represent the effects of plasma of all 10 patients and all 5 controls at various time points. T0: at 11–13 weeks of gestational age; T1: at 20–22 weeks of gestational age; T3: before Cesarean section or at 48–72 h after the delivery. The results are the median and range of repeated experiments. Untreated cells: non-treated hUMSCs. * $p < 0.05$ vs. untreated cells; a: $p < 0.05$ vs. T3 controls; square brackets indicate significance between the groups ($p < 0.05$).

As performed for plasma measurements of TBARS, GSH and NO, we executed a statistical analysis about response trends over time in measured variables of the in vitro experiments too. As shown in Table 4, it is of note that a significant interaction group × time emerged for in vitro data about NO and cell viability. This means that changes in response over time differed among AMA patients and controls.

Table 4. Response trends over time obtained by comparing groups averaged over time and measurement time within a group in the in vitro experiments.

	F Value	p Value	Outcomes
Group	53.03	<0.0001	NO
Time	1.87	0.20	
Interaction group × time	10.41	0.003	
Type of delivery	2.58	0.11	
Group	184.9	<0.0001	CELL VIABILITY
Time	9.34	0.004	
Interaction group × time	7.30	0.01	
Type of delivery	0.21	0.88	
Group	32.34	0.0001	ROS
Time	31.47	<0.0001	
Interaction group × time	0.45	0.65	
Type of delivery	1.33	0.31	

NO: nitric oxide; ROS: reactive oxygen species.

Concerning ROS, we observed an increased release by hUMSCs treated with plasma of AMA patients both vs. the untreated cells and vs. hUMSCs treated with plasma of the controls (Figure 2B).

Also, in hUMSCs treated with plasma of AMA patients at T3, cell viability and NO release were lower than those found in hUMSCs treated with plasma of controls, whereas ROS release was higher (Figure 2).

2.4. Clinical Pattern

The analysis of the data regarding AMA patients suffering from gestational diabetes, preterm birth, intrahepatic gestational cholestasis, gestational hypertension, pre-eclampsia, low-lying placenta, IUGR and deep vein thrombosis vs. controls, and even comparing patients with pregestational diseases vs. uncomplicated ones, did not show any statistically significant difference. Thus, the altered redox balance we found in AMA patients can be linked to age rather than to any comorbidities.

3. Discussion

Our results highlighted an altered redox balance in the plasma of AMA patients at diagnosis and throughout the duration of pregnancy up to 48 h postpartum. Moreover, we showed that unknown factors capable of affecting viability and ROS release in hUMSCs circulate in the plasma of the aforementioned patients.

In recent years, there has been a significant increase in AMA pregnancies, probably due to epidemiological and socio-economic reasons, as well as a progressive strengthening of medically assisted fertility techniques [2]. The percentage of women who decide to postpone pregnancy in developing countries is increasing. According to the recent Certificate of Assistance in Childbirth (CeDAP), the average age of mothers is 33.1 years for the Italian population, and the average age of having a first child is over 31 years, according to the global report [26,27].

In the past, an AMA gestation was almost always accidental and related to pluriparity; however, today it is more frequently associated with the first pregnancy and with changes in demographics of childbirth.

As previously reported, AMA pregnancy has been considered as an independent risk factor for several complications, varying across ages. In fact, these pregnancies are known to be linked to several comorbidities and adverse outcomes, such as pregnancy-related hypertensive disorders, blood transfusion, and maternal and fetal mortality [16,28]. In this regard, AMA has been reported to increase the risk of stillbirth (RR 2.16, 95% CI 1.86–2.51), perinatal mortality, intrauterine growth restriction, neonatal death, admission to neonatal intensive care unit, pre-eclampsia, preterm delivery, cesarean delivery and maternal mortality compared with women younger than 40 years old (RR 3.18, 95% CI 1.68–5.98) [29].

The relationship between AMA and stillbirth is not directly linked to maternal morbidity or assisted reproductive technology [30].

One of the possible explanations of those phenomena could be the increased oxidative stress. Pregnancy itself augments susceptibility to oxidative stress, with an elevated presence of ROS and RNS in the circulatory system. The amplified gestational redox imbalance can trigger several adverse outcomes such as abnormal placental function, and complications, including the forementioned pre-eclampsia, embryonic resorption, recurrent pregnancy loss, fetal developmental anomalies, intrauterine growth restriction and, in extreme instances, fetal death. The body's answer to control the increase in RNS/ROS levels requires nonenzymatic and enzymatic defense processes (i.e., acting on TBARS, GSH and NO). The literature describes a plausible association between compromised antioxidant enzyme function and the occurrence of these adverse gestational outcomes. Oxidative stress causes detrimental consequences on maternal physiology, pregnancy progression and fetal development by acting on placental function and compromising oxygen and nutrient delivery to the fetus [31].

Pregnancies complicated by conditions such as oxidative stress, hypoxia and inflammation are associated with alterations in placental vasculogenesis, trophoblast expression of transporters and hormone production contributing to alteration in fetal development [32].

Regarding this issue, it should be pointed out that, in physiological pregnancy, the increase in oxidants is counterbalanced by the production of antioxidants and the occurrence of oxidative stress is limited [33,34]. In particular, it was shown that there is a gradual reduction in lipoperoxide formation with the progress of gestation to protect the fetus against ROS [35].

Instead, in all conditions in which the systems for regulating the balance between oxidants and antioxidants are poorly effective, the development of oxidative stress is markedly greater. Certainly, one of those conditions predisposing the onset of oxidative stress is aging [36].

The primary endpoint of this study was the analysis of plasma redox balance at different time points during pregnancy in AMA vs. younger pregnant women.

We measured lipid peroxidation as MDA release by means of the TBARS assay, since it has been widely used in previous studies aimed at analyzing oxidants in biologic fluids during pregnancy too [34,37].

We found that patients and controls differed in terms of plasma TBARS and GSH levels throughout pregnancy. Hence, lipid peroxidation markers were higher in women with advanced age, while the antioxidant was higher in the control. Similar results were noted also in the experiments performed by using plasma from umbilical cords of AMA patients. Although both age and body mass index were different among groups, mean body mass index for AMA women was within the normal weight range, suggesting differences in redox potential could be attributable to age [38].

TBARS are the most reliable biomarkers of oxidative stress [39] and have been shown to have some clinical relevance in aging and age-related pathologies [40]. Our data about TBARS are in agreement with those found by de Lucca L et al. and Draganovic D et al. in complicated and uncomplicated pregnancies and in pregnancy-induced hypertension women, respectively [32,35]. Indeed, among the 45 pregnant women recruited in the study by de Lucca L et al., TBARS increased in the second trimester when compared to the first trimester of pregnancy in the uncomplicated group, whereas they increased in the third trimester when compared to the first trimester of pregnancy in the complicated group. Also, the 100 pregnant women with hypertension recruited by Draganovic D et al. had increased mean TBARS values over 20 µmol, whereas, in healthy pregnant women, only 1% experienced increased TBARS value.

Regarding GSH, it was suggested that the measurement of its plasma levels could provide a way to measure resilience of redox networks in aging and age-related disease [41], and it is strongly reduced in placental aging and in adverse pregnancy outcomes [18]. For this reason, we have performed GSH measurements in plasma of AMA patients throughout pregnancy and compared it with those of controls.

Also, although a role of the above reported pregnancy-related complications in the alteration of the balance between oxidants and antioxidants is conceivable [42,43], the statistical analysis showed that there were no significant differences regarding the parameters of redox balance and oxidative stress between AMA patients with complications and without complications. In fact, 28 patients presented no complications and 36 developed some gestational comorbidities. Indeed, as described in the Results section, the analysis of the data regarding the 36 AMA patients suffering from pathologies during pregnancy vs. controls, and even comparing patients with pregestational diseases vs. uncomplicated ones, did not show any statistically significant difference.

For this reason, we can attribute the increased values of plasma TBARS and reduced values of plasma GSH more to the advanced age of the patients than to their complications.

The data we obtained about the redox balance in AMA pregnant women up to 48 h after delivery and in umbilical blood could have clinical implications in those subjects and newborns. In this regard, there are data demonstrating that both children born from complicated pregnancies and AMA pregnant women are at increased risk of cardiovascular disease, which could be related to oxidative stress and endothelial dysfunction that had arisen during pregnancy [42–47].

In particular, our results could be helpful to increase knowledge about the role of oxidative stress in the physiopathology of AMA pregnancy and of systemic diseases that could arise in AMA pregnant women [44,45]. Also, it could stimulate studies on the efficacy of preventive or therapeutic treatments aimed at contrasting them.

As reported above, a key role as a pathogenic mechanism could be played by oxidative stress and endothelial dysfunction [12,34,46–48]. The data we obtained about plasma levels of NO would confirm this hypothesis.

Hence, in AMA pregnant women, we found statistical significance regarding group effect in the measurements of plasma TBARS, GSH and NO between AMA patients and controls.

Data about NO are of relevance for the physiologic regulation of pregnancy and for the onset of pregnancy-related pathologies. In the placental–umbilical unit, the NOS isoforms, namely the endothelial (eNOS) and the inducible ones, which are involved in NO release, are expressed in the syncytiotrophoblast (STB) and in the endothelium of umbilical vein and arteries and in placental microvascular endothelial cells [33]. NO, through its dilating effects, plays a pivotal role for maintaining endothelial function and is responsible for keeping vascular homeostasis, which ensures constant uterine blood flow [49]. Also, NO is engaged in STB endovascular invasion and development of the placenta through its angiogenic and vasculogenic effects [50].

For those reasons, changes in NO release, which can be observed in conditions of placental hypoperfusion and oxidative stress, may be involved in the onset of pregnancy-related diseases. In fact, in the presence of hypoxia and oxidative stress, the bioavailability of tetrahydrobiopterin, which is a cofactor of eNOS, is reduced and eNOS uncoupling can be observed. In this way, NO released from eNOS turns into peroxinitrites, which can reduce placental blood flow through vasoconstriction and activate inflammatory-response-associated signaling pathways [51].

In order to better understand the pathophysiology of pregnancy in AMA, both in relation to the effects on the mother and on the fetus, we performed some in vitro experiments on hUMSCs, which could play a fundamental role for the proper functioning of the placenta and the continuation of a pregnancy without complications [13]. The stimulation of those cells with plasma taken from AMA pregnant women was able to reduce cell viability and NO release and increase ROS release. It is noteworthy that the effects of plasma of AMA patients on hUMSCs persisted even after delivery.

It is also of note that a significant interaction group × time was found about NO and cell viability. This means that changes in response over time differed among AMA patients and controls. In particular, we observed an increase in cell viability of hUMSCs treated with plasma of controls, whereas no changes over time were observed for AMA patients.

Our data may have pathophysiological implications, since it is well known that hUMSCs can regulate trophoblast invasion, angiogenesis and placental function through the modulation of inflammation and the release of mediators like vascular endothelial growth factor, the placental growth factor, miRNAs or NO [16,21,52–54]. Alterations in the production of the aforementioned factors, with particular reference to NO, would, therefore, result in the loss of the role played by hUMSCs in the physiological regulation of pregnancy. For those reasons, our results, which highlight a possible mechanism underlying AMA pregnancy physiopathology, which could be related to unknown harmful circulating factors targeting hUMSCs, would add knowledge about this issue.

Regarding the nature of those "unknown circulating factors", among others, we could speculate that extracellular vesicles (EVs) could be good candidates, since they are released by STB into the maternal circulation in increasing amounts with advancing gestational age to target various organs, and may represent a mechanism by which the placenta orchestrates maternal responses in normal and pathological pregnancy [54]. Actually, the aim of this study was not to investigate the nature of those circulating factors, which could be the object of subsequent research.

Overall, we could hypothesize that, in AMA pregnant women, a condition of feto-placental dysfunction related to aging phenomena would cause an imbalance between oxidants/antioxidants in favor of the former, with an increase in lipid peroxidation, which would worsen endothelial dysfunction. Furthermore, circulating factors released by the endothelium itself, the immune system and the uteroplacental tissues and cells, including hUMSCs, could have detrimental effects on maternal physiology, pregnancy progression, and fetal development. Thus, we could speculate that the oxidative imbalance we found in our study and the release of those factors could contribute to miscarriages, fetal developmental abnormalities, preterm birth, low birth weight and the development of cardiovascular or other chronic disease, which could arise in AMA pregnant women and in children too. However, those issues will need to be explored in greater depth in order to define their role in AMA pregnancy and AMA-related diseases.

Thus, it would be crucial to explore this field of research also from a prevention perspective and for giving recommendation to AMA pregnant women to decrease the AMA pregnancy-associated risks. In this regard, it could be suggested to change lifestyle in order to increase the antioxidant power and keep endothelial function.

Anyway, for the development of effective interventions aimed at mitigating the detrimental effects of oxidative stress during pregnancy, there should be an advance in the knowledge of the underlying mechanisms.

Limitations

The results of this study could be implemented by increasing the sample size of both AMA pregnant women and controls as well and by carrying out analyses of plasma parameters even some months after delivery.

Although our study has possible relevant clinical implications, it should be, certainly, considered that the pathological conditions under examination could have many possible correlates, which have been investigated here and are not entirely well known.

Also, in our study, we used hUMSc from a depository since they are widely adopted to perform studies about inflammation, cancer and neurologic diseases [54–56]. In addition, the results we obtained were similar to those we found from preliminary experiments performed in primary hUMSCs isolated from umbilical cords. Anyway, the use of primary hUMSCs could be suggested for future studies about pregnancy and pregnancy-related diseases in order to better correlate the results obtained with the clinical ones and to increase the knowledge about their possible involvement in the physiopathology of AMA. In this regard, it could be interesting to perform crosstalk experiments between hUMSCs and other placental cell lines and analyze the role of EVs released by hUMSCs. As reported above, the nature of the "unknown circulating factors" should also be investigated as well by focusing on EVs.

Also, future analyses performed on hUMSCs through qPCR could be organized in order to investigate in detail the role of oxidants and antioxidants and of various NOS isoforms. Moreover, by means of confocal microscopy, it could be possible to implement knowledge on the cellular effects of the plasma of AMA pregnant women.

4. Materials and Methods

4.1. Patients

This observational case–control study was performed on 100 pregnant women older than 40 years (patients) and 20 pregnant women younger than 40 years (controls) enrolled at the Gynecology and Obstetrics Unit, Università del Piemonte Orientale, Azienda Ospedaliera Universitaria Maggiore della Carità, Novara, between 30 June 2021 and 30 June 2023, for a total of two years.

4.2. Clinical Evaluation

Patients and controls were subjected to anamnestic investigation. In each pregnant woman, blood samples were collected for the analysis of oxidants/antioxidants markers, such as TBARS, GSH and NO, as specified below. The inclusion criteria were singular or twin pregnancies; age $\geq$ 40 years for patients/<40 years for the controls; written informed consent; acceptance to the follow-up and giving birth at Gynecologic and Obstetrics Unit, Azienda Ospedaliera Universitaria Maggiore della Carità.

Women were excluded in cases of legal interdiction; childbirth in different centers; age < 40 years for patients and $\geq$40 years for the controls; positivity for SARS-CoV-2, anti-HIV or anti-HCV.

Plasma samples were collected at T0, T1, T2 and T3. T0 was taken at 11–13 weeks of gestational age, during the first trimester screening, when the patients were counseled and recruited. T1 was taken at 20–22 weeks of gestational age, during the second trimester morphological ultrasound. T2 was at the time of delivery and umbilical cord blood was collected. Finally, T3 was taken before Cesarean section or within 48–72 h after vaginal delivery.

Of the 134 AMA patients that visited in the selected study period, 34 women were excluded, as shown in Figure 3.

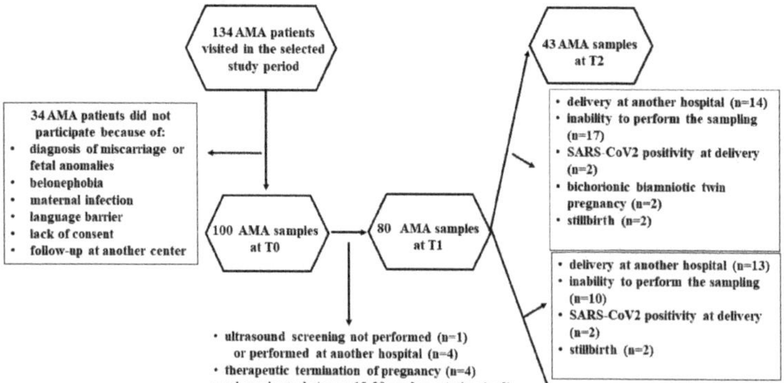

Figure 3. Flowchart describing advanced-maternal-age (AMA) pregnant women that underwent the analysis of plasma redox balance and NO at various time points.

Therefore, 100 patients and 20 controls were enrolled. At various time points, some patients were lost, as shown in Figures 3 and 4. Thus, the evaluations regarding the plasma redox balance and NO were executed on 100 patients at T0, 80 patients at T1, 43 patients at T2 and 53 patients at T3; and 20 controls at T0, 18 controls at T1, 16 controls at T2 and 18 controls at T3.

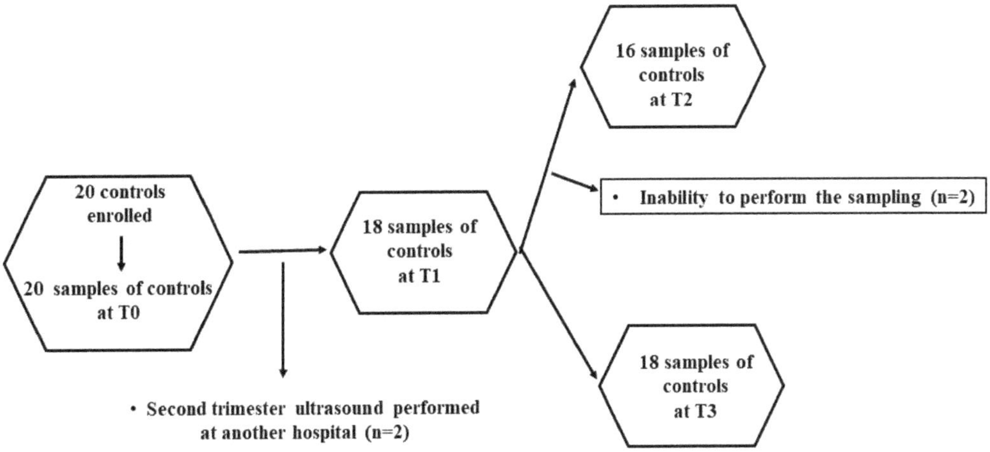

Figure 4. Flowchart describing the controls that underwent the analysis of plasma redox balance and NO at various time points.

4.3. Biological Sample Analysis

In each pregnant woman, 15 mL blood samples were taken at 9 am in fasting conditions by using BD Vacutainer tubes (sodium heparin or ethylenediaminetetraacetic acid as anticoagulant). Each sample was centrifuged for 15 min through a centrifuge (Eppendorf, mod. 5702 with rotor A-4-38, Milan, Italy) at 1600 rpm and 4 °C. The plasma was aliquoted into 5 tubes, which were stored at −80 °C at the Physiology laboratory, Università del Piemonte Orientale, until they were used for the quantification of TBARS, GSH and NO and to perform the in vitro experiments on hUMSCs. Plasma samples were handled in pseudonymized conditions and measurements were performed at least in triplicate.

4.4. Quantification of Plasma TBARS

Plasma TBARS were measured by using the TBARS assay Kit (Cayman Chemical, Ann Arbor, MI, USA), which evaluates the MDA release [23,24]. Briefly, 100 μL of each plasma sample was added to sodium dodecyl sulfate solution (100 μL) and Color Reagent (2 mL). Samples were boiled for 1 h and then put on ice for 10 min to stop the reaction. After centrifugation for 10 min at 1600 g at 4 °C, 150 μL of each sample was transferred to 96-well plates for malonyldialdeide detection through a spectrophotometer (VICTOR™ X Multilabel Plate Reader; Perkin Elmer; Waltham, MA, USA) at 530–540 nm excitation/emission. A standard curve with the MDA Standard was performed as a reference for TBARS in each sample (expressed as malonyldialdeide in μM).

4.5. Quantification of Plasma GSH

Plasma GSH levels were quantified through the Glutathione Assay Kit (Cayman Chemical) [23,24]. Briefly, each plasma sample was deproteinated through the addition of meta-phosphoric acid solution. After centrifugation (2000× g for 2 min), the supernatant of each sample was collected and 50 μL/mL of TEAM reagent was added. Thereafter, 50 μL of each sample was moved to 96-well plates, where the measurement of GSH was executed through a spectrophotometer (VICTOR™ X Multilabel Plate Reader; Perkin Elmer) at

405–414 nm excitation/emission. To perform an accurate GSH quantification (as μM), a standard curve was prepared as well by using the GSH Standard.

4.6. Quantification of Plasma NO

Plasma NO was quantified by the Griess assay (Promega Italia Srl, Milan, Italy) [23,25]. To achieve this, 5 mL plasma sample was deproteinated by adding 10 mL sulfosalicylic acid. The samples were then vortexed every 5 min and allowed to react for 30 min at room temperature. After centrifugation (10,000× *g* for 15 min), 50 μL of the supernatant was added to saline (1:2 dilution) for the subsequent analysis. The remaining microliters were used without dilution. In order to reduce nitrate to nitrite, the samples were passed through a copper–cadmium column of an autoanalyzer (Autoanalyzer; Technicon Instruments Corp., Tarrytown, NY, USA) and then they were mixed with an equal volume of Griess reagents. After 10 min, the absorbance was measured by a spectrometer (VICTOR™ X Multilabel Plate Reader; Perkin Elmer) at 570 nm. NO release was examined in comparison with a standard curve and expressed as nitrites (μM).

4.7. In Vitro Experiments

4.7.1. Effects of Plasma on hUMSCs

Cell Culture

The hUMSCs (StemBioSys, San Antonio, TX, USA) were cultured in α-Minimum Essential Medium Eagle (VWR International srl, Milan, Italy) supplemented with 2.5% heparin-free human pooled platelet lysate, 1% Antibiotic-Antimycotic (VWR International srl), and 1% GlutaMax (Euroclone, Pero, Italy).

Experiments were carried out using the plasma of 10 AMA patients at the time points, T0, T1 and T3, and of 5 patients belonging to the controls at the same time points. For the in vitro experiments, we selected patients/controls characterized by a physiological pregnancy, without the onset of any complications, and for whom samples were available and adequate at the three chosen time points.

Untreated cells were also included in the analysis. In particular, we used specific Transwell inserts in order to analyze cell viability, ROS and NO release in hUMSCs treated with plasma samples, as previously performed. As depicted in Figure 5, 10% plasma samples calculated in relation to total volume of each insert were positioned in the apical surface of the insert itself for 3 h, while hUMSCs were plated in the basal one. After 3 h stimulation with plasma, the inserts were removed and various assays were performed, as described below. Experiments were executed in triplicate and repeated three times on different pools of hUMSCs.

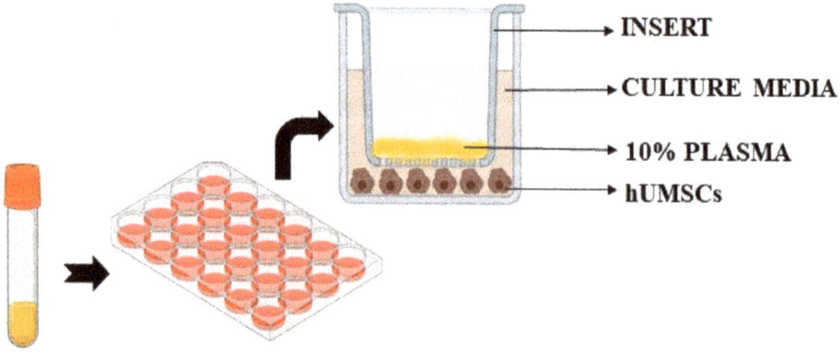

Figure 5. Experiments on hUMSCs. hUMSCs: human umbilical cord mesenchymal stem cells.

Cell Viability

Cell viability of hUMSCs was investigated by using 1% 3-[4,5-dimethylthiazol-2-yl]-2,5-diphenyl tetrazolium bromide (MTT assay; Cayman Chemical), as previously performed [25,56]. Briefly, after the 10% plasma stimulation of hUMSCs (50,000 cells/well in 24-well plates), the media were removed and 200 μL of the MTT solution diluted in Dulbecco's Modified Eagle Medium high glucose w/o phenol red, supplemented with 2 mM L-glutamine and 1% penicillin-streptomycin, was added to each well. After the addition of 100 μL of dimethyl sulfoxide (Sigma, Milan, Italy) to dissolve the formazan crystals, cell viability was determined by measuring the absorbance through a spectrophotometer (VICTOR™ X Multilabel Plate Reader; PerkinElmer; Waltham, MA, USA), with a wavelength of 570 nm. Cell viability was calculated by setting control cells (untreated cells) as 100%.

ROS Release

ROS release by hUMSCs was analyzed through the DCFDA assay, which evaluated the oxidation of 2,7-dichlorodihydrofluorescein diacetate into 2,7-dichlorodihydrofluorescein (Abcam, Cambridge, UK) [25,56]. To achieve this, 50,000 hUMSCs/well were plated in 24-Transwells plates in complete medium, as executed for MTT assay. After stimulation with 10% plasma for 3 h, the medium was removed and washing was performed with phosphate buffer saline, which was followed by staining with 10 μM 2,7-dichlorodihydrofluorescein diacetate for 20 min at 37 °C. The fluorescence intensity of 2,7-dichlorodihydrofluorescein was measured at 485 nm/530 nm excitation/emission by using a spectrophotometer (VICTOR™ X Multilabel Plate Reader; PerkinElmer). Results were expressed as 2,7-dichlorodihydrofluorescein fluorescence intensity, which was proportional to the amount of intracellular ROS.

NO Release

NO release by hUMSCs was quantified through the Griess assay (Promega) [25,56]. To achieve this, 50,000 hUMSCs were plated in 24-Transwells plates in complete medium and then the same experimental protocol followed for MTT and ROS assays was used. At the end of the stimulations, NO was examined in the sample's supernatants by adding Griess reagent following the manufacturer's instruction. Thereafter, the absorbance of each sample was read at 570 nm through a spectrometer (VICTOR™ X Multilabel Plate Reader; Perkin Elmer). A standard curve was prepared to quantify the NO production, which was expressed as nitrites (μM).

Statistical Analyses

All data were collected using the Research Electronic Data Capture software, v10.3.3 (RED-Cap, Vanderbilt University, Nashville, TN, USA). The mean of the multiple measurements taken on each patient/control was considered for the analysis. For plasma quantification and the in vitro experiments, the results were presented as median and interquartile (IQR) range. The differences in quantitative variables between the two groups were assessed using the Mann–Whitney test. Spearman's correlation coefficient was used to calculate the correlation between quantitative variables. The association between categorical variables was assessed using the chi-square test. Laboratory data were analyzed through repeated-measures ANOVA. This method of analysis was used because of the multiple responses taken in sequence on each experimental unit. The objective was to examine and compare response trends over time, comparing groups averaged over time and comparing measurement time within a group. The unstructured covariance matrix was used as a correlation pattern among responses to the same subject. A p-value < 0.05 was considered statistically significant. Statistical analysis was performed by using STATA version 18 (College Station, TX, USA: Stata Corp LLC), SAS 9.4 (SAS Institute Inc., Cary, NC, USA) and Graph PAD (GraphPad Prism 6 Software, San Diego, CA, USA).

5. Conclusions

Our results have highlighted an altered redox balance and a reduction in NO in plasma of AMA pregnant women and evidenced the existence of unknown circulating factors capable of inducing damages in hUMSCs. Those alterations were shown early and persisted up to 48 h after delivery. Beyond the possible obstetric complications, the increased plasma levels of TBARS, the reduced plasma levels of GSH and the harmful circulating factors could represent risk factors for the development of cardiovascular or other chronic disease, which could arise in women after advanced maternal age pregnancy and in children [13].

The results of this study about the role of oxidative stress in AMA pregnant women could have clinical implications regarding the usefulness of preventive or therapeutic treatments aimed at contrasting it.

Considering the role of the altered redox balance in the physiopathology of AMA pregnancy and its possible deleterious systemic effects, our research could correlate with the advice to act above all on lifestyle. In particular, a healthy and varied diet and moderate physical activity appear to be fundamental elements for the modulation of the redox state and the keeping of the mothers' and children's health.

Moreover, in the future, our data could have interesting clinical repercussions in terms of diagnosis, treatment and follow-up of several widespread diseases related to AMA pregnancy.

Author Contributions: Conceptualization: E.G., C.I.A., R.V., V.R. and D.S. Investigation and data curation: C.I.A., S.V., L.T., E.T. and F.F. Formal analysis: E.G., S.V., L.T., E.T., F.F., D.F., R.V., V.R. and D.S. Methodology: E.G., S.V., L.T., E.T., F.F., D.F., R.V., V.R. and D.S. Funding acquisition and resources: E.G., R.V., V.R. and D.S. Project administration: E.G., R.V., V.R. and D.S. Writing the original draft: E.G., C.I.A., S.V., D.F., R.V., V.R. and D.S. Writing, review and editing: E.G., C.I.A., S.V., L.T., E.T., F.F., D.F., R.V., V.R. and D.S. Supervision: E.G., R.V., V.R. and D.S. Validation: E.G., C.I.A., S.V., L.T., E.T., F.F., D.F., R.V., V.R. and D.S. Visualization: E.G., C.I.A., S.V., L.T., E.T., F.F., D.F., R.V., V.R. and D.S. All authors have read and agreed to the published version of the manuscript.

Funding: This research was supported by the Italian Ministry of Education, University and Research (MIUR) program "Departments of Excellence 2018–2022", Aging Project—Department of Translational Medicine, Università del Piemonte Orientale.

Institutional Review Board Statement: The study was approved by the Intercompany Ethics Committee n. CE 119/21 and complies with the Declaration of Helsinki and principles of Good Clinical Practice.

Informed Consent Statement: Informed consent was obtained from all subjects involved in the study.

Data Availability Statement: The data that support the findings of the present study are available from the corresponding author upon reasonable request.

Acknowledgments: We are thankful to all the patients for giving consent to provide blood samples and to the staff from the Gynecology and Obstetrics, Department of Translational Medicine, Università del Piemonte Orientale, Azienda Ospedaliera Universitaria (AOU) Maggiore della Carità, Novara, Italy, who were involved in patients' management and providing blood samples.

Conflicts of Interest: The authors declare no conflict of interest.

References

1. Frederiksen, L.E.; Ernst, A.; Brix, N.; Lauridsen, L.L.B.; Roos, L.; Ramlau-Hansen, C.H.; Ekelund, C.K. Risk of adverse pregnancy outcomes at advanced maternal age. *Obstet. Gynecol.* **2018**, *131*, 457–463. [CrossRef] [PubMed]
2. Pinheiro, R.L.; Areia, A.L.; Mota Pinto, A.; Donato, H. Advanced Maternal Age: Adverse Outcomes of Pregnancy, A Meta-Analysis. *Acta Med. Port.* **2019**, *32*, 219–226. [CrossRef] [PubMed]
3. Glick, I.; Kadish, E.; Rottenstreich, M. Management of Pregnancy in Women of Advanced Maternal Age: Improving Outcomes for Mother and Baby. *Int. J. Womens Health* **2021**, *13*, 751–759. [CrossRef] [PubMed]
4. Martin, J.A.; Hamilton, B.E.; Ventura, S.J.; Osterman, M.J.; Wilson, E.C.; Mathews, T.J. Births: Final data for 2010. *Natl. Vital Stat. Rep.* **2012**, *61*, 1–72. [PubMed]
5. Martin, J.A.; Hamilton, B.E.; Osterman, M.J.K.; Driscoll, A.K. Births: Final Data for 2019. In *National Vital Statistics Reports*; Centers for Disease Control and Prevention National Center for Health Statistics: Hyattsville, MD, USA, 2021; Volume 70, 51p.
6. Matthews, T.J.; Hamilton, B.E. First Births to Older Women Continue to Rise. In *NCHS Data Brief No. 152*; Centers for Disease Control and Prevention National Center for Health Statistics: Hyattsville, MD, USA, 2014; 8p.

7. Sauer, M.V. Reproduction at an advanced maternal age and maternal health. *Fertil. Steril.* **2015**, *103*, 1136–1143. [CrossRef] [PubMed]
8. Kawwass, J.F.; Badell, M.L. Maternal and Fetal Risk Associated With Assisted Reproductive Technology. *Obstet. Gynecol.* **2018**, *132*, 763–772. [CrossRef]
9. Cleary-Goldman, J.; Malone, F.D.; Vidaver, J.; Ball, R.H.; Nyberg, D.A.; Comstock, C.H.; Saade, G.R.; Eddleman, K.A.; Klugman, S.; Dugoff, L.; et al. Impact of maternal age on obstetric outcome. *Obstet. Gynecol.* **2005**, *105*, 983–990. [CrossRef]
10. Jacobsson, B.; Ladfors, L.; Milsom, I. Advanced maternal age and adverse perinatal outcome. *Obstet. Gynecol.* **2014**, *104*, 727–733. [CrossRef]
11. Di Fabrizio, C.; Giorgione, V.; Khalil, A.; Murdoch, C.E. Antioxidants in Pregnancy: Do We Really Need More Trials? *Antioxidants* **2022**, *11*, 812. [CrossRef]
12. Cooke, C.M.; Davidge, S.T. Advanced maternal age and the impact on maternal and offspring cardiovascular health. *Am. J. Physiol. Heart Circ. Physiol.* **2019**, *317*, H387–H394. [CrossRef]
13. Miriyala, S.; Spasojevic, I.; Tovmasyan, A.; Salvemini, D.; Vujaskovic, Z.; St Clair, D.; Batinic-Haberle, I. Manganese superoxide dismutase, MnSOD and its mimics. *Biochim. Biophys. Acta* **2012**, *1822*, 794–814. [CrossRef] [PubMed]
14. Volpe, C.M.O.; Villar-Delfino, P.H.; Dos Anjos, P.M.F.; Nogueira-Machado, J.A. Cellular death, reactive oxygen species (ROS) and diabetic complications. *Cell Death Dis.* **2018**, *9*, 119. [CrossRef] [PubMed]
15. Hussain, T.; Murtaza, G.; Metwally, E.; Kalhoro, D.H.; Kalhoro, M.S.; Rahu, B.A.; Sahito, R.G.A.; Yin, Y.; Yang, H.; Chughtai, M.I.; et al. The Role of Oxidative Stress and Antioxidant Balance in Pregnancy. *Mediators Inflamm.* **2021**, *2021*, 9962860. [CrossRef] [PubMed]
16. Surico, D.; Bordino, V.; Cantaluppi, V.; Mary, D.; Gentilli, S.; Oldani, A.; Farruggio, S.; Melluzza, C.; Raina, G.; Grossini, E. Preeclampsia and intrauterine growth restriction: Role of human umbilical cord mesenchymal stem cells-trophoblast cross-talk. *PLoS ONE* **2019**, *14*, e0218437. [CrossRef] [PubMed]
17. Jauniaux, E.; Poston, L.; Burton, G.J. Placental-related diseases of pregnancy: Involvement of oxidative stress and implications in human evolution. *Hum. Reprod. Update* **2006**, *12*, 747–755. [CrossRef] [PubMed]
18. Sultana, Z.; Maiti, K.; Aitken, J.; Morris, J.; Dedman, L.; Smith, R. Oxidative stress, placental ageing-related pathologies and adverse pregnancy outcomes. *Am. J. Reprod. Immunol.* **2017**, *77*, e12653. [CrossRef] [PubMed]
19. Kamath, U.; Rao, G.; Kamath, S.U.; Rai, L. Maternal and fetal indicators of oxidative stress during intrauterine growth retardation (IUGR). *Indian J. Clin. Biochem.* **2006**, *21*, 111–115. [CrossRef] [PubMed]
20. Liu, L.; Zhao, G.; Fan, H.; Zhao, X.; Li, P.; Wang, Z.; Hu, Y.; Hou, Y. Mesenchymal stem cells ameliorate Th1-induced pre-eclampsia-like symptoms in mice via the suppression of TNF-α expression. *PLoS ONE* **2014**, *9*, e88036. [CrossRef] [PubMed]
21. Choi, J.H.; Jung, J.; Na, K.H.; Cho, K.J.; Yoon, T.K.; Kim, G.J. Effect of mesenchymal stem cells and extracts derived from the placenta on trophoblast invasion and immune responses. *Stem Cells Dev.* **2014**, *23*, 132–145. [CrossRef]
22. Yang, W.; Zhang, J.; Xu, B.; He, Y.; Liu, W.; Li, J.; Zhang, S.; Lin, X.; Su, D.; Wu, T.; et al. HucMSC-Derived Exosomes Mitigate the Age-Related Retardation of Fertility in Female Mice. *Mol. Ther.* **2020**, *28*, 1200–1213. [CrossRef]
23. Grossini, E.; Garhwal, D.; Venkatesan, S.; Ferrante, D.; Mele, A.; Saraceno, M.; Scognamiglio, A.; Mandrioli, J.; Amedei, A.; De Marchi, F.; et al. The Potential Role of Peripheral Oxidative Stress on the Neurovascular Unit in Amyotrophic Lateral Sclerosis Pathogenesis: A Preliminary Report from Human and In Vitro Evaluations. *Biomedicines* **2022**, *10*, 691. [CrossRef] [PubMed]
24. Grossini, E.; Concina, D.; Rinaldi, C.; Russotto, S.; Garhwal, D.; Zeppegno, P.; Gramaglia, C.; Kul, S.; Panella, M. Association Between Plasma Redox State/Mitochondria Function and a Flu-Like Syndrome/COVID-19 in the Elderly Admitted to a Long-Term Care Unit. *Front. Physiol.* **2021**, *12*, 707587. [CrossRef] [PubMed]
25. Grossini, E.; Smirne, C.; Venkatesan, S.; Tonello, S.; D'Onghia, D.; Minisini, R.; Cantaluppi, V.; Sainaghi, P.P.; Comi, C.; Tanzi, A.; et al. Plasma Pattern of Extracellular Vesicles Isolated from Hepatitis C Virus Patients and Their Effects on Human Vascular Endothelial Cells. *Int. J. Mol. Sci.* **2023**, *24*, 10197. [CrossRef] [PubMed]
26. Mills, M.; Rindfuss, R.R.; McDonald, P.; Velde, E.T.; ESHRE Reproduction and Society Task Force. Why do people postpone parenthood? Reasons and social policy incentives. *Hum. Reprod. Update* **2011**, *17*, 848–860. [CrossRef] [PubMed]
27. Machado-Gédéon, A.; Badeghiesh, A.; Baghlaf, H.; Dahan, M.H. Adverse pregnancy, delivery and neonatal outcomes across different advanced maternal ages: A population-based retrospective cohort study. *Eur. J. Obstet. Gynecol. Reprod. Biol. X* **2023**, *17*, 100180. [CrossRef] [PubMed]
28. Lean, S.C.; Derricott, H.; Jones, R.L.; Heazell, A.E.P. Advanced maternal age and adverse pregnancy outcomes: A systematic review and meta-analysis. *PLoS ONE* **2017**, *12*, e0186287. [CrossRef] [PubMed]
29. Grzeszczak, K.; Łanocha-Arendarczyk, N.; Malinowski, W.; Ziętek, P.; Kosik-Bogacka, D. Oxidative Stress in Pregnancy. *Biomolecules* **2023**, *13*, 1768. [CrossRef]
30. Saccone, G.; Gragnano, E.; Ilardi, B.; Marrone, V.; Strina, I.; Venturella, R.; Berghella, V.; Zullo, F. Maternal and perinatal complications according to maternal age: A systematic review and meta-analysis. *Int. J. Gynaecol. Obstet.* **2022**, *159*, 43–55. [CrossRef]
31. Perrone, S.; Tataranno, M.L.; Negro, S.; Longini, M.; Toti, M.S.; Alagna, M.G.; Proietti, F.; Bazzini, F.; Toti, P.; Buonocore, G. Placental histological examination and the relationship with oxidative stress in preterm infants. *Placenta* **2016**, *46*, 72–78. [CrossRef]

32. Chiarello, D.I.; Abad, C.; Rojas, D.; Toledo, F.; Vázquez, C.M.; Mate, A.; Sobrevia, L.; Marín, R. Oxidative stress: Normal pregnancy versus preeclampsia. *Biochim. Biophys. Acta Mol. Basis Dis.* **2020**, *1866*, 165354. [CrossRef]
33. de Lucca, L.; Jantsch, L.B.; Vendrame, S.A.; de Paula, H.L.; Dos Santos Stein, C.; Gallarreta, F.M.P.; Moresco, R.N.; de Lima Gonçalves, T. Variation of the Oxidative Profile in Pregnant Women With and Without Gestational Complications. *Matern Child Health J.* **2022**, *26*, 2155–2168. [CrossRef] [PubMed]
34. Qanungo, S.; Mukherjea, M. Ontogenic profile of some antioxidants and lipid peroxidation in human placental and fetal tissues. *Mol. Cell Biochem.* **2000**, *215*, 11–19. [CrossRef]
35. Husain, S.; Hillmann, K.; Hengst, K.; Englert, H. Effects of a lifestyle intervention on the biomarkers of oxidative stress in non-communicable diseases: A systematic review. *Front. Aging* **2023**, *4*, 1085511. [CrossRef] [PubMed]
36. Draganovic, D.; Lucic, N.; Jojic, D. Oxidative Stress Marker and Pregnancy Induced Hypertension. *Med. Arch.* **2016**, *70*, 437–440. [CrossRef] [PubMed]
37. Malti, N.; Merzouk, H.; Merzouk, S.A.; Loukidi, B.; Karaouzene, N.; Malti, A.; Narce, M. Oxidative stress and maternal obesity: Feto-placental unit interaction. *Placenta* **2014**, *35*, 411–416. [CrossRef]
38. Tsikas, D. Assessment of lipid peroxidation by measuring malondialdehyde (MDA) and relatives in biological samples: Analytical and biological challenges. *Anal. Biochem.* **2017**, *524*, 13–30. [CrossRef]
39. Jacob, K.D.; Noren Hooten, N.; Trzeciak, A.R.; Evans, M.K. Markers of oxidant stress that are clinically relevant in aging and age-related disease. *Mech. Ageing Dev.* **2013**, *134*, 139–157. [CrossRef]
40. Go, Y.M.; Jones, D.P. Redox theory of aging: Implications for health and disease. *Clin. Sci.* **2017**, *131*, 1669–1688. [CrossRef]
41. Velazquez, M.A.; Smith, C.G.; Smyth, N.R.; Osmond, C.; Fleming, T.P. Advanced maternal age causes adverse programming of mouse blastocysts leading to altered growth and impaired cardiometabolic health in post-natal life. *Hum. Reprod.* **2016**, *31*, 1970–1980. [CrossRef]
42. Davidge, S.T.; Morton, J.S.; Rueda-Clausen, C.F. Oxygen and perinatal origins of adulthood diseases: Is oxidative stress the unifying element? *Hypertension* **2008**, *52*, 808–810. [CrossRef]
43. Rueda-Clausen, C.F.; Morton, J.S.; Davidge, S.T. The early origins of cardiovascular health and disease: Who, when, and how. *Semin. Reprod. Med.* **2011**, *29*, 197–210. [CrossRef] [PubMed]
44. Care, A.S.; Bourque, S.L.; Morton, J.S.; Hjartarson, E.P.; Davidge, S.T. Effect of advanced maternal age on pregnancy outcomes and vascular function in the rat. *Hypertension* **2015**, *65*, 1324–1330. [CrossRef] [PubMed]
45. Gluckman, P.D.; Hanson, M.A.; Cooper, C.; Thornburg, K.L. Effect of in utero and early-life conditions on adult health and disease. *N. Engl. J. Med.* **2008**, *359*, 61–73. [CrossRef] [PubMed]
46. Leiva, A.; Pardo, F.; Ramírez, M.A.; Farías, M.; Casanello, P.; Sobrevia, L. Fetoplacental vascular endothelial dysfunction as an early phenomenon in the programming of human adult diseases in subjects born from gestational diabetes mellitus or obesity in pregnancy. *Exp. Diabetes Res.* **2011**, *2011*, 349286. [CrossRef] [PubMed]
47. Choi, J.W.; Im, M.W.; Pai, S.H. Nitric oxide production increases during normal pregnancy and decreases in preeclampsia. *Ann. Clin. Lab. Sci.* **2002**, *32*, 257–263. [PubMed]
48. Tesfa, E.; Munshea, A.; Nibret, E.; Gizaw, S.T. Association of endothelial nitric oxide synthase gene variants in pre-eclampsia: An updated systematic review and meta-analysis. *J. Matern. Neonatal Med.* **2023**, *36*, 2290918. [CrossRef] [PubMed]
49. Dymara-Konopka, W.; Laskowska, M. The Role of Nitric Oxide, ADMA, and Homocysteine in The Etiopathogenesis of Preeclampsia-Review. *Int. J. Mol. Sci.* **2019**, *20*, 2757. [CrossRef]
50. Guerby, P.; Tasta, O.; Swiader, A.; Pont, F.; Bujold, E.; Parant, O.; Vayssiere, C.; Salvayre, R.; Negre-Salvayre, A. Role of oxidative stress in the dysfunction of the placental endothelial nitric oxide synthase in preeclampsia. *Redox. Biol.* **2021**, *40*, 101861. [CrossRef] [PubMed]
51. Castrechini, N.M.; Murthi, P.; Gude, N.M.; Erwich, J.J.; Gronthos, S.; Zannettino, A.; Brennecke, S.P.; Kalionis, B. Mesenchymal stem cells in human placental chorionic villi reside in a vascular Niche. *Placenta* **2010**, *31*, 203–212. [CrossRef]
52. Demir, R.; Kaufmann, P.; Castellucci, M.; Erbengi, T.; Kotowski, A. Fetal vasculogenesis and angiogenesis in human placental villi. *Acta Anat.* **1989**, *136*, 190–203. [CrossRef]
53. Awoyemi, T.; Cerdeira, A.S.; Zhang, W.; Jiang, S.; Rahbar, M.; Logenthiran, P.; Redman, C.; Vatish, M. Preeclampsia and syncytiotrophoblast membrane extracellular vesicles (STB-EVs). *Clin. Sci.* **2022**, *136*, 1793–1807. [CrossRef] [PubMed]
54. Liu, A.; Liang, X.; Wang, W.; Wang, C.; Song, J.; Guo, J.; Sun, D.; Wang, D.; Song, M.; Qian, J.; et al. Human umbilical cord mesenchymal stem cells ameliorate colon inflammation via modulation of gut microbiota-SCFAs-immune axis. *Stem Cell Res. Ther.* **2023**, *14*, 271. [CrossRef] [PubMed]
55. Qin, D.; Wang, C.; Li, D.; Guo, S. Exosomal miR-23a-3p derived from human umbilical cord mesenchymal stem cells promotes remyelination in central nervous system demyelinating diseases by targeting Tbr1/Wnt pathway. *J. Biol. Chem.* **2024**, *300*, 105487. [CrossRef] [PubMed]
56. Chao, K.C.; Yang, H.T.; Chen, M.W. Human umbilical cord mesenchymal stem cells suppress breast cancer tumourigenesis through direct cell-cell contact and internalization. *J. Cell Mol. Med.* **2012**, *16*, 1803–1815. [CrossRef] [PubMed]

International Journal of
Molecular Sciences

MDPI

Communication

Oxidative Stress Markers and Histopathological Changes in Selected Organs of Mice Infected with Murine Norovirus 1 (MNV-1)

Paulina Janicka [1,2,†], Dominika Stygar [3], Elżbieta Chełmecka [4], Piotr Kuropka [5], Arkadiusz Miążek [6], Aleksandra Studzińska [6], Aleksandra Pogorzelska [1], Katarzyna Pala [2] and Barbara Bażanów [1,*]

1 Department of Pathology, Wrocław University of Environmental and Life Sciences, Norwida 31, 50-375 Wrocław, Poland; paulina.janicka@upwr.edu.pl (P.J.); aleksandra.pogorzelska@upwr.edu.pl (A.P.)
2 Food4Future Technologies Sp. z o.o., Tarasa Szewczenki 24, 51-351 Wroclaw, Poland; kasia@f4ftech.com
3 Department of Physiology, Faculty of Medical Sciences in Zabrze, Medical University of Silesia, Jordana 19, 41-808 Zabrze, Poland; dstygar@sum.edu.pl
4 Department of Statistics, Department of Instrumental Analysis, Faculty of Pharmaceutical Sciences in Sosnowiec, Medical University of Silesia in Katowice, Ostrogórska 30, 41-200 Sosnowiec, Poland; echelmecka@sum.edu.pl
5 Division of Histology and Embryology, Department of Biostructure and Animal Physiology, Faculty of Veterinary Medicine, Wrocław University of Environmental and Life Sciences, Norwida 25, 50-635 Wroclaw, Poland; piotr.kuropka@upwr.edu.pl
6 Department of Biochemistry and Molecular Biology, Wroclaw University of Environmental and Life Sciences, Norwida 31, 50-375 Wroclaw, Poland; arkadiusz.miazek@upwr.edu.pl (A.M.); aleksandra.studzinska@upwr.edu.pl (A.S.)
* Correspondence: barbara.bazanow@upwr.edu.pl
† PhD Student in the 4th Edition of the Implementation Doctorate Programme—Ministry of Education and Science.

Abstract: This paper describes the effects of murine norovirus (MNV) infection on oxidative stress and histopathological changes in mice. This study uses histopathological assays, enzymatic and non-enzymatic antioxidant markers, and total oxidative status and capacity (TOS, TAC). The results suggest that MNV infection can lead to significant changes with respect to the above-mentioned parameters in various organs. Specifically, reduced superoxide dismutase (SOD), Mn superoxide dismutase (MnSOD), catalase (CAT), and glutathione reductase (GR) activities were observed in liver tissues, while higher MnSOD activity was observed in kidney tissues of MNV-infected mice when compared to the control. GR activity was lower in all tissues of MNV-infected mice tested, with the exception of lung tissue. This study also showed that norovirus infection led to increased TOS levels in the brain and liver and TAC levels in the brain, while TOS levels were significantly reduced in the kidneys. These changes may be due to the production of reactive oxygen species (ROS) caused by the viral infection. ROS can damage cells and contribute to oxidative stress. These studies help us to understand the pathogenesis of MNV infection and its potential effects on oxidative stress and histopathological changes in mice, and pave the way for further studies of the long-term effects of MNV infection.

Keywords: oxidative stress; bacterial and viral diseases

1. Introduction

Viral infections are one of the major causes of global health issues [1], where noroviruses (NoVs) are one of the most common causes of gastroenteritis in humans in all age groups worldwide [2]. These pathogens belong to the family *Caliciviridae* and are a type of non-enveloped, single-stranded RNA viruses characterized by being positive sense [3]. Noroviruses have also been identified in a variety of animal species, including pigs, cows, sheep, cats, dogs, rats, and mice, but murine norovirus (MNV) is the only representative of this family showing the ability to replicate in cell culture and laboratory animals [4], and at the same time, it is one of the most common pathogens of laboratory mice [5].

The initiation of effective therapy, as well as ensuring the removal of side effects caused by viral infections, requires a deep understanding of the pathogenesis and effects induced by norovirus infections at the cellular, tissue, and organ levels.

Wild-type mice (adults) infected with MNV show subclinical intestinal infection without gastric symptoms [6]. In contrast, MNV infection is lethal in immunocompromised mice (for example in immunodeficient mice with impaired innate immunity and negative expression of STAT1 and *Rag2* genes). There has been a steady increase in MNV infections in laboratories, where non-susceptible mouse strains infect susceptible mice. After administration of the virus to the immunocompromised animals, systemic disease develops—bloating, intestinal pathology, and diarrhea that result in weight loss [7–9]. Currently, MNV is the most widespread pathogen in animal facilities [10]. Mice infected with asymptomatic MNV are the source of the virus, and they can adversely affect other laboratory animals and undermine the reliability of studies conducted on animals [11]. Several histopathological changes were observed during MNV infection in different strains of laboratory mice. NOD.CB17-Prkdcscid/J (NOD-scid) immunodeficient mice showed no lesions, while liver and small intestine lesions were noted in STAT1-/- mice, and *Rag2*/STAT1-/- mice showed liver, lung, and brain lesions. The latter were also seen in infected animals of the IFNγR-/- strain [4,12,13]. Histopathological changes in the colon were observed in immunocompromised B6.129P2-Il10tm1Cgn/JZtm, C3Bir.129P2-Il10tm1Cgn/JZtm, and germ-free B6.129P2-Il10tm1Cgn/JZtm mice [11].

Pathological changes in organs observed over the course of infection may result from the disruption of metabolic pathways within the infected cell. Viral proliferation takes over the functions of host cells and causes a significant imbalance in the intracellular physiological processes and systems, including the redox system [10]. Oxidative stress, resulting from the imbalance between the antioxidant systems and toxic reactive oxygen species (ROS) [14], can be triggered by a wide variety of viral infections, including HIV 1, hepatitis B, C, and D viruses, herpes viruses, and respiratory viruses, such as coronaviruses [15]. However, oxidative stress and ROS generation are essential for numerous biological and physiological functions. Increased ROS concentration triggers enzymatic antioxidant systems, such as superoxide dismutase (SOD, MnSOD), glutathione peroxidase (GPx), glutathione reductase (GR), and catalase (CAT), considered the main oxidative stress markers, and non-enzymatic antioxidant systems including total oxidative status (TOS), total antioxidant capacity (TAC), and malondialdehyde (MDA) [16].

Superoxide dismutase (SOD), with its isoforms, e.g., MnSOD, is one of the main antioxidative enzymes that neutralize superoxide ions [17]. It mainly catalyzes the conversion of superoxide radicals (O2•−) to hydrogen peroxide (H_2O_2), which is subsequently converted into H_2O by GPx and CAT [18]. Oxidative stress caused by an imbalance in oxidant/antioxidant markers and disruption of endogenous antioxidant systems can be quantified using total antioxidant capacity (TAC), total oxidative status (TOS) [19], and MDA concentration, which is considered a marker of oxidative stress, more specifically of lipid peroxidation understood as lipid damage caused by ROS [20].

The comprehension of relevant antioxidant mechanisms and the relationship between viral infection pathological changes and oxidative stress may play a crucial role in understanding viral pathogenesis and identifying potential therapeutic targets, especially as antiviral therapies are often unavailable. Therefore, instead of acting directly on the virus, the effects of infection can be reduced by using therapy to limit the effects of oxidative stress.

The purpose of this project was to study the effects of MNV infection on oxidative stress and histopathological changes in C56Bl/6J mice.

2. Results

2.1. Assessing the Effectiveness of Infection

Due to the rapid rate of elimination of the virus from the body, RT-PCR tests showed such low titers that the result was not statistically significant. Successful infection of mice was confirmed by the detection of antibodies to norovirus in the serum of infected mice by the virus neutralization test (VNT) and interferon measurements.

2.2. Monitoring Histopathological Changes in Organs of MNV-Infected Mice

Despite the fact that tissue sections were taken from the small and large intestine, pancreas, liver, heart, lungs, kidneys, and brain fragments (forebrain, midbrain) after 3, 4, and 7 days of norovirus incubation, similar changes were found in the liver and brain in all individuals studied. In the other organs, there were either no changes (small and large intestine, pancreas, heart) or individual changes independently of the length of virus incubation (lungs, kidneys).

In the lung and kidney, small perivascular infiltrates within the blood vessels compared to a normal picture were observed. As infiltrates were found in the subcapsular zone as well as in the cortex and medulla of the kidney and in the lungs surrounding the bronchioles in the control group, it was decided not to consider changes in these organs as caused by the virus (Figure 1).

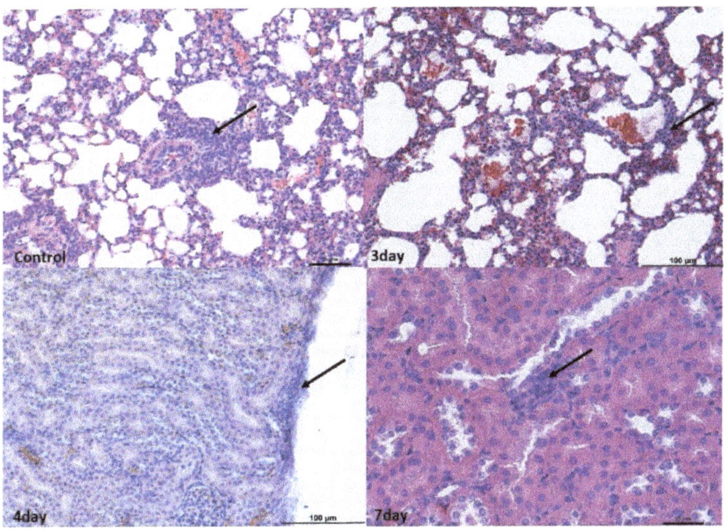

Figure 1. Histopathological changes in the lungs and kidneys. Lungs: Example pictures of lymphocyte infiltration in the control group and on the 3rd day. Note the presence of lymphoreticular tissue present (arrow) next to the bronchiole in the control group as well as in the alveoli on the 3rd day. Similar changes were found in individuals from other groups. Mag 100×. Kidneys: Example pictures of lymphocyte infiltration in animals on the 3rd, 4th, and 7th days. Mild lymphocyte infiltration (arrow) under the kidney capsule and between proximal tubules on the 4th day. Similar changes were observed in other time periods and in the control group. Fourth day—Mag 200×, seventh—Mag 400×. Scale bar—control and 7th day—25 µm, scale bar—3rd and 4th days—100 µm. These changes were considered not to be associated with norovirus.

In the liver, perivascular infiltrates were present surrounding both the hepatic triads and intralobular vessels. In the lobules, they often surrounded single hepatocytes; however, no features of necrosis or apoptosis were found. In the liver, numerous binucleated hepatocytes were present in all experimental groups, which, however, indicated enhanced repair processes (Figure 2).

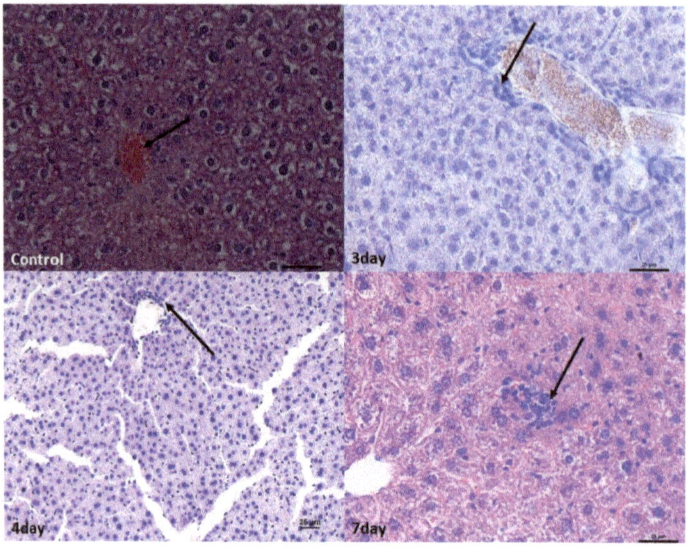

Figure 2. Histopathological changes in the liver. Control group—normal structure of hepatocytes surrounding the central vein (arrow). Mag 400×. Mild lymphocyte infiltration around the blood vessel (arrow) after 3rd (Mag 400×) and 4th days. ((Mag 200×).After the 7th day, lymphocyte infiltration in the hepatic stroma (arrow). Mag 400×. Scale bar—25 μm.

In the brain, infiltrates were localized in the meningeal vessels within the cerebellum and midbrain, mostly in white matter and pia mater. No such infiltrates were found in the forebrain. The infiltrates were often accompanied by edema, giving the white matter a foam-like character (Figures 3 and 4).

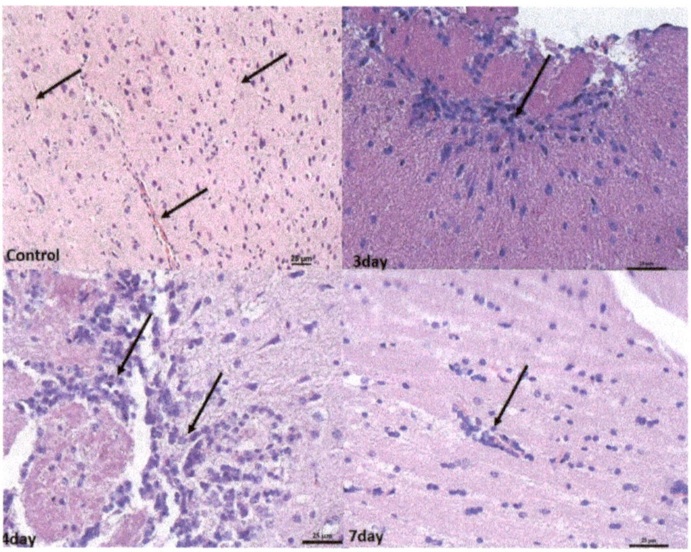

Figure 3. Histopathological changes in the midbrain. Control group—numerous blood vessels in the midbrain (arrow). Mag 200×. After 3rd and 4th days—perivascular infiltration of lymphocytes in the white matter (arrow). Mag 400×. (D) after the 7th day, mild perivascular lymphocyte infiltration in white matter (arrow). Mag 400×. Scale bar—25 μm.

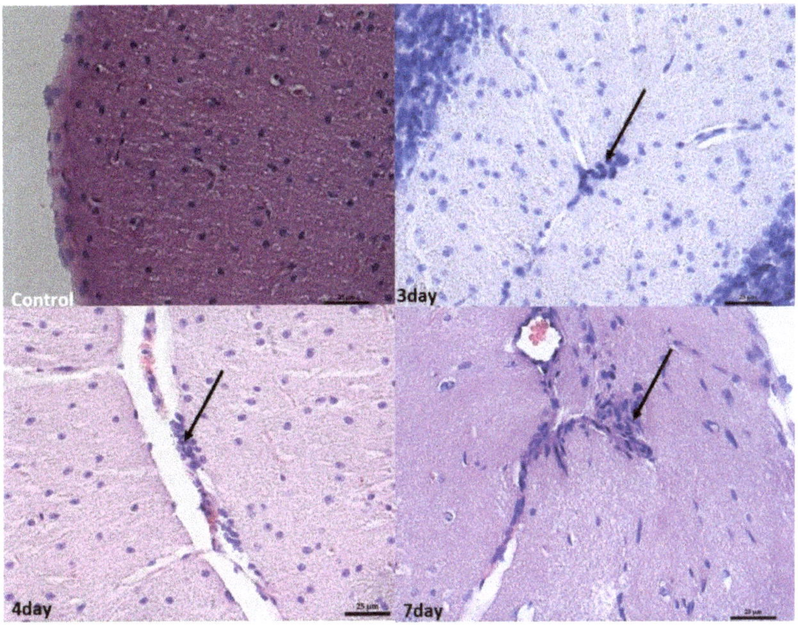

Figure 4. Histopathological changes in the cerebellum. Control group—normal image of unaltered cortex covered with pia matter. Mag 400×. On the 3rd day, 4th day, and 7th day—perivascular infiltration of lymphocytes in the white matter (arrow). Mag 400×. Scale bar—25 μm.

At 4 and 7 days after infection, the changes were broadly similar. Although there was less swelling in the cerebellum 7 days after infection compared to 3 days after infection, there were also lower levels of blood present in the vessels in the organs. In addition, an increased number of small infiltrates was observed in the liver and brain compared to 3 days post-infection. Nevertheless, no increased apoptosis or necrosis was observed in the organs examined (Figures 1–4).

2.3. Oxidative Stress Markers

We found statistically significant differences in SOD, MnSOD, GST, GPx, GR, and CAT activities, TAC and TOS levels, and MDA concentration between the sampled tissues in the control and norovirus-infected mice (Table 1).

We found higher SOD activity in the liver tissue of control mice when comparing it in individual tissues between the control and norovirus-infected mice (Table 1). Similarly, we found statistically significant differences in MnSOD activity when comparing it in individual tissues between the control and norovirus-infected mice. We observed higher MnSOD activity in the liver, but lower MnSOD activity in the kidney tissues of the control mice (Table 1). Further comparisons within individual tissues showed higher GR concentrations in the brain, cerebellum, and liver tissues of the norovirus-infected mice compared to control mice. We observed higher GR activity in the lung and kidney tissues of the control mice compared to norovirus-infected mice (Table 1). Comparisons within individual tissues showed higher CAT activity in the cerebellum, lung, and kidney tissues of the norovirus-infected mice compared to control mice. We observed increased CAT activity in the liver of the control mice compared to the norovirus-infected ones. (Table 1).

Table 1. Levels of oxidative stress markers in the brain, cerebellum, liver, lung, and kidney tissues collected from all control (n = 7) and norovirus-infected mice euthanized 7 days after inoculation (n = 7). Results are presented as median with lower and upper quartiles (Me (Q1–Q3)).

Oxidative Stress Marker	Group	Brain	Cerebellum	Liver	Lungs	Kidney	*p*
SOD (NU/mg protein)	control	11.1 (9.0–14.1)	15.0 (11.6–18.9)	3.6 (3.3–4.8)	4.1 (2.5–5.7)	5.4 (4.8–6.8)	<0.001
	norovirus	12.0 (11.2–13.4)	15.0 (11.5–15.5)	2.7 (2.4–2.8)	4.9 (3.2–6.3)	5.9 (5.0–6.1)	<0.001
	$p_{\text{control vs. norovirus}}$	0.443	0.443	**<0.01**	0.523	0.701	
MnSOD (NU/mg protein)	control	3.1 (2.1–4.1)	2.5 (1.4–3.7)	2.4 (2.2–2.9)	2.6 (2.3–3.0)	1.9 (1.8–2.0)	<0.05
	norovirus	2.7 (2.6–3.0)	2.6 (2.3–2.8)	2.0 (1.9–2.2)	2.6 (2.2–2.8)	2.2 (2.1–2.4)	<0.01
	$p_{\text{control vs. norovirus}}$	0.609	0.609	**<0.05**	0.701	**<0.05**	
GR (IU/g protein)	control	26.6 (25.7–28.0)	26.8 (24.9–27.5)	5.7 (5.4–7.4)	13.5 (12.4–15.9)	1.8 (1.5–2.4)	<0.001
	norovirus	15.4 (15.0–17.8)	21.9 (19.3–22.6)	0.3 (0.2–0.3)	18.3 (17.8–18.4)	3.6 (2.6–4.7)	<0.001
	$p_{\text{control vs. norovirus}}$	**<0.01**	**<0.01**	**<0.01**	**<0.01**	**<0.05**	
CAT (IU/g protein)	control	184.1 (141.2–187.7)	133.2 (108.1–147.1)	396.4 (382.7–437.5)	140.3 (133.4–144.3)	290.8 (264.3–298.1)	<0.001
	norovirus	154.0 (117.6–179.8)	193.5 (186.3–238.9)	225.6 (204.1–247.3)	180.8 (166.1–187.7)	358.2 (355.0–374.5)	<0.001
	$p_{\text{control vs. norovirus}}$	0.250	**<0.01**	**<0.01**	**<0.01**	**<0.01**	
GPx (IU/g protein)	control	446 (257–555)	488 (413–522)	65 (60–69)	496 (485–530)	270 (208–401)	<0.001
	norovirus	309 (271–358)	256 (205–373)	135 (113–150)	389 (384–399)	250 (245–270)	<0.001
	$p_{\text{control vs. norovirus}}$	0.201	**<0.01**	**<0.01**	**<0.01**	0.443	
GST (IU/g protein)	control	6.4 (6.2–7.0)	8.1 (6.8–8.8)	3.9 (3.1–4.6)	6.2 (5.2–7.6)	5.6 (5.2–6.6)	<0.001
	norovirus	6.5 (5.4–6.8)	7.5 (6.4–9.6)	5.9 (4.7–7.5)	6.4 (8.9–7.0)	5.1 (5.1–5.2)	<0.05
	$p_{\text{control vs. norovirus}}$	0.609	1.0	**<0.01**	0.798	0.196	
TAC (mmol/g protein)	control	0.34 (0.32–0.38)	0.41 (0.38–0.46)	0.32 (0.31–0.36)	0.26 (0.22–0.27)	0.33 (0.32–0.44)	<0.001
	norovirus	0.43 (0.41–0.53)	0.16 (0.14–0.18)	0.18 (0.11–0.24)	0.25 (0.21–0.34)	0.47 (0.44–0.48)	<0.001
	$p_{\text{control vs. norovirus}}$	**<0.01**	**<0.01**	**<0.01**	0.256	0.070	
TOS (IU/g protein)	control	0.45 (0.29–0.53)	0.46 (0.43–0.56)	0.40 (0.38–0.46)	2.48 (2.20–2.76)	0.62 (0.60–0.71)	<0.001
	norovirus	0.74 (0.67–0.90)	0.55 (0.51–0.59)	0.87 (0.86–1.07)	1.97 (1.54–2.14)	0.43 (0.40–0.46)	<0.001
	$p_{\text{control vs. norovirus}}$	**<0.01**	0.097	**<0.01**	0.055	**<0.01**	
MDA (IU/g protein)	control	7.1 (6.5–8.2)	14.3 (13.5–15.1)	2.9 (2.8–3.6)	3.3 (3.1–3.8)	2.7 (2.1–3.2)	<0.001
	norovirus	7.2 (6.7–8.4)	12.1 (11.2–12.3)	2.4 (2.1–2.6)	3.7 (3.2–4.6)	3.1 (2.8–3.8)	<0.001
	$p_{\text{control vs. norovirus}}$	0.898	0.055	**<0.05**	0.523	0.246	

Abbreviations: CAT—catalase activity; GPx—glutathione peroxidase activity; GR—glutathione reductase activity; GST—glutathione S-transferase; MDA—malondialdehyde concentration; MnSOD—Mn superoxide dismutase activity; SOD—superoxide dismutase; TAC—total antioxidative capacity; TOS—total oxidative status. Green indicates biochemical analysis results showing an increase relative to the control. Biochemical analysis results showing a decrease relative to the control are highlighted in red. *p*—statistical significance ($p \leq 0.05$).

When comparing GPx activity in individual tissues between control and norovirus-infected animals (Table 1), we found higher GPx activity in the cerebellum and lungs of control mice, as well as in the liver tissue of norovirus-infected mice. Additionally, this comparison revealed higher GST activity in the liver of norovirus-infected mice (Table 1). Furthermore, we observed higher TAC levels in brain tissue but lower levels in the cerebellum and liver tissues of norovirus-infected mice compared to control mice. Conversely, we noted higher TOS levels in the brain and liver, but lower TOS levels in the kidney tissues of norovirus-infected mice compared to controls. Finally, concentrations of MDA were higher in the liver tissue of control mice compared to norovirus-infected mice.

3. Discussion

The present study was designed to investigate how MNV infection affects oxidative stress and histological changes in mice.

It is known that MNV infection is associated with histopathological changes in immunocompetent hosts, but clinical disease is prevented by a STAT1-dependent interferon response [21]. Interferon gamma, along with other type I and III interferons, has previously been shown to play a role in limiting MNV macrophage infection [22].

We and others have used several inbred strains of mice as hosts for MNV infection [1,23,24]. But, since most of the genetically engineered mouse strains are from the C57Bl/6 genetic background, in this report, we decided to use this host strain. Moreover, a direct comparison of MNV-associated diarrhea in C57Bl/6 and Balb/c mice showed that both strains are comparably susceptible [25]. In our project, we also considered the use of different MNV strains. In studies of the pathogenesis of this virus in mice, the authors usually use MNV-1.CW 1, MNV-1.CW 3, and MNV-4, -5, and -6. These strains induce histopathological changes in many organs in some experiments, while in other studies, these changes are weak or not noted at all. This depends on many factors—the immunocompetence of the animals, their age, and, finally, the dose of virus [26]. Mumphrey et al. [27], comparing MNV-1.CW 3 and CW 1, reported that the MNV-1.CW 3 strain, being more virulent than CW 1, caused more severe histopathological changes. However, these studies, as well as those of other authors [5], do not apply to the neural tissue in which we wanted to demonstrate lesions. In a publication by Karst et al. [5], such changes were obtained using purified MNV-1 from IFNαβγR-/mouse brains. To obtain similar changes, we chose the MNV-1.CW1 (ATCC) strain of the same origin. Here. we also analyzed oxidative stress markers, which were tissue dependent, suggesting variation in oxidative stress associated with viral pathogenesis.

In the study by Hsu et al. [28], MNV-infected mice were analyzed for histopathological changes in various groups, including naturally MNV-infected mice; wild-type mice experimentally infected with MNV; immunodeficient mice experimentally infected with MNV; mice in gastrointestinal disease models experimentally infected with MNV; and mice in other disease models experimentally infected with MNV. Some naturally MNV-infected mice showed histopathological changes in the liver, lungs, mesenteric lymph nodes, lungs, and spleen. In some wild-type mice experimentally infected with MNV, histopathological changes occurred in the liver, small intestines, and spleen. In the histopathological examination of immunodeficient mice experimentally infected with MNV, changes were observed in the liver, lungs, small intestine, and spleen, and in the case of *Rag2-/-*, /STAT1-/-, and IfnαβγR-/- mice, changes were observed in the brain. Some mice in gastrointestinal disease models experimentally infected with MNV exhibited histopathological changes in mesenteric lymph nodes, small and large intestines, the spleen, and the stomach. In the case of some mice in other disease models experimentally infected with MNV, changes were observed in mesenteric lymph nodes and the aorta [29]. In our experiment, no severe changes in examined organs were found except mild lymphocyte infiltrations in the brain and liver without visible cell damage [21].

In fact, some of the most common causes of encephalitis are viral infections. Several viral agents have been described to cause this condition, such as arboviruses, rhabdoviruses, enteroviruses, herpesviruses, retroviruses, orthomyxoviruses, orthopneumoviruses, and coronaviruses, among others. However, MNV was not treated as an agent that can cause encephalitis [30].

The present study also showed that the enzymatic and non-enzymatic markers of oxidative stress were tissue dependent. Viral infection caused depletion of most of the assessed enzymes like SOD, MnSOD, CAT, and GR in liver tissue compared to the control mice, whereas the GPx and GST activities measured in liver tissue were significantly higher in the norovirus-infected mice. Superoxide dismutase, catalase, and glutathione-dependent enzymes, such as GPx, are the most important antioxidant enzymes. Superoxide dismutases, like total SOD and MnSOD, are enzymes that catalyze the conversion of superoxide into O_2 and H_2O_2. Catalase catalyzes the conversion of H_2O_2 into H_2O and O_2, and GPxs catalyze the conversion of H_2O_2 or hydroperoxides to H_2O and alcohols while oxidizing GSH to GSSG. Reactive oxygen species' role during viral infections may be unclear and may depend on how they are produced [31]. The activity of MnSOD was higher in the kidney tissue of the norovirus-infected mice when compared to control ones. The activity of GR was lower in all the studied tissues in norovirus-infected mice except for lung tissue. It is known that the activities and concentration of the selected antioxidative enzymes change during viral infection or different stages of infection. A study conducted on an in vitro model reported that during the first hours after infection, SOD, GST, CAT, and GPx were the main antioxidative enzymes induced in response to the infection [32]. As the infection progressed, only SOD concentration increased, leading to increased H_2O_2 production, whereas other antioxidant enzymes decreased, including those that are critical for neutralizing H_2O_2 [33].

The body's overall oxidative state is typically estimated using the total oxidant status (TOS), whereas the body's overall antioxidant level is determined by total antioxidant capacity (TAC) [34]. Other studies showed that TOS levels clearly increase with the aggravation of the COVID-19 disease, since the elevated level of oxidative stress has the capability to intensify the severity of COVID-19. Karkhanei et al. reported that serum TOS level, one of the oxidative stress biomarkers, was higher in patients with acute SARS-CoV-2 presentation. The authors showed that serum TOS levels in infected COVID-19 patients were related to other important factors, such as fever, hospitalization longer than one week, residency, educational status, and job type, and indirectly related to SpO2 (oxygen saturation). In the present study, the selected non-enzymatic antioxidative and oxidative parameters such as TOS, TAC, and MDA changed after the norovirus infection, depending on the analyzed tissue. Nevertheless, noroviral infection resulted in increased TOS levels in the brain and liver and increased TAC levels in brain tissues, while TOS levels were significantly decreased in the kidneys. The decrease in TAC may lead to reduced body resilience and ultimately to increased mortality [35]. The presented results prove that oxidative stress markers change rapidly depending on the conditions of the external environment. Selected evidence suggests that ROS overproduction and antioxidant system depletion play a significant role in the pathogenesis of viral infections and in the progression of the disease severity.

With high levels of oxidative stress in target organs including the brain, liver, kidney, and lung, oxidative stress plays a crucial role in a number of pathological disorders. We believe that this shows a strong connection between oxidative stress levels and the specific functions of each organ. Multiple intracellular signaling pathways are known to be activated by oxidative stress, resulting in subsequent organ failure. Targeting oxidative stress is thus thought to be useful in preventing organ damage, and assessing the level of oxidative stress may act as a biomarker in a variety of disease conditions. Here, we understand that the selected virus may have varied effects and, thus, create varying stresses in different organs—depending on the location of inflammation, the severity of the infection, and the corresponding immune response—which may also be influenced by other factors. In order to support this hypothesis further, biomolecular studies have to be performed.

4. Materials and Methods

4.1. Mice

Animal experimentation was carried out under permission no. 022/2022 from the Local Ethical Committee in Wroclaw, Poland. Mice were housed in individually ventilated cages and fed with standard sterilized rodent chow and water ad libitum. The specific pathogen-free, hygienic status of the animal facility is maintained by periodic sentinel screening according to FELASA recommendations. All measures were taken to minimize the number of animals in procedures and their suffering.

The study material consisted of healthy and norovirus-infected mice, which were divided into 4 study groups:

1. Healthy controls (control);
2. Euthanized 3 days after infection (3 days);
3. Euthanized 4 days after infection (4 days);
4. Euthanized 7 days after infection (7 days).

The method used to euthanize mice was cervical dislocation. Before the procedure, the mice were placed in the induction chamber of a veterinary anesthesia workstation (Équipement Vétérinaire Minerve, Esternay, France). The isoflurane flow was set to 2.5%. Two minutes after the mice were fully premedicated, they were transferred to pads with an inhalation mask and the anesthetic administration was continued. With a firm grip, placing the index finger and thumb directly behind the skull, and with the other hand grabbing the tail just at the base, one strong movement led to the dislocation of the cervical vertebrae. Pulling was continued until the skull was felt to be completely unattached to the spine. Each time, breathing stopped within a few seconds of the dislocation. Before starting the necropsy procedure, mice were decapitated.

4.2. Monitoring the Kinetics of MNV-1 Infection in C57Bl/6J Mice

The amount of virus in the suspension to be administered to the experimental animals was tested by RT-PCR. C57Bl/6J mice were infected per os with 50 microliters of MNV1 suspension (100TCID$_{50}$) on day 0. Mice were kept in isolation under specific pathogen-free (SPF) conditions to prevent accidental infection with another pathogen. After MNV was experimentally administered to the animals, a repeat RT-PCR test was performed after euthanasia. We used an indirect interferon gamma (IFNγ) test and a virus neutralization test to ascertain comparable MNV infection and antibody levels between individual mice.

4.3. Cell Culture

In vitro experiments were conducted using RAW 264.7 macrophage cells (TIB-7 TM, ATCC, Manassas, VA, USA). These cells were cultured in Dulbecco's Modified Eagle's Medium (DMEM) supplemented with nonessential amino acids and 10% fetal bovine serum (FBS), obtained from Biological Industries (Kibbutz Beit-Haemek, Israel). The cell cultures were maintained in 25 mL polystyrene flasks (Googlab Scientific, Rokocin, Poland) and incubated at 37 °C in a controlled environment with 5% CO_2 and 95% humidity. Passage of the cells was carried out using 0.25% trypsin–EDTA (Biological Industries).

4.4. Virus Propagation

Murine norovirus (ATCC, Los Angeles, CA, USA, VR-1973) was used for this project. When the RAW 264.7 (TIB-7^{TM}, ATCC, Manassas, VA, USA) cells reached 75–80% confluence, the cell culture medium was removed, and the cells were rinsed with phosphate-buffered saline (PBS). Subsequently, the virus was introduced into the flask and allowed to incubate for 3 h under conditions of 37 °C with 5% CO_2. Following this incubation, the virus suspension was removed, and the cell culture was washed again with PBS. Dulbecco's Modified Eagle's Medium (DMEM) was then added, and the cells were subjected to further incubation for a period of 4 to 6 days. Daily observations were carried out using an inverted microscope (Olympus Corp., Hamburg, Germany; Axio Observer, Carl Zeiss MicroImaging GmbH, Munich, Germany) to monitor the development of cytopathic effects (CPEs). Norovirus was titrated using a TCID$_{50}$ (50% Tissue Culture Infectious Dose) assay and then stored in a freezer at −80 °C

4.5. Histological Examination

For histopathological evaluation, small and large intestine, pancreas, liver, heart, lung, kidney, and brain fragments (forebrain, midbrain) were taken from all animals. The material was fixed in a solution of 4% buffered formalin (pH 7.2–7.4) for 48 h and then dehydrated in an alcoholic series after washing in tap water. The material was then embedded in paraffin. The 6 μm thick sections were stained with hematoxylin and eosin. The material was analyzed in a Nikon Eclipse 80i light microscope using NisElements Ar software https://www.microscope.healthcare.nikon.com/products/software/nis-elements/nis-elements-advanced-research, accesed on 1 June 2023 (Nikon, Tokyo, Japan).

4.6. Oxidative Stress

Seven days after viral inoculation, we harvested 100 mg of selected tissues: brain, cerebellum, liver, lungs, and kidneys from the infected animals (n = 7) and put it in 1 mL of a homogenizing buffer with protease inhibitors. The tissues were homogenized (1:10 *w*/*v*) in 0.9% NaCl with a glass homogenizer (Potter-Elvehjem PTFE, Sigma-Aldrich, Darmstadt, Germany) and then sonicated (Virsonic 100, VirTis, Gardiner, NY, USA). The tissue samples were then frozen and stored at −80 °C, until the analysis.

For this study, organs were taken from all control (n = 7) and infected mice were euthanized 7 days after inoculation (n = 7).

4.6.1. Superoxide Dismutase (SOD) (EC 1.15.1.1) Activity

Total SOD activity was measured using the Oyanagui method [35]. In this method, xanthine oxidase catalyzes the production of superoxide anion that reacts with hydroxylamine to produce nitroso ions. The latter combined with n-(1-naphthyl)ethylenediamine and sulfanilic acid gives a color combination that can be measured spectrophotometrically. The SOD and mnSOD activities were determined using the Oyanagui method [35]. The SOD activities were expressed in NU/mg of protein, with 1 NU (nitrate unit) equal to 50% inhibition of nitrite ion formation.

4.6.2. Catalase (CAT) Activity (EC 1.11.1.6)

CAT activity was assessed using the Aebi method [36]. In this method, the homogenate is mixed with perhydrol in 50 mM TRIS/HCl buffer, pH 7.4, and the reaction is started by adding freshly prepared hydrogen peroxide. The rate of decomposition of hydrogen peroxide can be measured spectrophotometrically at 240 nm. CAT activity is expressed as units per 1 g protein (IU/g protein).

The CAT activity was determined using the Aebi method [36] and expressed as units per 1 g protein (IU/g protein).

4.6.3. Glutathione Peroxidase (GPx) Activity (EC 1.11.1.9)

GPx activity was measured by using the kinetic method [37], with t-butyl peroxide as a substrate. In this reaction, oxidized glutathione (GSSG) is regenerated in the presence of glutathione reductase (GR) and NADPH. GPx activity was expressed as μmoles of NADPH oxidized in 1 min per 1 g of protein (IU/g protein) for selected tissues.

The GPx activity was measured using the kinetic method [37] with t-butyl peroxide as a substrate and is expressed as μmoles of oxidized NADPH in 1 min per 1 g of protein (IU/g protein).

4.6.4. Glutathione Reductase (GR) Activity (EC 1.8.1.7)

GR activity was determined by using the kinetic method and is expressed as μmoles of nicotinamide adenine dinucleotide phosphate (NADPH) utilized in 1 min per 1 g of protein (IU/g protein) for assessed samples [38]. This method is based on changes in the concentration of NADPH that reacts with oxidized glutathione. The changes in absorbance at 340 nm were measured with a PERKIN ELMER Victor X3 reader (PerkinElmer, Inc., Waltham, MA, USA).

The GR activity was determined using the kinetic method [38] and expressed in μmoles of NADPH utilized in 1 min by 1 g of protein (IU/g protein).

4.6.5. Glutathione-S Transferase (GST) Activity (EC 2.5.1.18)

GST activity was estimated using the Habig and Jakoby kinetic method [39]. The reaction mixture containing reduced glutathione was added to the samples. After initial stabilization, 1-chloro-2,3-dinitrobenzene (in ethyl alcohol solution) was added, and absorbance changes were monitored using a PERKIN ELMER Victor X3 reader, at 340 nm wavelength, for at least 3 min. GST activity was expressed as μmoles of thioether formed within 1 min per 1 g of protein (IU/g protein).

4.6.6. Total Oxidant Status (TOS) and Total Antioxidant Capacity (TAC)

The total oxidant status (TOS) and total antioxidant capacity (TAC) levels were determined using the Erel methods [40,41]. TOS was determined using the methodology provided by Erel based on a system containing xylene orange, o-dianisidine, and Fe^{+2} ions [35]. The determination of TAC in serum was based on the method of decolorization of oxidized ABTS (2,2-azinobis(3-ethylbenzothiazo-ine-6-sulfonate)) under the influence of antioxidants occurring in the tested material [35].

4.6.7. Malondialdehyde (MDA) Concentration

The malondialdehyde (MDA) concentration was measured using a reaction with thiobarbituric acid [42], calculated from a standard curve prepared with 1,1,3,3-tetraethoxypropane, and expressed in μmol/g protein.

4.6.8. Statistical Analysis of Antioxidant Stress Markers

The normality of distributions was assessed using the Shapiro–Wilk test. Kruskal–Wallis ANOVA was used for comparisons in the study and norovirus groups depending on the type of tissue. Within a single tissue, comparisons between the study group and the control group were made using the Mann–Whitney U test. The median (lower–upper quartile) (Me (Q1–Q3)) was used in the description. The tests performed were two-sided and the significance level was set at 0.05. The calculations were performed in the TIBCO Statistica® 13.3.0 program (data analysis software system), (TIBCO Software Inc., Santa Clara, CA, USA), accessed on 15 September 2023

Author Contributions: Conceptualization, P.J., P.K., D.S., A.M. and B.B.; methodology, P.J., D.S., P.K., A.M. and B.B.; investigation, P.J., P.K., A.M., A.S. and B.B.; data curation, E.C.; writing—original draft preparation, P.J., D.S., P.K., A.M., A.P. and B.B.; writing—review and editing, P.J., D.S., P.K., A.M., A.P., K.P. and B.B. All authors have read and agreed to the published version of the manuscript.

Funding: The article is part of a PhD dissertation titled "Development of a diagnostic kit for the detection of noroviruses in biological and environmental samples. Prevalence and epidemiological significance of noroviruses", prepared during Doctoral School at the Wrocław University of Environmental and Life Sciences. Project No. DWD/4/12/2020. The APC is financed by Wrocław University of Environmental and Life Sciences.

Institutional Review Board Statement: This study was conducted in accordance with the Declaration of Helsinki, and approved by the Local Ethics Committee in Wrocław (protocol code 022/2022/P2 and date 30 September 2022 of approval).

Data Availability Statement: The data presented in this study are available on request from the corresponding author. The data are not publicly available due to the large number of results, which would affect the clarity of the manuscript.

Conflicts of Interest: Authors Paulina Janicka and Katarzyna Pala were employed by the company Food4Future Technologies Ltd. The remaining authors declare that the research was conducted in the absence of any commercial or financial relationships that could be construed as a potential conflict of interest.

References

1. Tsikas, D. Assessment of lipid peroxidation by measuring malondialdehyde (MDA) and relatives in biological samples: Analytical and biological challenges. *Anal. Biochem.* **2017**, *524*, 13–30. [CrossRef]
2. Fernandes, I.G.; de Brito, C.A.; Dos Reis, V.M.S.; Sato, M.N.; Pereira, N.Z. SARS-CoV-2 and Other Respiratory Viruses: What Does Oxidative Stress Have to Do with It? *Oxid. Med. Cell. Longev.* **2020**, *2020*, 8844280. [CrossRef]
3. Buesa, J.; Collado, B.; Lopez-Andujar, P.; Abu-Mallouh, R.; Rodriguez Diaz, J.; Garcia Diaz, A.; Prat, J.; Guix, S.; Llovet, T.; Prats, G.; et al. Molecular epidemiology of caliciviruses causing outbreaks and sporadic cases of acute gastroenteritis in Spain. *J. Clin. Microbiol.* **2002**, *40*, 2854–2859. [CrossRef]
4. Green, K.Y. Caliciviridae: The noroviruses. In *Fields Virology*, 6th ed.; Knipe, D.M., Howley, P.M., Eds.; Lippincott Williams & Wilkins Health: Philadelphia, PA, USA, 2013; pp. 582–608.
5. Karst, S.M.; Wobus, C.E.; Lay, M.; Davidson, J.; Virgin, H.W., 4th. STAT1-dependent innate immunity to a Norwalk-like virus. *Science* **2003**, *299*, 1575–1578. [CrossRef]
6. Hsu, C.C.; Wobus, C.E.; Steffen, E.K.; Riley, L.K.; Livingston, R.S. Development of a microsphere-based serologic multiplexed fluorescent immunoassay and a reverse transcriptase PCR assay to detect murine norovirus 1 infection in mice. Clin. *Vaccine Immunol.* **2005**, *12*, 1145–1151. [CrossRef]
7. Kahan, S.M.; Liu, G.; Reinhard, M.K.; Hsu, C.C.; Livingston, R.S.; Karst, S.M. Comparative murine norovirus studies reveal a lack of correlation between intestinal virus titers and enteric pathology. *Virology* **2011**, *421*, 202–210. [CrossRef] [PubMed]
8. Strong, D.W.; Thackray, L.B.; Smith, T.J.; Virgin, H.W. Protruding domain of capsid protein is necessary and sufficient to determine murine norovirus replication and pathogenesis in vivo. *J. Virol.* **2012**, *86*, 2950–2958. [CrossRef] [PubMed]
9. Kim, J.R.; Seok, S.H.; Kim, D.J.; Baek, M.W.; Na, Y.R.; Han, J.H.; Kim, T.H.; Park, J.H.; Turner, P.V.; Chung, D.H.; et al. Prevalence of murine norovirus infection in Korean laboratory animal facilities. *J. Vet. Med. Sci.* **2011**, *73*, 687–691. [CrossRef] [PubMed]
10. Pritchett-Corning, K.R.; Cosentino, J.; Clifford, C.B. Contemporary prevalence of infectious agents in laboratory mice and rats. *Lab. Anim.* **2009**, *43*, 165–173. [CrossRef] [PubMed]
11. Muller, B.; Klemm, U.; Mas Marques, A.; Schreier, E. Genetic diversity and recombination of murine noroviruses in immunocompromised mice. *Arch. Virol.* **2007**, *152*, 1709–1719. [CrossRef] [PubMed]
12. Henderson, K.S. Murine norovirus, a recently discovered and highly prevalent viral agent of mice. *Lab. Anim.* **2008**, *37*, 314–320. [CrossRef]
13. Kitajima, M.; Oka, T.; Tohya, Y.; Katayama, H.; Takeda, N.; Katayama, K. Development of a broadly reactive nested reverse transcription-PCR assay to detect murine noroviruses, and investigation of the prevalence of murine noroviruses in laboratory mice in Japan. *Microbiol. Immunol.* **2009**, *53*, 531–534. [CrossRef]
14. Kelmenson, J.A.; Pomerleau, D.P.; Griffey, S.; Zhang, W.; Karolak, M.J.; Fahey, J.R. Kinetics of transmission, infectivity, and genome stability of two novel mouse norovirus isolates in breeding mice. *Comp. Med.* **2009**, *59*, 27–36.
15. McFadden, N.; Bailey, D.; Carrara, G.; Benson, A.; Chaudhry, Y.; Shortland, A.; Heeney, J.; Yarovinsky, F.; Simmonds, P.; Macdonald, A.; et al. Norovirus regulation of the innate immune response and apoptosis occurs via the product of the alternative open reading frame 4. *PLoS Pathog.* **2011**, *7*, e1002413. [CrossRef]
16. Basic, M.; Keubler, L.M.; Buettner, M.; Achard, M.; Breves, G.; Schröder, B.; Smoczek, A.; Jörns, A.; Wedekind, D.; Zschemisch, N.H.; et al. Norovirus triggered microbiota-driven mucosal inflammation in interleukin 10-deficient mice. *Inflamm. Bowel Dis.* **2014**, *20*, 431–443. [CrossRef]
17. Cecchini, R.; Cecchini, A.L. SARS-CoV-2 infection pathogenesis is related to oxidative stress as a response to aggression. *Med. Hypotheses* **2020**, *143*, 110102. [CrossRef]
18. Halliwell, B. Free radicals and antioxidants: Updating a personal view. *Nutr. Rev.* **2012**, *70*, 257–265. [CrossRef]
19. Ntyonga-Pono, M.P. COVID-19 infection and oxidative stress: An under-explored approach for prevention and treatment? *Pan Afr. Med. J.* **2020**, *35* (Suppl. S2), 12. [CrossRef]
20. Kazura, W.; Michalczyk, K.; Skrzep-Poloczek, B.; Chełmecka, E.; Zalejska-Fiolka, J.; Michalski, M.; Kukla, M.; Jochem, J.; Rutkowski, J.; Stygar, D. Liver Oxidative Status, Serum Lipids Levels after Bariatric Surgery and High-Fat, High-Sugar Diet in Animal Model of Induced Obesity. *Int. J. Mol. Sci.* **2023**, *24*, 16535. [CrossRef]
21. Kurutas, E.B. The importance of antioxidants which play the role in cellular response against oxidative/nitrosative stress: Current state. *Nutr. J.* **2015**, *15*, 71. [CrossRef] [PubMed]
22. Maloney, N.S.; Thackray, L.B.; Goel, G.; Hwang, S.; Duan, E.; Vachharajani, P.; Xavier, R.; Virgin, H.W. Essential cell-autonomous role for interferon (IFN) regulatory factor 1 in IFN-γ-mediated inhibition of norovirus replication in macrophages. *J. Virol.* **2012**, *86*, 12655–12664. [CrossRef]
23. Balmus, I.M.; Ciobica, A.; Antioch, I.; Dobrin, R.; Timofte, D. Oxidative stress implications in the affective disorders: Main biomarkers, animal models relevance, genetic perspectives, and antioxidant approaches. *Oxidative Med. Cell. Longev.* **2016**, *2016*, 25. [CrossRef]
24. Abedi, A.; Ghobadi, H.; Sharghi, A.; Iranpour, S.; Fazlzadeh, M.; Aslani, M.R. Effect of saffron supplementation on oxidative stress markers (MDA, TAC, TOS, GPx, SOD, and pro-oxidant/antioxidant balance): An updated systematic review and meta-analysis of randomized placebo-controlled trials. *Front. Med.* **2023**, *10*, 1071514. [CrossRef]

25. Mumphrey, S.M.; Changotra, H.; Moore, T.N.; Heimann-Nichols, E.R.; Wobus, C.E.; Reilly, M.J.; Moghadamfalahi, M.; Shukla, D.; Karst, S.M. Murine norovirus 1 infection is associated with histopathological changes in immunocompetent hosts, but clinical disease is prevented by STAT1-dependent interferon responses. *J. Virol.* **2007**, *81*, 3251–3263. [CrossRef]
26. Roth, A.N.; Helm, E.W.; Mirabelli, C.; Kirsche, E.; Smith, J.C.; Eurell, L.B.; Ghosh, S.; Altan-Bonnet, N.; Wobus, C.E.; Karst, S.M. Norovirus infection causes acute self-resolving diarrhea in wild-type neonatal mice. *Nat. Commun.* **2020**, *11*, 2968. [CrossRef]
27. Dycke, J.V.; Neyts, J.; Rocha-Pereira, J. Assessing the Efficacy of Small Molecule Inhibitors in a Mouse Model of Persistent Norovirus Infection. *Bio-Protocol* **2018**, *8*, e2831. [CrossRef]
28. Taube, S.; Kolawole, A.O.; Höhne, M.; Wilkinson, J.E.; Handley, S.A.; Perry, J.W.; Thackray, L.B.; Akkina, R.; Wobus, C.E. A mouse model for human norovirus. *mBio* **2013**, *4*, e00450-13. [CrossRef]
29. Peiper, A.M.; Helm, E.W.; Nguyen, Q.; Phillips, M.; Williams, C.G.; Shah, D.; Tatum, S.; Iyer, N.; Grodzki, M.; Eurell, L.B.; et al. Infection of neonatal mice with the murine norovirus strain WU23 is a robust model to study norovirus pathogenesis. *Lab. Anim* **2023**, *52*, 119–129. [CrossRef]
30. Hsu, C.C.; Piotrowski, S.L.; Meeker, S.M.; Smith, K.D.; Maggio-Price, L.; Treuting, P.M. histologic Lesions Induced by Murine Norovirus Infection in Laboratory Mice. *Vet. Pathol.* **2016**, *53*, 754–763. [CrossRef]
31. Wobus, C.E.; Thackray, L.B.; Virgin IV, H.W. Murine Norovirus: A Model System to Study Norovirus, Biology and Pathogenesis. *J. Virol.* **2006**, *80*, 5104–5112. [CrossRef]
32. Bohmwald, K.; Andrade, C.A.; Gálvez, N.M.S.; Mora, V.P.; Muñoz, J.T.; Kalergis, A.M. The Causes and Long-Term Consequences of Viral Encephalitis. *Front. Cell. Neurosci.* **2021**, *15*, 755875. [CrossRef]
33. Marrocco, I.; Altieri, F.; Peluso, I. Measurement and Clinical Significance of Biomarkers of Oxidative Stress in Humans. *Oxid. Med. Cell. Longev.* **2017**, *2017*, 6501046. [CrossRef]
34. Garofalo, R.P.; Kolli, D.; Casola, A. Respiratory syncytial virus infection: Mechanisms of redox control and novel therapeutic opportunities. *Antioxid. Redox Signal.* **2013**, *18*, 186–217. [CrossRef]
35. Erel, O. A novel automated direct measurement mettivehod for total antioxidant capacity using a new generation, more stable ABTS radical cation. *Clin. Biochem.* **2004**, *37*, 277–285. [CrossRef]
36. Karkhanei, B.; Talebi Ghane, E.; Mehri, F. Evaluation of oxidative stress level: Total antioxidant capacity, total oxidant status and glutathione activity in patients with COVID-19. *New Microbes New Infect.* **2021**, *42*, 100897. [CrossRef]
37. Oyanagui, Y. Reevaluation of assay methods and establishment of kit for superoxide dismutase activity. *Anal. Biochem.* **1984**, *142*, 290–296. [CrossRef]
38. Aebi, H. Catalase in vitro. *Methods Enzymol.* **1984**, *105*, 121–126.
39. Mannervik, B. Glutathione peroxidase. *Methods Enzymol.* **1985**, *113*, 490–495. [CrossRef]
40. Carlberg, I.; Mannervik, B. Glutathione reductase. *Methods Enzymol.* **1985**, *113*, 484–490. [CrossRef]
41. Erel, O. A new automated colorimetric method for measuring total oxidant status. *Clin. Biochem* **2005**, *38*, 1103–1111. [CrossRef]
42. Ohkawa, H.; Ohishi, N.; Yagi, K. Assay for lipid peroxides in animal tissues by thiobarbituric acid reaction. *Anal. Biochem.* **1979**, *95*, 351–358. [CrossRef] [PubMed]

International Journal of *Molecular Sciences*

MDPI

Article

Preliminary Report on the Influence of Acute Inflammation on Adiponectin Levels in Older Inpatients with Different Nutritional Status

Jakub Husejko [1], Marcin Gackowski [2], Jakub Wojtasik [3], Dominika Strzała [1], Maciej Pesta [1], Katarzyna Mądra-Gackowska [1], Jarosław Nuszkiewicz [4], Alina Woźniak [4,*], Mariusz Kozakiewicz [1] and Kornelia Kędziora-Kornatowska [1]

1 Department of Geriatrics, Faculty of Health Science, L. Rydygier Collegium Medicum in Bydgoszcz, Nicolaus Copernicus University in Torun, Skłodowska-Curie 9 Street, 85-094 Bydgoszcz, Poland; kubahusejko@gmail.com (J.H.); doominika05@gmail.com (D.S.); maciejpesta2000@gmail.com (M.P.); katarzyna.madra@cm.umk.pl (K.M.-G.); markoz@cm.umk.pl (M.K.); kornelia.kornatowska@cm.umk.pl (K.K.-K.)

2 Department of Toxicology and Bromatology, Faculty of Pharmacy, L. Rydygier Collegium Medicum in Bydgoszcz, Nicolaus Copernicus University in Torun, A. Jurasza 2 Street, 85-089 Bydgoszcz, Poland; marcin.gackowski@cm.umk.pl

3 Centre for Statistical Analysis, Nicolaus Copernicus University in Toruń, Chopina 12/18 Street, 87-100 Toruń, Poland; jwojtasik@umk.pl

4 Department of Medical Biology and Biochemistry, Faculty of Medicine, Ludwik Rydygier Collegium Medicum in Bydgoszcz, Nicolaus Copernicus University in Toruń, Karłowicza 24 Street, 85-092 Bydgoszcz, Poland; jnuszkiewicz@cm.umk.pl

* Correspondence: al1103@cm.umk.pl

Abstract: Inflammation can be triggered by a variety of factors, including pathogens, damaged cells, and toxic compounds. It is a biological response of the immune system, which can be successfully assessed in clinical practice using some molecular substances. Because adiponectin, a hormone released by adipose tissue, influences the development of inflammation, its evaluation as a potential measure of inflammation in clinical practice is justified. In the present contribution, statistical comparison of adiponectin concentration and selected molecular substances recognized in clinical practice as measures of inflammation were utilized to demonstrate whether adipose tissue hormones, as exemplified by adiponectin, have the potential to act as a measure of rapidly changing inflammation when monitoring older hospitalized patients in the course of bacterial infection. The study showed no statistically significant differences in adiponectin levels depending on the rapidly changing inflammatory response in its early stage. Interestingly, the concentration of adiponectin is statistically significantly higher in malnourished patients than in people with normal nutritional levels, assessed based on the MNA. According to the results obtained, adiponectin is not an effective measure of acute inflammation in clinical practice. However, it may serve as a biomarker of malnutrition in senile individuals.

Keywords: adiponectin; bacterial infection; inflammation; older inpatients; malnutrition; MNA; nutritional status; oxidative stress in infection

1. Introduction

Inflammation is the natural way in which the immune system responds to harmful stimuli such as pathogens, damaged cells, toxic compounds, or irradiation. It acts by removing harmful substances and initiating the healing process. Inflammation is, therefore, a defense mechanism that is crucial to maintaining good health. Various factors, such as infection, tissue injury, or cardiac infarction, can trigger inflammation by causing tissue damage [1]. The inflammation process causes changes in the patient's general clinical condition but is also reflected in the concentration levels of certain substances, which result, among other things, from processes related to oxidative stress. It is possible to successfully confirm the presence and monitor the course of the inflammatory reaction using several laboratory tests. In clinical practice, the levels of C-reactive protein (CRP), procalcitonin (PCT), and leukocytes (WBC) are usually determined for this purpose.

CRP is an acute-phase protein, a pro-inflammatory substance, the production of which is intensified during inflammation in the body. It is synthesized in the liver in response to increased interleukin-6 (IL-6) levels, which contributes to the activation of CRP transcription genes. Increased CRP levels may occur in both acute and chronic inflammation. However, this condition is most frequently observed in the case of infectious diseases [2,3]. Procalcitonin is a protein produced as a result of the transformation of pre-procalcitonin with the participation of endopeptidases in thyroid C cells. It is a precursor of a hormone naturally occurring in the body—calcitonin. Under homeostatic conditions, PCT concentrations are very low (typically < 0.05 ng/mL). The level of PCT is reactively increased as a result of the presence of heightened amounts of substances circulating in the blood, i.e., endotoxins, cytokines—interleukin-1b (IL-1b), and interleukin-6 (IL-6) or tumor necrosis factor (TNF-). A sudden increase in this marker is usually caused by an acute inflammatory response, most often caused by a bacterial infection [4,5]. Leukocytes (WBC) are a population of blood cells that participate in the immune system's response to the inflammatory reaction. A special role in acute inflammation is played by neutrophils, whose influx into the inflamed area is stimulated by pro-inflammatory cytokines, i.e., IL-6 or IL-1b. Normally, the level of white blood cells in the blood is $<10 \times 10^9$/L and a result exceeding this threshold is called leukocytosis. Leukocytosis with neutrophilia is the basis for suspecting an acute inflammatory reaction and suggests deepening of the diagnostics each time to detect its cause [6]. There are many examples of substances involved in the development of inflammation. Notably, uric acid is also more frequently measured in clinical practice. However, it does not serve as a marker for monitoring inflammatory changes. It has been shown that older people had atypically high CRP and Il-6 concentrations if the patient had hyperuricemia [7]. Therefore, when considering substances for their potential to monitor inflammation, it is worth taking into account their correlations with uric acid.

The process of inflammation is noticeably related to oxidative stress. During inflammation there are many substances produced, which also play a significant role in the oxidative stress. Higher concentrations of IL-6, IL-1b, or TNF-α can be noticed in the progress of both these cases. In general, inflammation and oxidative stress often occur simultaneously in the organism, which causes one of the processes to affect another, activating several biochemical pathways, using receptor mechanisms and effects of many chemicals on body cells. Reactive oxygen species (ROS), which are produced by the effect of oxidative stress, entail activation of the immune system. Immune cells, especially leukocytes, are activated and the course of inflammation in the organism begins. Consequence of the mentioned actions are pro-inflammation cytokine level upgrowth and an exacerbation of the inflammation [8].

Adipose tissue is a hormonally active organ that produces adipokines, which play essential roles, for instance, in the maintenance of whole-body energy and metabolic homeostasis at both organ and system levels, inflammation, obesity, and oxidative stress, which accompanies many age-related diseases. One of the most hormonally significant adipokines and the richest one in human plasma is adiponectin (APN). Numerous studies have demonstrated that it has insulin-sensitizing, anti-atherogenic, and anti-inflammatory effects. Furthermore, decreased serum levels of APN are associated with chronic inflammation of metabolic disorders, which include type 2 diabetes, obesity, and atherosclerosis. However, recent studies have shown that APN could have pro-inflammatory roles in patients with autoimmune diseases. Thus, it appears that adiponectin has both pro-inflammatory and anti-inflammatory effects, which indirectly suggests that adiponectin has different physiological roles based on the isoform and effector tissue [9]. Its production in excessive amounts by adipose tissue cells contributes to the development of unfavorable metabolic processes. APN stimulates the phosphorylation of the enzymes: adenosine monophosphate-activated protein kinase (AMPK) and acetyl-CoA carboxylase (ACC). This leads to the activation of key enzymes for oxidation of fatty acids. Additionally, it stimulates the functioning of the APN-AMPK-eNOS signaling pathway, as a result of which the production of free oxygen radicals (ROS) is intensified. Moreover, it has been shown that a low level of adiponectin is associated with an increased concentration of pro-inflammatory cytokines (including IL-6 and TNF-) in the blood, which favors the intensification of the inflammatory reaction and may be the basis of many diseases related to systemic inflammatory reaction, i.e., rheumatoid arthritis, type II diabetes, or systemic lupus erythematosus. Because adiponectin influences the development of chronic inflammation, its evaluation as a potential measure of inflammation in clinical practice is justified [9,10]. It should also be remembered that adiponectin increases under conditions of oxidative stress [11], which indicates its potential in measuring inflammation.

Senile patients admitted to the hospital for acute medical illness have a high prevalence of multimorbidity, malnutrition, and cognitive impairments [12]. In the case of such a patient, it is often difficult to make a quick and accurate diagnosis, so it takes time to implement appropriate treatment, and very often the patient's clinical condition worsens. Moreover, treating infections in older patients can be difficult as their symptoms and signs may not be specific, leading to both over and undertreatment. Additionally, they have a weaker immune response to infections, which can affect the kinetics of biomarkers of infection [13]. When interpreting admission biomarkers for older adults with infections, caution must be taken as their prognostic abilities are lower in comparison to younger adults [14,15]. For these reasons, searching for and validation of good biomarkers of disorders common in the senile population is imperative and may facilitate hospitalization, i.e., allow for faster diagnosis and application of effective treatment. Using infection markers such as PCT to guide personalized treatment decisions is particularly helpful for geriatric patients. PCT is a valuable biomarker for assessing the risk of septic complications and adverse outcomes in geriatric patients. It can also help make informed decisions about whether to use antibiotics or not.

In human research, leptin and APN are the two most extensively studied biomarkers among different molecules [16]. Many studies confirmed negative correlations between adiponectin and inflammatory biomarkers. However, the inverse correlation between the two variables was almost non-existent in participants who were older than 55 years of age. Adiponectin may be a possible predictor, among other things, of endothelial dysfunction [17], renal dysfunction [18], multiple sclerosis course [19], metabolic syndrome [20], osteoporosis [21], or even hair loss severity in alopecia areata [22]. Interestingly, data from patients with viral infections (chronic hepatitis C virus infections) [23] and human experimental endotoxemia [24] suggest altered patterns of APN release in these conditions. There is increasing evidence suggesting that adiponectin has an anti-inflammatory effect on the lungs. However, further research is needed to clarify the mechanism and roles of the APN pathway in protecting against lung diseases, including fungal, bacterial, and viral infections [25].

The study aimed to demonstrate whether adiponectin, as a prominent representative of adipose tissue hormones, has the potential to act as a measure of rapidly changing inflammation in the course of bacterial infection.

2. Results

2.1. The Influence of Nutritional Status on Evaluated Parameters

Table 1 illustrates the descriptive statistics of putative biomarkers divided by the MNA. According to the results, the mean uric acid concentration in hospitalized patients with normal nutritional status based on the MNA (N = 24) was 5.467 mg/dL. In patients at increased risk of malnutrition (N = 32), the mean uric acid concentration was 6.206 mg/dL, while in malnourished patients (N = 7), the mean concentration was 8.043 mg/dL. Therefore, there is a noticeable tendency for the uric acid level to increase as the nutritional status deteriorates, but the study group is too small to treat this as a rule.

The average adiponectin concentration in patients with a nutritional level within the normal range was 50.230 μg/mL, in patients with an increased risk of malnutrition it was 67.788 μg/mL, and in malnourished patients had a value of 79.626 μg/mL. As in the case of uric acid, here too there is a possible tendency to increase the concentration of adiponectin as the nutritional level deteriorates that needs further confirmation.

Similarly, the concentrations of inflammatory parameters, i.e., WBC and CRP, did not differ significantly in groups with different nutritional status. The average WBC at the time of admission to the hospital in patients with normal nutritional status was $9.195 \times 10^3/\mu L$, with the risk of malnutrition $10{,}295 \times 10^3/\mu L$, and in malnourished people $9374 \times 10^3/\mu L$. The mean WBC at discharge from the Clinic was, in properly nourished patients $8.078 \times 10^3/\mu L$, in patients at risk of malnutrition $7.504 \times 10^3/\mu L$, and in malnourished patients $7.266 \times 10^3/\mu L$. The mean CRP concentration upon admittance was 34.106 mg/dL in patients with normal nutritional status, 61.343 mg/dL in patients at risk of malnutrition, and 30.866 mg/dL in malnourished ones. The mean CRP concentration at the end of hospitalization was 24.519 mg/dL in patients with normal nutritional status, 18.800 mg/dL in patients at risk of malnutrition, and 7.757 mg/dL in malnourished people. It can be assumed that the obtained values result from factors other than nutritional level, which may be comorbidities or diseases that are the direct cause of hospitalization.

Due to the failure to meet the assumption of the equality of groups necessary for the parametric analysis of variance (ANOVA) (verification with the Chi-square test of equality, $p < 0.001$). Kruskal–Wallis tests were performed, which are its non-parametric equivalent. The results showed that there were statistically significant differences in adiponectin values in the individually analyzed groups (Table 2).

Post hoc tests showed differences in adiponectin concentration between the group of patients with normal nutritional levels and the group of malnourished patients ($p = 0.004$). No significant differences were found between the group of patients with normal nutritional status and those at risk of malnutrition ($p = 0.392$) and between patients at risk of malnutrition and patients with malnutrition ($p = 1.000$). The distributions of adiponectin measurement values in the groups are presented in Figure 1.

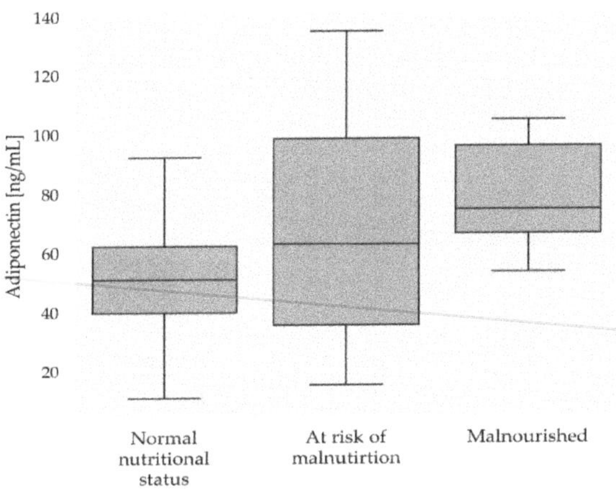

Figure 1. Boxplot of adiponectin divided by the MNA.

Table 1. Descriptive statistics of biomarkers divided by the MNA.

Nutritional Status	Normal Nutritional Status							At Risk of Malnutrition							Malnourished						
Variable	N	Mean	Std Error	Q1	Median	IQR	Q3	N	Mean	Std Error	Q1	Median	IQR	Q3	N	Mean	Std Error	Q1	Median	IQR	Q3
Uric acid [md/dL]	24	5.47	0.35	4.18	5.40	1.93	6.10	32	6.21	0.45	4.35	5.60	3.95	8.30	7	8.04	0.91	7.05	8.10	1.75	8.80
Adiponectin [μg/mL]	24	50.23	3.89	39.52	50.85	22.48	62.00	33	67.79	6.14	35.25	62.67	63.17	98.42	7	79.63	7.38	66.38	74.45	29.60	95.98
WBC adm. [10^3/μL]	24	9.20	1.16	5.84	8.08	4.84	10.68	33	10.30	0.95	6.70	8.44	5.92	12.65	7	9.37	1.42	6.49	9.84	4.84	11.32
WBC dis. [10^3/μL]	23	8.08	0.82	5.95	7.29	2.36	8.31	31	7.50	0.46	6.20	7.40	1.90	8.10	7	7.27	0.84	5.76	7.14	3.57	9.33
CRP adm. [mg/L]	23	34.11	10.00	1.41	5.73	46.72	48.13	32	61.34	14.49	5.79	28.49	72.75	78.54	7	30.87	16.79	1.29	21.79	31.35	32.64
CRP dis. [mg/L]	20	24.52	6.93	1.89	6.79	40.90	42.78	31	18.89	3.43	4.31	11.24	23.98	28.29	6	7.76	4.03	2.06	3.06	6.11	8.17
PCT adm. [ng/mL]	12	0.08	0.02	0.05	0.06	0.07	0.11	26	3.05	1.59	0.09	0.18	0.55	0.64	5	0.19	0.12	0.02	0.08	0.13	0.15
PCT dis. [ng/mL]	9	0.07	0.02	0.03	0.05	0.06	0.09	20	1.20	0.96	0.06	0.10	0.20	0.26	3	0.02	0.003	0.02	0.02	0.01	0.03

Table 2. Kruskal–Wallis test results.

Variable	p-Value	Decision	Test Power
Uric acid [md/dL]	0.055	no differences in subgroups	0.981
Adiponectin [μg/mL]	0.033	statistically significant differences in subgroups	1
WBC adm. [10^3/μL]	0.523	no differences in subgroups	0.9314
WBC dis. [10^3/μL]	0.978	no differences in subgroups	0.5108
CRP adm. [mg/L]	0.196	no differences in subgroups	1
CRP dis. [mg/L]	0.458	no differences in subgroups	1
PCT adm. [ng/mL]	0.063	no differences in subgroups	0.9985

2.2. The Influence of Rapidly Changing Inflammation in the Course of Infection on Discussed Parameters

Table 3 shows the descriptive statistics and results of statistical tests of biomarkers depending on the use of antibiotics. The average uric acid concentration was comparable in the group of people using antibiotics during hospitalization due to bacterial infection (N = 36) and those not using them (N = 27) and amounted to 5.897 and 6.437 mg/dL, respectively. Concerning adiponectin concentration, no significant differences were observed between the described groups. The average concentration was 65.458 μg/dL in patients taking antibiotics and 58.443 μg/dL in patients not taking antibiotics during hospitalization ($p = 0.361$). Statistically significant differences were found for WBC and CRP measurements, both at the time of admission and upon discharge from the Clinic, but this is due to the characteristics of the groups discussed: patients taking antibiotics had high parameters of inflammation resulting from infectious diseases, which quickly decreased as a result of proper antibiotic treatment.

2.3. The Influence of Gender on Discussed Parameters

Table 4 presents the descriptive statistics and p-values of biomarker tests among patients divided by gender. According to the analyses, the uric acid concentration did not differ significantly in women (N = 26) and men (N = 37) and amounted to an average of 6.058 mg/dL and 6.178 mg/dL, respectively ($p = 0.842$). There was also no statistically significant difference in adiponectin concentration between women and men, which averaged 52.275 μg/dL and 69.494 μg/dL, respectively ($p = 0.023$). The WBC concentrations, both upon admission to the Clinic and at the end of hospitalization did not differ depending on gender. A higher average CRP concentration at admission and the end of hospitalization was found in women, which amounted to 78.723 mg/dL at the beginning of hospitalization and 32.368 mg/dL at the end of the stay, respectively. While in men the average CRP concentration on admission was 25.463 mg/dL and at discharge 10.402 mg/dL. These differences, however, may be consequences of the diseases that led to the hospitalization of individual study participants, which led to different responses to inflammatory parameters.

2.4. Biomarker Correlation Analysis

Spearman's correlation coefficients were determined for the biomarkers analyzed, along with significance tests. The following graphs show the correlations among the entire study group, as well as by group of inpatients with normal nutritional status, at risk of malnutrition, and malnourished (Figure 2).

Among the entire study group, at the 0.05 significance level, significant correlations were found between WBC at the time of hospital admission and WBC upon discharge from the Clinic (R = 0.51), CRP at admission (R = 0.46), CRP at discharge (R = 0.35), and PCT on admission. CRP upon admittance is significantly correlated with CRP at discharge (R = 0.65) and PCT on admission (R = 0.47) and at discharge (R = 0.37). CRP upon discharge is also significantly correlated with PCT on admission and at discharge. Finally, PCT at the time of admission is correlated with PCT at discharge (R = 0.54) (Figure 2A).

For patients with normal nutritional status, statistically significant correlations at the 0.05 level were shown between WBC measurements on admission and upon discharge from the Clinic (R = 0.68), as well as CRP measurements at the time of admission and discharge (R = 0.82) and PCT upon admittance (R = 0.72) (Figure 2B).

Table 3. Descriptive statistics and biomarker statistical test results by antibiotic use.

	No Antibiotic					Antibiotic					Statistical Tests		
	N	Mean	Standard Error	Median	IQR	N	Mean	Standard Error	Median	IQR	Type	*p*-Value	Test Power
Uric acid [md/dL]	36	5.897	0.345	5.450	2.750	27	6.437	0.513	5.900	4.300	U Mann–Whitney test	0.466	0.519
Adiponectin [µg/mL]	37	65.458	5.339	60.111	52.300	27	58.443	5.111	53.350	39.740	Student's *t*-test	0.361	0.148
WBC adm. [10^3/µL]	37	8.180	0.569	8.120	4.390	27	11.978	1.274	11.370	7.905	U Mann–Whitney test	0.010	1.000
WBC dis. [10^3/µL]	34	7.052	0.439	6.535	1.873	27	8.500	0.682	7.960	2.320	U Mann–Whitney test	0.019	0.996
CRP adm. [mg/L]	35	25.466	8.543	3.480	28.07	27	76.746	14.907	49.140	118.510	U Mann–Whitney test	<0.001	1.000
CRP dis. [mg/L]	30	12.127	3.085	3.540	12.82	27	28.100	5.221	19.320	32.755	U Mann–Whitney test	0.004	1.000
PCT adm. [ng/mL]	19	0.213	0.065	0.080	0.145	24	3.215	1.724	0.140	0.552	Student's *t*-test	0.130	0.326
PCT dis. [ng/mL]	13	0.089	0.029	0.050	0.060	19	1.243	1.008	0.100	0.200	Student's *t*-test	0.354	0.149

Table 4. Descriptive statistics and *p*-values of biomarker tests among patients by gender.

	Male					Female					Statistical Tests		
	N	Mean	Standard Error	Median	IQR	N	Mean	Standard Error	Median	IQR	Type	*p*-Value	Test Power
Uric acid [md/dL]	37	6.178	0.386	5.600	3.400	26	6.058	0.465	5.600	3.300	Student's *t*-test	0.842	0.054
Adiponectin [µg/mL]	38	69.494	4.883	62.434	42.880	26	52.275	5.402	44.868	34.973	Student's *t*-test	0.023	0.629
WBC adm. [10^3/µL]	38	9.135	0.816	8.260	4.272	26	10.728	1.123	10.480	7.600	U Mann–Whitney test	0.234	0.999
WBC dis. [10^3/µL]	36	7.208	0.317	7.270	2.410	25	8.392	0.844	7.290	1.870	Student's *t*-test	0.143	0.309
CRP adm. [mg/L]	36	25.463	6.044	9.170	35.788	26	78.723	17.279	48.125	126.482	U Mann–Whitney test	0.016	1.000
CRP dis. [mg/L]	33	10.402	2.080	6.150	9.160	24	32.468	5.975	21.175	48.272	Student's *t*-test	<0.001	0.970
PCT adm. [ng/mL]	21	1.971	1.252	0.080	0.180	22	1.809	1.528	0.140	0.500	Student's *t*-test	0.935	0.051
PCT dis. [ng/mL]	17	1.246	1.130	0.070	0.190	15	0.239	0.108	0.080	0.150	Student's *t*-test	0.412	0.127

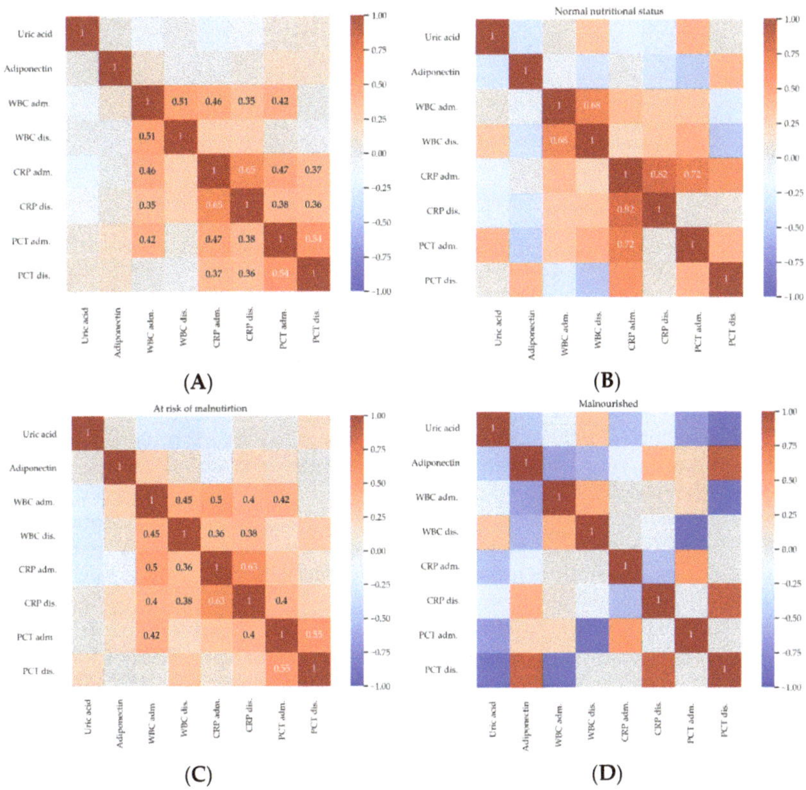

Figure 2. Spearman's correlation diagram for all subjects (**A**), for patients with normal nutritional status (**B**), for the group at risk of malnutrition (**C**), and for malnourished patients (**D**). Correlations significant at the 0.05 level are marked in black or white.

For patients at risk of malnutrition, significant correlations were found at the 0.05 significance level between WBC upon admittance and WBC at discharge (R = 0.45), CRP on admission (R = 0.5), CRP at discharge (R = 0.4), and PCT upon admittance (R = 0.42). In addition, WBC at discharge is statistically significantly correlated with CRP on admission and at discharge (R = 0.36 and R = 0.38, respectively). CRP at the time of hospital admission is significantly correlated with CRP at discharge (R = 0.63). CRP at discharge is statistically significantly correlated with PCT on admission (R = 0.4). Finally, PCT on admission is correlated with PCT at discharge (R = 0.55). (Figure 2C). Interestingly, there were no statistically significant correlations for biomarkers tested in malnourished patients (Figure 2D).

2.5. Effect of Antibiotic Therapy on the Change in WBC, CRP, and PCT Parameters between Admission and Discharge

To test the effect of antibiotic therapy on the change in WBC, CRP, and PCT parameters between admission and discharge, the differences in these measurements were determined and statistical tests were performed by groups defined by antibiotic intake. A negative value indicates a decrease in the value of the measurement at discharge relative to that taken at admission. Assuming the declared level of significance, there were no statistically significant differences in the values of the WBC, CRP, and PCT parameters (Table 5).

Table 5. Descriptive statistics and statistical test results of selected differences in biomarkers between admission and discharge from the Clinic by antibiotic use.

	No Antibiotic					Antibiotic					Statistical Tests		
	N	Mean	Standard Error	Median	IQR	N	Mean	Standard Error	Median	IQR	Type	*p*-Value	Test Power
WBC diff. [$10^3/\mu L$]	34	−1.311	0.576	−0.195	2.003	27	−3.477	1.145	−2.27	7.120	U Mann–Whitney test	0.152	1
CRP diff. [mg/L]	30	−17.407	8.157	−0.605	14.855	27	−48.644	14.481	−22.86	85.84	U Mann–Whitney test	0.054	1
PCT diff. [ng/mL]	13	−0.192	0.083	−0.010	0.3	19	−2.778	1.992	−0.03	0.305	Student's *t*-test	0.294	0.179

3. Discussion

A common reason for hospitalization, especially in older patients, is conditions involving a rapid increase in the inflammatory response. There are many causes of inflammatory reactions, including infectious diseases, rheumatological diseases, injuries, and cancer. During the treatment of patients with an inflammatory response, it is difficult to interpret commonly used inflammatory markers because they are characterized by the fact that their concentration increases in the general inflammatory response, which makes it difficult to specify what disease was the cause of the increase in inflammatory parameters [1]. For example, the treatment of infections in patients with advanced cancer can be carried out as effectively as possible. However, it is not always possible to achieve complete normalization of inflammatory parameters during hospitalization, because part of the inflammatory reaction will result from, for instance, the co-existing cancer [26]. The results of the present study confirm this phenomenon as substantial reductions in the WBC, CRP, and PCT (Table 5) values between the time of admission and discharge from the Clinic may be observed in most cases due to antibiotic treatment. However, the lack of statistically significant differences between the antibiotic and no antibiotic group suggests inflammation from causes different than bacterial infection. Therefore, it seems justified to search for further parameters measuring inflammation that could be used in clinical practice. The present study was conducted to demonstrate whether adipose tissue hormones, using the example of adiponectin, have the potential to act as a measure of rapidly changing inflammation when monitoring older patients in the course of infection.

Older adults are a very fragile and specific group of patients. They are at high risk of bacterial infections, often present atypically, and have weaker immune responses, which results in high morbidity and mortality [27]. Moreover, because of lower prognostic values of routinely assessed parameters in comparison to younger adults, finding a reliable biomarker of bacterial infections would accelerate the proper diagnosis and effective treatment. The other important matter is an inappropriate prescription of antimicrobials to senile patients, which not only contributes to antibiotic resistance but also potential adverse effects, especially due to multiple chronic diseases and polypharmacy.

The oxidation–inflammation theory of aging proposes that there is a close relationship between oxidative stress, inflammation, and aging. Aging is a loss of homeostasis caused by chronic oxidative stress, which mainly affects regulatory systems such as the immune system, nervous system, and endocrine system. This leads to activation of the immune system, inducing an inflammatory state, which creates a vicious cycle where chronic oxidative stress and inflammation reinforce each other. It leads to increased age-related morbidity and mortality [28]. There is a possible link between metabolic stress and oxidative stress in senile patients. It should be underlined that adiponectin plays a crucial role in maintaining metabolic balance and reducing oxidative stress related to obesity. Gradinaru et al. described significantly lower levels of adiponectin in senile patients with metabolic syndrome concomitantly with significantly higher levels of oxidative stress and cardiovascular disease risk markers [29]. Moreover, adiponectin levels were significantly positively associated with antioxidant capacity and significantly negatively associated with serum uric acid concentration. The results obtained in this study do not confirm the relationship between uric acid and adiponectin concentrations (Figure 2). The role of oxidative stress in acute and chronic infections, as well as in associated diseases, has been a topic of much debate. In bacterial infections, oxidative stress arises from altered metabolic pathways and has been linked to organ damage and the development of cancers. *Helicobacter pylori*, for instance, triggers enzymes that generate reactive oxygen species (ROS), such as spermine oxidase, and increases the expression of redox-regulated genes that are proinflammatory and potentially cancer causing, such as cyclooxygenase 2 [30].

Different effects of adiponectin in the human body have been already proven, including insulin-sensitizing, anti-atherosclerotic, and anti-inflammatory effects. This diversity is probably due to the existence of many isoforms of the hormone, which differ in their actions and the signaling pathways in which they participate. High concentrations of adiponectin are observed in many diseases caused by chronic inflammation. However, to understand and describe the effects of APN during the acute phase of inflammation, more research is necessary [9]. The most frequently described effect of APN is anti-inflammatory, even in diseases that do not specifically affect adipose tissue. It is so because receptors that may be influenced by adiponectin are located, for example, in the lungs. Therefore, therapies based on mechanisms involving APN may be used in the future in the treatment of inflammatory pulmonary diseases [25]. Naturally, due to the complex involvement of adiponectin in inflammatory processes, many researchers have been trying to identify the possibility of using it as a prognostic marker. So far, however, it has not been possible to link the dynamics of changes in adiponectin concentrations with ongoing acute inflammation. However, many studies have presented results suggesting that the measurement of a given protein can be used to detect and predict the course of some diseases, especially those in which there is chronic inflammation associated with obesity [31]. Determination of adiponectin concentration may probably be helpful in clinical practice in monitoring oncological diseases, diabetes, cardiovascular diseases, and Alzheimer's disease [10]. A meta-analysis by Yang et al. showed that elevated adiponectin levels are an independent predictor of cardiovascular death and mortality in patients with coronary artery disease. It follows that the described hormone could probably contribute to identifying patients with coronary artery disease at high risk of death [32]. Additionally, the study by Güven and co-authors proved the usefulness of measuring APN in monitoring the course of breast cancer and diagnosing patients at risk of more aggressive disease. However, detailed research on large databases is necessary to introduce the described method into clinical practice [33]. Interestingly, the disturbance of the adiponectin-to-leptin ratio in the body may also play a role in the development of inflammation. This is especially true in patients with metabolic syndrome (MS). As shown by Frühbeck et al., the total concentrations of both laboratory-determined adiponectin and its multimeric forms were significantly lower in MS patients. Moreover, the quantitative adiponectin/leptin ratio in this group of people was significantly lower compared to the general population. It also should be noted that the study group showed increased levels of inflammatory markers, including CRP. This constitutes the basis for the conclusion that abnormalities in the functioning of adipose tissue, manifested as disturbances in the adiponectin/leptin ratio among MS patients, may increase the effects of oxidative stress and significantly increase the unfavorable effects of ongoing inflammation [34]. Another study by Chen et al. describes that concentrations of adiponectin, hs-CRP (highly sensitive CRP), IL-6, and the level of activity of antioxidant enzymes, i.e., superoxide dismutase (SOD), may also have a predictive value for the occurrence of MS in a patient. In this study recruited patients had significantly higher concentrations of the above-mentioned inflammatory markers, as well as lower adiponectin concentrations and reduced SOD activity. It can therefore be concluded that high concentrations of inflammatory markers combined with reduced adiponectin levels significantly correlate with a higher risk of MS [35]. In addition to the above-mentioned observations in senile MS patients, Ma et al., in their investigation, demonstrated a higher body mass index (BMI), higher blood glycemia, significantly increased CRP levels, and significantly reduced APN levels compared to the group of patients without MS [36]. The above-mentioned investigations demonstrate the validity of considering the concentration of circulating APN as a predictive marker for MS in older patients. In the present study, statistically significant differences were found for WBC and CRP measurements both at the time of admission and upon discharge from the Clinic. Notably, patients taking antibiotics had high parameters of inflammation resulting from infectious diseases, which quickly decreased as a result of proper antibiotic treatment (Table 3). However, no statistical differences were found for the level of APN between these two groups of patients. Some statistically significant correlations were found for inflammatory parameters such as WBC and CRP, but not for adiponectin (Figure 2).

Some previous investigations indicate the role of APN in immunological response to infections. For instance, in the study by Behnes et al., it was shown that the level of the protein form of APN and the expression of the mRNA encoding it increase in response to the development of sepsis. Additionally, it has been proven that the deterioration of the patient's general condition and severe sepsis are related to the increase in the concentrations of the mentioned substances in the blood. In laboratory tests, an increase in adiponectin and mRNA levels was observed as the disease progressed. Additionally, the study revealed that the group of patients treated with drotrecogin a (DAA)-activated protein C therapy was characterized by increased adiponectin expression during the ongoing systemic inflammatory reaction. Therefore, the study results confirm that adiponectin may play an important role in the pathogenesis of inflammatory processes in sepsis [37]. An attempt to demonstrate the relationship between the concentration of adiponectin and leptin and the presence of acute inflammation was made by Ekström and co-authors [38]. This condition was attempted to be induced in one of the study groups by administering a vaccination against Salmonella typhi. In the second group, measurements were performed after undergoing open-heart cardiac surgery. Therefore, two validated methods of inducing the inflammatory process were used. However, summarizing the results of adiponectin and leptin determinations in plasma, no changes were noticed in samples taken before and after the intervention. In addition, it was decided to determine the expression of adiponectin genes in omental and subcutaneous fat tissue, and no changes were detected there either. It follows that quantitative changes in the production and release of adiponectin probably do not depend on the processes of the acute inflammatory response. This phenomenon is consistent with the results obtained in this study. Interestingly, a slightly lower median level of APN is observed among inpatients suffering from infection in comparison to the "no antibiotic" group, but the lack of statistical significance does not allow conclusions at this stage of research. The results obtained so far may be considered preliminary and need further investigations, especially considering many possible comorbidities characteristic of the senile population.

Adiponectin concentration is significantly influenced by body weight reduction, which deserves special attention in obese patients. It has been shown that reducing fat tissue while maintaining muscle mass in the weight-loss process in this group of patients leads to an increase in adiponectin concentration. This means that the assessment of serum adiponectin levels can serve as a predictor of the effects of weight loss in patients struggling with obesity [39]. The relationship between adiponectin concentration in the older population and their general condition, muscle mass, and the risk of mortality in this group of patients was investigated by Walowski et al. The results suggest that high levels of adiponectin are associated with lower muscle mass, reduced bone mineral content, and lower levels of fatty liver disease. This is most likely related to a decrease in the level of insulin-like growth factor 1 (IGF-1) in response to an increase in the concentration of adiponectin in the body [40]. The results of the above-mentioned studies are in agreement with the present investigation, which shows that a low nutritional status is associated with a higher concentration of adiponectin (Table 2). Significant differences in adiponectin concentration between the group of patients with normal nutritional status and the group of malnourished patients ($p = 0.004$) were observed (Figure 1). In contrast, Mądra-Gackowska et al. highlighted the influence of nutritional status according to the MNA on such adipokines as leptin and resistin [41]. No statistically significant differences between patients with normal nutritional status, at risk of malnutrition, and malnourished were found for APN. However, adiponectin concentration was inversely related to nutritional status, which was also observed in the current study. Indeed, the secretion of adiponectin by adipose tissue cells may be regulated by many factors, including diet and the patient's overall nutritional status. Important factors modulating the synthesis of adiponectin are the elements of the diet and the percentage of its components. The most beneficial in this respect seems to be the Mediterranean diet and the DASH diet, the main element of which is the supply of optimal amounts of unsaturated fatty acids, but also fiber and polyphenols. Thanks to their positive impact on maintaining the optimal level of APN, the above-mentioned dietary formulas may therefore support the prevention of some lifestyle diseases containing an inflammatory component, i.e., cardiovascular diseases or cancer [42]. The inflammatory process is one of the fundamental mechanisms of obesity. In its course, inflammatory mediators are released from adipocytes into the bloodstream, which affect other body tissues. APN has anti-inflammatory properties and, in the case of obesity, it counteracts the negative effects of pro-inflammatory cytokines. An unfavorable effect of obesity is the gradual development of resistance to adiponectin, which over time leads to a significant reduction in the beneficial effects of this APN, which may result in an exacerbation of inflammatory processes [43].

This study revealed some statistically significant differences in inflammatory parameters and adipokine levels between groups established based on nutritional status or antibiotic use. However, there are some limitations of the present investigation. The measurement of serum APN was performed only once while repeating the measurement at the end of hospitalization could shed more light on the behavior of APN. The group of recruited patients, in whom the measurements were carried out, was relatively small. Presumably, for a larger groups, the results could be more accurate. For this reason, the results presented in this study cannot be fully applied to the entire population of older people. The focus was also only on demonstrating the relationship between adiponectin concentration and the course of acute inflammation. Completely different correlations can probably be found in patients struggling with chronic diseases involving the inflammatory process, for example cancer or rheumatological diseases. Therefore, it is necessary to conduct more studies involving older people to confirm the validity of using adiponectin as an acute inflammatory marker.

4. Materials and Methods

4.1. Ethical Statement

The study was approved by the Bioethics Committee of the Collegium Medicum. L. Rydygier in Bydgoszcz. Nicolaus Copernicus University in Toruń on 14 December 2021. Consent number KB 675/2021.

4.2. Patient Recruitment

The study recruited 64 patients over 60 years of age, hospitalized in the Geriatrics Clinic of the University Hospital No. 1. Dr. A. Jurasz in Bydgoszcz in the period from July 2022 to April 2023, whose level of cognitive functioning allowed them to express informed consent to participate in the study. According to the organizational structure of the mentioned Clinic, patients can be hospitalized in two modes: planned, to perform a Comprehensive Geriatric Assessment (CGA), and urgent, when the health condition requires urgent hospitalization for general internal medicine reasons. Representatives of both groups took part in the study.

4.3. Nutritional Status Assessment

The Comprehensive Geriatric Assessment is a set of scales and tests used to assess independence in everyday functioning and to detect disorders that are part of the so-called Geriatric Giants such as falls, eating disorders, or frailty syndrome [44]. CGA is intended to indicate the patient's problems and deficits and draw attention to their strengths, which may become a starting point for improving everyday functioning. As part of the assessment, the nutritional status is also checked. It is particularly important in estimating the body's energy resources necessary to function while being exposed to multiple diseases [45]. A test that can be used to assess nutritional status is the Mini Nutritional Assessment (MNA). It includes the assessment of four basic parameters: anthropometric (weight, height, calf circumference, and arm circumference), food and fluid intake, general assessment (place of residence, physical activity, stressful situations or illness, medications taken, skin condition, neuropsychiatric disorders), and subjective assessment (self-assessment of health and nutritional status). During the assessment, points are awarded and then summed up: normal nutritional status is 24–30 points, a state at risk of malnutrition is 17–23.5 points, and malnutrition is a result below 17 points [46]. For the conducted research, all patients hospitalized both for a comprehensive geriatric assessment in a scheduled and urgent manner were assessed in terms of their nutritional status using the above-mentioned screening tool at the end of hospitalization, when their general condition allowed for assessment without a significant risk of the presence of factors interfering with the assessment, such as active, high inflammation. Depending on the number of points obtained according to the MNA, the study participants were divided into three groups:

- Group 0 (control) with normal nutritional level based on the MNA score.
- Group 1 at risk of malnutrition.
- Group 2 malnourished patients.

4.4. Taking into Account Comorbidities

The list of internal diseases that require hospitalization is long and varied. Patients who were able to give informed and voluntary consent to participate in the project were recruited for the study, with the following diseases: exacerbation of heart failure, complicated urinary tract infection, severe hyponatremia, severe anemia, pneumonia, acute kidney injury, decompensated diabetes, pulmonary embolism, and COVID-19. Since the number of diseases is large and they are often characterized by a different clinical course, further analyses took into account primarily the level of inflammation parameters and the use of any antibiotic therapy, bearing in mind that the analysis for specific disease entities individually may be irrelevant statistically with a limited number of research participants.

4.5. Evaluated Parameters

Uric acid is an example of a substance also involved in inflammatory reactions [47]. Its concentration was calculated at the beginning of patients' hospitalization in the Clinic, as part of routine laboratory parameter determinations. Uric acid concentration was determined in patients' serum using ELISA test. In analyses with other parameters, uric acid was included to demonstrate whether its concentration changes depending on changes in inflammatory parameters, such as WBC or CRP, or depending on the concentration of adiponectin, an example of an adipose tissue hormone.

Adiponectin was used as an adipokine that is relatively well known and can be commonly measured in medical care facilities. Serum APN concentration was also determined using ELISA test. Taking into account the fact that the studies involved patients with both increased inflammatory parameters and inflammation within the normal range, the level of adiponectin was determined once at the beginning of hospitalization, when discrepancies in inflammation parameters between individual patients were the largest.

In order to assess whether adiponectin concentrations change depending on the prevalence of acute inflammation, it was compared with other parameters of inflammation routinely assessed in hospitalized patients such as White Blood Cells (WBC), C-reactive protein, (CRP) and procalcitonin (PCT). These parameters are routinely measured in patients hospitalized at the Clinic upon admission and at the end of hospital treatment. The concentrations were determined in serum using ELISA test.

4.6. Consideration of Inflammation during Hospitalization

For further analyses, patients were also divided into two groups: patients with diagnosed infections who were treated with intravenous antibiotic therapy during hospitalization, and patients who did not require antibiotic therapy. The first group was always characterized by elevated inflammatory parameters that were subsequently decreased, while the second group could have chronically elevated inflammatory parameters (from comorbidities such as autoimmune diseases or cancer) or no inflammation.

4.7. Statistical Analysis

The Python programming language 3.8.10 (The Python Software Foundation, Fredericksburg, VA, USA) with the libraries pandas (1.4.3), matplotlib (3.1.3) and scipy (1.10.1) were used to perform the analyses. All analyses were performed assuming in advance a significance level of 0.05. After collecting the data described above, an attempt was made to demonstrate the correlation between the concentration of adiponectin, uric acid, WBC, and CRP on admission and at the end of hospitalization, and the nutritional status and the presence of acute inflammation, defined as inflammation caused by an infection requiring antibiotic therapy. Additionally, it was planned to demonstrate whether the mentioned parameters differed depending on gender. Verification was planned using the Chi-square test of equivalence and $p < 0.001$ to meet the assumption of equality of groups necessary for parametric analysis of variance (ANOVA). If verification is not possible, Kruskal–Wallis tests were planned to be performed. The Kruskal–Wallis test is used to compare two or more groups for a continuous or discrete variable. It is a non-parametric test assuming no particular distribution of the data and is analogous to the one-way analysis of variance (ANOVA) [48].

5. Conclusions

The present cross-sectional study revealed some statistically significant differences in inflammatory parameters and adipokine levels between groups of senile inpatients established based on nutritional status or antibiotic use. According to the results obtained, higher serum adiponectin levels are related to deterioration of nutritional status as significant differences in adiponectin concentration between the group of patients with normal nutritional levels and the group of malnourished patients were found. Despite bacterial infections being associated with oxidative stress rise, no statistically significant differences were found in the levels of APN in hospitalized patients diagnosed with acute infections who required antibiotic therapy compared to senile inpatients without diagnosed inflammation. Therefore, adiponectin may be a better biomarker of malnutrition in older adults than a parameter for monitoring the initial stage of acute inflammation. It should be noted that adiponectin was only measured once during hospital admission and commonly used markers of inflammation were tested throughout the hospitalization. Therefore, this study is preliminary and further research is needed. However, given the relationship between aging, infections, oxidative stress, and adipokines, it is possible that APN could be useful for monitoring later stages of the inflammatory response. As malnutrition is a serious and often underdiagnosed problem in the senile population, measuring adiponectin levels on hospital admission in the future may facilitate the identification of malnourished seniors. Nevertheless, further efforts should be made to search and validate potential biomarkers based on adipokines to translate them into clinical practice.

Author Contributions: Conceptualization J.H. and M.K.; methodology J.H. and M.G.; software J.H., J.W. and M.K.; validation J.H., M.G., K.M.-G. and J.W.; formal analysis J.H., D.S., M.P., M.G. and J.W.; investigation J.H.; resources J.H. and K.K.-K.; data curation J.H. and M.K.; writing—original draft preparation J.H., M.P. and D.S.; writing—review and editing J.H., D.S., M.P., J.N. and M.G.; visualization J.H., J.W., J.N. and M.G.; supervision M.K., A.W. and K.K.-K.; project administration J.H. All authors have read and agreed to the published version of the manuscript.

Funding: This research received no external funding.

Institutional Review Board Statement: The study was conducted in accordance with the Declaration of Helsinki and approved by the Bioethics Committee of Collegium Medicum im. Ludwik Rydygier in Bydgoszcz. Nicolaus Copernicus University in Toruń (No. KB 675/2021; 14 December 2021).

Informed Consent Statement: Informed consent was obtained from all subjects involved in the study.

Data Availability Statement: Data are available on request due to privacy/ethical restrictions.

Conflicts of Interest: The authors declare no conflicts of interest.

References

1. Chen, L.; Deng, H.; Cui, H.; Fang, J.; Zuo, Z.; Deng, J.; Li, Y.; Wang, X.; Zhao, L. Inflammatory Responses and Inflammation-Associated Diseases in Organs. *Oncotarget* **2018**, *9*, 7204–7218. [CrossRef]
2. Teissier, T.; Boulanger, E.; Cox, L.S. Interconnections between Inflammageing and Immunosenescence during Ageing. *Cells* **2022**, *11*, 359. [CrossRef]
3. Rizo-Téllez, S.A.; Sekheri, M.; Filep, J.G. C-Reactive Protein: A Target for Therapy to Reduce Inflammation. *Front. Immunol.* **2023**, *14*, 1237729. [CrossRef]
4. Kobayashi, H.; Amrein, K.; Lasky-Su, J.A.; Christopher, K.B. Procalcitonin Metabolomics in the Critically ill Reveal Relationships between Inflammation Intensity and Energy Utilization Pathways. *Sci. Rep.* **2021**, *11*, 23194. [CrossRef]
5. Matur, E.; Özcan, M.; Ergül Ekiz, E.; Ergen, E.; Erek, M.; Or, E.; Dokuzeylül, B.; Erhan, S.; Bilgiç, B. Use of Serum Procalcitonin (PCT) Level and PCT MRNA Expression as a Potential Clinical Biomarker in Cats with Bacterial and Viral Infections. *J. Feline Med. Surg.* **2022**, *24*, e595–e602. [CrossRef]
6. Margraf, A.; Lowell, C.A.; Zarbock, A. Neutrophils in Acute Inflammation: Current Concepts and Translational Implications. *Blood* **2022**, *139*, 2130–2144. [CrossRef]
7. Ruggiero, C.; Cherubini, A.; Ble, A.; Bos, A.J.G.; Maggio, M.; Dixit, V.D.; Lauretani, F.; Bandinelli, S.; Senin, U.; Ferrucci, L. Uric Acid and Inflammatory Markers. *Eur. Heart J.* **2006**, *27*, 1174–1181. [CrossRef]
8. Cannizzo, E.S.; Clement, C.C.; Sahu, R.; Follo, C.; Santambrogio, L. Oxidative Stress, Inflamm-Aging and Immunosenescence. *J. Proteomics* **2011**, *74*, 2313–2323. [CrossRef]
9. Choi, H.M.; Doss, H.M.; Kim, K.S. Multifaceted Physiological Roles of Adiponectin in Inflammation and Diseases. *Int. J. Mol. Sci.* **2020**, *21*, 1219. [CrossRef]
10. Khoramipour, K.; Chamari, K.; Hekmatikar, A.A.; Ziyaiyan, A.; Taherkhani, S.; Elguindy, N.M.; Bragazzi, N.L. Adiponectin: Structure, Physiological Functions, Role in Diseases, and Effects of Nutrition. *Nutrients* **2021**, *13*, 1180. [CrossRef]
11. Matsuda, M.; Shimomura, I. Roles of Adiponectin and Oxidative Stress in Obesity-Associated Metabolic and Cardiovascular Diseases. *Rev. Endocr. Metab. Disord.* **2014**, *15*, 1–10. [CrossRef] [PubMed]
12. van Dijk-Huisman, H.C.; Welters, M.H.P.; Bijnens, W.; van Kuijk, S.M.J.; Magdelijns, F.J.H.; de Bie, R.A.; Lenssen, A.F. Development and Internal Validation of a Prediction Model to Identify Older Adults at Risk of Low Physical Activity Levels during Hospitalisation: A Prospective Cohort Study. *BMC Geriatr.* **2022**, *22*, 479. [CrossRef] [PubMed]
13. Falcone, M.; Bauer, M.; Ferrer, R.; Gavazzi, G.; Gonzalez del Castillo, J.; Pilotto, A.; Schuetz, P. Biomarkers for Risk Stratification and Antibiotic Stewardship in Elderly Patients. *Aging Clin. Exp. Res.* **2023**, *35*, 925–935. [CrossRef]
14. Flamant, L.; Giordano Orsini, G.; Ramont, L.; Gornet, M.; De Ruffi, S.; Leroux, P.; Kanagaratnam, L.; Gennai, S. Association between Admission Biomarkers and Clinical Outcome in Older Adults Diagnosed with an Infection in the Emergency Department. *Acta Clin. Belg.* **2023**, *78*, 285–290. [CrossRef] [PubMed]
15. Gbinigie, O.A.; Onakpoya, I.J.; Richards, G.C.; Spencer, E.A.; Koshiaris, C.; Bobrovitz, N.; Heneghan, C.J. Biomarkers for Diagnosing Serious Bacterial Infections in Older Outpatients: A Systematic Review. *BMC Geriatr.* **2019**, *19*, 190. [CrossRef]
16. Graßmann, S.; Wirsching, J.; Eichelmann, F.; Aleksandrova, K. Association Between Peripheral Adipokines and Inflammation Markers: A Systematic Review and Meta-Analysis. *Obesity* **2017**, *25*, 1776–1785. [CrossRef]
17. Szydełko, J.; Trojanowska, P.; Dąbrowska, I.; Szydełko-Gorzkowicz, M.; Litwińczuk, M. Adiponectin as Novel Biomarker of Endothelial Dysfunction in Insulin Resistance and Obesity—A Narrative Review. *J. Educ. Health Sport* **2020**, *10*, 591–606. [CrossRef]
18. Song, S.H.; Oh, T.R.; Choi, H.S.; Kim, C.S.; Ma, S.K.; Oh, K.H.; Ahn, C.; Kim, S.W.; Bae, E.H. High Serum Adiponectin as a Biomarker of Renal Dysfunction: Results from the KNOW-CKD Study. *Sci. Rep.* **2020**, *10*, 5598. [CrossRef]
19. Galazka, J.K.; Polak, A.; Matyjaszek-Matuszek, B. Adiponectin as a Novel Predictive Biomarker of Multiple Sclerosis Course. *Curr. Issues Pharm. Med. Sci.* **2023**, *36*, 99–102. [CrossRef]
20. Wattanapol, P.; Vichinsartvichai, P.; Sakoonwatanyoo, P. Serum Adiponectin Is a Potential Biomarker for Metabolic Syndrome in Peri-and Postmenopausal Women. *Gynecol. Endocrinol.* **2020**, *36*, 620–625. [CrossRef]
21. Lubkowska, A.; Radecka, A.; Mieszkowski, J. *Adiponectin as Biomarker of Osteoporosis*; Springer: Berlin/Heidelberg, Germany, 2015; pp. 1–34.
22. Stochmal, A.; Waśkiel-Burnat, A.; Chrostowska, S.; Zaremba, M.; Rakowska, A.; Czuwara, J.; Rudnicka, L. Adiponectin as a Novel Biomarker of Disease Severity in Alopecia Areata. *Sci. Rep.* **2021**, *11*, 13809. [CrossRef]

23. Palmer, C.; Hampartzoumian, T.; Lloyd, A.; Zekry, A. A Novel Role for Adiponectin in Regulating the Immune Responses in Chronic Hepatitis C Virus Infection. *Hepatology* **2008**, *48*, 374–384. [CrossRef]
24. Keller, P.; Møller, K.; Krabbe, K.S.; Pedersen, B.K. Circulating Adiponectin Levels during Human Endotoxaemia. *Clin. Exp. Immunol.* **2003**, *134*, 107–110. [CrossRef]
25. Lim, J.Y.; Templeton, S.P. Regulation of Lung Inflammation by Adiponectin. *Front. Immunol.* **2023**, *14*, 1244586. [CrossRef]
26. Singh, N.; Baby, D.; Rajguru, J.; Patil, P.; Thakkannavar, S.; Pujari, V. Inflammation and Cancer. *Ann. Afr. Med.* **2019**, *18*, 121. [CrossRef]
27. Soraci, L.; Cherubini, A.; Paoletti, L.; Filippelli, G.; Luciani, F.; Laganà, P.; Gambuzza, M.E.; Filicetti, E.; Corsonello, A.; Lattanzio, F. Safety and Tolerability of Antimicrobial Agents in the Older Patient. *Drugs Aging* **2023**, *40*, 499–526. [CrossRef]
28. Gorni, D.; Finco, A. Oxidative Stress in Elderly Population: A Prevention Screening Study. *Aging Med.* **2020**, *3*, 205–213. [CrossRef]
29. Gradinaru, D.; Margina, D.; Borsa, C.; Ionescu, C.; Ilie, M.; Costache, M.; Dinischiotu, A.; Prada, G.-I. Adiponectin: Possible Link between Metabolic Stress and Oxidative Stress in the Elderly. *Aging Clin. Exp. Res.* **2017**, *29*, 621–629. [CrossRef]
30. Ivanov, A.V.; Bartosch, B.; Isaguliants, M.G. Oxidative Stress in Infection and Consequent Disease. *Oxid. Med. Cell. Longev.* **2017**, *2017*, 10–12. [CrossRef]
31. Ouchi, N.; Walsh, K. Adiponectin as an Anti-Inflammatory Factor. *Clin. Chim. Acta* **2007**, *380*, 24–30. [CrossRef]
32. Yang, L.; Li, B.; Zhao, Y.; Zhang, Z. Prognostic Value of Adiponectin Level in Patients with Coronary Artery Disease: A Systematic Review and Meta-Analysis. *Lipids Health Dis.* **2019**, *18*, 227. [CrossRef]
33. Guven, H.E.; Dogan, L.; Gulçelik, M.A.; Gulcelik, N.E. Adiponectin: A Predictor for Breast Cancer Survival? *Eur. J. Breast Health* **2019**, *15*, 13–17. [CrossRef]
34. Frühbeck, G.; Catalán, V.; Rodríguez, A.; Ramírez, B.; Becerril, S.; Salvador, J.; Portincasa, P.; Colina, I.; Gómez-Ambrosi, J. Involvement of the Leptin-Adiponectin Axis in Inflammation and Oxidative Stress in the Metabolic Syndrome. *Sci. Rep.* **2017**, *7*, 6619. [CrossRef]
35. Chen, S.J.; Yen, C.H.; Huang, Y.C.; Lee, B.J.; Hsia, S.; Lin, P.T. Relationships between inflammation, adiponectin, and oxidative stress in metabolic syndrome. *PLoS ONE* **2012**, *7*, e45693. [CrossRef] [PubMed]
36. Ma, L.; Xu, Y.; Zhang, Y.; Ji, T.; Li, Y. Lower Levels of Circulating Adiponectin in Elderly Patients with Metabolic Inflammatory Syndrome: A Cross-Sectional Study. *Diabetes Metab. Syndr. Obes. Targets Ther.* **2020**, *13*, 591–596. [CrossRef]
37. Behnes, M.; Brueckmann, M.; Lang, S.; Putensen, C.; Saur, J.; Borggrefe, M.; Hoffmann, U. Alterations of Adiponectin in the Course of Inflammation and Severe Sepsis. *Shock* **2012**, *38*, 243–248. [CrossRef]
38. Ekström, M.; Söderberg, S.; Tornvall, P. Acute Systemic Inflammation Is Unlikely to Affect Adiponectin and Leptin Synthesis in Humans. *Front. Cardiovasc. Med.* **2015**, *2*, 7. [CrossRef] [PubMed]
39. Takao, N.; Kurose, S.; Miyauchi, T.; Onishi, K.; Tamanoi, A.; Tsuyuguchi, R.; Fujii, A.; Yoshiuchi, S.; Takahashi, K.; Tsutsumi, H.; et al. The Relationship between Changes in Serum Myostatin and Adiponectin Levels in Patients with Obesity Undergoing a Weight Loss Program. *BMC Endocr. Disord.* **2021**, *21*, 147. [CrossRef]
40. Walowski, C.O.; Herpich, C.; Enderle, J.; Braun, W.; Both, M.; Hasler, M.; Müller, M.J.; Norman, K.; Bosy-Westphal, A. Analysis of the Adiponectin Paradox in Healthy Older People. *J. Cachexia. Sarcopenia Muscle* **2023**, *14*, 270–278. [CrossRef] [PubMed]
41. Mądra-Gackowska, K.; Szewczyk-Golec, K.; Gackowski, M.; Woźniak, A.; Kędziora-Kornatowska, K. Evaluation of Selected Parameters of Oxidative Stress and Adipokine Levels in Hospitalized Older Patients with Diverse Nutritional Status. *Antioxidants* **2023**, *12*, 569. [CrossRef]
42. Janiszewska, J.; Ostrowska, J.; Szostak-Węgierek, D. The Influence of Nutrition on Adiponectin—A Narrative Review. *Nutrients* **2021**, *13*, 1394. [CrossRef] [PubMed]
43. Febriza, A.; Ridwan, R.; As'ad, S.; Kasim, V.N.; Idrus, H.H. Adiponectin and Its Role in Inflammatory Process of Obesity. *Mol. Cell. Biomed. Sci.* **2019**, *3*, 60. [CrossRef]
44. Welsh, T.J.; Gordon, A.L.; Gladman, J.R. Comprehensive Geriatric Assessment—A Guide for the Non-specialist. *Int. J. Clin. Pract.* **2014**, *68*, 290–293. [CrossRef] [PubMed]
45. Norman, K.; Haß, U.; Pirlich, M. Malnutrition in Older Adults—Recent Advances and Remaining Challenges. *Nutrients* **2021**, *13*, 2764. [CrossRef]
46. Vellas, B.; Guigoz, Y.; Garry, P.J.; Nourhashemi, F.; Bennahum, D.; Lauque, S.; Albarede, J.-L. The Mini Nutritional Assessment (MNA) and Its Use in Grading the Nutritional State of Elderly Patients. *Nutrition* **1999**, *15*, 116–122. [CrossRef]
47. Shi, Y. Caught Red-Handed: Uric Acid Is an Agent of Inflammation. *J. Clin. Investig.* **2010**, *120*, 1809–1811. [CrossRef]
48. Ostertagová, E.; Ostertag, O.; Kováč, J. Methodology and Application of the Kruskal-Wallis Test. *Appl. Mech. Mater.* **2014**, *611*, 115–120. [CrossRef]

International Journal of *Molecular Sciences*

MDPI

Review

Inhibition of Thioredoxin-Reductase by Auranofin as a Pro-Oxidant Anticancer Strategy for Glioblastoma: In Vitro and In Vivo Studies

Nelly Chmelyuk [1,2], Maria Kordyukova [3], Maria Sorokina [1,3], Semyon Sinyavskiy [1], Valeriya Meshcheryakova [1], Vsevolod Belousov [1,3] and Tatiana Abakumova [1,*]

1 Department of Synthetic Neurotechnologies, Pirogov Russian National Research Medical University, 117997 Moscow, Russia
2 Laboratory of Biomedical nanomaterials, National Research Technological University "MISIS", Leninskiy Prospekt 4, 119049 Moscow, Russia
3 Neurotechnology Laboratory, Federal Center of Brain Research and Neurotechnologies, Federal Medical Biological Agency, 117513 Moscow, Russia
* Correspondence: sandalovato@gmail.com

Abstract: Reactive oxygen species (ROS) play a key role in cancer progression and antitumor therapy. Glioblastoma is a highly heterogeneous tumor with different cell populations exhibiting various redox statuses. Elevated ROS levels in cancer cells promote tumor growth and simultaneously make them more sensitive to anticancer drugs, but further elevation leads to cell death and apoptosis. Meanwhile, various subsets of tumor cells, such a glioblastoma stem cells (GSC) or the cells in tumor microenvironment (TME), demonstrate adaptive mechanisms to excessive ROS production by developing effective antioxidant systems such as glutathione- and thioredoxin-dependent. GSCs demonstrate higher chemoresistance and lower ROS levels than other glioma cells, while TME cells create a pro-oxidative environment and have immunosuppressive effects. Both subpopulations have become an attractive target for developing therapies. Increased expression of thioredoxin reductase (TrxR) is often associated with tumor progression and poor patient survival. Various TrxR inhibitors have been investigated as potential anticancer therapies, including nitrosoureas, flavonoids and metallic complexes. Gold derivatives are irreversible inhibitors of TrxR. Among them, auranofin (AF), a selective TrxR inhibitor, has proven its effectiveness as a drug for the treatment of rheumatoid arthritis and its efficacy as an anticancer agent has been demonstrated in preclinical studies in vitro and in vivo. However, further clinical application of AF could be challenging due to the low solubility and insufficient delivery to glioblastoma. Different delivery strategies for hydrophobic drugs could be used to increase the concentration of AF in the brain. Combining different therapeutic approaches that affect the redox status of various glioma cell populations could become a new strategy for treating brain tumor diseases.

Keywords: auranofin; glioblastoma; thioredoxin reductase; redox system; glioma stem cells

1. Redox Balance in Glioblastoma and the Role of ROS in Glioma Stem Cells and Tumor Microenvironment

Glioma is the most common primary malignant brain tumor in adults. This condition generally has a poor prognosis, especially for the most aggressive form of glioma, glioblastoma (GBM), where the median survival time after diagnosis is approximately 14 months despite standard modern treatments, including surgical tumor removal followed by radiation and chemotherapy [1]. Therefore, the development of new therapeutic strategies to treat this disease is urgently needed.

One of the key reasons for chemoresistance in GBM cells is their insusceptibility to oxidative stress caused by elevation of cellular ROS level. ROS participate in mediating the cytotoxic effects of various treatments [2]. For example, the main chemotherapeutic agent for GBM, temozolomide (TMZ), has been shown to induce ROS production in addition to its alkylating action [3–5]. Furthermore, rapid and repeated increases in ROS caused by TMZ can activate autophagy [5], apoptosis, and cell death [6].

The primary ROS in cells include superoxide radicals, hydrogen peroxide, and hydroxyl radicals. Cellular antioxidant enzymes such as superoxide dismutase, catalase, thioredoxin reductase, glutathione reductase, peroxiredoxins, and glutathione peroxidases, as well as cofactors such as nicotinamide adenine dinucleotide (NADH) and nicotinamide adenine dinucleotide phosphate (NADPH), regulate physiological ROS concentrations [7]. Notably, many cellular antioxidant systems are functionally interconnected and often compensate for one another. The balance between antioxidant system activity and intracellular ROS concentrations is referred to as the redox status of the cell [8].

Cancer cells contain higher levels of ROS than normal cells. Sources of increased ROS production in tumor cells have been attributed to oncogene- and damage-stimulated production of O_2 by mitochondria, associated with altered assembly of the electron transport chain, hypoxia; O_2 by NADPH oxidases (NOX); H_2O_2 by 5-lipoxygenase; and H_2O_2 within the endoplasmic reticulum [9]. A mild increase in ROS stimulates the proliferation of cancer cells by inhibition of PTEN [10] and MAPK phosphatases [11]. To avoid cell death due to such an increased ROS level, cancer cells activate antioxidant defense systems. Since the toxic threshold of ROS levels in cancer cells is higher than in normal cells, a further increase in ROS and/or a decrease in antioxidant proteins may be potential strategies for glioblastoma treatment [12].

One of the most critical antioxidant systems in cells is the thioredoxin system [7], which includes thioredoxin (Trx) and thioredoxin reductase (TrxR). Trx is a peptide containing three SH groups that are sensitive to oxidation. Mammalian TrxR, a selenium-containing enzyme, catalyzes the reduction of oxidized Trx. TrxR exists in three isoforms: TrxR1 in the cytoplasm and extracellular space, TrxR2 in mitochondria, and TrxR3 in the testes [13]. While this enzyme participates in numerous cellular processes, including cell growth and differentiation [7], its primary function is to protect cells from oxidative stress. Furthermore, TrxR is highly expressed in mammalian brain cells [14].

Another important antioxidant system is the glutathione system, which consists of more components than the thioredoxin system does. Glutathione reductase (GR), the functional analog of TrxR, catalyzes the reduction of oxidized glutathione. Glutathione (GSH) is a water-soluble tripeptide composed of glutamine, cysteine, and glycine. The thiol group of cysteine can act as a reducing agent in redox reactions, making GSH the most abundant intracellular thiol, with concentrations reaching several millimoles in some tissues. GSH is synthesized de novo in cells through the sequential actions of two enzymes: γ-glutamylcysteine synthetase and glutathione synthetase [7]. Another component of the glutathione system is glutaredoxin. Unlike Trx, which is reduced by TrxR, glutaredoxin is nonenzymatically reduced via glutathione oxidation, with glutathione subsequently regenerated by GR [8].

Both antioxidant systems use electrons provided by NADPH, transferring them to glutaredoxins and terminal peroxidases, such as peroxiredoxins and glutathione peroxidases, which reduce ROS to nontoxic compounds [7,15]. Numerous studies have investigated the characteristics of these antioxidant systems in tumor cells, documenting the roles of both the glutathione [16–18] and thioredoxin systems [19–21] in conferring chemotherapy resistance. It was shown that increased activity of GSH related enzymes protects cancer cells and helps them to adjust to chemotherapy. In particular, it has been demonstrated that high intracellular GSH concentration in glioma cells is associated with chemoresistance and its depletion substantially enhances the cytotoxicity of cisplatin and TMZ [22,23]. Thioredoxin system is also important for supporting chemoresistance in gliomas. For patients with high-grade gliomas, both high cytoplasmic TrxR and Trx expression are associated with poor overall survival [24]. TrxR inhibition has been shown to sensitize glioma cells to chemotherapy leading to the development of novel treatments. Recent studies have demonstrated that TrxR inhibition by novel small molecules inhibitors can enhance the effectiveness of chemotherapy [25,26].

Affecting both antioxidant systems involves the transcription factor nuclear factor erythroid 2-related 2 (NRF2) pathway which is well known as the master regulator of antioxidant response, maintaining redox homeostasis in the cells [27,28]. NRF2 controls the expression of components of glutathione and thioredoxin systems [29]. NRF2 expression has a robust negative correlation with glioblastoma patient survival rates [30]. It limits the extent of autophagy induced by TMZ and protects tumor cells from the death induced by the drug [31]. The NRF2 induces GSH synthesis as a protective mechanism during TMZ treatment, silencing NRF2 potentiated TMZ-induced cell death [32]. Genome-Wide CRISPR screen has revealed that the NRF2 pathway contributes to TMZ resistance [30]. NRF2 has also been reported to be important for maintaining the stemness of glioblastoma stem cells (GSCs), which are one of the factors that contribute to GBM resistance to therapy [33].

The main challenge of GBM treatment is the heterogeneity of cell populations within this tumor type, which requires a multidrug approach. Recent studies in new drug development have mainly focused on targeting GSCs or cells in the tumor microenvironment, such as tumor-associated macrophages (TAM), T cells, or cancer-associated fibroblasts (CAF).

GSCs is a is a unique cell population within the tumor that were first identified as a subset of cells capable of initiating tumor growth in vivo [34,35]. These cells exhibit high transcriptional plasticity, proliferation capacity, and self-renewal potential [36,37]. Moreover, compared with more differentiated tumor cells, GSCs are characterized by an enhanced ability to invade surrounding tissues and have greater resistance to radiation and chemotherapy [38,39]. Numerous mechanisms underlying the increased resistance of GSCs to therapy have been described, including their altered energy profile, which allows them to survive increased intracellular ROS concentrations [40]. GSCs, however, are better capable of maintaining lower levels of ROS than are more differentiated tumor cells, which helps preserve genome integrity and supports their self-renewal [41,42]. Consequently, standard doses of TMZ may fail to induce cytotoxic effects in GSCs, necessitating additional strategies targeting the antioxidant systems that regulate ROS levels in these cells. Moreover, GSCs exhibit greater transcriptional plasticity than other tumor cells, making them better equipped to adapt to oxidative stress. These properties make GSCs highly resistant to therapy and the primary drivers of tumor recurrence. Elevated activity of antioxidant systems has been shown in cancer stem cells. It has been demonstrated that TrxR inhibitor could effectively eradicate cancer stem cells and particularly GSC by increasing ROS [43]. So, to overcome the therapeutic resistance of GSC inhibition of key cellular antioxidant systems particularly thioredoxin system could be a promising strategy.

Another important factor in glioma chemo- and radioresistance and a potential target for developing effective therapeutic strategies is a tumor microenvironment (TME). TME plays a vital role in the survival of cancer cells and their response to therapy. TME niche is comprised of many cell types, including endothelial cells, neural cells, CAFs [44], and resident and circulating immune cells and forms an environment that promotes and supports tumorigenesis [45].

ROS have a major impact on the functioning of TME. Elevated ROS level in particular CAFs, TAMs and T cells [9], have profound effects on tumor biology and demonstrated significant immunosuppressive effects [46]. First, CAFs and TAMs themselves can contribute to the formation of a pro-oxidant environment. It has been shown that CAFs express the enzyme NOX4 and generate intracellular ROS which are also necessary for maintaining their phenotype. CAFs suppress immune response by specifically excluding CD8+ T-cells from tumors and NOX4 is necessary for this activity [47]. Also, CAFs can recruit monocytes to TME and promote their differentiation into immunosuppressive TAMs. TAM generates ROS through activation of NOX-2 and also suppress CD8+ T-cell proliferation [48]. Also, macrophages contribute to drug resistance and relapse after chemotherapy treatment by promoting tumor revascularization and activating anti-apoptotic programs in cancer cells [49]. TAM are divided into tumor-suppressive M1-like (classic activation of macrophages) and tumor-supportive M2-like (alternatively activated macrophages) polarized cells [50]. The data on the role of ROS in induction of M1-M2 macrophage phenotypes are controversial. There are some data that ROS can stimulate both activation statuses in tumor-associated macrophages [51]. Increased ROS level may participate in inducing the polarization of macrophages from the phenotype M1 to M2 except for its involvement in tumorigenesis, thus causing suppressive tumor immune microenvironment [52,53]. It has been shown that in glioblastoma elevated oxidative stress and SOD3 are associated with increased M2-like pro-tumoral macrophages [54].

Despite the differences in the functioning of the TME during oxidative stress, certain adaptive mechanisms of the TME cell populations are required to avoid excessive ROS production and cell death. It was shown that thioredoxin system can support and protect the NK and T cells from elevated ROS levels [55]. Additionally, GSH depletion has been successfully used to overcome the immunosuppressive function of regulatory T cells (T regs) [56]. Wang et al. also demonstrated that the survival benefit of TME cells is accompanied by elevated expression of Trx [57]. So, the modulation of the ROS level by inhibiting the antioxidant system in cancer could be beneficial for glioblastoma therapy. This approach could have an impact not only on the tumor cells, but also on GSCs and the TME cells. However, there are dose-dependent effects of ROS on tumor biology. While a primary increase in ROS levels can lead to cancer cell death, it can also stimulate tumor progression in GSC or immunosuppressive mechanisms in the TME cells. All aspects of the redox status of different glioma cell populations should be considered when designing a multidrug treatment approach.

2. Inhibition of TrxR in Cancer

The thioredoxin-dependent system, in particular Trx, is considered a potential target for the development of effective anticancer drugs. Many TrxR inhibitors have been discovered and developed thus far, some of which have shown pronounced antitumor effects. Because of its high reactivity, TrxR can be inhibited by many electrophilic compounds, such as nitrosoureas, flavonoids, and metal-containing compounds, which are used in clinical practice. According to previous studies, TrxR inhibitors have high potential for improving the treatment of oncological diseases. They can potentiate the action of other cytostatic agents and overcome drug resistance in tumors, making them highly effective in combination therapy.

Alkylating agents—nitrosoureas—are irreversible inhibitors of TrxR. The nitrosoureas carmustine, lomustine and fotemustine covalently deactivate thioredoxin reductase, glutathione reductase and ribonucleotide reductase by alkylating their thiolate active sites [58,

59]. Nitrogen mustards (chlorambucil and melphalan), cyclophosphamide [60,61] and alkyl sulfonates (busulfan) efficiently inhibit TrxR but do not inhibit glutathione reductase [59].

Arsenic trioxide is another irreversible inhibitor of TrxR with a mechanism involving both the C-terminal and N-terminal redox active sites of the enzyme [62]. Arsenic trioxide is a first-line treatment for acute promyelocytic leukemia and is also effective for other types of leukemia [63]. Arsenic trioxide has been shown to have anticancer activity in gliomas [64], breast cancer [65], hepatocellular carcinoma [66], lung cancer [67] and pancreatic cancer [68]. Phase I and pharmacodynamic study of arsenic trioxide showed promising results for patients with newly diagnosed GBM. The median overall survival for all patients in this trial was 17.7 months vs. 12 months for patients without the use of concurrent and adjuvant temozolomide [69].

Motexafin gadolinium (MGd) is a member of a class of rationally designed porphyrin-like molecules called texaphyrins. MGd preferentially spreads to cancer cells in animal models and in human clinical trials [70,71]. MGd acts as a substrate of cytosolic TrxR, and MGd and NADPH react in the presence of TrxR, producing ROS, which generate oxidative stress in cells [72]. In phase III clinical trials, MGd significantly prolonged the interval to neurologic progression in non–small-cell lung cancer patients with brain metastases receiving prompt whole-brain radiation therapy [73].

The natural compounds that exhibit inhibitory activity toward TrxR1 can be grouped as follows: phenylpropanoids, polyphenols, quinones, terpenoids, and chromones [74]. Curcumine—one of the most studied natural TrxR inhibitors causes alkylation of both residues in the catalytically active site (Cys(496)/Sec(497)) of TrxR [75]. It induces ROS production and elevates oxidative stress in cells [76]. These compound have an antitumor effect on xenografts in vivo [77].

Piperlongumine is a compound derived from long pepper plants. Piperlongumine inhibits cell growth and induces ROS accumulation in hepatocellular carcinoma cells. It induces a lethal endoplasmic reticulum stress response in HCC cells by targeting TrxR1 and increasing intracellular ROS levels [78].

Platinum compounds are potent TrxR inhibitors. They induce covalently stable intracellular TrxR1–Trx1 and TrxR1–TRP14 complexes [79]. TrxR, but not GR, is efficiently inhibited by cisplatin, carboplatin and oxaliplatin [59,80]. However, platinum-based compounds are only effective in a limited number of cancer types and have severe adverse effects [81]. Therefore, other metallodrugs, such as gold, mercury, copper and silver have been extensively studied.

Mercury (Hg) compounds such as thimerosal are rather quite effective because of their high affinity for binding to thiols and selenols. TrxR is highly sensitive to Hg inhibition because of the reactivity and position of the Sec residue in the open C-terminus of the TrxR active site [82,83].

Numerous gold (I) derivatives have been considered TrxR inhibitors including Au(I) complexes with phosphines (auranofin, triphenyl phosphine gold chloride, among others [84–86]), phospholes [87], thiomalate [84], thiosulfate [84], N-heterocyclic carbenes [88–90], azolates and phosphanes [91,92].

Gold (I) complexes sometimes also inhibit glutathione reductase [87] and show non-specific reactions with thiols [93,94], which leads to cytotoxicity to normal cells.

Au(I) N-heterocyclic carbene (NHC) complexes are more cytotoxic to cancer cells than to normal cells and do not inhibit glutathione reductase [95]. A series of caffeine-based gold(I) NHC complexes have been synthesized and tested for their antiproliferative activities in different cancerous and nontumorigenic cell lines in vitro. The bis-carbene complex [Au(caffein-2-ylidene)2][BF4] appeared to be selective for human ovarian cancer cell lines and poorly toxic in healthy organs [88].

A bis-chelated tetrahedral gold(I) phosphine complex [Au(d2pype)2]Cl (where d2pype is 1,2-bis(di-2-pyridylphosphino)ethane) was designed to improve the gold(I) compound selectivity toward selenol- and thiol-containing proteins, such as TrxR. This compound is stable in the presence of serum proteins, thiols, or disulfides and exhibits high stability in solution [96]. The related bis-chelated Au(I) bidentate phosphine complex, 1,3-bis(di-2-pyridylphosphino)propane (d2pypp), also has unusually low thiol reactivity. It is selectively toxic to breast cancer cells but not normal breast cells, resulting in apoptotic programmed cell death [Au(d2pypp)2]Cl inhibits Trx and TrxR activity more in cancer cells than in normal cells. This difference appears to be in part a consequence of the increased uptake of the complex into cancer cells, particularly into their mitochondria [85]. It has been shown to exhibit anticancer effects against lymphoma [97], multiple myeloma [98] and chronic myeloid leukemia [99].

A series of water-soluble gold(I) compounds have been developed. Gold (I) complexes bearing water-soluble phosphine ligands, including 1,3,5-triaza-7-phosphaadamantane (PTA), 3,7-diacetyl-1,3,7-triaza-5-phosphabicyclo[3.3.1]nonane (DAPTA), and sodium triphenylphosphine trisulfonate (TPPTS), efficiently inhibited TrxRs at concentrations that did not affect glutathione reductase. These compounds are cytotoxic against cancer cell lines, particularly the cisplatin-resistant cell line [100]. Azolate gold (I) phosphane compounds have deactivating groups on the azole rings, which increases the solubility of these complexes in aqueous solutions. This gold (I) compound also has high cytotoxic activity for several human cancer cell lines in the low micromolar range and has significant selectivity for TrxR inhibition over GR inhibition [91]. Among other anti-TrxR inhibitors, AF plays a pivotal role in anticancer studies. It selectively inhibits Trx and TrxR and disrupts redox homeostasis leading to apoptosis in cancer cells.

3. The Anticancer Activity of AF

Numerous studies have demonstrated the anti-inflammatory, immunomodulatory, and antibacterial effects of AF. However, several studies have also highlighted the antitumor effects of AF in various models, both as monotherapy and in combination with other known antitumor agents. A few examples are presented in Table 1. Research papers have explored AF for the treatment of several types of blood cancer cells, including multiple myeloma [101], chronic lymphocytic leukemia [102], and chronic myeloid leukemia [103]. There is a constant search for new approaches to treating breast cancer (BC) due to its prevalence and rapid development of tumor resistance, which is characteristic of triple-negative BC (in which there is an absence of estrogen receptor, progesterone receptor, and receptor tyrosine-protein kinase gene expression). These tumors are not amenable to hormonal therapy, making it essential to suppress the development of drug resistance by identifying novel therapeutic strategies. One promising approach involves the use of a combination of antitumor drugs or mechanisms of action. For example, the combination of AF with antibodies targeting programmed death-ligand 1 (PD-L1) has been shown to significantly slow tumor growth [104]. For colon cancer therapy, AF (0.5 μM) was combined with 5Z-7-oxozeaenol (5 μM), a molecule capable of inhibiting transforming growth factor β-activated kinase 1 (TAK1). This combination led to cell death in colon carcinoma cells both in vitro and in vivo [105]. The activity of AF has also been reported in other cancer types, both in combination with known drugs and as monotherapy [106]. Of course, the most impressive results are from in vivo studies. Evaluating a series of data on the results of therapy in various tumor models, the following trends can be noted: the most effective therapy is therapy at doses ranging from 5 to 15 mg/kg with daily administration for 2 weeks. These methods make it possible to significantly reduce tumor growth and increase the survival of animals.

Table 1. Potent anticancer activity of AF in preclinical tumor models. Abbreviations: i.p. (intraperitoneal injection), i.v. (intravenous injection), p.o. (per os, oral administration).

Compound	Solvent (for AF)	Model	Cells	Treatment Protocol (Groups and Doses)	Effect of Tumor Growth Inhibition	Mechanism	Ref.
AF	(1) 4% (*v*/*v*) DMSO/10% (*v*/*v*) ethanol (2) 50% (*v*/*v*) DMSO, 40% (*v*/*v*) PEG300 and 10% (*v*/*v*) absolute ethanol	C57BL/6J and 129S2/SvPasCrl	SB28 and 344SQ	Doses: 2, 5, 10 and 15 mg/kg (1) i.p. injections over a period of 14 consecutive days; (2) s.c. via osmotic minipumps for 14 days; (3) p.o. for a period of 14 days	Daily i.p. injections of 10 mg/kg AF induced weight loss and gastro-intestinal problems. High doses of AF were able to inhibit TrxR activity in SB28 tumors after 14 days. A solvent consisting of 50% DMSO, 40% PEG300 and 10% ethanol provided optimal solubility of AF for p.o. administration to mice	inhibited TrxR activity	[107]
AF	Non-metioned	Athymic nude mice (Foxn1nu)	LN229, U251 and P3	AF (5 or 10 mg/kg/day) or PBS, p.o. starting at day 7 after implantation	AF inhibited GBM progression in vivo	Inhibited SC35 agglomerates, prevented NONO; binding to pre-mRNA and promoted the degradation of NONO protein	[108]
AF	DMSO and corn oil (1:4 ratio).	BALB/c nu/nu mice	TGS-01	Nigericin (4 mg/kg/day, i.p. injection, every 2 days) or AF (12 mg/kg/day for 2 days, i.p. injection)	AF inhibited GBM progression in vivo	Decreased ATP levels, induced abnormality in mitochondrial membrane potential	[109]
AF and AF with saracatinib	DMSO	Pten; Trp53; NOD-SCID	Human GBM	12 mg/kg by i.p. injection, 6 days per week and or AF + saracatinib (12 mg/kg by i.p. injection 6 days/week; 17.5 mg/kg by p.o. 5 days/week, respectively)	Increased median survival compared to the placebo group	Inhibition PKCι	[110]

Table 1. *Cont.*

Compound	Solvent (for AF)	Model	Cells	Treatment Protocol (Groups and Doses)	Effect of Tumor Growth Inhibition	Mechanism	Ref.
AF	5% DMSO, 10% Cremophor EL, 12.5% PEG, and 15% Ethyl Alcohol, and 57.5% H_2O	BALB/c nude mice	MiaPaCa-2	5–15 mg/kg AF, i.p. injection, 5 times per week for 21 days	15 mg/kg is the optimal dose: median survival increased to 12 days (versus 8 days in control); suppression of gross abdominal metastasis and a low occurrence of ascites	TrxR1 inhibition	[111]
AF and adriamycin (ADM)	DMSO	BALB/c nude mice	A549	(1) 10 mg/kg AF, i.p. injection, five times per week for 6 weeks, (2) ADM 5 mg/kg, i.v. injection, once per week for 6 weeks. (3) combination of AF and ADM	treatment with AF or ADM alone exhibited a moderate inhibitory effect on tumor growth. Combination of both drugs resulted in a significant decrease in tumor growth.	ROS increase, consistent with inhibition of TrxRs, inhibition of glycolytic hexokinase	[112]
AF and anti-PD-1 antibody	DMSO	BALB/c nude mice (human JeKo-1) and immunocompetent Balb/c (A20)	human JeKo-1 and A20	Jeko-1 model: 10 mg/kg AF, i.p. injection, 5 times per week. A20 model: (1) control, (2) anti-PD-1 antibody, (100 µg/mouse, i.p. every 5 days); (3) AF (10 mg/kg/day, i.p.); (4) combination of AF and anti-PD1 antibody	AF and anti-PD-1 antibody induced significant apoptosis in lymphoma tissues	Elevated ROS prouction, inhibiting ATP generation and induction of PD-L1 expression	[113]
AF	DMSO	BALB/c nude mice	NCI-H460	10 mg/kg AF, i.p. injection, every 2 days for 14 days	Tumor growth inhibition; AF increased the expression of apoptosis-related proteins	TrxR1 expression inhibition	[114]
AF	DMSO	BALB/c nude mice	PSN1	12.5 mg/kg AF, i.p. injection, 5 times weekly for 5 weeks	AF suppressed tumor growth of ~40% of vehicle control	Inhibited TrxR activity	[115]
AF and piper-longumine (PL)	DMSO	BALB/c nude mice	SGC-7901	i.p. injection once per day, groups: (1) 4 mg/kg PL, (2) 2 mg/kg AF (3) combination AF and PL (total 14 days)	combined treatment with AF and PL significantly inhibited tumor volume and tumor weight	Inhibited TrxR activity	[116]

Table 1. *Cont.*

Compound	Solvent (for AF)	Model	Cells	Treatment Protocol (Groups and Doses)	Effect of Tumor Growth Inhibition	Mechanism	Ref.
AF and cold atmospheric plasma (CAT)	DMSO	C57BL/6J mice	SB28 glioma	(1) 15 mg/kg AF, p.o., daily for 14 days (2) CAT for 5 days (3) combination of AF and CAT	the sequential combination regimen resulted in a significantly decreased tumor volume and significantly increased survival of the SB28-bearing mice	Inhibited TrxR activity	[117]
AF and celecoxib (CE)	olive oil	BALB/c nude mice	DLD-1	Six groups of mice were treated as follows (p.o., 30 days): (1) solvent control (olive oil); (2) AF 10 mg/kg; (3) CE 20 mg/kg; (4) CE 60 mg/kg; (5) AF 10 mg/kg + CE 20 mg/kg; (6) AF 10 mg/kg + CE 60 mg/kg.	The combination of AF and CE showed significant inhibition of tumor growth	Inhibition of hexokinase enzyme activity; increase in ROS accumulation	[118]
AF, carboplatin and buthionine-sulfoximine (BSO)	ethanol and Cremaphor EL	BALB/c nude mice	A549; H292	(1) 450 mg/kg BSO (2) 1.6 mg/kg AF (3) combination of AF and BSO i.p. daily for two weeks. (4) combination AF, BSO and carboplatin 5 days a week for 2 weeks.	Au+BSO+carbo resulted in a highly significant decrease in tumor growth rate when compared to control, carboplatin or combination of AF+BSO	inhibited TrxR activity	[119]
AF and cisplatin	DMSO	BALB/c nude mice	H69	Mice treated by i.p. injection every two days for 28 days with (1) DMSO as control, (2) AF (10 mg/kg), (3) cisplatin (2 mg/kg) and (4) combination of AF with cisplatin	The combination treatment with AF and cisplatin significantly inhibited tumor volume of H69 xenografts as compared with vehicle treatment. AF or cisplatin single treatment did not significantly inhibit tumor growth.	Enhanced cisplatin-induced S-phase cell cycle arrest in cisplatin-resistant H69 and H196 cells; increase of ROS accumulation	[120]

Table 1. *Cont.*

Compound	Solvent (for AF)	Model	Cells	Treatment Protocol (Groups and Doses)	Effect of Tumor Growth Inhibition	Mechanism	Ref.
AF and ganetespib	20% Cremophor RH40 and 80% D5W i.v. injection and 0.5% hydroxypropyl methylcellulose in 5% dextrose	BALB/c nude mice	A673	(1) 12 mg/kg AF, i.p. injection, 5 days per week, (2) 150 mg/kg ganetespib, i.v. injection, and (3) combination of AF and ganetispib once weekly for up to 18 days	The survival rate in the combination group was nearly doubled in this extremely aggressive EWS tumor animal model compared to the control group	Inhibited TrxR activities	[121]
AF and 5Z-7-oxozeaenol (TAK1 inhibitor)	sterile 2.5% DMSO in vegetable oil	BALB/c nude mice	SW 620	(1) vehicle control (2) 1.6 mg/kg AF (3) 5Z-7-oxozeaenol 15 mg/kg, (4) combination of AF and 5Z-7-oxozeaenol, The drugs were i.p. injected for five days, followed by 2 days of AF only, and then four days of combination treatment.	Auranofin alone did not significantly slow tumor growth as compared to the control group. The combination of 5Z-7-oxozeaenol plus AF showed the slowest tumor growth, however this did not reach statistical significance as compared to TAK1 inhibitor alone	Inhibited TrxR activities	[105]
AF and 5-fluorouracil (5-FU)	DMSO	BALB/c nude mice	SW620/5-FU	(1) 5-FU: 50 mg/kg, once every 4 days, i.p.; (2) 6 mg/kg AF, daily, intragastric (3) combination of FU + AF for 24 days	Animals in the 5-FU+AF treatment group had fewer lung metastatic nodules than in the 5-FU or AF groups, less necrosis was observed in the drug combination group than in either of the single-drug groups.	Inhibited TrxR activities and FoxO3-promotion	[122]
AF and selenocystine (SeC)	DMSO	BALB/c nude mice	A549	(1) 5 mg/kg SeC (2) 2 mg/kg AF (3) combination of SeC and AF, i.v. injection, daily for 16 days	Combined treatment with SeC and AF significantly inhibited the tumor weight and tumor volume but did not affect the body weight of mice.	inhibited TrxR activities; induction of mitochondria-mediated apoptosis confirmed by caspases activities	[123]

Table 1. *Cont.*

Compound	Solvent (for AF)	Model	Cells	Treatment Protocol (Groups and Doses)	Effect of Tumor Growth Inhibition	Mechanism	Ref.
AF	2% DMSO, 8.5% ethanol, and 5% PEG-400	BALB/c nude mice	Calu3	(1) AF (10 mg/kg/day) and (2) solvent, i.p. injection, daily for 14 days	Treatment with AF resulted in significant growth suppression of Calu3 tumors and led to 67% inhibition of tumor growth compared with control. No weight loss was detected.	PI3K/Akt/mTOR pathway and NRF2-mediated oxidative stress response inhibition	[124]
AF	DMSO (10%)	BALB/c mice	CT26	AF 10 mg/kg/day i.p. 3 times a week for 3 weeks	The treatment reduced tumor growth and preserved body weight in mice. Also, administration of auranofin before irradiation enhanced the radiation response of colon tumors while providing radioprotection to normal tissues, including the small intestine.	Activated the p53/p21 pathway	[125]
AF	DMSO (5 mmol/L)	TCL-1 transgenic mice (Tcl1-tg:p53−/−)	CLL (chronic lymphocytic leukemia) cells	AF 10 mg/kg i.p. daily, 5 times per week for 2 weeks	Treatment with AF resulted in a reduction in leukemia cell burden in every TCL-1 mouse tested and improved mouse survival	Inhibited TrxR activity	[126]

No more than 30% of the papers are devoted to GBM therapy by AF in Table 1. However, AF remains a promising drug for glioma therapy, which is confirmed by a significant number of in vitro studies on a variety of cell lines. For example, Ferraz et al. examined in detail the effect of gold and platinum complexes, where AF demonstrated one of the best abilities to inhibit TrxR and has a high cytotoxicity effect on the cells (U87, T-98, MRC-5) [127]. In another work, it was shown that the pro-oxidant Trx and/or GSH systems in U87MG and T98G cells were inhibited by AF, which caused increasing ROS, activating p53 and cell death [128]. The same results were demonstrated by Szeliga et al. on the U87MG, LN229, U87MG, LUB17, and LUB20 cells [129] and by Martinez-Jaramillo on the U87MG [130].

Notably, AF is being studied separately as an antitumor therapeutic agent and is currently in clinical trials. There are a number of clinical trials where AF was used as a potential drug for ovarian cancer (NCT03456700, NCT01747798), chronic lymphocytic leukemia (NCT01419691), non-small cell lung cancer (NCT02126527, NCT01737502), but the results of the clinical trials have not been posted. Furthermore, AF is being tested in combination with TMZ for the treatment of recurrent glioblastoma. This treatment protocol is called CUSP9v3, and in addition to AF, it includes eight other repurposed drugs in combination with TMZ. According to the CUSP9v3 protocol, after enrollment, the participant goes into the induction cycle, which lasts 35 days. The induction cycle consists of a drug-by-drug addition and up-dosing process. Hereafter, the subject will enter the treatment cycles (up to 12). During the induction cycle and the first 2 treatment cycles, regimen adjustments (dropping of certain drugs, dose modification of certain drugs) may be executed to accommodate to the patients' individual toxicity reactions that may occur during this period. At present, the detailed results of the clinical trial (NCT02770378) of the CUSP9v3 protocol have not been posted. Although it is not the highest anti-cancer efficacy of AF, it has been shown to be a safe drug. In addition, AF can inhibit the Trx system in GBM cells, which makes AF a promising drug for GBM therapy in the near future.

4. AF Inhibits Thioredoxin Reductase Followed by the Activation of Other Pathways

AF (Figure 1), which first demonstrated clinical efficacy in 1976 [131], belongs to the group of gold-containing disease-modifying antirheumatic drugs.

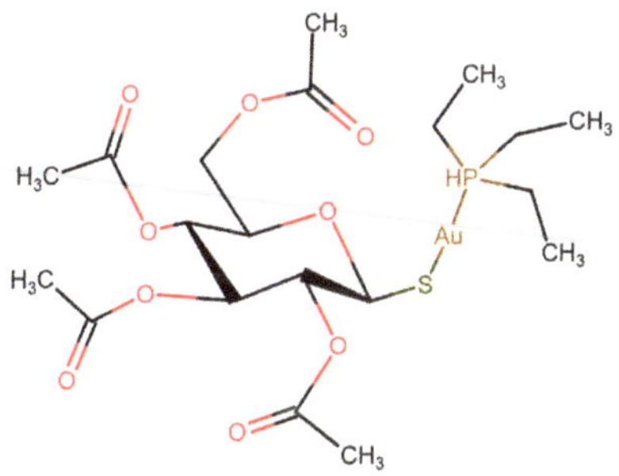

Figure 1. Chemical structure of auranofin.

The direct inhibition of TrxR by AF was first demonstrated by Stefan Gromer (Heidelberg University, Germany) and coauthors in 1998, and this mechanism is now widely accepted [132]. TrxR reduces Trx, a 12-kDa protein with a disulfide bond as its core functional group. The reduced form of Trx [Trx-(SH)2] restores oxidized protein substrates, which typically contain disulfide groups. The oxidized Trx form (Trx-[SS]) is regenerated by TrxR in an NADPH-dependent manner [133,134].

The thioredoxin reductase system, which includes Trx and TrxR plays a role in numerous biological processes, including redox homeostasis, antioxidant defense, transcription factor regulation, and cellular proliferation and division [135]. The inhibition of TrxR1, which leads to increased ROS levels, is currently recognized as the primary mechanism of AF cytotoxicity [106]. In vitro studies have also shown that AF inhibits other thioredoxin reductases, such as TrxR2 and TrxR3 [136,137].

Researchers have discovered that TrxR provides protective effects against various cellular stresses, including growth inhibition and cell death induced by hydrogen peroxide, tumor necrosis factor-α, and chemotherapeutic agents [138–140]. The literature indicates that not all cancer tumors exhibit the same level of TrxR expression, affecting their sensitivity to AF. For example, cisplatin-resistant human bladder cancer cells and PC-3 prostate cancer cells are characterized by elevated TrxR expression levels [138,141,142].

Conversely, tumor cells with high ROS levels may be more vulnerable to cell death when antioxidant systems are inhibited [143]. Therefore, a potential cancer treatment strategy involves selectively disrupting the redox balance of pathological cells by suppressing their antioxidant systems while sparing healthy cells [144].

Research, including bioinformatics approaches, is ongoing to identify alternative mechanisms of action of AF. For example, researchers from Peking University have shown in vitro and in vivo that AF inhibits the catalytic activity of TET1 (ten-eleven translocation 1). TET1 is a key epigenetic regulator of DNA that catalyzes the conversion of 5-methylcytosine (5mC) to 5-hydroxymethylcytosine (5hmC) to modulate gene expression. Analysis of public datasets (GSE5820) revealed that TET1 expression is elevated in patients with T-cell acute lymphoblastic leukemia (T-ALL) and serves as a biomarker for decreased survival. The competitive binding of AF to the substrate pocket of TET1 reduces 5hmC levels, leading to epigenetic reprogramming of the oncogene c-Myc. Thus, AF was shown to inhibit TET1 in T-ALL models via the TET1/5hmC/c-Myc signaling pathway [145].

Using a large transcriptomic dataset (9 cell lines: A375, A549, HCC515, HEPG2, HT29, MCF7, PC3, HA1E, and VCAP before and after AF exposure) and Clue software (clue.io/cmap, [146]), 978 key genes were identified that characterize the inhibitory effects on TrxR. Gene expression changes were compared to all available gene signatures in the database (over 1 million profiles), generating a "tau ball" that predicted the effect.

NF-κB inhibition showed the greatest predictive power. In 2000, Jeon et al. demonstrated that NF-κB is inhibited by 5–10 μM AF over 4 h in macrophages stimulated with lipopolysaccharide, blocking IκB kinase activity [147]. Nakaya et al. reported that 0.05 μM AF inhibits NF-κB DNA binding and reduces nuclear NF-κB levels in U266 multiple myeloma cells [101]. AF is thought to bind to the Cys179 residue of IκB kinase, blocking its activity. IκB kinase phosphorylates IκB, which binds to NF-κB, causing dissociation of the IκB + NF-κB complex. Once dissociated, phosphorylated IκB detaches from NF-κB, which migrates to the nucleus to activate transcription [148].

Interestingly, IκB is not the only substrate of IκB kinase. Hu et al. reported that it also phosphorylates the tumor suppressor FOXO3, which becomes inactive upon phosphorylation and translocates to the cytosol [149]. Thus, AF-induced IκB kinase inhibition may produce various responses, with cytotoxicity potentially dependent on specific cell types and tumor microenvironments in vivo.

Another proposed mechanism based on bioinformatics analysis involves the inhibition of deubiquitinating enzymes (DUBs), which remove ubiquitin degradation tags from proteins, preventing their proteasomal degradation. AF inhibits DUBs (UCHL5 and USP14) in the 16S proteasome regulatory subunit at 0.5 μM over 3 h, as demonstrated in HEPG2 and MCF7 cells [150]. These findings were later confirmed in prostate cancer models in vitro and in vivo [151]. However, recent proteomic research suggests that AF-dependent

DUB and proteasome inhibition occur only at high doses (>2.5 μM) and may represent off-target effects [152].

In 2013, Wang et al. proposed a direct interaction between AF and protein kinase C-iota (PKCiota), as AF analogs (aurothiomalate and aurothioglucose) inhibit this enzyme by binding critical cysteine residues [153,154]. PKCiota mRNA levels are elevated in various cancers, but the inhibitory effect of AF on this kinase requires further study [155].

Hou et al. demonstrated that AF inhibits hexokinase at 4–6 μM, reducing glucose uptake, lactate production, and ATP synthesis in vivo [112]. Since hexokinase is a key glycolytic enzyme that is overexpressed in many cancers, its inhibition could decrease ATP production and impair the ABCG2 pump, which contributes to multidrug resistance in chemotherapy [156]. The targeting of hexokinase is considered a promising anticancer strategy, but further data on the role of AF in this process are needed.

In conclusion, on the basis of research conducted in various cancer models, several mechanisms, including dose-dependent effects, have been proposed to explain the antitumor activity of AF. The main direct and indirect targets of AF are summarized in Figure 2 [106].

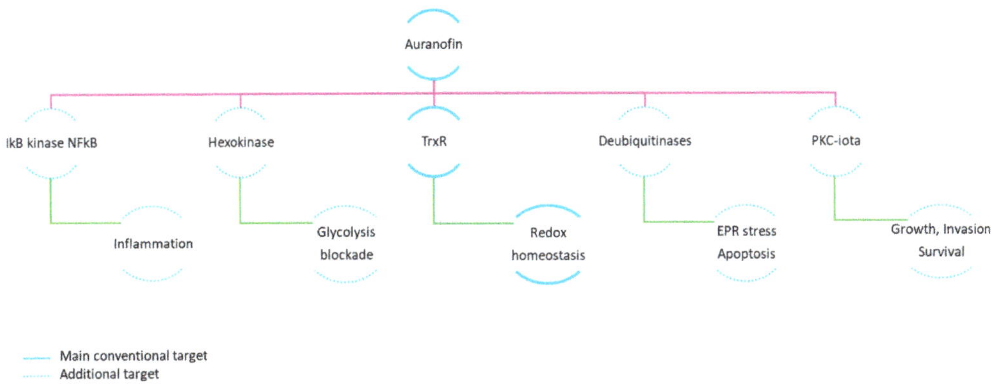

Figure 2. Summary of the main molecular targets affected by auranofin. Abbreviations: NFkB—Nuclear factor kappa-light-chain-enhancer of activated B cells; PKC-iota—Protein kinase C iota; TrxR—thioredoxin reductase; EPR—endoplasmic reticulum. Solid line stands for the conventional mechanism, dotted for additional.

5. Formulation and Effective Delivery of TrxR Inhibitors (AF) for Brain Tumors

The development of new drug forms for glioma treatment remains a complex challenge. There are three main issues associated with new drugs: stability, solubility and delivery.

An AF, an inhibitor of TrxR, is soluble in organic solvents such as ethanol and DMSO but insoluble in aqueous solutions, which can hinder its bioavailability in vivo and lead to complex formulation protocols [100]. The formulation of AFs is critical for treating brain pathologies, as it has another drawback: it reacts nonspecifically with protein thiols [93,94], which can make it toxic to normal cells. To address this issue, several approaches can be employed, such as increasing solubility by solubilizing agents or encapsulating the active pharmaceutical ingredient (API) in nanoparticles. Encapsulating AF into nanoparticles or combining it with other pharmaceutical excipients can minimize its potential adverse effects, preserve its interaction with serum proteins, and result in greater stability. This approach reduces the amount of free drug circulating in the body, thereby preventing unwanted tissue delivery and improving therapeutic efficacy [157].

The first approach (using solubilizing agents) involves the use of different expedients to improve the stability and solubility of AF. Typical excipients for improving the solubility of poorly soluble drugs for oral and intravenous administration include the following classes of compounds [158]: (1) water-soluble organic solvents (PEG 300 or 400, ethanol, propylene glycol, glycerol, N-methyl-2-pyrrolidone, dimethylacetamide and DMSO); (2) nonionic surfactants (Cremophor EL or RH 40 or RH 60, polysorbate 20 or 80, poloxamer 407 and mono- and differential acid esters of PEG); (3) water-insoluble lipids (castor oil, corn oil, olive oil, etc.); (4) organic liquids/semicolids (beeswax, d-α-tocopherol, oleic acid, medium-chain mono- and diglycerides), and (5) various cyclodextrins (β-cyclodextrin, α-cyclodextrin, hydroxypropyl-β-cyclodextrin). However, at present, few studies have described drug forms based on AF via the abovementioned substances for i.v. injection.

Nanoparticles are another form of excipients that should protect the API from degradation, ensure controlled release, enhance solubility, and reduce overall toxicity. Over the past 20 years, various drug delivery methods, including gels, micelles, polymer particles, PLGA particles, and inorganic particles (e.g., transition metal oxides or noble metals), have been proposed. These approaches enable the delivery of small molecules such as cytostatic or larger entities such as proteins, antibodies, RNA, or viruses. Owing to their relatively simple synthesis and availability of clinical trial data, liposomes and polymer nanoparticles are the most common carriers. Other promising delivery systems include gold, magnetic, and mesoporous silica nanoparticles. Magnetic and gold nanoparticles rely on coatings for controlled drug release, whereas mesoporous silica nanoparticles utilize their nanopores for controlled release. The addition of vector molecules and surface chemistry modifications further enhances penetration and therapeutic outcomes, as demonstrated in studies using magnetic nanoparticles loaded with TMZ [159]. However, the most popular method for brain delivery is the encapsulation of API into cationic liposomes [160].

There are several examples of formulations of AF using nanoparticles. One interesting example of nanoparticle use involves chitosan-PEG copolymers. Maame Abena O. Afrifa et al. demonstrated improved solubility and greater efficacy at lower doses of AF (3 mg/kg vs. 5 mg/kg) for the treatment of triple-negative breast cancer [161]. Another study by Marta Pérez-Lloret showed that silk fibroin nanoparticles loaded with AF improved its bioavailability, cellular absorption, and resistance to degradation, successfully inhibiting the growth of 3D colorectal carcinoma cultures [162]. AF also exhibit antimicrobial activity. Encapsulating it in PLGA nanoparticles enhanced its bactericidal effects, virtually sterilizing multiresistant pneumococcal strains (e.g., *Streptococcus pneumoniae* and *Streptococcus pyogenes*) after six hours of treatment at 0.25 μM. This potent effect was also observed in biofilms, where AF-NPs reduced bacterial populations by approximately four logarithms more than free AF did [163]. Interestingly, PLGA nanoparticles loaded with AF have also been used as neuroprotective agents for Alzheimer's disease treatment. These nanoparticles significantly reversed cognitive deficits, biochemical changes, neuroinflammatory markers, and neurotransmitter imbalances in a streptozotocin-induced model [164]. Notably, similar particles loaded with other drugs have already been delivered to the brain, which points to the possibility of delivering AF to the brain tumors in the form of a nanoparticle formulation. However, several approaches can improve the delivery of nanoparticles.

Although AF and AF-loaded nanoparticles have shown significant efficacy in preclinical studies on non-brain tumors, there are still limited data regarding their effectiveness in GBM studies. AF has already been shown to be effective as a treatment for rheumatoid arthritis, but the anticancer activity of AF requires higher doses. A major challenge in developing new drugs for GBM is achieving sufficient therapeutic concentrations in the brain, which is often hindered by the low solubility of the drugs and the limited BBB permeability. To achieve the maximum concentration, several approaches could be used: (1) the addition of lipid-like molecules through modification of the hydrophilic groups on the drug structure [165]; (2) the addition of hydrophobic groups to the drug molecule [166]; (3) active targeting; (4) opening of the BBB (FUS, osmotic opening); and (5) intraarterial/intrathecal administration.

In the first case, drug permeability across the BBB is favored by their lipophilicity, low molecular weight, and lack of ionization at physiological pH. Lipid-soluble molecules with a molecular weight of <500 Da may cross the BBB through the small pores that form transiently within the lipid bilayer [165]. An example of this type of modification is the addition of methyl groups to various drugs in the barbiturate class, which results in increased lipophilicity and brain penetration in animals [166]. A more recent approach is to enclose already known API in liposomes (mostly cationic). For example, improved drug delivery has been demonstrated for encapsulated doxorubicin in the treatment of experimental brain tumors [167].

The use of targeting moieties such as transferrin, insulin, apolipoproteins, and antibodies attached to polymeric nanoparticles for maximum accumulation has been demonstrated [168]. For example, transferrin strongly influences the binding of magnetic PLGA particles to glioma cells, and the results of in vivo experiments revealed that particles with transferrin loaded with doxorubicin and paclitaxel significantly suppress tumor growth [169]. Increased permeability and improved therapeutic efficacy due to the presence of transferrin on the nanoparticles have also been reported in other studies. Thus, the efficacy of zoledronic acid in the treatment of GBM was increased by encapsulating it in self-assembling nanoparticles with transferrin, which allowed the drug to have improved solubility and imparted the ability to penetrate the BBB [170].

While active targeting could enhance the accumulation of drugs in brain pathology, it also presents some challenges (difficulties in manufacturing and scaling). Other methods to increase the therapeutic concentration of a drug in brain tissue include temporary BBB opening via clinically approved focused ultrasound (FUS) [171] or osmotic opening (mannitol) [172]. Temporal BBB opening after exposure to ultrasound is possible after the injection of gas-filled microbubbles (e.g., sulfur hexafluoride). The use of microbubbles safely opens the BBB [173] and the absence or minimal damage to nervous tissue [174]. To date, more than 10 microbubbles have been registered for clinical practice, and several clinical trials have focused on the efficacy of drug therapy via focused ultrasound to improve the penetration of drugs (TMZ, doxorubicin, carboplatin, etc.) into brain tumors [175]. Additionally, this approach has been shown to be effective for the delivery of nanoparticles in preclinical trials. Thus, exposure to ultrasound leads to a 6–40-fold increase in the accumulation of particles with cisplatin (60-nm particles) [176], a 1.4–6.9-fold increase in the accumulation of particles with paclitaxel (170-nm particles) [177], and a 2-fold increase in the accumulation of nanoparticles with carmustine (10–20-nm particles).

In conclusion, there are various approaches thar could improve the efficacy of AF in clinical application. There are numerous methods for enhancing the delivery of AF across the BBB for the treatment of GBM. These methods can be categorized into two types: first, improving the formulation of the substance itself, either by adding specific functional groups or encapsulating it in various novel drug forms; second, temporarily increasing the permeability of the BBB through artificial means. Also, the antitumor efficacy of AF could be improved by modulation of dynamic thiol exchange of gold ions, which leads to a fast formation of covalent albumin-gold adducts. Researchers from Sun Yat-Sen University report that AF supplemented with its own thiol ligand, TGTA (1-thio-β-D-glucose tetraacetate) significantly restored anticancer activity in cells and patient-derived xenograft models [178]. Screening a library of ligand fragments and conducting machine learning analysis afterward may be a promising strategy for enhancing the effectiveness of AF even at a reduced dosage. In addition, using AF to sensitize cancer cells to other approved drugs or in combination with other anti-cancer therapies could be a promising approach. Combination therapy with AF demonstrated good results in therapy, including GBM [109,110]. Taken together, these approaches provide hope for the potential success of GBM treatment via AF, as AF has excellent potential to inhibit TrxR.

6. Conclusions

GBM is a challenging disease that has various cell populations with different redox statuses. GSCs and TME cells have become attractive targets for developing anticancer therapies while playing a key role in tumor relapse. These cell populations benefit from elevated ROS production and develop adaptive mechanisms to survive through high ROS level, such as strong antioxidant systems. One approach to targeting cancer cells is by suppressing their antioxidant system in the cell by inhibiting TrxR. AF has shown promising results as a selective inhibitor of TrxR, demonstrating anticancer activity against various tumor types, including GBM. However, there are some challenges in further clinical application for GBM that should be overcome, such as the low solubility of AF and poor delivery. While its solubility in water is limited, studies have explored the development of new formulations of AF suitable for intravenous administration. These formulations could potentially allow AF to cross the blood-brain barrier and be used in the treatment of GBM in the future.

Author Contributions: Conceptualization, V.B. and T.A.; writing—original draft preparation, N.C., M.K., M.S., S.S., V.M. and T.A.; writing—review and editing, V.B.; visualization, S.S.; supervision, V.B. and T.A.; project administration, T.A.; funding acquisition, T.A. All authors have read and agreed to the published version of the manuscript.

Funding: This research was funded by grants from the Ministry of Health of the Russian Federation 124021400005-3 (redox status of GBM, inhibition of TrxR in cancers, mechanisms) and RSF grant 22-75-10151 (anticancer activity of AF, brain delivery methods).

Institutional Review Board Statement: Not applicable.

Informed Consent Statement: Not applicable.

Data Availability Statement: Not applicable.

Conflicts of Interest: The authors declare no conflicts of interest.

References

1. Mohammed, S.; Dinesan, M.; Ajayakumar, T. Survival and Quality of Life Analysis in Glioblastoma Multiforme with Adjuvant Chemoradiotherapy: A Retrospective Study. *Rep. Pract. Oncol. Radiother.* **2022**, *27*, 1026–1036. [CrossRef] [PubMed]
2. Zhao, Y.; Ye, X.; Xiong, Z.; Ihsan, A.; Ares, I.; Martínez, M.; Lopez-Torres, B.; Martínez-Larrañaga, M.-R.; Anadón, A.; Wang, X.; et al. Cancer Metabolism: The Role of ROS in DNA Damage and Induction of Apoptosis in Cancer Cells. *Metabolites* **2023**, *13*, 796. [CrossRef] [PubMed]
3. Pawlowska, E.; Szczepanska, J.; Szatkowska, M.; Blasiak, J. An Interplay between Senescence, Apoptosis and Autophagy in Glioblastoma Multiforme—Role in Pathogenesis and Therapeutic Perspective. *Int. J. Mol. Sci.* **2018**, *19*, 889. [CrossRef]
4. Lo Dico, A.; Salvatore, D.; Martelli, C.; Ronchi, D.; Diceglie, C.; Lucignani, G.; Ottobrini, L. Intracellular Redox-Balance Involvement in Temozolomide Resistance-Related Molecular Mechanisms in Glioblastoma. *Cells* **2019**, *8*, 1315. [CrossRef]
5. Lin, C.-J.; Lee, C.-C.; Shih, Y.-L.; Lin, T.-Y.; Wang, S.-H.; Lin, Y.-F.; Shih, C.-M. Resveratrol Enhances the Therapeutic Effect of Temozolomide against Malignant Glioma in Vitro and in Vivo by Inhibiting Autophagy. *Free Radic. Biol. Med.* **2012**, *52*, 377–391. [CrossRef]

6. Knizhnik, A.V.; Roos, W.P.; Nikolova, T.; Quiros, S.; Tomaszowski, K.-H.; Christmann, M.; Kaina, B. Survival and Death Strategies in Glioma Cells: Autophagy, Senescence and Apoptosis Triggered by a Single Type of Temozolomide-Induced DNA Damage. *PLoS ONE* **2013**, *8*, e55665. [CrossRef]
7. Checa, J.; Aran, J.M. Reactive Oxygen Species: Drivers of Physiological and Pathological Processes. *J. Inflamm. Res.* **2020**, *13*, 1057–1073. [CrossRef] [PubMed]
8. Le Gal, K.; Schmidt, E.E.; Sayin, V.I. Cellular Redox Homeostasis. *Antioxidants* **2021**, *10*, 1377. [CrossRef] [PubMed]
9. Hayes, J.D.; Dinkova-Kostova, A.T.; Tew, K.D. Oxidative Stress in Cancer. *Cancer Cell* **2020**, *38*, 167–197. [CrossRef] [PubMed]
10. Kwon, J.; Lee, S.-R.; Yang, K.-S.; Ahn, Y.; Kim, Y.J.; Stadtman, E.R.; Rhee, S.G. Reversible Oxidation and Inactivation of the Tumor Suppressor PTEN in Cells Stimulated with Peptide Growth Factors. *Proc. Natl. Acad. Sci. USA* **2004**, *101*, 16419–16424. [CrossRef]
11. Kim, S.J.; Kim, H.S.; Seo, Y.R. Understanding of ROS-Inducing Strategy in Anticancer Therapy. *Oxid. Med. Cell. Longev.* **2019**, *2019*, 5381692. [CrossRef]
12. Zhou, Y.; Wang, L.; Wang, C.; Wu, Y.; Chen, D.; Lee, T.H. Potential Implications of Hydrogen Peroxide in the Pathogenesis and Therapeutic Strategies of Gliomas. *Arch. Pharm. Res.* **2020**, *43*, 187–203. [CrossRef]
13. Hasan, A.A.; Kalinina, E.; Tatarskiy, V.; Shtil, A. The Thioredoxin System of Mammalian Cells and Its Modulators. *Biomedicines* **2022**, *10*, 1757. [CrossRef] [PubMed]
14. Silva-Adaya, D.; Gonsebatt, M.E.; Guevara, J. Thioredoxin System Regulation in the Central Nervous System: Experimental Models and Clinical Evidence. *Oxid. Med. Cell. Longev.* **2014**, *2014*, 590808. [CrossRef] [PubMed]
15. Snezhkina, A.V.; Kudryavtseva, A.V.; Kardymon, O.L.; Savvateeva, M.V.; Melnikova, N.V.; Krasnov, G.S.; Dmitriev, A.A. ROS Generation and Antioxidant Defense Systems in Normal and Malignant Cells. *Oxid. Med. Cell. Longev.* **2019**, *2019*, 6175804. [CrossRef]
16. Kennedy, L.; Sandhu, J.K.; Harper, M.-E.; Cuperlovic-Culf, M. Role of Glutathione in Cancer: From Mechanisms to Therapies. *Biomolecules* **2020**, *10*, 1429. [CrossRef]
17. Navarro, J.; Obrador, E.; Carretero, J.; Petschen, I.; Aviñó, J.; Perez, P.; Estrela, J.M. Changes in Glutathione Status and the Antioxidant System in Blood and in Cancer Cells Associate with Tumour Growth in Vivo. *Free Radic. Biol. Med.* **1999**, *26*, 410–418. [CrossRef] [PubMed]
18. Elda Valenti, G.; Tasso, B.; Traverso, N.; Domenicotti, C.; Marengo, B. Glutathione in Cancer Progression and Chemoresistance: An Update. *Redox Exp. Med.* **2023**, *2023*, e220023. [CrossRef]
19. Raninga, P.V.; Di Trapani, G.; Tonissen, K.F. Cross Talk between Two Antioxidant Systems, Thioredoxin and DJ-1: Consequences for Cancer. *Oncoscience* **2014**, *1*, 95–110. [CrossRef] [PubMed]
20. Seitz, R.; Tümen, D.; Kunst, C.; Heumann, P.; Schmid, S.; Kandulski, A.; Müller, M.; Gülow, K. Exploring the Thioredoxin System as a Therapeutic Target in Cancer: Mechanisms and Implications. *Antioxidants* **2024**, *13*, 1078. [CrossRef]
21. Jovanović, M.; Podolski-Renić, A.; Krasavin, M.; Pešić, M. The Role of the Thioredoxin Detoxification System in Cancer Progression and Resistance. *Front. Mol. Biosci.* **2022**, *9*, 883297. [CrossRef]
22. Rocha, C.R.R.; Garcia, C.C.M.; Vieira, D.B.; Quinet, A.; de Andrade-Lima, L.C.; Munford, V.; Belizário, J.E.; Menck, C.F.M. Glutathione Depletion Sensitizes Cisplatin- and Temozolomide-Resistant Glioma Cells in Vitro and in Vivo. *Cell Death Dis.* **2014**, *5*, e1505. [CrossRef] [PubMed]
23. Chen, L.; Li, X.; Liu, L.; Yu, B.; Xue, Y.; Liu, Y. Erastin Sensitizes Glioblastoma Cells to Temozolomide by Restraining XCT and Cystathionine-γ-Lyase Function. *Oncol. Rep.* **2015**, *33*, 1465–1474. [CrossRef]
24. Yao, A.; Storr, S.J.; Al-hadyan, K.; Rahman, R.; Smith, S.; Grundy, R.; Paine, S.; Martin, S.G. Thioredoxin System Protein Expression Is Associated with Poor Clinical Outcome in Adult and Paediatric Gliomas and Medulloblastomas. *Mol. Neurobiol.* **2020**, *57*, 2889–2901. [CrossRef] [PubMed]
25. Jovanović, M.; Dragoj, M.; Zhukovsky, D.; Dar'in, D.; Krasavin, M.; Pešić, M.; Podolski-Renić, A. Novel TrxR1 Inhibitors Show Potential for Glioma Treatment by Suppressing the Invasion and Sensitizing Glioma Cells to Chemotherapy. *Front. Mol. Biosci.* **2020**, *7*, 586146. [CrossRef] [PubMed]
26. Yao, A.; Storr, S.J.; Inman, M.; Barwell, L.; Moody, C.J.; Martin, S.G. Cytotoxic and Radiosensitising Effects of a Novel Thioredoxin Reductase Inhibitor in Brain Cancers. *Mol. Neurobiol.* **2022**, *59*, 3546–3563. [CrossRef]
27. Ngo, V.; Duennwald, M.L. Nrf2 and Oxidative Stress: A General Overview of Mechanisms and Implications in Human Disease. *Antioxidants* **2022**, *11*, 2345. [CrossRef] [PubMed]
28. Almeida Lima, K.; Osawa, I.Y.A.; Ramalho, M.C.C.; de Souza, I.; Guedes, C.B.; Souza Filho, C.H.D.D.; Monteiro, L.K.S.; Latancia, M.T.; Rocha, C.R.R. Temozolomide Resistance in Glioblastoma by NRF2: Protecting the Evil. *Biomedicines* **2023**, *11*, 1081. [CrossRef] [PubMed]
29. Jaganjac, M.; Milkovic, L.; Sunjic, S.B.; Zarkovic, N. The NRF2, Thioredoxin, and Glutathione System in Tumorigenesis and Anticancer Therapies. *Antioxidants* **2020**, *9*, 1151. [CrossRef]

30. Rocha, C.R.R.; Rocha, A.R.; Silva, M.M.; Gomes, L.R.; Latancia, M.T.; Andrade-Tomaz, M.; de Souza, I.; Monteiro, L.K.S.; Menck, C.F.M. Revealing Temozolomide Resistance Mechanisms via Genome-Wide CRISPR Libraries. *Cells* **2020**, *9*, 2573. [CrossRef] [PubMed]
31. Zhou, Y.; Wang, H.-D.; Zhu, L.; Cong, Z.-X.; Li, N.; Ji, X.-J.; Pan, H.; Wang, J.-W.; Li, W.-C. Knockdown of Nrf2 Enhances Autophagy Induced by Temozolomide in U251 Human Glioma Cell Line. *Oncol. Rep.* **2013**, *29*, 394–400. [CrossRef] [PubMed]
32. Rocha, C.R.R.; Kajitani, G.S.; Quinet, A.; Fortunato, R.S.; Menck, C.F.M. NRF2 and Glutathione Are Key Resistance Mediators to Temozolomide in Glioma and Melanoma Cells. *Oncotarget* **2016**, *7*, 48081–48092. [CrossRef] [PubMed]
33. Zhu, J.; Wang, H.; Fan, Y.; Hu, Y.; Ji, X.; Sun, Q.; Liu, H. Knockdown of Nuclear Factor Erythroid 2-Related Factor 2 by Lentivirus Induces Differentiation of Glioma Stem-like Cells. *Oncol. Rep.* **2014**, *32*, 1170–1178. [CrossRef]
34. Hemmati, H.D.; Nakano, I.; Lazareff, J.A.; Masterman-Smith, M.; Geschwind, D.H.; Bronner-Fraser, M.; Kornblum, H.I. Cancerous Stem Cells Can Arise from Pediatric Brain Tumors. *Proc. Natl. Acad. Sci. USA* **2003**, *100*, 15178–15183. [CrossRef] [PubMed]
35. Singh, S.K.; Hawkins, C.; Clarke, I.D.; Squire, J.A.; Bayani, J.; Hide, T.; Henkelman, R.M.; Cusimano, M.D.; Dirks, P.B. Identification of Human Brain Tumour Initiating Cells. *Nature* **2004**, *432*, 396–401. [CrossRef]
36. Kim, H.-J.; Jeon, H.-M.; Batara, D.C.; Lee, S.; Lee, S.J.; Yin, J.; Park, S.-I.; Park, M.; Seo, J.B.; Hwang, J.; et al. CREB5 Promotes the Proliferation and Self-Renewal Ability of Glioma Stem Cells. *Cell Death Discov.* **2024**, *10*, 103. [CrossRef] [PubMed]
37. Safa, A.R.; Saadatzadeh, M.R.; Cohen-Gadol, A.A.; Pollok, K.E.; Bijangi-Vishehsaraei, K. Glioblastoma Stem Cells (GSCs) Epigenetic Plasticity and Interconversion between Differentiated Non-GSCs and GSCs. *Genes Dis.* **2015**, *2*, 152–163. [CrossRef]
38. Auffinger, B.; Spencer, D.; Pytel, P.; Ahmed, A.U.; Lesniak, M.S. The Role of Glioma Stem Cells in Chemotherapy Resistance and Glioblastoma Multiforme Recurrence. *Expert Rev. Neurother.* **2015**, *15*, 741–752. [CrossRef]
39. Pasqualetti, F.; Miniati, M.; Gonnelli, A.; Gadducci, G.; Giannini, N.; Palagini, L.; Mancino, M.; Fuentes, T.; Paiar, F. Cancer Stem Cells and Glioblastoma: Time for Innovative Biomarkers of Radio-Resistance? *Biology* **2023**, *12*, 1295. [CrossRef]
40. Singer, E.; Judkins, J.; Salomonis, N.; Matlaf, L.; Soteropoulos, P.; McAllister, S.; Soroceanu, L. Reactive Oxygen Species-Mediated Therapeutic Response and Resistance in Glioblastoma. *Cell Death Dis.* **2015**, *6*, e1601. [CrossRef] [PubMed]
41. Olivier, C.; Oliver, L.; Lalier, L.; Vallette, F.M. Drug Resistance in Glioblastoma: The Two Faces of Oxidative Stress. *Front. Mol. Biosci.* **2021**, *7*, 620677. [CrossRef]
42. Vlashi, E.; Lagadec, C.; Vergnes, L.; Matsutani, T.; Masui, K.; Poulou, M.; Popescu, R.; Della Donna, L.; Evers, P.; Dekmezian, C.; et al. Metabolic State of Glioma Stem Cells and Nontumorigenic Cells. *Proc. Natl. Acad. Sci. USA* **2011**, *108*, 16062–16067. [CrossRef] [PubMed]
43. Peng, G.; Tang, Z.; Xiang, Y.; Chen, W. Glutathione Peroxidase 4 Maintains a Stemness Phenotype, Oxidative Homeostasis and Regulates Biological Processes in Panc-1 Cancer Stem-like Cells. *Oncol. Rep.* **2019**, *41*, 1264–1274. [CrossRef]
44. Galbo, P.M.; Madsen, A.T.; Liu, Y.; Peng, M.; Wei, Y.; Ciesielski, M.J.; Fenstermaker, R.A.; Graff, S.; Montagna, C.; Segall, J.E.; et al. Functional Contribution and Clinical Implication of Cancer-Associated Fibroblasts in Glioblastoma. *Clin. Cancer Res.* **2024**, *30*, 865–876. [CrossRef] [PubMed]
45. Sharma, P.; Aaroe, A.; Liang, J.; Puduvalli, V.K. Tumor Microenvironment in Glioblastoma: Current and Emerging Concepts. *Neuro-Oncol. Adv.* **2023**, *5*, vdad009. [CrossRef]
46. Weinberg, F.; Ramnath, N.; Nagrath, D. Reactive Oxygen Species in the Tumor Microenvironment: An Overview. *Cancers* **2019**, *11*, 1191. [CrossRef]
47. Ford, K.; Hanley, C.J.; Mellone, M.; Szyndralewiez, C.; Heitz, F.; Wiesel, P.; Wood, O.; Machado, M.; Lopez, M.-A.; Ganesan, A.-P.; et al. NOX4 Inhibition Potentiates Immunotherapy by Overcoming Cancer-Associated Fibroblast-Mediated CD8 T-Cell Exclusion from Tumors. *Cancer Res.* **2020**, *80*, 1846–1860. [CrossRef] [PubMed]
48. Xiang, H.; Ramil, C.P.; Hai, J.; Zhang, C.; Wang, H.; Watkins, A.A.; Afshar, R.; Georgiev, P.; Sze, M.A.; Song, X.S.; et al. Cancer-Associated Fibroblasts Promote Immunosuppression by Inducing ROS-Generating Monocytic MDSCs in Lung Squamous Cell Carcinoma. *Cancer Immunol. Res.* **2020**, *8*, 436–450. [CrossRef] [PubMed]
49. Larionova, I.; Cherdyntseva, N.; Liu, T.; Patysheva, M.; Rakina, M.; Kzhyshkowska, J. Interaction of Tumor-Associated Macrophages and Cancer Chemotherapy. *Oncoimmunology* **2019**, *8*, e1596004. [CrossRef]
50. Ren, J.; Xu, B.; Ren, J.; Liu, Z.; Cai, L.; Zhang, X.; Wang, W.; Li, S.; Jin, L.; Ding, L. The Importance of M1-and M2-Polarized Macrophages in Glioma and as Potential Treatment Targets. *Brain Sci.* **2023**, *13*, 1269. [CrossRef] [PubMed]
51. Kennel, K.B.; Greten, F.R. Immune Cell—Produced ROS and Their Impact on Tumor Growth and Metastasis. *Redox Biol.* **2021**, *42*, 101891. [CrossRef] [PubMed]
52. Xing, J.; Cai, H.; Lin, Z.; Zhao, L.; Xu, H.; Song, Y.; Wang, Z.; Liu, C.; Hu, G.; Zheng, J.; et al. Examining the Function of Macrophage Oxidative Stress Response and Immune System in Glioblastoma Multiforme through Analysis of Single-Cell Transcriptomics. *Front. Immunol.* **2024**, *14*, 1288137. [CrossRef] [PubMed]
53. Kuo, C.-L.; Ponneri Babuharisankar, A.; Lin, Y.-C.; Lien, H.-W.; Lo, Y.K.; Chou, H.-Y.; Tangeda, V.; Cheng, L.-C.; Cheng, A.N.; Lee, A.Y.-L. Mitochondrial Oxidative Stress in the Tumor Microenvironment and Cancer Immunoescape: Foe or Friend? *J. Biomed. Sci.* **2022**, *29*, 74. [CrossRef] [PubMed]

54. Liang, X.; Wang, Z.; Dai, Z.; Liu, J.; Zhang, H.; Wen, J.; Zhang, N.; Zhang, J.; Luo, P.; Liu, Z.; et al. Oxidative Stress Is Involved in Immunosuppression and Macrophage Regulation in Glioblastoma. *Clin. Immunol.* **2024**, *258*, 109802. [CrossRef]
55. Yang, Y.; Neo, S.Y.; Chen, Z.; Cui, W.; Chen, Y.; Guo, M.; Wang, Y.; Xu, H.; Kurzay, A.; Alici, E.; et al. Thioredoxin Activity Confers Resistance against Oxidative Stress in Tumor-Infiltrating NK Cells. *J. Clin. Investig.* **2020**, *130*, 5508–5522. [CrossRef]
56. Kurniawan, H.; Franchina, D.G.; Guerra, L.; Bonetti, L.; Baguet, L.S.; Grusdat, M.; Schlicker, L.; Hunewald, O.; Dostert, C.; Merz, M.P.; et al. Glutathione Restricts Serine Metabolism to Preserve Regulatory T Cell Function. *Cell Metab.* **2020**, *31*, 920–936.e7. [CrossRef] [PubMed]
57. Wang, X.; Dong, H.; Li, Q.; Li, Y.; Hong, A. Thioredoxin Induces Tregs to Generate an Immunotolerant Tumor Microenvironment in Metastatic Melanoma. *Oncoimmunology* **2015**, *4*, e1027471. [CrossRef] [PubMed]
58. Schallreuter, K.U.; Gleason, F.K.; Wood, J.M. The Mechanism of Action of the Nitrosourea Anti-Tumor Drugs on Thioredoxin Reductase, Glutathione Reductase and Ribonucleotide Reductase. *Biochim. Biophys. Acta-Mol. Cell Res.* **1990**, *1054*, 14–20. [CrossRef]
59. Witte, A.-B.; Anestål, K.; Jerremalm, E.; Ehrsson, H.; Arnér, E.S.J. Inhibition of Thioredoxin Reductase but Not of Glutathione Reductase by the Major Classes of Alkylating and Platinum-Containing Anticancer Compounds. *Free Radic. Biol. Med.* **2005**, *39*, 696–703. [CrossRef]
60. Zhang, J.; Ma, K.; Wang, H. Cyclophosphamide Suppresses Thioredoxin Reductase in Bladder Tissue and Its Adaptive Response via Inductions of Thioredoxin Reductase and Glutathione Peroxidase. *Chem. Biol. Interact.* **2006**, *162*, 24–30. [CrossRef] [PubMed]
61. Wang, X.; Zhang, J.; Xu, T. Cyclophosphamide as a Potent Inhibitor of Tumor Thioredoxin Reductase in Vivo. *Toxicol. Appl. Pharmacol.* **2007**, *218*, 88–95. [CrossRef] [PubMed]
62. Lu, J.; Chew, E.-H.; Holmgren, A. Targeting Thioredoxin Reductase Is a Basis for Cancer Therapy by Arsenic Trioxide. *Proc. Natl. Acad. Sci. USA* **2007**, *104*, 12288–12293. [CrossRef]
63. Jiang, Y.; Shen, X.; Zhi, F.; Wen, Z.; Gao, Y.; Xu, J.; Yang, B.; Bai, Y. An Overview of Arsenic Trioxide-Involved Combined Treatment Algorithms for Leukemia: Basic Concepts and Clinical Implications. *Cell Death Discov.* **2023**, *9*, 284. [CrossRef] [PubMed]
64. Wu, J.; Ji, Z.; Liu, H.; Liu, Y.; Han, D.; Shi, C.; Shi, C.; Wang, C.; Yang, G.; Chen, X.; et al. Arsenic Trioxide Depletes Cancer Stem-like Cells and Inhibits Repopulation of Neurosphere Derived from Glioblastoma by Downregulation of Notch Pathway. *Toxicol. Lett.* **2013**, *220*, 61–69. [CrossRef] [PubMed]
65. Chow, S.K.Y.; Chan, J.Y.W.; Fung, K.P. Inhibition of Cell Proliferation and the Action Mechanisms of Arsenic Trioxide (As_2O_3) on Human Breast Cancer Cells. *J. Cell. Biochem.* **2004**, *93*, 173–187. [CrossRef]
66. Yang, Y.; Zhang, X.; Jia, S.; Yang, S.; Yang, Y.; Yang, T. Arsenic Trioxide Induces G2/M Arrest in Hepatocellular Carcinoma Cells by Increasing the Tumor Suppressor PTEN Expression. *J. Cell. Biochem.* **2012**, *113*, 3528–3535.
67. Walker, A.M.; Stevens, J.J.; Ndebele, K.; Tchounwou, P.B. Evaluation of Arsenic Trioxide Potential for Lung Cancer Treatment: Assessment of Apoptotic Mechanisms and Oxidative Damage. *J. Cancer Sci. Ther.* **2015**, *8*, 1–9. [CrossRef]
68. Tian, Z.; Tan, Y.; Lin, X.; Su, M.; Pan, L.; Lin, L.; Ou, G.; Chen, Y. Arsenic Trioxide Sensitizes Pancreatic Cancer Cells to Gemcitabine through Downregulation of the TIMP1/PI3K/AKT/MTOR Axis. *Transl. Res.* **2023**, *255*, 66–76. [CrossRef] [PubMed]
69. Ryu, S.; Ye, X.; Olson, J.J.; Mikkelsen, T.; Bangiyev, L.; Lesser, G.J.; Batchelor, T.; Nabors, B.; Desideri, S.; Walbert, T.; et al. Phase I and Pharmacodynamic Study of Arsenic Trioxide plus Radiotherapy in Patients with Newly Diagnosed Glioblastoma. *Neuro-Oncol. Adv.* **2024**, *6*, vdae089. [CrossRef]
70. Rosenthal, D.I.; Nurenberg, P.; Becerra, C.R.; Frenkel, E.P.; Carbone, D.P.; Lum, B.L.; Miller, R.; Engel, J.; Young, S.; Miles, D.; et al. A Phase I Single-Dose Trial of Gadolinium Texaphyrin (Gd-Tex), a Tumor Selective Radiation Sensitizer Detectable by Magnetic Resonance Imaging. *Clin. Cancer Res.* **1999**, *5*, 739–745. [PubMed]
71. Young, S.W.; Qing, F.; Harriman, A.; Sessler, J.L.; Dow, W.C.; Mody, T.D.; Hemmi, G.W.; Hao, Y.; Miller, R.A. Gadolinium(III) Texaphyrin: A Tumor Selective Radiation Sensitizer That Is Detectable by MRI. *Proc. Natl. Acad. Sci. USA* **1996**, *93*, 6610–6615. [CrossRef]
72. Hashemy, S.I.; Ungerstedt, J.S.; Avval, F.Z.; Holmgren, A. Motexafin Gadolinium, a Tumor-Selective Drug Targeting Thioredoxin Reductase and Ribonucleotide Reductase. *J. Biol. Chem.* **2006**, *281*, 10691–10697. [CrossRef] [PubMed]
73. Mehta, M.P.; Shapiro, W.R.; Phan, S.C.; Gervais, R.; Carrie, C.; Chabot, P.; Patchell, R.A.; Glantz, M.J.; Recht, L.; Langer, C.; et al. Motexafin Gadolinium Combined With Prompt Whole Brain Radiotherapy Prolongs Time to Neurologic Progression in Non–Small-Cell Lung Cancer Patients With Brain Metastases: Results of a Phase III Trial. *Int. J. Radiat. Oncol.* **2009**, *73*, 1069–1076. [CrossRef] [PubMed]
74. Chupakhin, E.G.; Krasavin, M.Y. Natural Compounds as Inhibitors of Thioredoxin Reductase (TrxR1). *Russ. Chem. Bull.* **2022**, *71*, 443–448. [CrossRef]
75. Fang, J.; Lu, J.; Holmgren, A. Thioredoxin Reductase Is Irreversibly Modified by Curcumin. *J. Biol. Chem.* **2005**, *280*, 25284–25290. [CrossRef] [PubMed]
76. Cai, W.; Zhang, B.; Duan, D.; Wu, J.; Fang, J. Curcumin Targeting the Thioredoxin System Elevates Oxidative Stress in HeLa Cells. *Toxicol. Appl. Pharmacol.* **2012**, *262*, 341–348. [CrossRef]

77. Larasati, Y.A.; Yoneda-Kato, N.; Nakamae, I.; Yokoyama, T.; Meiyanto, E.; Kato, J. Curcumin Targets Multiple Enzymes Involved in the ROS Metabolic Pathway to Suppress Tumor Cell Growth. *Sci. Rep.* **2018**, *8*, 2039. [CrossRef]
78. Zhang, Q.; Chen, W.; Lv, X.; Weng, Q.; Chen, M.; Cui, R.; Liang, G.; Ji, J. Piperlongumine, a Novel TrxR1 Inhibitor, Induces Apoptosis in Hepatocellular Carcinoma Cells by ROS-Mediated ER Stress. *Front. Pharmacol.* **2019**, *10*, 1180. [CrossRef]
79. Prast-Nielsen, S.; Cebula, M.; Pader, I.; Arnér, E.S.J. Noble Metal Targeting of Thioredoxin Reductase—Covalent Complexes with Thioredoxin and Thioredoxin-Related Protein of 14 kDa Triggered by Cisplatin. *Free Radic. Biol. Med.* **2010**, *49*, 1765–1778. [CrossRef]
80. Cheng, P.; Liu, H.; Li, Y.; Pi, P.; Jiang, Y.; Zang, S.; Li, X.; Fu, A.; Ren, X.; Xu, J.; et al. Inhibition of Thioredoxin Reductase 1 Correlates with Platinum-Based Chemotherapeutic Induced Tissue Injury. *Biochem. Pharmacol.* **2020**, *175*, 113873. [CrossRef] [PubMed]
81. Massai, L.; Cirri, D.; Marzo, T.; Messori, L. Auranofin and Its Analogs as Prospective Agents for the Treatment of Colorectal Cancer. *Cancer Drug Resist.* **2022**, *5*, 1–14. [CrossRef]
82. Wagner, C.; Sudati, J.H.; Nogueira, C.W.; Rocha, J.B.T. In Vivo and in Vitro Inhibition of Mice Thioredoxin Reductase by Methylmercury. *BioMetals* **2010**, *23*, 1171–1177. [CrossRef] [PubMed]
83. Branco, V.; Canário, J.; Holmgren, A.; Carvalho, C. Inhibition of the Thioredoxin System in the Brain and Liver of Zebra-Seabreams Exposed to Waterborne Methylmercury. *Toxicol. Appl. Pharmacol.* **2011**, *251*, 95–103. [CrossRef]
84. Omata, Y.; Folan, M.; Shaw, M.; Messer, R.L.; Lockwood, P.E.; Hobbs, D.; Bouillaguet, S.; Sano, H.; Lewis, J.B.; Wataha, J.C. Sublethal Concentrations of Diverse Gold Compounds Inhibit Mammalian Cytosolic Thioredoxin Reductase (TrxR1). *Toxicol. Vitr.* **2006**, *20*, 882–890. [CrossRef]
85. Rackham, O.; Nichols, S.J.; Leedman, P.J.; Berners-Price, S.J.; Filipovska, A. A Gold(I) Phosphine Complex Selectively Induces Apoptosis in Breast Cancer Cells: Implications for Anticancer Therapeutics Targeted to Mitochondria. *Biochem. Pharmacol.* **2007**, *74*, 992–1002. [CrossRef] [PubMed]
86. Meyer, A.; Bagowski, C.P.; Kokoschka, M.; Stefanopoulou, M.; Alborzinia, H.; Can, S.; Vlecken, D.H.; Sheldrick, W.S.; Wölfl, S.; Ott, I. On the Biological Properties of Alkynyl Phosphine Gold(I) Complexes. *Angew. Chem. Int. Ed.* **2012**, *51*, 8895–8899. [CrossRef]
87. Urig, S.; Fritz-Wolf, K.; Réau, R.; Herold-Mende, C.; Tóth, K.; Davioud-Charvet, E.; Becker, K. Undressing of Phosphine Gold(I) Complexes as Irreversible Inhibitors of Human Disulfide Reductases. *Angew. Chem. Int. Ed.* **2006**, *45*, 1881–1886. [CrossRef]
88. Bertrand, B.; Stefan, L.; Pirrotta, M.; Monchaud, D.; Bodio, E.; Richard, P.; Le Gendre, P.; Warmerdam, E.; de Jager, M.H.; Groothuis, G.M.M.; et al. Caffeine-Based Gold(I) N -Heterocyclic Carbenes as Possible Anticancer Agents: Synthesis and Biological Properties. *Inorg. Chem.* **2014**, *53*, 2296–2303. [CrossRef]
89. Rubbiani, R.; Kitanovic, I.; Alborzinia, H.; Can, S.; Kitanovic, A.; Onambele, L.A.; Stefanopoulou, M.; Geldmacher, Y.; Sheldrick, W.S.; Wolber, G.; et al. Benzimidazol-2-Ylidene Gold(I) Complexes Are Thioredoxin Reductase Inhibitors with Multiple Antitumor Properties. *J. Med. Chem.* **2010**, *53*, 8608–8618. [CrossRef]
90. Schmidt, C.; Albrecht, L.; Balasupramaniam, S.; Misgeld, R.; Karge, B.; Brönstrup, M.; Prokop, A.; Baumann, K.; Reichl, S.; Ott, I. A Gold (i) Biscarbene Complex with Improved Activity as a TrxR Inhibitor and Cytotoxic Drug: Comparative Studies with Different Gold Metallodrugs. *Metallomics* **2019**, *11*, 533–545. [CrossRef] [PubMed]
91. Galassi, R.; Burini, A.; Ricci, S.; Pellei, M.; Rigobello, M.P.; Citta, A.; Dolmella, A.; Gandin, V.; Marzano, C. Synthesis and Characterization of Azolate Gold(i) Phosphane Complexes as Thioredoxin Reductase Inhibiting Antitumor Agents. *Dalt. Trans.* **2012**, *41*, 5307. [CrossRef] [PubMed]
92. Serebryanskaya, T.V.; Lyakhov, A.S.; Ivashkevich, L.S.; Schur, J.; Frias, C.; Prokop, A.; Ott, I. Gold(I) Thiotetrazolates as Thioredoxin Reductase Inhibitors and Antiproliferative Agents. *Dalt. Trans.* **2015**, *44*, 1161–1169. [CrossRef] [PubMed]
93. Crooke, S.T.; Snyder, R.M.; Butt, T.R.; Ecker, D.J.; Allaudeen, H.S.; Monia, B.; Mirabelli, C.K. Cellular and Molecular Pharmacology of Auranofin and Related Gold Complexes. *Biochem. Pharmacol.* **1986**, *35*, 3423–3431. [CrossRef] [PubMed]
94. Radenkovic, F.; Holland, O.; Vanderlelie, J.J.; Perkins, A.V. Selective Inhibition of Endogenous Antioxidants with Auranofin Causes Mitochondrial Oxidative Stress Which Can Be Countered by Selenium Supplementation. *Biochem. Pharmacol.* **2017**, *146*, 42–52. [CrossRef] [PubMed]
95. Salmain, M.; Gaschard, M.; Baroud, M.; Lepeltier, E.; Jaouen, G.; Passirani, C.; Vessières, A. Thioredoxin Reductase and Organometallic Complexes: A Pivotal System to Tackle Multidrug Resistant Tumors? *Cancers* **2023**, *15*, 4448. [CrossRef]
96. Berners-Price, S.J.; Mirabelli, C.K.; Johnson, R.K.; Mattern, M.R.; McCabe, F.L.; Faucette, L.F.; Sung, C.M.; Mong, S.M.; Sadler, P.J.; Crooke, S.T. In Vivo Antitumor Activity and in Vitro Cytotoxic Properties of Bis [1,2-Bis(Diphenylphosphino)Ethane]Gold(I) Chloride. *Cancer Res.* **1986**, *46*, 5486–5493.
97. Wang, S.; Lu, Y.; Woods, K.; Di Trapani, G.; Tonissen, K.F. Investigating the Thioredoxin and Glutathione Systems' Response in Lymphoma Cells after Treatment with [Au(D2pype)2]Cl. *Antioxidants* **2021**, *10*, 104. [CrossRef]
98. Sze, J.H.; Raninga, P.V.; Nakamura, K.; Casey, M.; Khanna, K.K.; Berners-Price, S.J.; Di Trapani, G.; Tonissen, K.F. Anticancer Activity of a Gold(I) Phosphine Thioredoxin Reductase Inhibitor in Multiple Myeloma. *Redox Biol.* **2020**, *28*, 101310. [CrossRef]

99. Clapper, E.; Wang, S.; Raninga, P.V.; Di Trapani, G.; Tonissen, K.F. Cross-Talk between Bcr-Abl and the Thioredoxin System in Chronic Myeloid Leukaemia: Implications for CML Treatment. *Antioxidants* **2020**, *9*, 207. [CrossRef]
100. Vergara, E.; Casini, A.; Sorrentino, F.; Zava, O.; Cerrada, E.; Rigobello, M.P.; Bindoli, A.; Laguna, M.; Dyson, P.J. Anticancer Therapeutics That Target Selenoenzymes: Synthesis, Characterization, in Vitro Cytotoxicity, and Thioredoxin Reductase Inhibition of a Series of Gold(I) Complexes Containing Hydrophilic Phosphine Ligands. *ChemMedChem* **2010**, *5*, 96–102. [CrossRef] [PubMed]
101. Nakaya, A.; Sagawa, M.; Muto, A.; Uchida, H.; Ikeda, Y.; Kizaki, M. The Gold Compound Auranofin Induces Apoptosis of Human Multiple Myeloma Cells through Both Down-Regulation of STAT3 and Inhibition of NF-KB Activity. *Leuk. Res.* **2011**, *35*, 243–249. [CrossRef] [PubMed]
102. Bichi, R.; Shinton, S.A.; Martin, E.S.; Koval, A.; Calin, G.A.; Cesari, R.; Russo, G.; Hardy, R.R.; Croce, C.M. Human Chronic Lymphocytic Leukemia Modeled in Mouse by Targeted TCL1 Expression. *Proc. Natl. Acad. Sci. USA* **2002**, *99*, 6955–6960. [CrossRef] [PubMed]
103. Chen, X.; Shi, X.; Zhao, C.; Li, X.; Lan, X.; Liu, S.; Huang, H.; Liu, N.; Liao, S.; Zang, D.; et al. Anti-Rheumatic Agent Auranofin Induced Apoptosis in Chronic Myeloid Leukemia Cells Resistant to Imatinib through Both Bcr/Abl-Dependent and -Independent Mechanisms. *Oncotarget* **2014**, *5*, 9118–9132. [CrossRef] [PubMed]
104. Mittendorf, E.A.; Philips, A.V.; Meric-Bernstam, F.; Qiao, N.; Wu, Y.; Harrington, S.; Su, X.; Wang, Y.; Gonzalez-Angulo, A.M.; Akcakanat, A.; et al. PD-L1 Expression in Triple-Negative Breast Cancer. *Cancer Immunol. Res.* **2014**, *2*, 361–370. [CrossRef] [PubMed]
105. Hrabe, J.E.; O'Leary, B.R.; Fath, M.A.; Rodman, S.N.; Button, A.M.; Domann, F.E.; Spitz, D.R.; Mezhir, J.J. Disruption of Thioredoxin Metabolism Enhances the Toxicity of Transforming Growth Factor β-Activated Kinase 1 (TAK1) Inhibition in KRAS-Mutated Colon Cancer Cells. *Redox Biol.* **2015**, *5*, 319–327. [CrossRef] [PubMed]
106. Gamberi, T.; Chiappetta, G.; Fiaschi, T.; Modesti, A.; Sorbi, F.; Magherini, F. Upgrade of an Old Drug: Auranofin in Innovative Cancer Therapies to Overcome Drug Resistance and to Increase Drug Effectiveness. *Med. Res. Rev.* **2022**, *42*, 1111–1146. [CrossRef] [PubMed]
107. Freire Boullosa, L.; Van Loenhout, J.; Hermans, C.; Lau, H.W.; Merlin, C.; Marcq, E.; Takhsha, F.S.; Martinet, W.; De Meyer, G.R.Y.; Lardon, F.; et al. Optimization of the Solvent and In Vivo Administration Route of Auranofin in a Syngeneic Non-Small Cell Lung Cancer and Glioblastoma Mouse Model. *Pharmaceutics* **2022**, *14*, 2761. [CrossRef]
108. Wang, X.; Han, M.; Wang, S.; Sun, Y.; Zhao, W.; Xue, Z.; Liang, X.; Huang, B.; Li, G.; Chen, A.; et al. Targeting the Splicing Factor NONO Inhibits GBM Progression through GPX1 Intron Retention. *Theranostics* **2022**, *12*, 5451–5469. [CrossRef] [PubMed]
109. Hegazy, A.M.; Yamada, D.; Kobayashi, M.; Kohno, S.; Ueno, M.; Ali, M.A.E.; Ohta, K.; Tadokoro, Y.; Ino, Y.; Todo, T.; et al. Therapeutic Strategy for Targeting Aggressive Malignant Gliomas by Disrupting Their Energy Balance. *J. Biol. Chem.* **2016**, *291*, 21496–21509. [CrossRef] [PubMed]
110. Kenchappa, R.S.; Liu, Y.; Argenziano, M.G.; Banu, M.A.; Mladek, A.C.; West, R.; Luu, A.; Quiñones-Hinojosa, A.; Hambardzumyan, D.; Justilien, V.; et al. Protein Kinase Cι and SRC Signaling Define Reciprocally Related Subgroups of Glioblastoma with Distinct Therapeutic Vulnerabilities. *Cell Rep.* **2021**, *37*, 110054. [CrossRef] [PubMed]
111. Rios Perez, M.V.; Roife, D.; Dai, B.; Pratt, M.; Dobrowolski, R.; Kang, Y.; Li, X.; Augustine, J.J.; Zielinski, R.; Priebe, W.; et al. Antineoplastic Effects of Auranofin in Human Pancreatic Adenocarcinoma Preclinical Models. *Surg. Open Sci.* **2019**, *1*, 56–63. [CrossRef] [PubMed]
112. Hou, G.-X.; Liu, P.-P.; Zhang, S.; Yang, M.; Liao, J.; Yang, J.; Hu, Y.; Jiang, W.-Q.; Wen, S.; Huang, P. Elimination of Stem-like Cancer Cell Side-Population by Auranofin through Modulation of ROS and Glycolysis. *Cell Death Dis.* **2018**, *9*, 89. [CrossRef]
113. Yang, M.; Liu, J.; Li, J.; Wen, S.; Hu, Y.; Lu, W.; Liu, J.; Huang, P.; Liu, P. The Rheumatoid Arthritis Drug Auranofin Exerts Potent Anti-Lymphoma Effect by Stimulating TXNRD-Mediated ROS Generation and Inhibition of Energy Metabolism. *Redox Biol.* **2024**, *75*, 103245. [CrossRef] [PubMed]
114. Wang, R.; Zhong, L.; Wang, T.; Sun, T.; Yang, J.; Liu, X.; Wu, Y.; Guo, Q.; Gao, Y.; Zhao, K. Inducing Ubiquitination and Degradation of TrxR1 Protein by LW-216 Promotes Apoptosis in Non-Small Cell Lung Cancer via Triggering ROS Production. *Neoplasia* **2024**, *53*, 101004. [CrossRef] [PubMed]
115. Onodera, T.; Momose, I.; Adachi, H.; Yamazaki, Y.; Sawa, R.; Ohba, S.; Kawada, M. Human Pancreatic Cancer Cells under Nutrient Deprivation Are Vulnerable to Redox System Inhibition. *J. Biol. Chem.* **2020**, *295*, 16678–16690. [CrossRef] [PubMed]
116. Zou, P.; Chen, M.; Ji, J.; Chen, W.; Chen, X.; Ying, S.; Zhang, J.; Zhang, Z.; Liu, Z.; Yang, S.; et al. Auranofin Induces Apoptosis by ROS-Mediated ER Stress and Mitochondrial Dysfunction and Displayed Synergistic Lethality with Piperlongumine in Gastric Cancer. *Oncotarget* **2015**, *6*, 36505–36521. [CrossRef]
117. Van Loenhout, J.; Freire Boullosa, L.; Quatannens, D.; De Waele, J.; Merlin, C.; Lambrechts, H.; Lau, H.W.; Hermans, C.; Lin, A.; Lardon, F.; et al. Auranofin and Cold Atmospheric Plasma Synergize to Trigger Distinct Cell Death Mechanisms and Immunogenic Responses in Glioblastoma. *Cells* **2021**, *10*, 2936. [CrossRef] [PubMed]

118. Han, Y.; Chen, P.; Zhang, Y.; Lu, W.; Ding, W.; Luo, Y.; Wen, S.; Xu, R.; Liu, P.; Huang, P. Synergy between Auranofin and Celecoxib against Colon Cancer In Vitro and In Vivo through a Novel Redox-Mediated Mechanism. *Cancers* **2019**, *11*, 931. [CrossRef] [PubMed]
119. Fath, M.A.; Ahmad, I.M.; Smith, C.J.; Spence, J.; Spitz, D.R. Enhancement of Carboplatin-Mediated Lung Cancer Cell Killing by Simultaneous Disruption of Glutathione and Thioredoxin Metabolism. *Clin. Cancer Res.* **2011**, *17*, 6206–6217. [CrossRef]
120. Liu, X.; Wang, W.; Yin, Y.; Li, M.; Li, H.; Xiang, H.; Xu, A.; Mei, X.; Hong, B.; Lin, W. A High-Throughput Drug Screen Identifies Auranofin as a Potential Sensitizer of Cisplatin in Small Cell Lung Cancer. *Investig. New Drugs* **2019**, *37*, 1166–1176. [CrossRef] [PubMed]
121. Pessetto, Z.Y.; Chen, B.; Alturkmani, H.; Hyter, S.; Flynn, C.A.; Baltezor, M.; Ma, Y.; Rosenthal, H.G.; Neville, K.A.; Weir, S.J.; et al. In Silico and in Vitro Drug Screening Identifies New Therapeutic Approaches for Ewing Sarcoma. *Oncotarget* **2017**, *8*, 4079–4095. [CrossRef] [PubMed]
122. Liu, C.; Zhao, Y.; Wang, J.; Yang, Y.; Zhang, Y.; Qu, X.; Peng, S.; Yao, Z.; Zhao, S.; He, B.; et al. FoxO3 Reverses 5-Fluorouracil Resistance in Human Colorectal Cancer Cells by Inhibiting the Nrf2/TR1 Signaling Pathway. *Cancer Lett.* **2020**, *470*, 29–42. [CrossRef] [PubMed]
123. Fan, C.; Zheng, W.; Fu, X.; Li, X.; Wong, Y.-S.; Chen, T. Enhancement of Auranofin-Induced Lung Cancer Cell Apoptosis by Selenocystine, a Natural Inhibitor of TrxR1 in Vitro and in Vivo. *Cell Death Dis.* **2014**, *5*, e1191. [CrossRef] [PubMed]
124. Li, H.; Hu, J.; Wu, S.; Wang, L.; Cao, X.; Zhang, X.; Dai, B.; Cao, M.; Shao, R.; Zhang, R.; et al. Auranofin-Mediated Inhibition of PI3K/AKT/MTOR Axis and Anticancer Activity in Non-Small Cell Lung Cancer Cells. *Oncotarget* **2016**, *7*, 3548–3558. [CrossRef] [PubMed]
125. Nag, D.; Bhanja, P.; Riha, R.; Sanchez-Guerrero, G.; Kimler, B.F.; Tsue, T.T.; Lominska, C.; Saha, S. Auranofin Protects Intestine against Radiation Injury by Modulating P53/P21 Pathway and Radiosensitizes Human Colon Tumor. *Clin. Cancer Res.* **2019**, *25*, 4791–4807. [CrossRef]
126. Fiskus, W.; Saba, N.; Shen, M.; Ghias, M.; Liu, J.; Gupta, S.D.; Chauhan, L.; Rao, R.; Gunewardena, S.; Schorno, K.; et al. Auranofin Induces Lethal Oxidative and Endoplasmic Reticulum Stress and Exerts Potent Preclinical Activity against Chronic Lymphocytic Leukemia. *Cancer Res.* **2014**, *74*, 2520–2532. [CrossRef] [PubMed]
127. Ferraz, K.S.O.; Da Silva, J.G.; Costa, F.M.; Mendes, B.M.; Rodrigues, B.L.; dos Santos, R.G.; Beraldo, H. N(4)-Tolyl-2-Acetylpyridine Thiosemicarbazones and Their Platinum(II,IV) and Gold(III) Complexes: Cytotoxicity against Human Glioma Cells and Studies on the Mode of Action. *BioMetals* **2013**, *26*, 677–691. [CrossRef]
128. Jamali, F.; Lan, K.; Daniel, P.; Petrecca, K.; Sabri, S.; Abdulkarim, B. Synergistic Dual Targeting of Thioredoxin and Glutathione Systems Irrespective of P53 in Glioblastoma Stem Cells. *Antioxidants* **2024**, *13*, 1201. [CrossRef] [PubMed]
129. Szeliga, M.; Rola, R. Menadione Potentiates Auranofin-Induced Glioblastoma Cell Death. *Int. J. Mol. Sci.* **2022**, *23*, 15712. [CrossRef]
130. Martinez-Jaramillo, E.; Jamali, F.; Abdalbari, F.H.; Abdulkarim, B.; Jean-Claude, B.J.; Telleria, C.M.; Sabri, S. Pro-Oxidant Auranofin and Glutathione-Depleting Combination Unveils Synergistic Lethality in Glioblastoma Cells with Aberrant Epidermal Growth Factor Receptor Expression. *Cancers* **2024**, *16*, 2319. [CrossRef] [PubMed]
131. Finkelstein, A.; Walz, D.; Batista, V.; Mizraji, M.; Roisman, F.; Misher, A. Auranofin. New Oral Gold Compound for Treatment of Rheumatoid Arthritis. *Ann. Rheum. Dis.* **1976**, *35*, 251–257. [CrossRef]
132. Gromer, S.; Arscott, L.D.; Williams, C.H.; Schirmer, R.H.; Becker, K. Human Placenta Thioredoxin Reductase. *J. Biol. Chem.* **1998**, *273*, 20096–20101. [CrossRef] [PubMed]
133. Powis, G.; Montfort, W.R. Properties and Biological Activities of Thioredoxins. *Annu. Rev. Biophys. Biomol. Struct.* **2001**, *30*, 421–455. [CrossRef] [PubMed]
134. Karlenius, T.C.; Tonissen, K.F. Thioredoxin and Cancer: A Role for Thioredoxin in All States of Tumor Oxygenation. *Cancers* **2010**, *2*, 209–232. [CrossRef]
135. Onodera, T.; Momose, I.; Kawada, M. Potential Anticancer Activity of Auranofin. *Chem. Pharm. Bull.* **2019**, *67*, 186–191. [CrossRef]
136. Zhang, H.; Du, Y.; Zhang, X.; Lu, J.; Holmgren, A. Glutaredoxin 2 Reduces Both Thioredoxin 2 and Thioredoxin 1 and Protects Cells from Apoptosis Induced by Auranofin and 4-Hydroxynonenal. *Antioxid. Redox Signal.* **2014**, *21*, 669–681. [CrossRef] [PubMed]
137. Liu, X.; Zhang, Y.; Lu, W.; Han, Y.; Yang, J.; Jiang, W.; You, X.; Luo, Y.; Wen, S.; Hu, Y.; et al. Mitochondrial TXNRD3 Confers Drug Resistance via Redox-Mediated Mechanism and Is a Potential Therapeutic Target in Vivo. *Redox Biol.* **2020**, *36*, 101652. [CrossRef]
138. Zhao, F.; Yan, J.; Deng, S.; Lan, L.; He, F.; Kuang, B.; Zeng, H. A Thioredoxin Reductase Inhibitor Induces Growth Inhibition and Apoptosis in Five Cultured Human Carcinoma Cell Lines. *Cancer Lett.* **2006**, *236*, 46–53. [CrossRef] [PubMed]
139. Myers, J.M.; Myers, C.R. The Effects of Hexavalent Chromium on Thioredoxin Reductase and Peroxiredoxins in Human Bronchial Epithelial Cells. *Free Radic. Biol. Med.* **2009**, *47*, 1477–1485. [CrossRef]
140. Eriksson, S.E.; Prast-Nielsen, S.; Flaberg, E.; Szekely, L.; Arnér, E.S.J. High Levels of Thioredoxin Reductase 1 Modulate Drug-Specific Cytotoxic Efficacy. *Free Radic. Biol. Med.* **2009**, *47*, 1661–1671. [CrossRef] [PubMed]

141. Cassidy, P.B.; Edes, K.; Nelson, C.C.; Parsawar, K.; Fitzpatrick, F.A.; Moos, P.J. Thioredoxin Reductase Is Required for the Inactivation of Tumor Suppressor P53 and for Apoptosis Induced by Endogenous Electrophiles. *Carcinogenesis* **2006**, *27*, 2538–2549. [CrossRef] [PubMed]
142. Anestål, K.; Prast-Nielsen, S.; Cenas, N.; Arnér, E.S.J. Cell Death by SecTRAPs: Thioredoxin Reductase as a Prooxidant Killer of Cells. *PLoS ONE* **2008**, *3*, e1846. [CrossRef] [PubMed]
143. Toyokuni, S.; Okamoto, K.; Yodoi, J.; Hiai, H. Persistent Oxidative Stress in Cancer. *FEBS Lett.* **1995**, *358*, 1–3. [CrossRef] [PubMed]
144. Trachootham, D.; Alexandre, J.; Huang, P. Targeting Cancer Cells by ROS-Mediated Mechanisms: A Radical Therapeutic Approach? *Nat. Rev. Drug Discov.* **2009**, *8*, 579–591. [CrossRef] [PubMed]
145. Chen, L.; Ren, A.; Zhao, Y.; Chen, H.; Wu, Q.; Zheng, M.; Zhang, Z.; Zhang, T.; Zhong, W.; Lin, J.; et al. Direct Inhibition of Dioxygenases TET1 by the Rheumatoid Arthritis Drug Auranofin Selectively Induces Cancer Cell Death in T-ALL. *J. Hematol. Oncol.* **2023**, *16*, 113. [CrossRef]
146. Subramanian, A.; Narayan, R.; Corsello, S.M.; Peck, D.D.; Natoli, T.E.; Lu, X.; Gould, J.; Davis, J.F.; Tubelli, A.A.; Asiedu, J.K.; et al. A Next Generation Connectivity Map: L1000 Platform and the First 1,000,000 Profiles. *Cell* **2017**, *171*, 1437–1452.e17. [CrossRef]
147. Jeon, K.-I.; Jeong, J.-Y.; Jue, D.-M. Thiol-Reactive Metal Compounds Inhibit NF-KB Activation by Blocking IκB Kinase. *J. Immunol.* **2000**, *164*, 5981–5989. [CrossRef] [PubMed]
148. Mitchell, S.; Vargas, J.; Hoffmann, A. Signaling via the NFκB System. *WIREs Syst. Biol. Med.* **2016**, *8*, 227–241. [CrossRef]
149. Hu, M.C.-T.; Lee, D.-F.; Xia, W.; Golfman, L.S.; Ou-Yang, F.; Yang, J.-Y.; Zou, Y.; Bao, S.; Hanada, N.; Saso, H.; et al. IκB Kinase Promotes Tumorigenesis through Inhibition of Forkhead FOXO3a. *Cell* **2004**, *117*, 225–237. [CrossRef] [PubMed]
150. Liu, N.; Li, X.; Huang, H.; Zhao, C.; Liao, S.; Yang, C.; Liu, S.; Song, W.; Lu, X.; Lan, X.; et al. Clinically Used Antirheumatic Agent Auranofin Is a Proteasomal Deubiquitinase Inhibitor and Inhibits Tumor Growth. *Oncotarget* **2014**, *5*, 5453–5471. [CrossRef]
151. Huang, H.; Liao, Y.; Liu, N.; Hua, X.; Cai, J.; Yang, C.; Long, H.; Zhao, C.; Chen, X.; Lan, X.; et al. Two Clinical Drugs Deubiquitinase Inhibitor Auranofin and Aldehyde Dehydrogenase Inhibitor Disulfiram Trigger Synergistic Anti-Tumor Effects in Vitro and in Vivo. *Oncotarget* **2016**, *7*, 2796–2808. [CrossRef] [PubMed]
152. Zhang, X.; Selvaraju, K.; Saei, A.A.; D'Arcy, P.; Zubarev, R.A.; Arnér, E.S.; Linder, S. Repurposing of Auranofin: Thioredoxin Reductase Remains a Primary Target of the Drug. *Biochimie* **2019**, *162*, 46–54. [CrossRef] [PubMed]
153. Wang, Y.; Hill, K.S.; Fields, A.P. PKCι Maintains a Tumor-Initiating Cell Phenotype That Is Required for Ovarian Tumorigenesis. *Mol. Cancer Res.* **2013**, *11*, 1624–1635. [CrossRef] [PubMed]
154. Erdogan, E.; Lamark, T.; Stallings-Mann, M.; Jamieson, L.; Pellechia, M.; Thompson, E.A.; Johansen, T.; Fields, A.P. Aurothiomalate Inhibits Transformed Growth by Targeting the PB1 Domain of Protein Kinase Cι. *J. Biol. Chem.* **2006**, *281*, 28450–28459. [CrossRef]
155. Murray, N.R.; Kalari, K.R.; Fields, A.P. Protein Kinase Cι Expression and Oncogenic Signaling Mechanisms in Cancer. *J. Cell. Physiol.* **2011**, *226*, 879–887. [CrossRef]
156. Westover, D.; Li, F. New Trends for Overcoming ABCG2/BCRP-Mediated Resistance to Cancer Therapies. *J. Exp. Clin. Cancer Res.* **2015**, *34*, 159. [CrossRef]
157. Merlino, A. Metallodrug Binding to Serum Albumin: Lessons from Biophysical and Structural Studies. *Coord. Chem. Rev.* **2023**, *480*, 215026. [CrossRef]
158. Strickley, R.G. Solubilizing Excipients in Oral and Injectable Formulations. *Pharm. Res.* **2004**, *21*, 201–230. [CrossRef] [PubMed]
159. Afzalipour, R.; Khoei, S.; Khoee, S.; Shirvalilou, S.; Raoufi, N.J.; Motevalian, M.; Karimi, M.Y. Thermosensitive Magnetic Nanoparticles Exposed to Alternating Magnetic Field and Heat-Mediated Chemotherapy for an Effective Dual Therapy in Rat Glioma Model. *Nanomed. Nanotechnol. Biol. Med.* **2021**, *31*, 102319. [CrossRef]
160. Simonis, B.; Vignone, D.; Gonzalez Paz, O.; Donati, E.; Falchetti, M.L.; Bombelli, C.; Cellucci, A.; Auciello, G.; Fini, I.; Galantini, L.; et al. Transport of Cationic Liposomes in a Human Blood Brain Barrier Model: Role of the Stereochemistry of the Gemini Amphiphile on Liposome Biological Features. *J. Colloid Interface Sci.* **2022**, *627*, 283–298. [CrossRef] [PubMed]
161. Afrifa, M.A.O.; Kim, J.H.; Pitton, K.A.; Olelewe, C.; Arojojoye, A.S.; Strachan, D.R.; Suckow, M.A.; Awuah, S.G. Auranofin-Loaded Chitosan Nanoparticles Demonstrate Potency against Triple-Negative Breast Cancer. *ACS Appl. Bio Mater.* **2024**, *7*, 2012–2022. [CrossRef]
162. Pérez-Lloret, M.; Reidy, E.; Lozano-Pérez, A.A.; Marchal, J.A.; Lens, P.N.L.; Ryan, A.E.; Erxleben, A. Auranofin Loaded Silk Fibroin Nanoparticles for Colorectal Cancer Treatment. *Drug Deliv. Transl. Res.* **2024**. [CrossRef] [PubMed]
163. Díez-Martínez, R.; García-Fernández, E.; Manzano, M.; Martínez, Á.; Domenech, M.; Vallet-Regí, M.; García, P. Auranofin-Loaded Nanoparticles as a New Therapeutic Tool to Fight Streptococcal Infections. *Sci. Rep.* **2016**, *6*, 19525. [CrossRef] [PubMed]
164. Kushawaha, S.K.; Ashawat, M.S.; Baldi, A. Auranofin-Loaded PLGA Nanoparticles Alleviate Cognitive Deficit Induced by Streptozotocin in Rats Model: Modulation of Oxidative Stress, Neuroinflammatory Markers, and Neurotransmitters. *Naunyn. Schmiedebergs. Arch. Pharmacol.* **2024**, *397*, 10031–10047. [CrossRef] [PubMed]
165. Misra, A.; Ganesh, S.; Shahiwala, A.; Shah, S.P. Drug Delivery to the Central Nervous System: A Review. *J. Pharm. Pharm. Sci.* **2003**, *6*, 252–273.

166. Temsamani, J.; Scherrmann, J.M.; Rees, A.R.; Kaczorek, M. Brain Drug Delivery Technologies: Novel Approaches for Transporting Therapeutics. *Pharm. Sci. Technol. Today* **2000**, *3*, 155–162. [CrossRef] [PubMed]
167. Lesniak, M.S. Novel Advances in Drug Delivery to Brain Cancer. *Technol. Cancer Res. Treat.* **2005**, *4*, 417–428. [CrossRef]
168. Kabanov, A.; Batrakova, E. New Technologies for Drug Delivery Across the Blood Brain Barrier. *Curr. Pharm. Des.* **2005**, *10*, 1355–1363. [CrossRef]
169. Cui, Y.; Xu, Q.; Chow, P.K.-H.; Wang, D.; Wang, C.-H. Transferrin-Conjugated Magnetic Silica PLGA Nanoparticles Loaded with Doxorubicin and Paclitaxel for Brain Glioma Treatment. *Biomaterials* **2013**, *34*, 8511–8520. [CrossRef] [PubMed]
170. Porru, M.; Zappavigna, S.; Salzano, G.; Luce, A.; Stoppacciaro, A.; Balestrieri, M.L.; Artuso, S.; Lusa, S.; De Rosa, G.; Leonetti, C.; et al. Medical Treatment of Orthotopic Glioblastoma with Transferrin-Conjugated Nanoparticles Encapsulating Zoledronic Acid. *Oncotarget* **2014**, *5*, 10446–10459. [CrossRef] [PubMed]
171. Narsinh, K.H.; Perez, E.; Haddad, A.F.; Young, J.S.; Savastano, L.; Villanueva-Meyer, J.E.; Winkler, E.; de Groot, J. Strategies to Improve Drug Delivery Across the Blood–Brain Barrier for Glioblastoma. *Curr. Neurol. Neurosci. Rep.* **2024**, *24*, 123–139. [CrossRef] [PubMed]
172. Rapoport, S.I. Osmotic Opening of the Blood-Brain Barrier: Principles, Mechanism, and Therapeutic Applications. *Cell. Mol. Neurobiol.* **2000**, *20*, 217–230. [CrossRef]
173. Dauba, A.; Delalande, A.; Kamimura, H.A.S.; Conti, A.; Larrat, B.; Tsapis, N.; Novell, A. Recent Advances on Ultrasound Contrast Agents for Blood-Brain Barrier Opening with Focused Ultrasound. *Pharmaceutics* **2020**, *12*, 1125. [CrossRef] [PubMed]
174. Meng, Y.; Pople, C.B.; Lea-Banks, H.; Abrahao, A.; Davidson, B.; Suppiah, S.; Vecchio, L.M.; Samuel, N.; Mahmud, F.; Hynynen, K.; et al. Safety and Efficacy of Focused Ultrasound Induced Blood-Brain Barrier Opening, an Integrative Review of Animal and Human Studies. *J. Control. Release* **2019**, *309*, 25–36. [CrossRef] [PubMed]
175. Bouakaz, A.; Michel Escoffre, J. From Concept to Early Clinical Trials: 30 Years of Microbubble-Based Ultrasound-Mediated Drug Delivery Research. *Adv. Drug Deliv. Rev.* **2024**, *206*, 115199. [CrossRef] [PubMed]
176. Timbie, K.F.; Afzal, U.; Date, A.; Zhang, C.; Song, J.; Wilson Miller, G.; Suk, J.S.; Hanes, J.; Price, R.J. MR Image-Guided Delivery of Cisplatin-Loaded Brain-Penetrating Nanoparticles to Invasive Glioma with Focused Ultrasound. *J. Control. Release* **2017**, *263*, 120–131. [CrossRef] [PubMed]
177. Li, Y.; North, R.Y.; Rhines, L.D.; Tatsui, C.E.; Rao, G.; Edwards, D.D.; Cassidy, R.M.; Harrison, D.S.; Johansson, C.A.; Zhang, H.; et al. DRG Voltage-Gated Sodium Channel 1.7 Is Upregulated in Paclitaxel-Induced Neuropathy in Rats and in Humans with Neuropathic Pain. *J. Neurosci.* **2018**, *38*, 1124–1136. [CrossRef] [PubMed]
178. Wang, Y.; Cao, B.; Wang, Q.; Fang, X.; Wang, J.; Chan, A.S.C.; Xiong, X.; Zou, T. Ligand Supplementation Restores the Cancer Therapy Efficacy of an Antirheumatic Drug Auranofin from Serum Inactivation. *bioRxiv* **2024**. [CrossRef]

International Journal of
Molecular Sciences

MDPI

Review

Retinal Pigment Epithelium Under Oxidative Stress: Chaperoning Autophagy and Beyond

Yuliya Markitantova * and Vladimir Simirskii

Koltzov Institute of Developmental Biology, Russian Academy of Sciences, 119334 Moscow, Russia; simir@mail.ru
* Correspondence: yuliya.mark@gmail.com; Tel.: +7-(499)1350052

Abstract: The structural and functional integrity of the retinal pigment epithelium (RPE) plays a key role in the normal functioning of the visual system. RPE cells are characterized by an efficient system of photoreceptor outer segment phagocytosis, high metabolic activity, and risk of oxidative damage. RPE dysfunction is a common pathological feature in various retinal diseases. Dysregulation of RPE cell proteostasis and redox homeostasis is accompanied by increased reactive oxygen species generation during the impairment of phagocytosis, lysosomal and mitochondrial failure, and an accumulation of waste lipidic and protein aggregates. They are the inducers of RPE dysfunction and can trigger specific pathways of cell death. Autophagy serves as important mechanism in the endogenous defense system, controlling RPE homeostasis and survival under normal conditions and cellular responses under stress conditions through the degradation of intracellular components. Impairment of the autophagy process itself can result in cell death. In this review, we summarize the classical types of oxidative stress-induced autophagy in the RPE with an emphasis on autophagy mediated by molecular chaperones. Heat shock proteins, which represent hubs connecting the life supporting pathways of RPE cells, play a special role in these mechanisms. Regulation of oxidative stress-counteracting autophagy is an essential strategy for protecting the RPE against pathological damage when preventing retinal degenerative disease progression.

Keywords: retinal pigment epithelium; phagocytosis; oxidative stress; proteostasis system; mitochondrial dysfunction; ubiquitin–proteasome system; autophagy; heat shock proteins; chaperone mediated autophagy; lysosomes; programmed cell death

1. Introduction

The state of retinal pigment epithelium (RPE) cells plays a central role in the viability and homeostasis of the neural retina. The RPE provides functional support for the retinal neurons and the choroid, thanks to the coordinated components in the endogenous defense system [1–3]. Interactions between the RPE and neighboring tissues such as the choroid and photoreceptors maintain the local RPE and vision functions [4,5].

RPE cells and photoreceptors are characterized by high metabolic activity, due to their involvement in the mechanisms of the visual cycle. During the visual cycle and metabolism of rhodopsin, the isomerization of 11-cis-retinal to trans-retinal leads to chain of biochemical reactions and as a result to the production of a large amount of reactive oxygen species (ROS), causing lipid peroxidation (LPO) [6,7]. The high metabolism rate in the RPE and retina induces oxidative stress (OS) in these tissues. The key physiological function of the RPE is associated with protecting photoreceptors from excess light and ROS by means of endogenous defense systems [7–9]. The endogenous defense systems of the RPE provide antioxidant protection and redox homeostasis and promote the remodeling of damaged proteins and their repair and/or degradation through autophagy or the ubiquitin–proteasome system [10,11].

Structural, metabolic, and genetic disorders of RPE cells and neighboring tissue structures (the choroid, Bruch's membrane, and photoreceptors) lead to the accumulation of toxic components, which creates a risk of redox disbalance and OS development [12,13]. RPE cell defects and dysfunction are associated with disorders in the key components of the endogenous defense system in the RPE and contribute to the development of retinal diseases, such as age-related macular degeneration (AMD) and *proliferative vitreoretinopathy* (PVR) [14–16].

Maintaining the structural and functional integrity and homeostatic balance in RPE cells constitutes an urgent problem in biomedicine and ophthalmology. The regulatory network of the endogenous defense system defines the physiological functions of the RPE cell layer, ensuring retinal neuron viability and homeostasis and enabling adaptation in response to cellular stress [17].

A number of therapeutic approaches are aimed at maintaining the stability of intercellular contacts and the functionality of RPE cells. Thus, the use of factors such as nicotinamide or lysophosphatidic acid contributes to the formation of the tight junctions in the RPE [18,19]. Several proposed therapeutic strategies targeting RPE cells are aimed at activating the antioxidant defense and alleviating both OS and mitochondrial and lysosomal dysfunction, which are the main mechanisms of RPE dysfunction resulting in AMD. Photobiomodulation agents can also remove drusen in AMD by activating RPE phagocytosis [15,20–22].

Regulation of RPE cell dysfunction aimed at preventing RPE-related neurodegenerative pathologies is a priority for modern research. Potential therapeutic strategies involve targeting autophagy as an important biological process to maintain homeostasis in RPE cells [23–25].

In this review, we have performed an analysis of the current state of the research focusing on the endogenous antioxidant defense system in the RPE with a focus on autophagy and molecular chaperones, highlighting them as potential targets as therapies for RPE-related cell death and retinal degenerative pathologies. The molecular changes and signaling pathways involving heat shock proteins (HSPs) that control the RPE response to injury remain largely undefined. Given the important roles of HSPs in the RPE in retinal disease and health, we reviewed recent data on their roles in the molecular protection of RPE cells obtained using modern molecular genetic methods. It would be of interest to fully characterize their roles in the molecular defense responses of RPE cells to injury as a result of trauma or disease affecting the RPE and the retina. Autophagy is a key example of such molecular defense responses.

2. RPE Functions for Maintaining the Viability of Photoreceptors

RPE cells form a monolayer of polygonal pigmented cells. RPE cells are located between the choroidal vasculature of the retina and the outer segments of retinal photoreceptors. RPE cells play various roles in the retina and the choroid, maintaining the homeostasis of these tissues [17]. On the basal side, RPE cells interact with Bruch's membrane, which is closely associated with the choroid. On the apical side, RPE cells interdigitate with photoreceptor outer segments (POS), forming tight intercellular contacts [17,26,27].

Numerous vital RPE functions include (1) the utilization of shed POS discs (by phagocytosis); (2) maintaining the visual cycle in retina (by degrading or recycling components of the visual cascade such as visual chromophore 11-cis-retinal); (3) the absorption of light energy (by melanosome bleaching); (4) the maintenance of redox homeostasis (by free radicals and ROS scavenging as well as the activities of antioxidant enzymes and chaperone systems); (5) the blood–retinal barrier (through the formation of Bruch's membrane and tight junctions); and (6) the transport of nutrients and metabolites to the photoreceptors and choroid (through secretion, pinocytosis, endocytosis, and exocytosis) [17,20,28–32].

RPE cells constantly renew the "exhausted" photoreceptor discs by phagocytosis of the used POS, which are subjected to autophagic and lysosomal degradation. POS discs are enriched in ROS and lipofuscin (LF), which is the main product of LPO [9]. Superoxide radicals generated by LF under the action of visible light have a destructive effect on RPE melanosomes, which perform screening and antioxidant functions [33].

A high metabolism rate of RPE cells, high oxygen consumption and intensity of energy metabolism processes, a constant renewal of POS membrane discs using phagocytosis, high levels of polyunsaturated fatty acids and exposure to light are the main RPE characteristics needed to maintain a normal physiological state of the photoreceptors and the retinal function [1]. On the apical side of RPE cells, abundant melanosomes, which absorb stray light during the visual function, are distinguished from the nucleus and melano-LF granules on the basal side [34–36].

Intense mitochondrial metabolism, the phagocytosis of POS discs, the phototoxic activity of LF, and the photosensitization of hemoglobin precursors are the main sources of ROS, both in the RPE and photoreceptors [12]. In differentiated RPE cells, redox homeostasis largely depends on autophagic clearance and the intensity of the accumulation of cellular debris, including LF. With age, the intracellular redox balance shifts toward the intensification of oxidative reactions, which enhances the production of ROS [37].

3. Major Intercellular and Molecular Events That Contribute to Oxidative Damage in RPE Cells

In retinal cells, ROS are produced by lysosomes (phagosomes), peroxisomes, melanosomes, and intracellular membrane-bound NADPH oxidases [38,39].

Disruption of RPE polarity as a result of a disrupted interactions with the photoreceptors is accompanied by a decrease in intercellular adhesion, reduced barrier function, increased ROS production, migration of RPE cells, and retinal detachment, leading to several RPE-related diseases, such as PVR and AMD [40–42] (Figure 1).

The destructive influence of ROS manifests itself in the oxidation of proteins and membrane lipids and can result in DNA damage [43]. Excessive intracellular ROS production leads to the loss of the inherent structure of some proteins and to their aggregation. This can contribute to endoplasmic reticulum (ER) stress and cause a response to unstructured proteins. The accumulation of aggregated proteins increases OS, leading to lysosome damage and inflammation [44,45].

The structural disorders and dysfunction of RPE cells impair photoreceptor metabolism and are caused by an excessive accumulation of LF, releasing lysosomal-degrading enzymes into the cytosol, which can lead to cell death [46,47]. The process of LF formation is accompanied by an inhibition of proteasome activity in the RPE [48]. In turn, a decrease in proteasome activity is accompanied by the accumulation of LF in the RPE and a decrease in the intensity of autophagy [10,49].

ROS overproduction in the mitochondria contributes to ROS production and can result in mitochondrial DNA damage [50]. Under OS, the mitochondria may aggravate the production of ROS, which can lead to cell apoptosis [37].

RPE cells have abundant mitochondria to provide for the energy needs of the outer retina cells [51]. Damaged respiration due to mitochondrial dysfunction in RPE cells is one of the main links to AMD pathogenesis [22,52]. ROS levels can increase as a result of interactions between the mitochondria and cytochrome C oxidase [12,53].

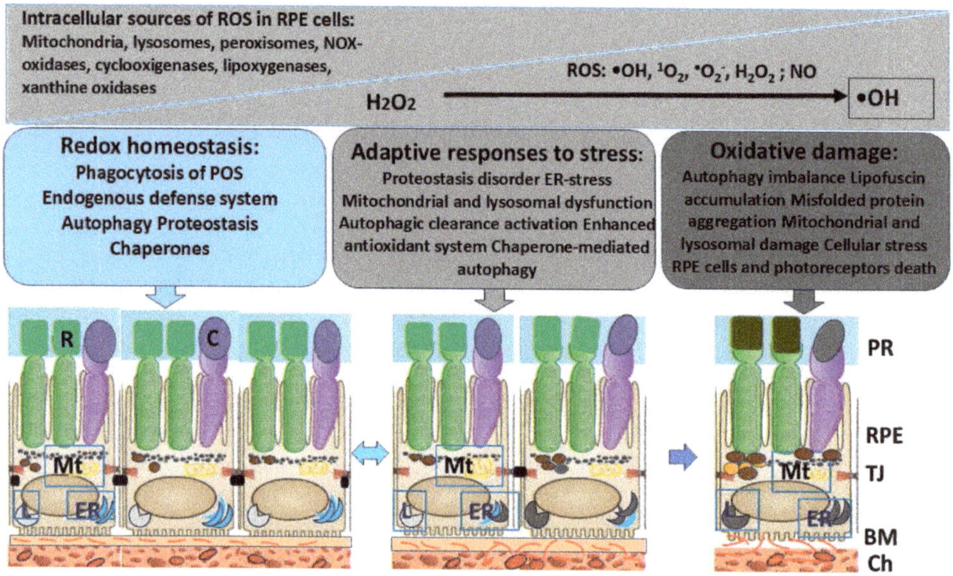

Figure 1. The main endogenous defense system that maintains RPE cell homeostasis. Given that the RPE is vital for key photoreceptor function, RPE dysfunction may lead to photoreceptor degeneration and severe visual impairment. RPE cells are characterized by a high metabolic activity and a predisposition for oxidative damage. The balance of RPE cell homeostasis is achieved through the coordinated work of endogenous regulatory systems, where the proteolysis and antioxidant protection systems play an important role. Disruption of redox homeostasis by endogenous and exogenous sources of ROS can lead to autophagy clearance activation, enhancement of the antioxidant system, chaperone-mediated autophagy activation, and, as a result, to an adaptive cell response. If these systems fail, oxidative stress can lead to the development of pathological processes and to the death of the RPE and photoreceptors. Abbreviations: TJ—tight junctions; BM—Bruch's membrane; Ch—choroid; PR—photoreceptors; R—rod; C—cone; ER—endoplasmic reticulum; Mt—mitochondria; L—lysosomes; POS—photoreceptor outer segments; ROS—reactive oxygen species.

Exposure of LF from melano-LF granules to light manifests itself in the production of ROS, which destroys the melanin component. With age and the development of degenerative diseases of the RPE and photoreceptors, a large number of melano-LF granules are formed. Excessive accumulation of melano-LF granules leads to a decrease in and even disappearance of melanin in pigmented RPE cells, thereby reducing the protection of the cells from ROS [34].

It has been shown that impaired lysosomal degradation owing to the accumulation of LF is closely related to autophagy disorders in AMD. Activation of lysosomal proteases from the cathepsin class in the RPE promotes the destruction of POS, as well as the formation of end products of LPO and oxidized low-density lipoproteins [54]. Their accumulation leads to the excessive accumulation of LF. In turn, excessive intracellular deposition of LF enhances OS and the formation of autolysosomes and drusen in the RPE [55]. The impaired lysosomal degradation due to the accumulation of LF is closely related to autophagy disorders and apoptosis, as shown during AMD. Inflammasome priming during inflammation development in retinal pigment epithelial cells increases their susceptibility to LF phototoxicity, thereby switching from the cell death mechanism via apoptosis to the pyroptosis pathway [54,56].

4. Endogenous Defense System in RPE Activated in Response to OS

In response to OS conditions, the intensity of phagocytosis in RPE cells increases. This also manifests itself in the increased phagocytosis of the apoptotic fragments of dying cells and the utilization of Bruch's membrane metabolic products [9].

The endogenous defense systems of RPE cells include the antioxidant system, different kinds of autophagy, and a multilevel system of chaperones. Antioxidant enzymes (superoxide dismutase [SOD], catalase, and cytochrome P450 monooxygenase) and non-enzymatic molecules (thioredoxin, glutathione ascorbate, and β-carotene) [57] take part in maintaining redox homeostasis. Among them, SOD2 is a primary enzyme activated in response to OS and protects the cells from damage by removing ROS and maintains the function of mitochondria [53]. Under normal conditions, these systems maintain cellular homeostasis, but they can activate any kind of programmed cell death under stress conditions (Figure 2).

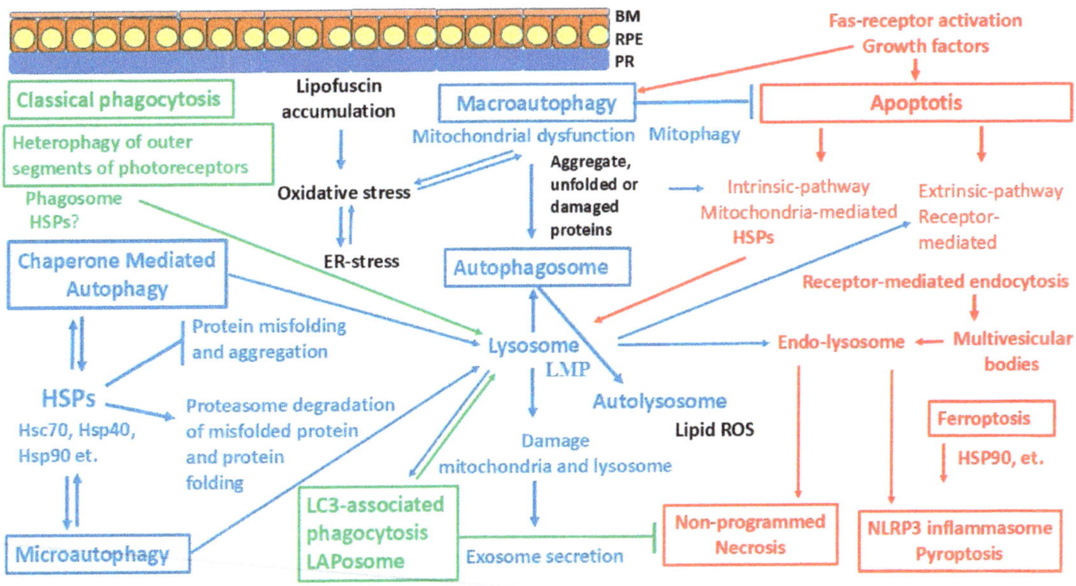

Figure 2. Schematic presentation of the intersection between degradative pathways in RPE cells. HSP involvement in phagocytosis, different types of autophagy, and cell death. RPE cells are constantly subjected to phagocytosis of photoreceptor outer segment discs and exposed to oxidative stress. Degradative pathways in RPE cells and associated processes are indicated by arrows with the corresponding colors: types of phagocytosis (green), types of autophagy (blue), and forms of cell death (red). Photoreceptor outer segment discs are phagocytosed by the RPE (heterophagy) and are degraded in the lysosomes in the process of LAP. LAP bridges the phagocytic and canonical autophagic pathways. Abbreviations: RPE—retinal pigment epithelium; ROS—reactive oxygen species; ER—endoplasmic reticulum; LMP—lysosome membrane permeabilization; LAP—LC3-associated phagocytosis; HSPs—heat shock proteins.

RPE pigments absorb light, protecting the retina against oxidative damage, continuous exposure to light, and the photo-oxidation of LF [58]. Melanin is a key component of the antioxidant system. The density of melanin is relatively high in the RPE cells adjacent to the central zone of the retina, while the highest density is found in the fovea [59]. A significant age-related decrease in the amount of melanin, as well as in the number of melanosomes that act as screening light filters and antioxidants in RPE cells, may lead to an increase in the risk of oxidative and photo-OS in the structures of the eye [34]. Melanin decreases the photo-oxidation of LF by shielding from the harmful bright light in the RPE [60] and removes ROS, thus contributing to protection from OS [35].

Autophagy is the key biological process necessary for maintaining cellular homeostasis by lysosomal degradation of unused and damaged cellular components [61]. In RPE cells, autophagy is closely related to the regulation of proteostasis and redox homeostasis [62].

5. Chaperone Defense System of Heat Shock Proteins in the RPE

Chaperone proteins play a critical role in all taxonomic groups and are involved in ensuring both the folding of newly synthesized peptides into their mature conformation, the refolding of misfolded proteins, and the movement of proteins to subcellular compartments [63]. They function under normal conditions and can be activated in response to internal and external stressors.

The most common and studied group of chaperones are HSPs. They can be categorized into several families based on their molecular size, structure, and function: small HSPs (HSPB), HSP40 (DNAJ), HSP60/HSP10 (HSPD/E), HSP70 (HSPA), HSP90 (HSPC), and HSP100/110 (HSPH) [64,65]. To date, up to 83 HSP genes have been identified in the human genome: 4 for HSP110, 4 for HSP90, 13 for HSP70, 1 for HSP60, 50 for HSP40, and 11 for small HSPs [65].

RPE cells are exposed to chronic OS from three sources: high levels of oxygen consumption, LPO from phagocytized POS, and light stimuli. Cells constitutively express many types of HSPs but increase their expression to stabilize and restore cellular homeostasis under stress (OS, hypoxia, high glucose, and others).

HSPs have been demonstrated to play a key role in cellular responses to stress, serving as an adaptive response center and stabilizing cellular structures [66]. Many HSPs are upregulated during cellular stress and are expressed with high tissue specificity. There is a strong connection between the stress regulation systems in cells and the mechanisms that control their growth and differentiation. HSPs not only act as molecular chaperones, but also participate in cell homeostasis and cell viability, as well as cell differentiation processes. The ability of HSPs to protect cells is a key aspect of their functioning to support cell viability [67,68].

αA-crystallin contributes to the activation of phosphorylation reactions in the PI3K/Akt signaling pathway, which ensures the resistance of RPE cells to the action of OS [69]. In the RPE, αB-crystallin protects proteins against aggregation and unfolding and stimulates the production of vascular endothelial growth factor (VEGF) in response to inflammation, which contributes to neovascularization in AMD [70].

In the RPE, OS induces the activation of redox-dependent antioxidants, DJ-1 chaperones and alpha-1 microglobulin. The latter can directly bind and neutralize ROS [71,72].

Almost all the proteins of the HSP family, with the exception of sHSPs, need ATP (adenosine 5′-triphosphate) hydrolysis for their activity. The interactions of Hsp60 with the co-chaperone Hsp10 control the substrate-binding and ATPase activities of Hsp60 [73]. Despite the fact that the Hsp60 protein is localized mainly in mitochondria, it can, like other stress proteins, change its intracellular localization in response to OS and is expressed on the cell membrane [74].

Each member of the HSP70 family can respond specifically in cells and tissues, depending on the type of stress [67]. For example, the human Hsp70B protein is strictly expressed in response to heat shock. This distinguishes Hsp70B from Hsp70, which is characterized by a basal level of expression and activation under different stress conditions [75].

HSP110 is involved in the proteolysis of α-synuclein and/or the prevention of associated neurodegeneration in mammals. The Hsp110 protein has a high structural and functional similarity to Hsp70 and performs cytoprotective functions against the negative effects of OS. It is assumed that this function of HSP110 as a co-chaperone is associated with the regulation of HSP70 expression [76].

Hsp90 proteins control the quality of proteins, ensuring their transport to the proteasomes for destruction in cases of protein defects that prevent them from performing their functions. Normally, Hsp90 protein expression is insignificant in the total content of all proteins in the cells, but its amount increases during stress [77]. During heat shock, HSP90 activity is necessary to restore the functions of misfolded and denatured proteins. The synthesis of Hsp90 also increases under the influence of other stress factors such as inflammation, which cause OS [77].

Neurodegenerative pathologies are associated with dysfunction of sHSPs [78]. Disorders in the expression of HSPs are also associated with inflammatory reactions, as well as with the death of pigment epithelial cells and retinal neurons in various eye pathologies [79,80]. A decrease in the amount of HSP proteins (HSP60, HSP70, and αA-crystallin) occurs against the background of increased OS in the human retina in AMD [81]. Intracellular aggregates formed by incorrectly folded proteins can form β-amyloid structures as a result of dysfunction of sHSPs. These structures are characterized by their high stability to proteolytic cleavage, leading to their excessive accumulation, which causes cell death. sHSPs are actively involved in the formation of elaborate protein complexes that can affect the activity of cytoskeletal proteins. It is noted that in Parkinson's disease, Lewy bodies consisting of sHSPs and neurofilaments are found in a complex of protein aggregates in the neurons, but the mechanisms of the accumulation of these complexes have not been clarified [82]. HSP27 also plays an important role in the organization of microfilaments. It interacts with actin filaments, is involved in programmed cell death (apoptosis), and is important for cell survival under stress conditions [83,84].

In general, HSPs (HSP70, HSP90 and others) primarily act as molecular chaperones, preventing cell damage from proteins with disrupted conformation caused by OS. HSPs also participate in the mechanisms of redox balance regulation in cells. In a recent study, a polymorphic variant of HSP70 was discovered, namely, HSP70-2, which is a sensor of the redox balance of cells and changes in OS. The activity of HSP70-2 is associated with the concentration of ROS in the cell, which plays a role in various functional disorders [85].

6. Proteolytic and Autophagy Defense Systems

As a result of LPO, which is rich in POS, and the metabolism of trans-retinol, RPE cells produce excess amounts of ROS, which initiate OS. An additional source of ROS is constituted by the products of photo-oxidation of LF [6,36]. RPE cells have numerous antioxidant systems that ensure the removal of excess ROS and the restoration of redox homeostasis [20,59]. One of the most universal systems for maintaining and restoring proteostasis is the chaperone system. The most studied chaperones are HSPs that provide a unique system to regulate the traffic of newly synthetized proteins between cellular compartments, to promote the refolding of misfolded proteins, and to inhibit the formation of toxic protein aggregates [86,87]. Among the defense mechanisms responsible for maintaining cellular homeostasis, HSPs provide the only way of restoring the configuration of unfolded or misfolded proteins [88]. Once the capacity of HSPs is exceeded, misfolded, aggregated, and damaged proteins, as well as damaged organelles, undergo degradation using proteolytic systems. Four different well-coordinated systems are responsible for protein degradation in eukaryotic cells: (1) the proteasome-based UPS, which degrades most long- and short-lived normal and abnormal intracellular proteins; (2) mitochondrial proteases, which degrade mitochondrial proteins; (3) calcium-activated calpains, which degrade membrane and cytoskeletal proteins and several membrane-associated enzymes; and (4) lysosome-based autophagy, which degrades cellular organelles, membrane proteins, and endocytosed proteins [89,90]. Both autophagy and proteasomal clearance are especially crucial for cell types with no proliferation, such as RPE cells. These systems regulate cellular homeostasis, control the quality of mitochondria and the production of ROS under conditions of normal functioning, and provide antioxidant protection to RPE cells under OS conditions [91,92].

Proteasomes are efficient in the degradation of small and short-lived proteins, whereas larger and longer-lived substrates (including lipids) are degraded by autophagy [93]. Despite the unique role of proteasomes in the regulation of cellular homeostasis, they are functionally related and can act together, which is especially pronounced under OS conditions. Proteasome inhibition may be compensated by increased autophagy [94].

In both systems, molecular chaperones play important roles in the recognition and selection of the proteins or organelles to be degraded [64,95,96].

6.1. *Ubiquitin–Proteasome System (UPS)*

UPS-mediated protein degradation is a multistep process involving various proteins. In particular, a protein to be degraded is first labeled with ubiquitin and then recognized and degraded by a proteasome [97]. Proteasomes are multi-protein complexes responsible for the selective degradation of misfolded and high-turnover proteins in the cytoplasm. These proteins have many proteolytically active sites that break down proteins into peptides. The 26S proteasome is highly conserved throughout eukaryotes, where it is found in both the nucleus and cytoplasm [98]. The proteasome is composed of 33 subunits assembled in two sub-complexes, as well as the 20S core particle, which bears the actual protease active sites and is flanked at one or both ends by the 19S regulatory particle to form the 26S proteasome [99]. Proteasome assembly requires the assistance of proteasome assembly chaperones. In mammals, four evolutionarily conserved 19S regulatory particle assembly chaperones (PACs), including p27, p28, S5b, and Rpn14/PAAF1, are needed for regulatory particle assembly [98,100]. Formation of the proteasomal 20S core complex relies on the function of the other five chaperones PAC1-PAC4 (proteasome assembly chaperones 1–4) and POMP (proteasome maturation protein) [101,102].

Involvement of αB-crystallin in protein degradation pathways has also been discovered. αB-Crystallin interaction with C8/a7, one of the 14 subunits of the 20S proteasome, has been reported both in vitro and in vivo. This interaction is highly specific since C8/a7 does not bind to αA-crystallin or HSP27 [103]. A mutation in αB-crystallin decreased Atg7 expression, which is a mediator of autophagosomal biogenesis [104]. αB-crystallin can also promote the degradation of certain misfolded proteins that cannot be converted to their native state after repeated cycling through the chaperone systems by the proteasome [105,106]. The opposite role is played by HSP90, which interacts with and protects the transcription factor SP1 from degradation in the ubiquitin–proteasome pathway [107].

Degradation of a protein by the UPS is typically mediated by ubiquitination of the target protein, which involves the energy-dependent covalent attachment of the small protein ubiquitin (Ub) to one or more lysines within the protein via the concerted action of three substrate-specific E1-E3 Ub-protein ligases [98]. Normally, the cell maintains a balance between the formation of ROS and their neutralization [108]. Disruption of redox homeostasis can lead to OS and the formation of oxidation products of proteins, lipids, and other macromolecules [109]. Proteasomal degradation of oxidized proteins mainly occurs in the cytosol, which is zone of greatest production of oxidized proteins. On the other hand, the nucleus is well protected against the formation of oxidized proteins and their accumulation, as a result the high proteasome content [110]. Among these proteins, only slightly oxidized forms of proteins re suitable substrates for the proteasome. Highly oxidized proteins are likely to exist in stable aggregates due to covalent cross-links, disulfide bonds, or hydrophobic interactions. Such proteins aggregate, together with oxidized lipids, thereby forming LF granules. They are no longer suitable as a substrate for proteasomes and are degraded by lysosomes [111]. Human AMD donor RPE in which OS is activated exhibited a significantly higher content of the proteasome as well as HSP27 and HSP90 [112].

6.2. Phagocytosis

POS discs are constantly shed and subsequently phagocytosed by the RPE before new outer segments are constructed at the cilium. RPE cells are the most active phagocytic cells in the human body, and defects in the phagocytic process lead to impaired retinal function [113].

The process of POS phagocytosis is triggered by ligand secretion by the RPE, which provides specific binding of the POS discs to the RPE apical membrane. One of these ligands, milk fat globule-EGF8 (MFGE8), binds to exposed phosphatidylserines on POS fragments, bridging them to aVb5 integrin receptors on the apical surface of the RPE [114]. This binding initiates two downstream signaling cascades. On the one hand, it stimulates Mer receptor tyrosine kinase (MERTK) via aVb5 integrin-associated focal adhesion kinase [115]. On the other hand, it activates RAC1-GTPase [116], which leads to F-actin recruitment to the phagocytic cup.

The rearrangement of cytoskeletal filamentous actin RPE (F-actin) is required for POS internalization [117]. The phagocytic cup, formed by the ordered aggregation of F-actin in combination with POS, is the key to the initiation of phagocytosis [118]. Cyclic TCP-1 complex, also known as chaperonin-containing TCP-1 (CCT), consists of eight parallel subunits (CCT1–8) [119]. A recent study showed that CCT is required for efficient assembly of actin myofilaments [120], and CCT5-specific ATP binding is required for efficient actin folding [121]. In addition, CCT5 controls lysosome biogenesis through the actin cytoskeleton [122]. The association of HSP90 with F-actin, but not with α-tubulin, is important for phagosome formation. Silencing of HSP90 (siHSP90) reduced expression of cytoskeletal proteins and the phagosome marker (Rab5) and successfully diminished phagocytosis in U937-derived macrophages [123].

Mesoderm development candidate 2 (Mesd or Mesdc2) has been identified as facilitating phagocytosis in the RPE. Mesd is an ER-located chaperone that facilitates the folding of the low-density lipoprotein receptor-related protein (LRP) family [124]. Mesd is predominantly expressed in POS fragments, which are an ER-free cellular compartment [125]. It has been demonstrated that Mesd can be released from shedding POS fragments and stimulates their phagocytosis by RPE cells binding to phagosomes [126].

It is known that most lysosomal RPE enzymes function within a narrow pH range in the acidic environment of the lysosomal lumen [127]. The acidic luminal environment is primarily created by the vacuolar-type H+-ATPase (V-ATPase), which pumps protons into the lumen [128]. In RPE cells, CRYBA1/βA3/A1-crystallin is found in lysosomes, where it functions as a regulator of endolysosomal acidification by modulating V-ATPase, to control of both phagocytosis and autophagy. CRYBA/βA3/A1-crystallin 1 directly influences lysosomal V-ATPase activity. Crosstalk between V-ATPase and mTORC1 is required for the regulation of autophagy, and CRYBA1 regulates the activity of the mTORC1 signaling pathway [129].

6.3. LC3-Associated Phagocytosis

A new type of phagocytosis has been described in the RPE, where the maturing phagosome acquires microtubule-associated protein 1 light chain 3B (LC3B) classically associated with macroautophagy. This hybrid autophagy–phagocytosis degradation pathway is termed LC3-associated phagocytosis (LAP) [130–132]. Microtubule-associated protein 1 light chain 3B (LC3B) mediates the physical interactions between microtubules and components of the cytoskeleton. LAP and macroautophagy both rely on the lipidation of LC3 to LC3II, which ensures its membrane link with double-membrane autophagosomes in autophagy or single-membrane phagosomes in LAP. Lipidation of LC3 involves a complex of Atg5-12 and Atg16L; these factors act in cooperation as an E3-like enzyme to transfer phosphatidylethanolamine (PE) to LC3 [133]. The lipidated form of the autophagy protein LC3 binds to the phagosome in a manner dependent on Atg5 and Beclin1, but independent of the autophagy preinitiation complex including the Ulk1/Atg13/Fip200 [134]. The composition of the PI3K complex consisting of subunits involved in LAP also differs from that of autophagy. The LAP complex contains the RUN domain protein Rubicon, while the autophagy complex includes Atg14 [131]. Phagosomes require Atg5 to move through the RPE and enter the lysosomal compartment for degradation. In addition to the degradation of shed POS discs, this noncanonical form of autophagy supports optimal visual function by supplying a portion of the retinoids required for chromophore synthesis [132].

LAP coexists in the RPE with canonical phagocytosis [132]. New adapter proteins have been found that bind LC3, targeting specific organelles to lysosomes. One of them is melanoregulin, which is specific for phagosomes [130]. Notably, only part of the POS-phagosome is degraded by LAP. A significant portion of POS phagosomes are ubiquitinated and, as a result, can be included in phagosomes that have LC3B, melanoregulin, or SQSTM1 as receptors.

In RPE cells, phagocytosis starts with the capture and ensheathment of POS by the apical processes of the RPE. Ensheathment is stimulated by MERTK ligands, GAS6 and PROS1, but not by V5 integrin receptor ligands, MFG-E8 and vitronectin. Remarkably, the ensheathment participates in POS fragmentation before their internalization. It is suggested that MERTK activation is required for ensheathment-mediated POS fragmentation before internalization [135].

In the RPE, LAP plays a necessary role in cell homeostasis through the optimal clearance of phagocytosed POS fragments. LAP supports (1) the visual cycle by recycling retinoids, (2) retinal metabolism (e.g., by metabolizing lipids to generate ketones), (3) lipid homeostasis, and (4) the synthesis of anti-inflammatory lipids [132,136,137].

RPE cells use Rubicon (RUN domain and cysteine-rich domain containing Beclin 1-interacting protein) similarly to macrophages to stimulate LAP and efficiently degrade phagocytosed cargo. It was found that Rubicon expression is highest in the morning during the time of maximal POS degradation. The phagocytosis of outer segments activates EGFR (epidermal growth factor receptor), suppressing autophagy. When exposed to starvation stress, RPE cells activate autophagy, which impairs phagocytic degradation. Thus, RPE cells maintain a balance between phagocytosis and autophagy, ensuring their long-term functions and retinal homeostasis [138]. ARPE-19 cells are efficient at phagocytizing rod POS under both normal and high-glucose conditions. However, under high-glucose conditions, ARPE-19 cells treated with oxidized rod POS fragments accumulated malondialdehyde and LF and displayed altered LC3, GRP78, and caspase-8 expression compared with untreated and unoxidized rod POS-treated cells [139]. Thus, phagocytizing cells grown in high glucose appear more prone to suffer a permanent oxidative insult than those grown under normal conditions.

6.4. *Autophagy*

Autophagy is the catabolism of intracellular material in specialized cellular structures, namely, lysosomes. Autophagy is involved in maintaining cellular homeostasis, adaptation to stress, immune responses, and regulation of inflammatory processes.

In mammals, three main autophagic pathways target substrates for lysosomal degradation. Macroautophagy involves the formation of autophagosomes, double-membrane vesicles, to transport substrates to lysosomes for degradation. In chaperone-mediated autophagy, lysosomal degradation is regulated by the interaction between lysosomal receptor LAMP2 (lysosome-associated membrane protein 2) and chaperone Hsc70, facilitating the selective access of substrates with the target motif to the lysosome. In microautophagy, lysosomal degradation is the simplest, involving the direct invagination and sequestration of substrates into the lysosome [140].

6.4.1. Macroautophagy

Protein aggregates, cellular organelles, and protein complexes in signaling cascades are typically degraded through selective autophagy. Selective autophagy uses autophagy receptors that bind to specific structural elements of the protein or organelle that must be eliminated [141]. Ubiquitin modification is also part of the signal that marks cellular components for destruction. Autophagy receptors, such as SQSTM1/p62, NBR1, and optineurin, contain domains that recognize ubiquitin and unfolded protein structural elements in cargo [142–144]. After receptor binding, the cargo is chaperoned to the autophagosome where it is degraded. The autophagosome is decorated by ATG8 (autophagy-associated protein 8)/LC3 (microtubule-associated protein 1A/1B-light chain 3), which are recognized by the LIR (LC3-interacting regions) motif that is present on all autophagy receptors [30]. The RPE phagocytizes and degrades the POS in autophagosomes. The process of autophagosome formation includes initiation and elongation phases, during which an insulating membrane is formed. This membrane grows during the nucleation phase to form the mature autophagosome. Fusion of the autophagosome with lysosomes will allow the engulfed material to be degraded during the degradation phase [145]. All stages of the process from induction to autophagosome formation and its fusion with the lysosome are regulated by members of the ATG family of proteins. More than 35 ATG genes have been identified that control the autophagy process [146].

The RPE maintains basal autophagy for cellular homeostasis, with autophagic variations common in aging and diseased cells [55,62,147]. Hypoxia, OS, the unfolded protein response, and inflammation are typical inducers of autophagy [11,147–149]. These conditions are also associated with RPE aging and AMD. In the RPE, p62 promotes autophagy and simultaneously enhances an Nrf2-mediated antioxidant response to protect against acute OS [150]. Members of the Bcl-2-associated anthanogen (BAGs) may be also cytoprotective by inducing autophagy [151].

6.4.2. Chaperone-Mediated Autophagy

Chaperone-mediated autophagy (CMA) degrades the proteins that have a specific tag related to pentapeptide KFERQ [152]. These proteins first bind to cytosolic chaperone hsc70 and its co-chaperones to unfold, then bind to lysosomal membrane protein LAMP-2A, and are directly translocated across the lysosomal membrane, without requiring the formation of intermediate vesicles or membrane deformation [153,154].

Molecular chaperones (Hsp70, Hsp40, and Hsp90) can form a complex with their transcription factor HSF1 in the cytosol. Upon binding misfolded proteins Hsp70, Hsp40, and Hsp90 dissociate from HSF1, which is activated via phosphorylation, and are trafficked to the nucleus to increase chaperone expression. The substrate proteins in the cytosol are selectively bound by the KFERQ-motif with chaperone proteins and transported to the lysosomal lumen via CMA. In the lysosome, the substrate proteins are digested and degraded by lysosomal enzymes. The KFERQ-motif is a conserved peptide sequence in the target proteins recognized by cytosolic Hsc70 and chaperoned into the lysosome via the LAMP-2A channel. LAMP-2A is the key component of chaperone-mediated autophagy. LAMP-2A organizes into specific protein complexes at the lysosomal membrane. HSP90 regulates the stability of lysosome-associated membrane protein-2a (Lamp-2a) in the process of CMA. The central regulator of this degradation pathway is constitutively expressed heat shock protein 70 (HSC70). It recognizes the KFERQ motif in protein sequences and stimulates protein translocation across membranes [155]. There is a group of chaperone proteins that interact with HSC70 and regulate its activity or may themselves act as chaperones. They include heat shock protein 40 (HSP40), which stimulates the ATPase activity of HSC70; HSC70-interacting protein (HIP), which stimulates complex assembly; heat shock protein 90 (HSP90), which can refold unfolded proteins and/or prevents the degradation of unfolded proteins from aggregating; and organizing protein HSC70–HSP90 (HOP), which binds the HSC70–HSP90 chaperones [156,157]. The chaperone complex is associated with the target protein and LAMP-2A on the cytosolic surface of the lysosomal membrane. The main function of chaperones is unfolding the target protein before its transport through the lysosomal membrane [158]. Lys-hsc70 induces disassembly of LAMP-2A from the 700 kDa complex once the substrate has crossed the membrane, and lys-hsp90 stabilizes LAMP-2A during its transition from monomeric to multimeric forms [159].

The main integral transmembrane protein of lysosomes, LAMP-2, is represented by three isoforms. All of them have a short C-terminal cytoplasmic domain (11 amino acids), one transmembrane domain, and a large, highly glycosylated luminal domain. LAMP-2A, B, and C, share an identical lumenal region, but differ in their trans-membrane and cytosolic domain [160,161]. Only the cytosolic domain of LAMP-2A is recognized by the CMA substrates [162]. Organization of LAMP-2A into multimeric complexes, required for CMA activation, only occurs outside the cholesterol-rich microdomains of the lysosomal membrane, whereas the LAMP-2A located within these regions is susceptible to proteolytic cleavage and degradation [163].

CMA is activated under OS, prolonged nutrient deprivation, or exposure to toxic compounds that induce protein damage [164,165]. The first CMA regulation point is constituted by the target proteins that become more accessible after oxidation [166]. This is at least partly due to the fact that oxidized proteins are more easily unfolded, which is necessary for their translocation across the lysosomal membrane [167]. Post-translation modifications on substrate proteins not only generate a KFERQ-like motif, but also change the state of exposure of the KFERQ-like motif so that Hsc70 can recognize or de-identify the substrate proteins [168]. Another CMA regulation point is LAMP2A. Oxidation makes lysosomes more active in CMA due to increased LAMP-2A in the lysosomal membrane and

HSC70 in the lysosomal lumen [166]. It is LAMP-2A that regulates CMA activity [165,169]. A portion of full-sized LAMP-2A resides within the lysosomal lumen, perhaps complexed with lipids. These molecules are able to reinsert into the lysosomal membrane when CMA is activated [170]. In addition, the degradation rate of LAMP-2A can be regulated through the activity of lysosomal cathepsin A [169]. Lys-Hsp90 plays a role in maintaining LAMP2A stability, while Lys-Hsc70 induces LAMP2A to disassemble for a new cycle [159]. In rainbow trout, activation of CMA upon high-glucose exposure was mediated by generation of mitochondrial ROS and involved the antioxidant transcription factor Nrf2 [171]. It was demonstrated that the signaling mediated by P2X7, a member of the purinergic family of receptors, can increase Hsc70 and LAMP2A mRNA levels, allowing LAMP2A association with the lysosomal membrane under inflammation conditions [172]. In humans and mice, NRF2 performs a positive regulatory function for CMA through binding to the regulatory elements in the LAMP2 gene and enhancing its transcription [173]. However, signaling mediated by the RARα (retinoic acid receptor-α) has been shown to negatively regulate the transcription of CMA components, including LAMP2A [174].

In addition, the lysosome-associated form of GFAP (glial fibrillary acidic protein) has been shown to stabilize the multimeric translocation complex in response to starvation or OS. In this case, GTP mediates the release of EF1α from the lysosomal membrane, promoting the self-association of GFAP and disassembly of the CMA translocation complex, thereby reducing CMA [175].

Moreover, lysosomal Akt and kinase TORC2 (target of rapamycin complex 2) regulate the activity of CMA by controlling the dynamics of assembly and disassembly of the CMA translocation complex on the lysosomal membrane [176].

6.4.3. Microautophagy

Microautophagy is a catabolic process, in which the dysfunctional or superfluous proteins and organelles are delivered directly to the endosomal or lysosomal lumen [177,178]. Microautophagy is classified as non-selective bulk degradation, which occurs under conditions of starvation and provides essential nutrients for cell survival [141]. This process largely depends on the endosomal sorting complexes required for transport (ESCRT) I and III systems and the protein chaperone, hsc70 [179].

Only microER-phagy has been identified in mammalian systems [180]. Mammalian microautophagy pathways all direct their cargo to late endosomes/multivesicular bodies (LEs/MVBs) [181]. The degradation of cytosolic proteins in LEs/MVBs is referred to endosomal microautophagy. As in the case of CMA, endosomal microautophagy may require recognition of a KFERQ-like motif by HSC70 [182]. HSC70 binds to phosphatidylserine in LE/MVB membranes, triggering substrate internalization into the lumen via membrane invaginations that form in an ESCRT-dependent manner. Cargo protein degradation can occur in the LE/MVB compartment itself, although the bulk of degradation occurs after LE/MVB-lysosome fusion [183].

7. Programmed Cell Death and Chaperones

The regulation of cell death and survival is under strict control in eukaryotic development and tissue homeostasis. It is distinguished between accidental and programmed cell death (PCD). PCD is mediated genetically and is tightly controlled [184,185]. Cell death can be induced by the genetically programmed suicide mechanisms of apoptosis, necroptosis, and pyroptosis, or it can be a consequence of dysregulated metabolism, as in ferroptosis [186,187].

The death of RPE and other eye tissues occurs through apoptosis, necrosis, pyroptosis, and ferroptosis pathways (Table 1) under stress conditions, with aging, or as a result of pathological processes [188–191].

Table 1. Main features of major types of programmed cell death (modified from [184,192–195]).

Characteristics		Apoptosis	Necroptosis	Ferroptosis	Pyroptosis
Morphology	Cell	Shrinkage	Swelling	Rounding up	Bubbling
	Plasmatic membrane	Blebbing	Pore formation	Lack of rupture and blebbing	Rupture
	Subcellular structures	Pseudopod retraction	Necrotizing bodies	Rupture of outer mitochondrial membrane	Pyroptotic bodies
	Nuclei	Fragmentation	Moderate chromatin condensation	Lack of condensation	Chromatin condensation
Activated/increased markers		Caspase3/7 Cytochrome C release	RIPK1,3 MLKL PS exposure	Iron and ROS accumulation Lipid peroxidation MAP kinases	NLRP3 ASC Pro-caspase-1 Gasdermin D
Outcome		Engulf cells by phagocytes without inflammation	Cell lysis Inflammation	Inflammation	Cell lysis Inflammation

Abbreviations: ASC—apoptosis-associated speck-like protein containing a CARD; MLKL—pseudokinase mixed lineage kinase-like; NLRP3—inflammasome; PS—phosphatidylserine; RIPK—receptor-interacting kinase.

7.1. Apoptosis

The development of apoptosis is associated with an inhibition of growth and division, leading to controlled death without a leakage of its contents into the environment. Apoptosis is initiated by the activation of a chain of caspases, which belong to the class of cysteine-aspartic proteases [196].

Cell damage activates initiator caspases (caspases 8 and 9), which trigger the activation of effector caspases (caspases 3, 6, and 7). Apoptosis leads to DNA and nuclear fragmentation, cytoskeletal destruction, and the formation of apoptotic bodies. This process can be triggered through intrinsic and extrinsic signaling pathways. The intrinsic pathway, which is OS dependent, is associated with mitochondrial dysfunction and depends on factors secreted by mitochondria [196]. The extrinsic apoptosis pathway is activated after binding of the TNF1 to Fas-associated death receptors. The chain of reactions then activates caspase-8, which induces caspase-3-based apoptosis [197]. Bax translocation to mitochondria is mediated by signals from c-Jun N-terminal kinases and p38 mitogen-activated protein kinase in OS [198,199].

RPE cells constitutively express members of the sHSP family αA- and αB-crystallin, which can function as anti-apoptotic proteins induced during OS [200,201]. Under OS conditions, the antioxidant activity of chaperone proteins αA and αB crystallins increases, which prevents OS-mediated apoptotic cell death of RPE cells. In the RPE cells of mice with the αA-crystallin or αB-crystallin gene knocked out, the accumulation of ROS was followed by the degeneration of retinal photoreceptors under conditions of experimentally simulated OS [202,203].

It is proposed that αB-crystallin may inhibit apoptosis through interaction with p53, preventing its translocation to mitochondria and blocking the apoptotic signaling [204]. p53 is involved in the initiation of the calcium-activated RAF/MEK/ERK signaling pathway of apoptosis, which can be suppressed by αB-crystallin via inhibition of RAS activation [201]. αB-Crystallin interacts directly with the pro-apoptotic members Bax and Bcl-XS and P53 polypeptides in vitro and in vivo, with sequestration of these proteins preventing the translocation to mitochondria and hence suppressing apoptosis [205,206]. Exogenously added recombinant human αB-crystallin was taken up by stressed cells and protected these cells from apoptosis by inhibiting caspase-3 and through poly (ADP-ribose) polymerase activation [207]. A 20-mer functional chaperone peptide (αB-crystallin peptide) derived from the amino acid residues 73–92 (DRFSVNLDVKHFSPEELKVK) of αB-crystallin protects RPE cells from OS-induced cell death by inhibiting caspase-3 activation [70,208].

The functions of ATP-independent chaperone HspB1 in RPE cells consist of blocking signaling pathways that trigger caspase-dependent apoptosis [209]. HspB1 is among the first chaperones to be activated in RPE cells under OS conditions, which decreases the level of ATP. These events are accompanied by the blocking of the external receptor-dependent pathway of cell death mediated by tumor necrosis factor receptors (TNFRs) and the internal mitochondrial signaling pathway. Then, after partial RPE cell recovery, ATP-dependent chaperone HSP70 is activated [209]. Hsp70 inhibits the formation of a functional apoptosome by interactions with Apaf-1. Hsp70 protects against the forced destructive action of caspase-3 and prevents the translocation of Bax from the cytoplasm to the mitochondria [210,211].

Hsp90 can prevent the formation of the apoptosome complex by inhibiting the oligomerization of Apaf-1 [210]. HSP90 regulates Akt activated by vascular endothelial growth factor (VEGF) and neuroprotectin D1 (NPD1) in the RPE in response to OS by inhibiting the signal cascade of dephosphorylation [212,213]. Akt is involved in the I3K-AktmTOR pathway, and its activation mainly occurs in the Nrf2-related response to OS and blocks autophagy. Proteasome inhibitors can enhance the autophagy process by inhibition of the PI3K–Akt–mTOR pathway, which makes it possible for them to be considered as therapeutic agents for enhancing anti-oxidative defense in human RPE cells [214–216].

The Hsp27 low-molecular-weight chaperone protein can maintain mitochondrial stability and redox homeostasis in cells and interacts with the apoptotic signaling pathways at many stages. Its activation leads to the blocking of Ca^{2+}-induced apoptosis, which is a result of the suppression of caspase-3 functions and the prevention of cytochrome C release from Bcl-xS in the cytoplasm. HSP27 is also involved in the stabilization of Akt [217,218].

In addition to HSPs, other chaperones with anti-apoptotic functions are also known, for example, members of the Bcl-2-associated anthanogen (BAG) and clusterin (Clu) families [219,220]. Some of these proteins, such as BAGs likely serve to inhibit apoptosis by acting as co-chaperones for other proteins like HSPs, although there is recent evidence that BAGs may be cytoprotective by inducing autophagy [151,221].

7.2. Necroptosis (Regulated Necrosis)

Necrosis is activated by the production of proinflammatory factors, which leads to the destruction of the cell membrane. The necrotic pathway is initiated by the binding of TNF (tumor necrosis factor) to death receptors on the cell membrane, which is mainly controlled by RIP (receptor-interacting protein kinases) in the absence of caspase-8. Autophosphorylation of RIPK1 and RIPK3 leads to the formation of necrosomes, which are associated with mitochondrial dysfunction, leading to cell death [190,222,223]. In the RPE, the main characteristics of necrosis are ATP depletion and RIPK3 protein aggregation, cell swelling, and loss of cell membrane integrity under OS conditions [222,223].

Necroptosis is considered as regulated necrosis dependent on RIPK3 and MLKL (mixed lineage kinase domain) proteins. Caspase-8 activity can suppress this type of cell death by cleaving RIPK1 and RIPK3 [224]. The elimination of caspase-8 and FADD leads to autonomous activation of RIPK3 and MLKL that initiates the process of necroptosis [186,225]. It is interesting that RIP-mediated necroptosis becomes the predominant form of cell death after caspase inhibition. Therefore, necroptosis may serve as a backup mechanism along with apoptosis in various retinal diseases [226]. RPE cells respond to necrosis by enhanced producing of inflammatory cytokines that cause increased cell permeability. Necrosis-induced production of inflammatory cytokines in RPE cells have been demonstrated partially mimicked by recombinant HSP90 [227]. Inhibition of HSP90 by CDDO (synthetic triterpenoid, 2-cyano-3,12-dioxoolean-1,9-dien-28-oic acid) blocked necroptosis by inhibiting the activation of RIPK1 kinase [228].

7.3. Pyroptosis

Pyroptosis is an inflammatory form of cell death executed by gasdermins (GSDMs), a family of transmembrane pore-forming proteins activated via inflammasome-dependent or inflammasome-independent pathways [229].

Inflammasome activation is related to lysosomal destabilization. Inflammasomes contain the NOD-like receptor family pyrin domains, which are involved in initiating immune cell death by activating apoptotic or pyroptotic pathways in RPE cells [230–232]. Inflammasome priming by IL-1α, C5a, or medium conditioned by pyroptotic cells increased the susceptibility of RPE cells to photooxidative damage-mediated cell death and switched the mechanism of induced cell death from apoptosis to pyroptosis [56]. Two types of cell death, including pyroptosis and apoptosis, were activated in RPE cells after prolonged inflammasome activation induced by the drusen component of amyloid-beta (Aβ) [231]. All-trans retinal-derived A2E (*bis*-retinoid *N*-retinyl-*N*-retinylidene ethanolamine) of LF granules activates NLRP3 inflammasome to trigger pyroptosis or apoptosis of ARPE-19 cells [233].

Inflammasomes in the caspase-1-dependent pathway play an important role in triggering pyroptosis. The process is initiated by binding to proteins of the NLPR1, NLPR3, and NLPR4 (NOD receptor family) by inflammasomes, which activates caspase-1 with ASC as an adaptor protein. Caspase-1 promotes the cleavage of propyroptotic factor gasdermin D, producing an N-terminal fragment that induces cell death. Pyroptosis can be also activated independently of caspase-1. In this case, human caspase-4/5 and mouse caspase-11 promote the cleavage of gasdermin D to activate pyroptosis [191]. The NLRP3 inflammasome is a central regulator of inflammation and its activation has been associated with several age-related diseases, such as AMD, diabetic retinopathy, uveitis, and others [234]. 4-Hydroxynonenalis is one of the primary end products during LPO, which accumulates in RPE cells in AMD [235]. 4-HNE induces IL-6, IL-1β, and TNF-α production by promoting the extracellular efflux of HSP70 [236]. HSP70 affects inflammation by functioning as a negative regulator of the NLRP3 inflammasome [237].

HSP90 is a crucial chaperone protecting NLRP3 from destruction, keeping it intact but ready to be activated [238]. The inhibition of Hsp90 by TAS-116 (4-(1H-pyrazolo[3,4-b]pyridine-1-yl)benzamide) could prevent NLRP3 activation-dependent IL-1 release from RPE cells [239]. It is speculated that Hsp90 can assist in NLRP3 activation by stabilizing the P2X7 receptor [240]. Previously, it has been shown that Hsp90 inhibition can reduce NLRP3 inflammasome activation in RPE cells and that this effect relies on the activation of autophagy [241].

7.4. Ferroptosis

Ferroptosis is a cell death mode associated with LPO of polyunsaturated fatty acids leading to plasma membrane rupture. This form of death is characteristic of RPE cells due to their participation in the phagocytosis of POS, which constitute a major source of intracellular ROS and polyunsaturated fatty acids. The core metabolic mechanisms of ferroptosis are LPO and an imbalance of iron homeostasis [242]. Autophagy regulates ferroptosis by regulating cellular iron homeostasis and cell ROS production [243]. Ferroptosis is a major pathological process in OS-mediated RPE degeneration in cases of AMD, diabetic retinopathy, and others [244–246].

Recently, the involvement of chaperones in the induction and regulation of ferroptosis has been extensively reviewed [247]. The HSP90 family may act on GPX4 (glutathione peroxidase 4) and inhibit its antioxidant capacity [248]. The HSP90 family then participates in the regulation of ferroptosis through GSH/GPX4 pathway and inhibits LPO, therefore influencing ferroptosis [249]. HSP90 has also been identified as an important molecular chaperone that mediates the degradation of Gpx4 during ferroptosis, while suppressing ferroptosis in mouse neuronal HT-22 cells (*mouse hippocampal neuronal cell* line) [228]. A recent study has also demonstrated that overexpression of HSPA5 can negatively regulate ferroptosis by limiting Gpx4 degradation and LPO [250]. The upregulation of HSPA5 increases the expression and activity of GPX4, while GPX4 protects glioma cells from ferroptosis by neutralizing DHA-induced LPO [251]. The IP3R–HSPA9 (also known as GRP75)-voltage dependent anion channel 1 (VDAC1) complex bridges the gap in the mitochondria-associated membranes and establishes a platform for the transmission of ferroptosis signals from the ER to the mitochondria [252]. These findings establish a direct connection between the ER and the mitochondria mediated by calcium signals, with the ER acting as the initiator and the mitochondria as the effector of ferroptosis.

Overexpression of DNAJB6 (HSP40 family member) enhances the degradation of GSH, downregulates GPX4, enhances LPO, and promotes ferroptosis in esophageal squamous adenocarcinoma [247,253].

HO-1(HSP32), member of the sHSP family, has an anti-ferroptosis effect in human renal tubular epithelial cells, protecting AKI from ferroptosis by promoting GSH depletion [254]. Another member of this family HSPB1 also plays an integral role in ferroptosis. HSPB1 is considered to be a negative regulator of iron accumulation and uptake in fibroblasts or cardiac cells [227,255].

Under conditions of ferroptotic stress, there is an augmentation in the interaction between sigma non-opioid intracellular receptor 1 (SIGMAR1, also known as σ1R), a molecular chaperone situated in the mitochondria-associated membranes, and inositol 1,4,5-trisphosphate receptor (ITPR). This enhancement prompts an exchange of calcium ions between the ER and the mitochondria, thereby intensifying sensitivity to ferroptosis [252].

8. Extracellular Vesicles and Chaperones

RPE is a polarized epithelium that performs a barrier function, being located between the neural retina and the choroid. Extracellular vesicles (EVs) play an important role in the RPE interaction with adjacent tissues. They represent heterogeneous group of the extracellular particles that are delimited by a lipid bilayer and cannot replicate on their own. There are a number of EV subtypes (in particular, exosomes, which are a kind of small EV). EVs carry RNA, lipids, and proteins, such as heat shock proteins [256]. Exosomes released from the apical surface of the RPE cells contain αB-crystallin which may provide neuroprotection to neighboring photoreceptor cells [203]. Another function of αB-crystallin is inhibition of b-amyloid fibril formation. Minipeptides derived from αB-crystallin have been identified as anti-apoptotic agents in addition to their chaperone function [79]. αB-Crystallin is also a modulator of angiogenesis and vascular endothelial growth factor [79]. It is supposed that αB-crystallin controls the fusion of multivesicular bodies with the plasma membrane and the release of the exosomes [257].

In AMD, the secretion of RPE-derived EVs is enhanced to mediate OS, inflammation, angiogenesis, amyloid fibril formation, and drusen accumulation [258]. Extracellular microparticles enhanced senescence and interrupted phagocytic activity of RPE to lead to its degeneration [192]. In addition, stressed RPE cells released exosomes with a higher expression of VEGFR in the membrane and enclosed extra cargo of VEGFR mRNA. These exosomes may stimulate angiogenesis in choroids and during the healing of diabetic wounds [259]. Exosomes can also transport anti-inflammatory drugs to microglia, inhibit neuroinflammatory responses, and play a neuroprotective role in photoreceptor cells [260]. The cumulative evidence presented in recent review underscores the pivotal role of EVs in the onset and progression of retinal degenerative diseases (AMD, DR) [261].

9. Conclusions

The universal function of the RPE is the constant phagocytosis of the photoreceptor outer segment (POS) discs, which has led to the emergence of a unique form of phagocytosis without the formation of autophagosomes. The homeostatic balance of RPE cells is ensured by close interaction of endogenous defense systems indispensable to the health of the neural retina. Fine regulation of proteostasis and autophagy processes is primary to guarantee RPE homeostasis, protecting it from oxidative damage and protein accumulation. The redox balance in the RPE is largely dependent on autophagic clearance, which is a pleiotropic process by which cells deliver cytoplasmic components to the lysosome for degradation. Impaired autophagy significantly contributes into the development of RPE dysfunctions and photoreceptors degeneration in neural retina.

It is obvious that autophagy involves multiple mechanisms that use different pathways to maintaining cellular homeostasis, adaptation to stress, and regulation of immune responses and inflammatory processes. Analysis of scientific data demonstrates the multifaceted role of autophagy and much more specific endogenous selectivity of this process for the intracellular degradation and removal of destroyed and harmful components such as mitochondria (mitophagy), endoplasmic reticulum (ERphagy), peroxisomes (pexophagy), liposomes (lipophagy), and aggregated proteins (aggregophagy). Autophagy balance can act as an important switch between programmed and unprogrammed cell death.

Most of the chaperones are components of the molecular network of HSPs and their accessory proteins. HSPs promote cell survival through protection against changes in the cellular redox homeostasis, are implicated in junctional biogenesis, and can be responsible for the selective assembly of different junctional complexes. Recent studies have shown that various chaperones and co-chaperones are involved in the organization of the actin cytoskeleton, which is important for maintaining the stability of cell differentiation. HSPs can trigger or modulate multiple signals. HSP signaling pathways largely depend on the intensity of OS and on the state of RPE cells and demonstrates the possible use of alternative signal pathways.

The most of the molecular targets and signaling pathways of HSPs remain largely obscure. The extensive cross-talk between HSP signaling pathways and OS producers and mediators that control cellular response is problematic. This is due to the fact that HSPs integrate various mechanisms of endogenous defense systems into an overlapping molecular and genetic regulatory network and can perform oppositely directed functions. These proteins are involved in the regulation of proteostasis and maintenance of RPE cell viability, as well as in cell death signaling pathways. This makes it difficult to target the signaling to modulate autophagy processes and largely explains the controversial questions regarding insufficient selectivity and side effects of the HSPs activators or blockers for neuroprotection. Targeting HSPs may negatively affect other cell life support mechanisms and should be aimed at dissecting downstream signaling pathways. The multifaceted functions of molecular chaperones, HSPs, and their isoforms, co-chaperones, and HSP coinducers, as well as the spectrum of specific client proteins, at each step of proteostasis are still poorly understood. Understanding how HSPs in the RPE are regulated by both transcriptional and post-transcriptional mechanisms will promote the details of pathogenetic pathways of different forms of autophagy in RPE proteostasis in retinal disorders and shed light on potential strategies to treat visual impairments. The impact on molecular targets for autophagy by blocking the production of ROS or activating an endogenic defense requires knowledge on how to fine tune their participation in the regulation of the vital functions of RPE cells.

Inflammation and oxidative stress are thought to play major causative roles in the pathogenesis of many retinal degenerative diseases, such as age-related macular degeneration (AMD), diabetic retinopathy (DR), retinal vein occlusion, and retinitis pigmentosa [262,263]. This is evidenced by the abundance of clinical and experimental data. The current arsenal of strategies for neuroprotection uses antioxidant and anti-inflammatory therapies that are aimed at reducing oxidative stress and its consequences in RPE cells, which is usually accompanied by inflammatory processes, to prevent tissue damage.

Despite the common links in the pathogenesis of these diseases, they have their own specific development of inflammatory processes, which is largely associated with the intensity and duration of oxidative stress.

Alterations in autophagy processes are involved in the development of these ocular pathologies. In most ocular diseases (AMD, glaucoma, and cataract), proteolytic and autophagic capacity is attenuated, but excessive autophagic activity may accelerate the development of DR. Therefore, treatments focused on the modulation of autophagy processes in ocular RPE-related diseases could constitute a perspective therapeutic intervention [264,265]. It is important to emphasize that the effectiveness of autophagy-targeting drugs may have low specificity [266] since the targets are involved in multiple signaling pathways. In addition, the cross-talk between these pathways is not fully understood. Future studies using metabolomic approaches could identify and explore specific novel molecular targets that could pave the way to the prevention and targeted treatment of RPE-dependent retinal pathologies of specific diseases that lead to irreversible vision loss and blindness.

Author Contributions: Both authors of the review contributed equally to the writing—original draft preparation, final version preparation, editing, and funding acquisition. All authors have read and agreed to the published version of the manuscript.

Funding: This study has been funded by a Government Program of Basic Research at the Koltzov Institute of Developmental Biology of the Russian Academy of Sciences in 2024 (No. 0088-2024-0014).

Institutional Review Board Statement: Protocol Code: 23, 4 December 2024.

Informed Consent Statement: Not applicable.

Data Availability Statement: Not applicable.

Acknowledgments: The authors express their gratitude to the Collective Use Center of the Koltsov Institute of Developmental Biology RAS. We thank Natasha Grigorian for her help with proofreading the manuscript.

Conflicts of Interest: The authors declare no conflict of interest. The funders had no role in the design of the study; in the collection, analyses, or interpretation of data; in the writing of the manuscript; or in the decision to publish the results. The authors state that the manuscript has not been published previously.

References

1. Lakkaraju, A.; Umapathy, A.; Tan, L.X.; Daniele, L.; Philp, N.J.; Boesze-Battaglia, K.; Williams, D.S. The cell biology of the retinal pigment epithelium. *Prog. Retin. Eye Res.* **2020**, *78*, 100846. [CrossRef] [PubMed]
2. Lutty, G.A.; McLeod, D.S. Development of the hyaloid, choroidal and retinal vasculatures in the fetal human eye. *Prog. Retin. Eye Res.* **2018**, *62*, 58–76. [CrossRef] [PubMed]
3. Stern, J.; Temple, S. Retinal pigment epithelial cell proliferation. *Exp. Biol. Med.* **2015**, *240*, 1079–1086. [CrossRef]
4. George, S.M.; Lu, F.; Rao, M.; Leach, L.L.; Gross, J.M. The retinal pigment epithelium: Development, injury responses, and regenerative potential in mammalian and non-mammalian systems. *Prog. Retin. Eye Res.* **2021**, *85*, 100969. [CrossRef] [PubMed]
5. Booij, J.C.; Baas, D.C.; Beisekeeva, J.; Gorgels, T.G.; Bergen, A.A. The dynamic nature of Bruch's membrane. *Prog. Retin. Eye Res.* **2010**, *29*, 1–18. [CrossRef] [PubMed]
6. Pan, W.W.; Wubben, T.J.; Besirli, C.G. Photoreceptor metabolic reprogramming: Current understanding and therapeutic implications. *Commun. Biol.* **2021**, *4*, 245. [CrossRef] [PubMed]
7. Masutomi, K.; Chen, C.; Nakatani, K.; Koutalos, Y. All-trans retinal mediates light-induced oxidation in single living rod photoreceptors. *Photochem. Photobiol.* **2012**, *88*, 1356–1361. [CrossRef] [PubMed]
8. Adijanto, J.; Banzon, T.; Jalickee, S.; Wang, N.S.; Miller, S.S. CO_2-induced ion and fluid transport in human retinal pigment epithelium. *J. Gen. Physiol.* **2009**, *133*, 603–622. [CrossRef] [PubMed]
9. Sun, M.; Finnemann, S.C.; Febbraio, M.; Shan, L.; Annangudi, S.P.; Podrez, E.A.; Hoppe, G.; Darrow, R.; Organisciak, D.T.; Salomon, R.G.; et al. Light-induced oxidation of photoreceptor outer segment phospholipids generates ligands for CD36-mediated phagocytosis by retinal pigment epithelium: A potential mechanism for modulating outer segment phagocytosis under oxidant stress conditions. *J. Biol. Chem.* **2006**, *281*, 4222–4230. [CrossRef] [PubMed]
10. Kaarniranta, K.; Koskela, A.; Felszeghy, S.; Kivinen, N.; Salminen, A.; Kauppinen, A. Fatty acids and oxidized lipoproteins contribute to autophagy and innate immunity responses upon the degeneration of retinal pigment epithelium and development of age-related macular degeneration. *Biochimie* **2019**, *159*, 49–54. [CrossRef]
11. Mitter, S.K.; Song, C.; Qi, X.; Mao, H.; Rao, H.; Akin, D.; Lewin, A.; Grant, M.; Dunn, W.; Ding, J.; et al. Dysregulated autophagy in the RPE is associated with increased susceptibility to oxidative stress and AMD. *Autophagy* **2014**, *10*, 1989–2005. [CrossRef] [PubMed]
12. Godley, B.F.; Shamsi, F.A.; Liang, F.Q.; Jarrett, S.G.; Davies, S.; Boulton, M. Blue light induces mitochondrial DNA damage and free radical production in epithelial cells. *J. Biol. Chem.* **2005**, *280*, 21061–21066. [CrossRef] [PubMed]
13. Miceli, M.V.; Liles, M.R.; Newsome, D.A. Evaluation of oxidative processes in human pigment epithelial cells associated with retinal outer segment phagocytosis. *Exp. Cell Res.* **1994**, *214*, 242–249. [CrossRef] [PubMed]
14. Kim, J.; Lee, Y.J.; Won, J.Y. Molecular Mechanisms of Retinal Pigment Epithelium Dysfunction in Age-Related Macular Degeneration. *Int. J. Mol. Sci.* **2021**, *22*, 12298. [CrossRef]
15. Wong, W.L.; Su, X.; Li, X.; Cheung, C.M.; Klein, R.; Cheng, C.Y.; Wong, T.Y. Global prevalence of age-related macular degeneration and disease burden projection for 2020 and 2040: A systematic review and meta-analysis. *Lancet Glob. Health* **2014**, *2*, e106–e116. [CrossRef]
16. Bhutto, I.; Lutty, G. Understanding age-related macular degeneration (AMD): Relationships between the photoreceptor/retinal pigment epithelium/Bruch's membrane/choriocapillaris complex. *Mol. Aspects Med.* **2012**, *33*, 295–317. [CrossRef] [PubMed]
17. Strauss, O. The retinal pigment epithelium in visual function. *Physiol. Rev.* **2005**, *85*, 845–881. [CrossRef] [PubMed]
18. Hazim, R.A.; Volland, S.; Yen, A.; Burgess, B.L.; Williams, D.S. Rapid differentiation of the human RPE cell line, ARPE-19, induced by nicotinamide. *Exp. Eye Res.* **2019**, *179*, 18–24. [CrossRef] [PubMed]
19. Lidgerwood, G.E.; Morris, A.J.; Conquest, A.; Daniszewski, M.; Rooney, L.A.; Lim, S.Y.; Hernández, D.; Liang, H.H.; Allen, P.; Connell, P.P.; et al. Role of lysophosphatidic acid in the retinal pigment epithelium and photoreceptors. *Biochim. Biophys. Acta Mol. Cell Biol. Lipids* **2018**, *1863*, 750–761. [CrossRef] [PubMed]
20. Markitantova, Y.; Simirskii, V. Endogenous and Exogenous Regulation of Redox Homeostasis in Retinal Pigment Epithelium Cells: An Updated Antioxidant Perspective. *Int. J. Mol. Sci.* **2023**, *24*, 10776. [CrossRef] [PubMed]
21. Handa, J.T.; Bowes Rickman, C.; Dick, A.D.; Gorin, M.B.; Miller, J.W.; Toth, C.A.; Ueffing, M.; Zarbin, M.; Farrer, L.A. A systems biology approach towards understanding and treating non-neovascular age-related macular degeneration. *Nat. Commun.* **2019**, *10*, 3347. [CrossRef] [PubMed]

22. Terluk, M.R.; Kapphahn, R.J.; Soukup, L.M.; Gong, H.; Gallardo, C.; Montezuma, S.R.; Ferrington, D.A. Investigating mitochondria as a target for treating age-related macular degeneration. *J. Neurosci.* **2015**, *35*, 7304–7311. [CrossRef] [PubMed]
23. Caruso, L.; Fields, M.; Rimondi, E.; Zauli, G.; Longo, G.; Marcuzzi, A.; Previati, M.; Gonelli, A.; Zauli, E.; Milani, D. Classical and Innovative Evidence for Therapeutic Strategies in Retinal Dysfunctions. *Int. J. Mol. Sci.* **2024**, *25*, 2124. [CrossRef] [PubMed]
24. Wang, X.L.; Gao, Y.X.; Yuan, Q.Z.; Zhang, M. NLRP3 and autophagy in retinal ganglion cell inflammation in age-related macular degeneration: Potential therapeutic implications. *Int. J. Ophthalmol.* **2024**, *17*, 1531–1544. [CrossRef] [PubMed]
25. Li, C.P.; Yao, J.; Tao, Z.F.; Li, X.M.; Jiang, Q.; Yan, B. Epigallocatechin-gallate (EGCG) regulates autophagy in human retinal pigment epithelial cells: A potential role for reducing UVB light-induced retinal damage. *Biochem. Biophys. Res. Commun.* **2013**, *438*, 739–745. [CrossRef]
26. Beranova-Giorgianni, S.; Giorgianni, F. Proteomics of Human Retinal Pigment Epithelium (RPE) Cells. *Proteomes* **2018**, *6*, 22. [CrossRef] [PubMed]
27. Fuhrmann, S.; Zou, C.; Levine, E.M. Retinal pigment epithelium development, plasticity, and tissue homeostasis. *Exp. Eye Res.* **2014**, *123*, 141–150. [CrossRef] [PubMed]
28. Yang, S.; Zhou, J.; Li, D. Functions and Diseases of the Retinal Pigment Epithelium. *Front. Pharmacol.* **2021**, *12*, 727870. [CrossRef]
29. Naylor, A.; Hopkins, A.; Hudson, N.; Campbell, M. Tight Junctions of the Outer Blood Retina Barrier. *Int. J. Mol. Sci.* **2019**, *21*, 211. [CrossRef] [PubMed]
30. Ferrington, D.A.; Sinha, D.; Kaarniranta, K. Defects in retinal pigment epithelial cell proteolysis and the pathology associated with age-related macular degeneration. *Prog. Retin. Eye Res.* **2016**, *51*, 69–89. [CrossRef] [PubMed]
31. Baehr, W.; Wu, S.M.; Bird, A.C.; Palczewski, K. The retinoid cycle and retina disease. *Vision Res.* **2003**, *43*, 2957–2958. [CrossRef] [PubMed]
32. Bazan, N.G.; Gordon, W.C.; Rodriguez de Turco, E.B. Docosahexaenoic acid uptake and metabolism in photoreceptors: Retinal conservation by an efficient retinal pigment epithelial cell-mediated recycling process. *Adv. Exp. Med. Biol.* **1992**, *318*, 295–306. [CrossRef]
33. Dontsov, A.E.; Sakina, N.L.; Golubkov, A.M.; Ostrovsky, M.A. Light-induced release of A2E photooxidation toxic products from lipofuscin granules of human retinal pigment epithelium. *Dokl. Biochem. Biophys.* **2009**, *425*, 98–101. [CrossRef] [PubMed]
34. Dontsov, A.; Ostrovsky, M. Retinal Pigment Epithelium Pigment Granules: Norms, Age Relations and Pathology. *Int. J. Mol. Sci.* **2024**, *25*, 3609. [CrossRef] [PubMed]
35. Hu, D.N.; Simon, J.D.; Sarna, T. Role of ocular melanin in ophthalmic physiology and pathology. *Photochem. Photobiol.* **2008**, *84*, 639–644. [CrossRef] [PubMed]
36. Boulton, M.; Dayhaw-Barker, P. The role of the retinal pigment epithelium: Topographical variation and ageing changes. *Eye* **2001**, *15*, 384–389. [CrossRef] [PubMed]
37. Kaarniranta, K.; Pawlowska, E.; Szczepanska, J.; Jablkowska, A.; Blasiak, J. Role of Mitochondrial DNA Damage in ROS-Mediated Pathogenesis of Age-Related Macular Degeneration (AMD). *Int. J. Mol. Sci.* **2019**, *20*, 2374. [CrossRef] [PubMed]
38. Markitantova, Y.V.; Simirskii, V.N. Role of the Redox System in Initiation of a Regenerative Response of Neural Eye Tissues in Vertebrates. *Russ. J. Dev. Biol.* **2020**, *51*, 16–30. [CrossRef]
39. Kaemmerer, E.; Schutt, F.; Krohne, T.U.; Holz, F.G.; Kopitz, J. Effects of lipid peroxidation-related protein modifications on RPE lysosomal functions and POS phagocytosis. *Investig. Ophthalmol. Vis. Sci.* **2007**, *48*, 1342–1347. [CrossRef] [PubMed]
40. Campbell, H.K.; Maiers, J.L.; DeMali, K.A. Interplay between tight junctions & adherens junctions. *Exp. Cell Res.* **2017**, *358*, 39–44. [CrossRef]
41. Sluysmans, S.; Vasileva, E.; Spadaro, D.; Shah, J.; Rouaud, F.; Citi, S. The role of apical cell-cell junctions and associated cytoskeleton in mechanotransduction. *Biol. Cell.* **2017**, *109*, 139–161. [CrossRef] [PubMed]
42. González-Mariscal, L.; Domínguez-Calderón, A.; Raya-Sandino, A.; Ortega-Olvera, J.M.; Vargas-Sierra, O.; Martínez-Revollar, G. Tight junctions and the regulation of gene expression. *Semin. Cell Dev. Biol.* **2014**, *36*, 213–223. [CrossRef] [PubMed]
43. Sies, H.; Berndt, C.; Jones, D.P. Oxidative Stress. *Annu. Rev. Biochem.* **2017**, *86*, 715–748. [CrossRef]
44. Zhang, S.X.; Sanders, E.; Fliesler, S.J.; Wang, J.J. Endoplasmic reticulum stress and the unfolded protein responses in retinal degeneration. *Exp. Eye Res.* **2014**, *125*, 30–40. [CrossRef]
45. He, S.; Yaung, J.; Kim, Y.H.; Barron, E.; Ryan, S.J.; Hinton, D.R. Endoplasmic reticulum stress induced by oxidative stress in retinal pigment epithelial cells. *Graefes Arch. Clin. Exp. Ophthalmol.* **2008**, *246*, 677–683. [CrossRef] [PubMed]
46. Yao, J.; Tao, Z.F.; Li, C.P.; Li, X.M.; Cao, G.F.; Jiang, Q.; Yan, B. Regulation of autophagy by high glucose in human retinal pigment epithelium. *Cell. Physiol. Biochem.* **2014**, *33*, 107–116. [CrossRef] [PubMed]
47. Brunk, U.T.; Terman, A. Lipofuscin: Mechanisms of age-related accumulation and influence on cell function. *Free Radic. Biol. Med.* **2002**, *33*, 611–619. [CrossRef]

48. Terman, A.; Sandberg, S. Proteasome inhibition enhances lipofuscin formation. *Ann. N. Y. Acad. Sci.* **2002**, *973*, 309–312. [CrossRef] [PubMed]
49. Lakkaraju, A.; Finnemann, S.C.; Rodriguez-Boulan, E. The lipofuscin fluorophore A2E perturbs cholesterol metabolism in retinal pigment epithelial cells. *Proc. Natl. Acad. Sci. USA* **2007**, *104*, 11026–11031. [CrossRef]
50. Liang, F.Q.; Godley, B.F. Oxidative stress-induced mitochondrial DNA damage in human retinal pigment epithelial cells: A possible mechanism for RPE aging and age-related macular degeneration. *Exp. Eye Res.* **2003**, *76*, 397–403. [CrossRef]
51. Toms, M.; Burgoyne, T.; Tracey-White, D.; Richardson, R.; Dubis, A.M.; Webster, A.R.; Futter, C.; Moosajee, M. Phagosomal and mitochondrial alterations in RPE may contribute to KCNJ13 retinopathy. *Sci. Rep.* **2019**, *9*, 3793. [CrossRef] [PubMed]
52. Brown, E.E.; DeWeerd, A.J.; Ildefonso, C.J.; Lewin, A.S.; Ash, J.D. Mitochondrial oxidative stress in the retinal pigment epithelium (RPE) led to metabolic dysfunction in both the RPE and retinal photoreceptors. *Redox Biol.* **2019**, *24*, 101201. [CrossRef] [PubMed]
53. Candas, D.; Li, J.J. MnSOD in oxidative stress response-potential regulation via mitochondrial protein influx. *Antioxid. Redox Signal.* **2014**, *20*, 1599–1617. [CrossRef]
54. Kaarniranta, K.; Tokarz, P.; Koskela, A.; Paterno, J.; Blasiak, J. Autophagy regulates death of retinal pigment epithelium cells in age-related macular degeneration. *Cell Biol. Toxicol.* **2017**, *33*, 113–128. [CrossRef] [PubMed]
55. Krohne, T.U.; Stratmann, N.K.; Kopitz, J.; Holz, F.G. Effects of lipid peroxidation products on lipofuscinogenesis and autophagy in human retinal pigment epithelial cells. *Exp. Eye Res.* **2010**, *90*, 465–471. [CrossRef] [PubMed]
56. Brandstetter, C.; Patt, J.; Holz, F.G.; Krohne, T.U. Inflammasome priming increases retinal pigment epithelial cell susceptibility to lipofuscin phototoxicity by changing the cell death mechanism from apoptosis to pyroptosis. *J. Photochem. Photobiol. B* **2016**, *161*, 177–183. [CrossRef] [PubMed]
57. Kurtz, J.; Jones, D.; Sternberg, P.; Wu, M.; Olsen, W. Antioxidant functions of glutathione in human retinal pigment epithelium in relation to age-related macular degeneration. In *Retinal Pigment Epithelium and Macular Diseases*; Coscas, G., Piccolino, F.C., Eds.; Springer: Berlin/Heidelberg, Germany, 1998; pp. 47–57.
58. Kim, H.J.; Montenegro, D.; Zhao, J.; Sparrow, J.R. Bisretinoids of the Retina: Photo-Oxidation, Iron-Catalyzed Oxidation, and Disease Consequences. *Antioxidants* **2021**, *10*, 1382. [CrossRef] [PubMed]
59. Handa, J.T. How does the macula protect itself from oxidative stress? *Mol. Aspects Med.* **2012**, *33*, 418–435. [CrossRef] [PubMed]
60. Rózanowska, M.; Jarvis-Evans, J.; Korytowski, W.; Boulton, M.E.; Burke, J.M.; Sarna, T. Blue light-induced reactivity of retinal age pigment. In vitro generation of oxygen-reactive species. *J. Biol. Chem.* **1995**, *270*, 18825–18830. [CrossRef] [PubMed]
61. Chan, C.M.; Huang, D.Y.; Sekar, P.; Hsu, S.H.; Lin, W.W. Reactive oxygen species-dependent mitochondrial dynamics and autophagy confer protective effects in retinal pigment epithelial cells against sodium iodate-induced cell death. *J. Biomed. Sci.* **2019**, *26*, 40. [CrossRef]
62. Mitter, S.K.; Rao, H.V.; Qi, X.; Cai, J.; Sugrue, A.; Dunn, W.A.; Grant, M.B.; Boulton, M.E. Autophagy in the retina: A potential role in age-related macular degeneration. *Adv. Exp. Med. Biol.* **2012**, *723*, 83–90. [CrossRef] [PubMed]
63. Zhao, H.; Raines, L.N.; Huang, S.C. Molecular Chaperones: Molecular Assembly Line Brings Metabolism and Immunity in Shape. *Metabolites* **2020**, *10*, 394. [CrossRef] [PubMed]
64. Kaarniranta, K.; Salminen, A.; Eskelinen, E.L.; Kopitz, J. Heat shock proteins as gatekeepers of proteolytic pathways-Implications for age-related macular degeneration (AMD). *Ageing Res. Rev.* **2009**, *8*, 128–139. [CrossRef]
65. Kampinga, H.H.; Hageman, J.; Vos, M.J.; Kubota, H.; Tanguay, R.M.; Bruford, E.A.; Cheetham, M.E.; Chen, B.; Hightower, L.E. Guidelines for the nomenclature of the human heat shock proteins. *Cell Stress Chaperones* **2009**, *14*, 105–111. [CrossRef]
66. Balchin, D.; Hayer-Hartl, M.; Hartl, F.U. Recent advances in understanding catalysis of protein folding by molecular chaperones. *FEBS Lett.* **2020**, *594*, 2770–2781. [CrossRef] [PubMed]
67. Szyller, J.; Bil-Lula, I. Heat Shock Proteins in Oxidative Stress and Ischemia/Reperfusion Injury and Benefits from Physical Exercises: A Review to the Current Knowledge. *Oxid. Med. Cell. Longev.* **2021**, *2021*, 6678457. [CrossRef]
68. Kregel, K.C. Heat shock proteins: Modifying factors in physiological stress responses and acquired thermotolerance. *J. Appl. Physiol.* **2002**, *92*, 2177–2186. [CrossRef] [PubMed]
69. De, S.; Rabin, D.M.; Salero, E.; Lederman, P.L.; Temple, S.; Stern, J.H. Human retinal pigment epithelium cell changes and expression of alphaB-crystallin: A biomarker for retinal pigment epithelium cell change in age-related macular degeneration. *Arch. Ophthalmol.* **2007**, *125*, 641–645. [CrossRef]
70. Kannan, R.; Sreekumar, P.G.; Hinton, D.R. Alpha crystallins in the retinal pigment epithelium and implications for the pathogenesis and treatment of age-related macular degeneration. *Biochim. Biophys. Acta* **2016**, *1860*, 258–268. [CrossRef] [PubMed]
71. Bonilha, V.L. Oxidative Stress Regulation and DJ-1 Function in the Retinal Pigment Epithelium: Implications for AMD. *Adv. Exp. Med. Biol.* **2018**, *1074*, 3–9. [CrossRef]
72. Bergwik, J.; Kristiansson, A.; Allhorn, M.; Gram, M.; Åkerström, B. Structure, Functions, and Physiological Roles of the Lipocalin α1-Microglobulin (A1M). *Front. Physiol.* **2021**, *12*, 645650. [CrossRef] [PubMed]

73. Gomez-Llorente, Y.; Jebara, F.; Patra, M.; Malik, R.; Nisemblat, S.; Chomsky-Hecht, O.; Parnas, A.; Azem, A.; Hirsch, J.A.; Ubarretxena-Belandia, I. Structural basis for active single and double ring complexes in human mitochondrial Hsp60-Hsp10 chaperonin. *Nat. Commun.* **2020**, *11*, 1916. [CrossRef]
74. Meng, Q.; Li, B.X.; Xiao, X. Toward Developing Chemical Modulators of Hsp60 as Potential Therapeutics. *Front. Mol. Biosci.* **2018**, *5*, 35. [CrossRef] [PubMed]
75. Leung, T.K.; Rajendran, M.Y.; Monfries, C.; Hall, C.; Lim, L. The human heat-shock protein family. Expression of a novel heat-inducible HSP70 (HSP70B′) and isolation of its cDNA and genomic DNA. *Biochem. J.* **1990**, *267*, 125–132. [CrossRef]
76. Polier, S.; Dragovic, Z.; Hartl, F.U.; Bracher, A. Structural basis for the cooperation of Hsp70 and Hsp110 chaperones in protein folding. *Cell* **2008**, *133*, 1068–1079. [CrossRef] [PubMed]
77. Prodromou, C. Mechanisms of Hsp90 regulation. *Biochem. J.* **2016**, *473*, 2439–2452. [CrossRef]
78. Benndorf, R.; Martin, J.L.; Kosakovsky Pond, S.L.; Wertheim, J.O. Neuropathy- and myopathy-associated mutations in human small heat shock proteins: Characteristics and evolutionary history of the mutation sites. *Mutat. Res. Rev. Mutat. Res.* **2014**, *761*, 15–30. [CrossRef] [PubMed]
79. Daugaard, M.; Rohde, M.; Jäättelä, M. The heat shock protein 70 family: Highly homologous proteins with overlapping and distinct functions. *FEBS Lett.* **2007**, *581*, 3702–3710. [CrossRef] [PubMed]
80. Arrigo, A.P.; Simon, S. Expression and functions of heat shock proteins in the normal and pathological mammalian eye. *Curr. Mol. Med.* **2010**, *10*, 776–793. [CrossRef]
81. Martindale, J.L.; Holbrook, N.J. Cellular response to oxidative stress: Signaling for suicide and survival. *J. Cell. Physiol.* **2002**, *192*, 1–15. [CrossRef] [PubMed]
82. Cox, D.; Carver, J.A.; Ecroyd, H. Preventing α-synuclein aggregation: The role of the small heat-shock molecular chaperone proteins. *Biochim. Biophys. Acta* **2014**, *1842*, 1830–1843. [CrossRef] [PubMed]
83. Zou, Y.; Shi, H.; Liu, N.; Wang, H.; Song, X.; Liu, B. Mechanistic insights into heat shock protein 27, a potential therapeutic target for cardiovascular diseases. *Front. Cardiovasc. Med.* **2023**, *10*, 1195464. [CrossRef]
84. Lee, G.J.; Roseman, A.M.; Saibil, H.R.; Vierling, E. A small heat shock protein stably binds heat-denatured model substrates and can maintain a substrate in a folding-competent state. *EMBO J.* **1997**, *16*, 659–671. [CrossRef] [PubMed]
85. Pistono, C.; Monti, M.C.; Boiocchi, C.; Berzolari, F.G.; Osera, C.; Mallucci, G.; Cuccia, M.; Pascale, A.; Montomoli, C.; Bergamaschi, R. Response to oxidative stress of peripheral blood mononuclear cells from multiple sclerosis patients and healthy controls. *Cell Stress Chaperones* **2020**, *25*, 81–91. [CrossRef] [PubMed]
86. Richter, K.; Haslbeck, M.; Buchner, J. The heat shock response: Life on the verge of death. *Mol. Cell* **2010**, *40*, 253–266. [CrossRef] [PubMed]
87. Penke, B.; Bogár, F.; Crul, T.; Sántha, M.; Tóth, M.E.; Vígh, L. Heat Shock Proteins and Autophagy Pathways in Neuroprotection: From Molecular Bases to Pharmacological Interventions. *Int. J. Mol. Sci.* **2018**, *19*, 325. [CrossRef]
88. Amirkavei, M.; Plastino, F.; Kvanta, A.; Kaarniranta, K.; André, H.; Koskelainen, A. Hormetic Heat Shock Enhances Autophagy through HSF1 in Retinal Pigment Epithelium Cells. *Cells* **2022**, *11*, 1778. [CrossRef]
89. Klaips, C.L.; Jayaraj, G.G.; Hartl, F.U. Pathways of cellular proteostasis in aging and disease. *J. Cell. Biol.* **2018**, *217*, 51–63. [CrossRef]
90. Ciechanover, A. Proteolysis: From the lysosome to ubiquitin and the proteasome. *Nat. Rev. Mol. Cell. Biol.* **2005**, *6*, 79–87. [CrossRef] [PubMed]
91. Shang, F.; Taylor, A. Ubiquitin-proteasome pathway and cellular responses to oxidative stress. *Free Radic. Biol. Med.* **2011**, *51*, 5–16. [CrossRef] [PubMed]
92. Giordano, S.; Darley-Usmar, V.; Zhang, J. Autophagy as an essential cellular antioxidant pathway in neurodegenerative disease. *Redox Biol.* **2013**, *2*, 82–90. [CrossRef] [PubMed]
93. Nam, T.; Han, J.H.; Devkota, S.; Lee, H.W. Emerging Paradigm of Crosstalk between Autophagy and the Ubiquitin-Proteasome System. *Mol. Cells* **2017**, *40*, 897–905. [CrossRef]
94. Ji, C.H.; Kwon, Y.T. Crosstalk and Interplay between the Ubiquitin-Proteasome System and Autophagy. *Mol. Cells* **2017**, *40*, 441–449. [CrossRef] [PubMed]
95. Reggiori, F.; Komatsu, M.; Finley, K.; Simonsen, A. Autophagy: More than a nonselective pathway. *Int. J. Cell. Biol.* **2012**, *2012*, 219625. [CrossRef] [PubMed]
96. Kaarniranta, K.; Hyttinen, J.; Ryhanen, T.; Viiri, J.; Paimela, T.; Toropainen, E.; Sorri, I.; Salminen, A. Mechanisms of protein aggregation in the retinal pigment epithelial cells. *Front. Biosci.* **2010**, *2*, 1374–1384. [CrossRef] [PubMed]
97. Galves, M.; Rathi, R.; Prag, G.; Ashkenazi, A. Ubiquitin Signaling and Degradation of Aggregate-Prone Proteins. *Trends Biochem. Sci.* **2019**, *44*, 872–884. [CrossRef]

98. Tomko, R.J.; Hochstrasser, M. Molecular architecture and assembly of the eukaryotic proteasome. *Annu. Rev. Biochem.* **2013**, *82*, 415–445. [CrossRef]
99. Finley, D. Recognition and processing of ubiquitin-protein conjugates by the proteasome. *Annu. Rev. Biochem.* **2009**, *78*, 477–513. [CrossRef]
100. Kaneko, T.; Hamazaki, J.; Iemura, S.; Sasaki, K.; Furuyama, K.; Natsume, T.; Tanaka, K.; Murata, S. Assembly pathway of the Mammalian proteasome base subcomplex is mediated by multiple specific chaperones. *Cell* **2009**, *137*, 914–925. [CrossRef] [PubMed]
101. Le Tallec, B.; Barrault, M.B.; Courbeyrette, R.; Guérois, R.; Marsolier-Kergoat, M.C.; Peyroche, A. 20S proteasome assembly is orchestrated by two distinct pairs of chaperones in yeast and in mammals. *Mol. Cell* **2007**, *27*, 660–674. [CrossRef]
102. Kusmierczyk, A.R.; Hochstrasser, M. Some assembly required: Dedicated chaperones in eukaryotic proteasome biogenesis. *Biol. Chem.* **2008**, *389*, 1143–1151. [CrossRef] [PubMed]
103. Boelens, W.C.; Croes, Y.; de Jong, W.W. Interaction between alphaB-crystallin and the human 20S proteasomal subunit C8/alpha7. *Biochim. Biophys. Acta* **2001**, *1544*, 311–319. [CrossRef] [PubMed]
104. Pattison, J.S.; Robbins, J. Autophagy and proteotoxicity in cardiomyocytes. *Autophagy* **2011**, *7*, 1259–1260. [CrossRef]
105. den Engelsman, J.; Bennink, E.J.; Doerwald, L.; Onnekink, C.; Wunderink, L.; Andley, U.P.; Kato, K.; de Jong, W.W.; Boelens, W.C. Mimicking phosphorylation of the small heat-shock protein alphaB-crystallin recruits the F-box protein FBX4 to nuclear SC35 speckles. *Eur. J. Biochem.* **2004**, *271*, 4195–4203. [CrossRef] [PubMed]
106. Lin, D.I.; Barbash, O.; Kumar, K.G.; Weber, J.D.; Harper, J.W.; Klein-Szanto, A.J.; Rustgi, A.; Fuchs, S.Y.; Diehl, J.A. Phosphorylation-dependent ubiquitination of cyclin D1 by the SCF(FBX4-alphaB crystallin) complex. *Mol. Cell* **2006**, *24*, 355–366. [CrossRef] [PubMed]
107. Chen, D.D.; Liu, B.; Wang, Y.; Jiang, M.; Shang, G.; Xue, M.; Jia, X.; Lang, Y.; Zhou, G.; Zhang, F.; et al. The downregulation of HSP90-controlled CRALBP expression is associated with age-related vision attenuation. *FASEB J.* **2023**, *37*, e22832. [CrossRef]
108. Nickel, A.; Kohlhaas, M.; Maack, C. Mitochondrial reactive oxygen species production and elimination. *J. Mol. Cell. Cardiol.* **2014**, *73*, 26–33. [CrossRef]
109. Dunlop, R.A.; Brunk, U.T.; Rodgers, K.J. Oxidized proteins: Mechanisms of removal and consequences of accumulation. *IUBMB Life* **2009**, *61*, 522–527. [CrossRef]
110. Jung, T.; Bader, N.; Grune, T. Oxidized proteins: Intracellular distribution and recognition by the proteasome. *Arch. Biochem. Biophys.* **2007**, *462*, 231–237. [CrossRef]
111. Jung, T.; Bader, N.; Grune, T. Lipofuscin: Formation, distribution, and metabolic consequences. *Ann. N. Y. Acad. Sci.* **2007**, *1119*, 97–111. [CrossRef] [PubMed]
112. Decanini, A.; Nordgaard, C.L.; Feng, X.; Ferrington, D.A.; Olsen, T.W. Changes in select redox proteins of the retinal pigment epithelium in age-related macular degeneration. *Am. J. Ophthalmol.* **2007**, *143*, 607–615. [CrossRef] [PubMed]
113. Mazzoni, F.; Safa, H.; Finnemann, S.C. Understanding photoreceptor outer segment phagocytosis: Use and utility of RPE cells in culture. *Exp. Eye Res.* **2014**, *126*, 51–60. [CrossRef] [PubMed]
114. Nandrot, E.F.; Anand, M.; Almeida, D.; Atabai, K.; Sheppard, D.; Finnemann, S.C. Essential role for MFG-E8 as ligand for alphavbeta5 integrin in diurnal retinal phagocytosis. *Proc. Natl. Acad. Sci. USA* **2007**, *104*, 12005–12010. [CrossRef] [PubMed]
115. Nandrot, E.F.; Kim, Y.; Brodie, S.E.; Huang, X.; Sheppard, D.; Finnemann, S.C. Loss of synchronized retinal phagocytosis and age-related blindness in mice lacking alphavbeta5 integrin. *J. Exp. Med.* **2004**, *200*, 1539–1545. [CrossRef]
116. Mao, Y.; Finnemann, S.C. Essential diurnal Rac1 activation during retinal phagocytosis requires αvβ5 integrin but not tyrosine kinases focal adhesion kinase or Mer tyrosine kinase. *Mol. Biol. Cell* **2012**, *23*, 1104–1114. [CrossRef]
117. Bulloj, A.; Maminishkis, A.; Mizui, M.; Finnemann, S.C. Semaphorin4D-PlexinB1 Signaling Attenuates Photoreceptor Outer Segment Phagocytosis by Reducing Rac1 Activity of RPE Cells. *Mol. Neurobiol.* **2018**, *55*, 4320–4332. [CrossRef] [PubMed]
118. Freeman, S.A.; Goyette, J.; Furuya, W.; Woods, E.C.; Bertozzi, C.R.; Bergmeier, W.; Hinz, B.; van der Merwe, P.A.; Das, R.; Grinstein, S. Integrins Form an Expanding Diffusional Barrier that Coordinates Phagocytosis. *Cell* **2016**, *164*, 128–140. [CrossRef] [PubMed]
119. Cong, Y.; Baker, M.L.; Jakana, J.; Woolford, D.; Miller, E.J.; Reissmann, S.; Kumar, R.N.; Redding-Johanson, A.M.; Batth, T.S.; Mukhopadhyay, A.; et al. 4.0-A resolution cryo-EM structure of the mammalian chaperonin TRiC/CCT reveals its unique subunit arrangement. *Proc. Natl. Acad. Sci. USA* **2010**, *107*, 4967–4972. [CrossRef] [PubMed]
120. Berger, J.; Berger, S.; Li, M.; Jacoby, A.S.; Arner, A.; Bavi, N.; Stewart, A.G.; Currie, P.D. In Vivo Function of the Chaperonin TRiC in α-Actin Folding during Sarcomere Assembly. *Cell Rep.* **2018**, *22*, 313–322. [CrossRef]
121. Sergeeva, O.A.; Tran, M.T.; Haase-Pettingell, C.; King, J.A. Biochemical characterization of mutants in chaperonin proteins CCT4 and CCT5 associated with hereditary sensory neuropathy. *J. Biol. Chem.* **2014**, *289*, 27470–27480. [CrossRef] [PubMed]

122. Pavel, M.; Imarisio, S.; Menzies, F.M.; Jimenez-Sanchez, M.; Siddiqi, F.H.; Wu, X.; Renna, M.; O'Kane, C.J.; Crowther, D.C.; Rubinsztein, D.C. CCT complex restricts neuropathogenic protein aggregation via autophagy. *Nat. Commun.* **2016**, *7*, 13821. [CrossRef] [PubMed]
123. Singhto, N.; Sintiprungrat, K.; Thongboonkerd, V. Alterations in macrophage cellular proteome induced by calcium oxalate crystals: The association of HSP90 and F-actin is important for phagosome formation. *J. Proteome Res.* **2013**, *12*, 3561–3572. [CrossRef] [PubMed]
124. Hsieh, J.C.; Lee, L.; Zhang, L.; Wefer, S.; Brown, K.; DeRossi, C.; Wines, M.E.; Rosenquist, T.; Holdener, B.C. Mesd encodes an LRP5/6 chaperone essential for specification of mouse embryonic polarity. *Cell* **2003**, *112*, 355–367. [CrossRef] [PubMed]
125. Pearring, J.N.; Salinas, R.Y.; Baker, S.A.; Arshavsky, V.Y. Protein sorting, targeting and trafficking in photoreceptor cells. *Prog. Retin. Eye Res.* **2013**, *36*, 24–51. [CrossRef] [PubMed]
126. Chen, X.; Guo, F.; LeBlanc, M.E.; Ding, Y.; Zhang, C.; Shakya, A.; Li, W. Mesd extrinsically promotes phagocytosis by retinal pigment epithelial cells. *Cell Biol. Toxicol.* **2016**, *32*, 347–358. [CrossRef] [PubMed]
127. Liu, J.; Lu, W.; Reigada, D.; Nguyen, J.; Laties, A.M.; Mitchell, C.H. Restoration of lysosomal pH in RPE cells from cultured human and ABCA4$^{-/-}$ mice: Pharmacologic approaches and functional recovery. *Investig. Ophthalmol. Vis. Sci.* **2008**, *49*, 772–780. [CrossRef]
128. Mindell, J.A. Lysosomal acidification mechanisms. *Annu. Rev. Physiol.* **2012**, *74*, 69–86. [CrossRef]
129. Valapala, M.; Wilson, C.; Hose, S.; Bhutto, I.A.; Grebe, R.; Dong, A.; Greenbaum, S.; Gu, L.; Sengupta, S.; Cano, M.; et al. Lysosomal-mediated waste clearance in retinal pigment epithelial cells is regulated by CRYBA1/βA3/A1-crystallin via V-ATPase-MTORC1 signaling. *Autophagy* **2014**, *10*, 480–496. [CrossRef]
130. Frost, L.S.; Lopes, V.S.; Bragin, A.; Reyes-Reveles, J.; Brancato, J.; Cohen, A.; Mitchell, C.H.; Williams, D.S.; Boesze-Battaglia, K. The Contribution of Melanoregulin to Microtubule-Associated Protein 1 Light Chain 3 (LC3) Associated Phagocytosis in Retinal Pigment Epithelium. *Mol. Neurobiol.* **2015**, *52*, 1135–1151. [CrossRef]
131. Martinez, J.; Malireddi, R.K.; Lu, Q.; Cunha, L.D.; Pelletier, S.; Gingras, S.; Orchard, R.; Guan, J.L.; Tan, H.; Peng, J.; et al. Molecular characterization of LC3-associated phagocytosis reveals distinct roles for Rubicon, NOX2 and autophagy proteins. *Nat. Cell Biol.* **2015**, *17*, 893–906. [CrossRef] [PubMed]
132. Kim, J.Y.; Zhao, H.; Martinez, J.; Doggett, T.A.; Kolesnikov, A.V.; Tang, P.H.; Ablonczy, Z.; Chan, C.C.; Zhou, Z.; Green, D.R.; et al. Noncanonical autophagy promotes the visual cycle. *Cell* **2013**, *154*, 365–376. [CrossRef] [PubMed]
133. Mizushima, N.; Yoshimori, T.; Levine, B. Methods in mammalian autophagy research. *Cell* **2010**, *140*, 313–326. [CrossRef]
134. Fazeli, G.; Wehman, A.M. Safely removing cell debris with LC3-associated phagocytosis. *Biol. Cell* **2017**, *109*, 355–363. [CrossRef] [PubMed]
135. Almedawar, S.; Vafia, K.; Schreiter, S.; Neumann, K.; Khattak, S.; Kurth, T.; Ader, M.; Karl, M.O.; Tsang, S.H.; Tanaka, E.M. MERTK-Dependent Ensheathment of Photoreceptor Outer Segments by Human Pluripotent Stem Cell-Derived Retinal Pigment Epithelium. *Stem Cell Rep.* **2020**, *14*, 374–389. [CrossRef] [PubMed]
136. Dhingra, A.; Bell, B.A.; Peachey, N.S.; Daniele, L.L.; Reyes-Reveles, J.; Sharp, R.C.; Jun, B.; Bazan, N.G.; Sparrow, J.R.; Kim, H.J.; et al. Microtubule-Associated Protein 1 Light Chain 3B, (LC3B) Is Necessary to Maintain Lipid-Mediated Homeostasis in the Retinal Pigment Epithelium. *Front. Cell. Neurosci.* **2018**, *12*, 351. [CrossRef] [PubMed]
137. Singh, R.; Kaushik, S.; Wang, Y.; Xiang, Y.; Novak, I.; Komatsu, M.; Tanaka, K.; Cuervo, A.M.; Czaja, M.J. Autophagy regulates lipid metabolism. *Nature* **2009**, *458*, 1131–1135. [CrossRef] [PubMed]
138. Muniz-Feliciano, L.; Doggett, T.A.; Zhou, Z.; Ferguson, T.A. RUBCN/rubicon and EGFR regulate lysosomal degradative processes in the retinal pigment epithelium (RPE) of the eye. *Autophagy* **2017**, *13*, 2072–2085. [CrossRef] [PubMed]
139. Ravera, S.; Bertola, N.; Puddu, A.; Bruno, S.; Maggi, D.; Panfoli, I. Crosstalk between the Rod Outer Segments and Retinal Pigmented Epithelium in the Generation of Oxidative Stress in an In Vitro Model. *Cells* **2023**, *12*, 2173. [CrossRef]
140. Morel, E.; Mehrpour, M.; Botti, J.; Dupont, N.; Hamaï, A.; Nascimbeni, A.C.; Codogno, P. Autophagy: A Druggable Process. *Annu. Rev. Pharmacol. Toxicol.* **2017**, *57*, 375–398. [CrossRef] [PubMed]
141. Boya, P.; Reggiori, F.; Codogno, P. Emerging regulation and functions of autophagy. *Nat. Cell Biol.* **2013**, *15*, 713–720. [CrossRef] [PubMed]
142. Birgisdottir, Å.B.; Lamark, T.; Johansen, T. The LIR motif - crucial for selective autophagy. *J. Cell Sci.* **2013**, *126*, 3237–3247. [CrossRef] [PubMed]
143. Rogov, V.; Dötsch, V.; Johansen, T.; Kirkin, V. Interactions between autophagy receptors and ubiquitin-like proteins form the molecular basis for selective autophagy. *Mol. Cell* **2014**, *53*, 167–178. [CrossRef] [PubMed]
144. Svenning, S.; Johansen, T. Selective autophagy. *Essays Biochem.* **2013**, *55*, 79–92. [CrossRef] [PubMed]
145. Boya, P.; Esteban-Martínez, L.; Serrano-Puebla, A.; Gómez-Sintes, R.; Villarejo-Zori, B. Autophagy in the eye: Development, degeneration, and aging. *Prog. Retin. Eye Res.* **2016**, *55*, 206–245. [CrossRef] [PubMed]

146. Ktistakis, N.T.; Tooze, S.A. Digesting the Expanding Mechanisms of Autophagy. *Trends Cell Biol.* **2016**, *26*, 624–635. [CrossRef]
147. Kaarniranta, K.; Sinha, D.; Blasiak, J.; Kauppinen, A.; Veréb, Z.; Salminen, A.; Boulton, M.E.; Petrovski, G. Autophagy and heterophagy dysregulation leads to retinal pigment epithelium dysfunction and development of age-related macular degeneration. *Autophagy* **2013**, *9*, 973–984. [CrossRef] [PubMed]
148. Blasiak, J.; Petrovski, G.; Veréb, Z.; Facskó, A.; Kaarniranta, K. Oxidative stress, hypoxia, and autophagy in the neovascular processes of age-related macular degeneration. *Biomed. Res. Int.* **2014**, *2014*, 768026. [CrossRef] [PubMed]
149. Valapala, M.; Edwards, M.; Hose, S.; Grebe, R.; Bhutto, I.A.; Cano, M.; Berger, T.; Mak, T.W.; Wawrousek, E.; Handa, J.T.; et al. Increased Lipocalin-2 in the retinal pigment epithelium of Cryba1 cKO mice is associated with a chronic inflammatory response. *Aging Cell* **2014**, *13*, 1091–1094. [CrossRef] [PubMed]
150. Wang, L.; Cano, M.; Handa, J.T. p62 provides dual cytoprotection against oxidative stress in the retinal pigment epithelium. *Biochim. Biophys. Acta* **2014**, *1843*, 1248–1258. [CrossRef]
151. Gurusamy, N.; Lekli, I.; Gherghiceanu, M.; Popescu, L.M.; Das, D.K. BAG-1 induces autophagy for cardiac cell survival. *Autophagy* **2009**, *5*, 120–121. [CrossRef] [PubMed]
152. Majeski, A.E.; Dice, J.F. Mechanisms of chaperone-mediated autophagy. *Int. J. Biochem. Cell. Biol.* **2004**, *36*, 2435–2444. [CrossRef]
153. Martinez-Vicente, M.; Sovak, G.; Cuervo, A.M. Protein degradation and aging. *Exp. Gerontol.* **2005**, *40*, 622–633. [CrossRef] [PubMed]
154. Terman, A.; Gustafsson, B.; Brunk, U.T. Autophagy, organelles and ageing. *J. Pathol.* **2007**, *211*, 134–143. [CrossRef] [PubMed]
155. Terlecky, S.R.; Chiang, H.L.; Olson, T.S.; Dice, J.F. Protein and peptide binding and stimulation of in vitro lysosomal proteolysis by the 73-kDa heat shock cognate protein. *J. Biol. Chem.* **1992**, *267*, 9202–9209. [CrossRef]
156. Richter, K.; Buchner, J. hsp90: Twist and fold. *Cell* **2006**, *127*, 251–253. [CrossRef]
157. Agarraberes, F.A.; Dice, J.F. A molecular chaperone complex at the lysosomal membrane is required for protein translocation. *J. Cell Sci.* **2001**, *114*, 2491–2499. [CrossRef] [PubMed]
158. Dice, J.F. Chaperone-mediated autophagy. *Autophagy* **2007**, *3*, 295–299. [CrossRef]
159. Bandyopadhyay, U.; Kaushik, S.; Varticovski, L.; Cuervo, A.M. The chaperone-mediated autophagy receptor organizes in dynamic protein complexes at the lysosomal membrane. *Mol. Cell. Biol.* **2008**, *28*, 5747–5763. [CrossRef] [PubMed]
160. Eskelinen, E.L.; Cuervo, A.M.; Taylor, M.R.; Nishino, I.; Blum, J.S.; Dice, J.F.; Sandoval, I.V.; Lippincott-Schwartz, J.; August, J.T.; Saftig, P. Unifying nomenclature for the isoforms of the lysosomal membrane protein LAMP-2. *Traffic* **2005**, *6*, 1058–1061. [CrossRef]
161. Eskelinen, E.L.; Tanaka, Y.; Saftig, P. At the acidic edge: Emerging functions for lysosomal membrane proteins. *Trends Cell Biol.* **2003**, *13*, 137–145. [CrossRef]
162. Cuervo, A.M.; Dice, J.F. Regulation of lamp2a levels in the lysosomal membrane. *Traffic* **2000**, *1*, 570–583. [CrossRef] [PubMed]
163. Kaushik, S.; Massey, A.C.; Cuervo, A.M. Lysosome membrane lipid microdomains: Novel regulators of chaperone-mediated autophagy. *EMBO J.* **2006**, *25*, 3921–3933. [CrossRef]
164. Bandyopadhyay, U.; Cuervo, A.M. Chaperone-mediated autophagy in aging and neurodegeneration: Lessons from alpha-synuclein. *Exp. Gerontol.* **2007**, *42*, 120–128. [CrossRef]
165. Massey, A.C.; Zhang, C.; Cuervo, A.M. Chaperone-mediated autophagy in aging and disease. *Curr. Top. Dev. Biol.* **2006**, *73*, 205–235. [CrossRef]
166. Kiffin, R.; Christian, C.; Knecht, E.; Cuervo, A.M. Activation of chaperone-mediated autophagy during oxidative stress. *Mol. Biol. Cell* **2004**, *15*, 4829–4840. [CrossRef] [PubMed]
167. Salvador, N.; Aguado, C.; Horst, M.; Knecht, E. Import of a cytosolic protein into lysosomes by chaperone-mediated autophagy depends on its folding state. *J. Biol. Chem.* **2000**, *275*, 27447–27456. [CrossRef] [PubMed]
168. Liao, Z.; Wang, B.; Liu, W.; Xu, Q.; Hou, L.; Song, J.; Guo, Q.; Li, N. Dysfunction of chaperone-mediated autophagy in human diseases. *Mol. Cell. Biochem.* **2021**, *476*, 1439–1454. [CrossRef] [PubMed]
169. Cuervo, A.M.; Mann, L.; Bonten, E.J.; d'Azzo, A.; Dice, J.F. Cathepsin A regulates chaperone-mediated autophagy through cleavage of the lysosomal receptor. *EMBO J.* **2003**, *22*, 47–59. [CrossRef]
170. Cuervo, A.M.; Dice, J.F. Unique properties of lamp2a compared to other lamp2 isoforms. *J. Cell Sci.* **2000**, *113*, 4441–4450. [CrossRef] [PubMed]
171. Vélez, E.J.; Schnebert, S.; Goguet, M.; Balbuena-Pecino, S.; Dias, K.; Beauclair, L.; Fontagné-Dicharry, S.; Véron, V.; Depincé, A.; Beaumatin, F.; et al. Chaperone-mediated autophagy protects against hyperglycemic stress. *Autophagy* **2024**, *20*, 752–768. [CrossRef] [PubMed]
172. Das, S.; Seth, R.K.; Kumar, A.; Kadiiska, M.B.; Michelotti, G.; Diehl, A.M.; Chatterjee, S. Purinergic receptor X7 is a key modulator of metabolic oxidative stress-mediated autophagy and inflammation in experimental nonalcoholic steatohepatitis. *Am. J. Physiol. Gastrointest. Liver Physiol.* **2013**, *305*, G950–G963. [CrossRef]

173. Pajares, M.; Rojo, A.I.; Arias, E.; Díaz-Carretero, A.; Cuervo, A.M.; Cuadrado, A. Transcription factor NFE2L2/NRF2 modulates chaperone-mediated autophagy through the regulation of LAMP2A. *Autophagy* **2018**, *14*, 1310–1322. [CrossRef] [PubMed]
174. Anguiano, J.; Garner, T.P.; Mahalingam, M.; Das, B.C.; Gavathiotis, E.; Cuervo, A.M. Chemical modulation of chaperone-mediated autophagy by retinoic acid derivatives. *Nat. Chem. Biol.* **2013**, *9*, 374–382. [CrossRef] [PubMed]
175. Bandyopadhyay, U.; Sridhar, S.; Kaushik, S.; Kiffin, R.; Cuervo, A.M. Identification of regulators of chaperone-mediated autophagy. *Mol. Cell* **2010**, *39*, 535–547. [CrossRef]
176. Arias, E.; Koga, H.; Diaz, A.; Mocholi, E.; Patel, B.; Cuervo, A.M. Lysosomal mTORC2/PHLPP1/Akt Regulate Chaperone-Mediated Autophagy. *Mol. Cell* **2015**, *59*, 270–284. [CrossRef] [PubMed]
177. Fleming, A.; Bourdenx, M.; Fujimaki, M.; Karabiyik, C.; Krause, G.J.; Lopez, A.; Martín-Segura, A.; Puri, C.; Scrivo, A.; Skidmore, J.; et al. The different autophagy degradation pathways and neurodegeneration. *Neuron* **2022**, *110*, 935–966. [CrossRef] [PubMed]
178. Galluzzi, L.; Baehrecke, E.H.; Ballabio, A.; Boya, P.; Bravo-San Pedro, J.M.; Cecconi, F.; Choi, A.M.; Chu, C.T.; Codogno, P.; Colombo, M.I.; et al. Molecular definitions of autophagy and related processes. *EMBO J.* **2017**, *36*, 1811–1836. [CrossRef]
179. Sahu, R.; Kaushik, S.; Clement, C.C.; Cannizzo, E.S.; Scharf, B.; Follenzi, A.; Potolicchio, I.; Nieves, E.; Cuervo, A.M.; Santambrogio, L. Microautophagy of cytosolic proteins by late endosomes. *Dev. Cell* **2011**, *20*, 131–139. [CrossRef] [PubMed]
180. Loi, M.; Raimondi, A.; Morone, D.; Molinari, M. ESCRT-III-driven piecemeal micro-ER-phagy remodels the ER during recovery from ER stress. *Nat. Commun.* **2019**, *10*, 5058. [CrossRef] [PubMed]
181. Krause, G.J.; Cuervo, A.M. Assessment of mammalian endosomal microautophagy. *Meth. Cell Biol.* **2021**, *164*, 167–185. [CrossRef]
182. Kirchner, P.; Bourdenx, M.; Madrigal-Matute, J.; Tiano, S.; Diaz, A.; Bartholdy, B.A.; Will, B.; Cuervo, A.M. Proteome-wide analysis of chaperone-mediated autophagy targeting motifs. *PLoS Biol.* **2019**, *17*, e3000301. [CrossRef] [PubMed]
183. Morozova, K.; Clement, C.C.; Kaushik, S.; Stiller, B.; Arias, E.; Ahmad, A.; Rauch, J.N.; Chatterjee, V.; Melis, C.; Scharf, B.; et al. Structural and Biological Interaction of hsc-70 Protein with Phosphatidylserine in Endosomal Microautophagy. *J. Biol. Chem.* **2016**, *291*, 18096–18106. [CrossRef] [PubMed]
184. Gudipaty, S.A.; Conner, C.M.; Rosenblatt, J.; Montell, D.J. Unconventional Ways to Live and Die: Cell Death and Survival in Development, Homeostasis, and Disease. *Annu. Rev. Cell Dev. Biol.* **2018**, *34*, 311–332. [CrossRef]
185. Galluzzi, L.; Bravo-San Pedro, J.M.; Vitale, I.; Aaronson, S.A.; Abrams, J.M.; Adam, D.; Alnemri, E.S.; Altucci, L.; Andrews, D.; Annicchiarico-Petruzzelli, M.; et al. Essential versus accessory aspects of cell death: Recommendations of the NCCD 2015. *Cell Death Differ.* **2015**, *22*, 58–73. [CrossRef] [PubMed]
186. Newton, K.; Strasser, A.; Kayagaki, N.; Dixit, V.M. Cell death. *Cell* **2024**, *187*, 235–256. [CrossRef] [PubMed]
187. Kroemer, G.; Galluzzi, L.; Vandenabeele, P.; Abrams, J.; Alnemri, E.S.; Baehrecke, E.H.; Blagosklonny, M.V.; El-Deiry, W.S.; Golstein, P.; Green, D.R.; et al. Nomenclature Committee on Cell Death 2009. Classification of cell death: Recommendations of the Nomenclature Committee on Cell Death 2009. *Cell Death Differ.* **2009**, *16*, 3–11. [CrossRef]
188. Tong, Y.; Wu, Y.; Ma, J.; Ikeda, M.; Ide, T.; Griffin, C.T.; Ding, X.Q.; Wang, S. Comparative mechanistic study of RPE cell death induced by different oxidative stresses. *Redox Biol.* **2023**, *65*, 102840. [CrossRef] [PubMed]
189. Yang, M.; So, K.F.; Lam, W.C.; Lo, A.C.Y. Novel Programmed Cell Death as Therapeutic Targets in Age-Related Macular Degeneration? *Int. J. Mol. Sci.* **2020**, *21*, 7279. [CrossRef] [PubMed]
190. Hanus, J.; Anderson, C.; Wang, S. RPE necroptosis in response to oxidative stress and in AMD. *Ageing Res. Rev.* **2015**, *24*, 286–298. [CrossRef]
191. Man, S.M.; Karki, R.; Kanneganti, T.D. Molecular mechanisms and functions of pyroptosis, inflammatory caspases and inflammasomes in infectious diseases. *Immunol. Rev.* **2017**, *277*, 61–75. [CrossRef]
192. Yang, C.; Shani, S.; Tahiri, H.; Ortiz, C.; Gu, M.; Lavoie, J.C.; Croteau, S.; Hardy, P. Extracellular microparticles exacerbate oxidative damage to retinal pigment epithelial cells. *Exp. Cell Res.* **2020**, *390*, 111957. [CrossRef]
193. Guo, D.; Liu, Z.; Zhou, J.; Ke, C.; Li, D. Significance of Programmed Cell Death Pathways in Neurodegenerative Diseases. *Int. J. Mol. Sci.* **2024**, *25*, 9947. [CrossRef] [PubMed]
194. Pierzynowska, K.; Rintz, E.; Gaffke, L.; Węgrzyn, G. Ferroptosis and Its Modulation by Autophagy in Light of the Pathogenesis of Lysosomal Storage Diseases. *Cells* **2021**, *10*, 365. [CrossRef]
195. Xie, Y.; Hou, W.; Song, X.; Yu, Y.; Huang, J.; Sun, X.; Kang, R.; Tang, D. Ferroptosis: Process and function. *Cell Death Differ.* **2016**, *23*, 369–379. [CrossRef] [PubMed]
196. Elmore, S. Apoptosis: A review of programmed cell death. *Toxicol. Pathol.* **2007**, *35*, 495–516. [CrossRef] [PubMed]
197. Kaneko, H.; Dridi, S.; Tarallo, V.; Gelfand, B.D.; Fowler, B.J.; Cho, W.G.; Kleinman, M.E.; Ponicsan, S.L.; Hauswirth, W.W.; Chiodo, V.A.; et al. DICER1 deficit induces Alu RNA toxicity in age-related macular degeneration. *Nature* **2011**, *471*, 325–330. [CrossRef] [PubMed]

198. Gomez-Lazaro, M.; Galindo, M.F.; Melero-Fernandez de Mera, R.M.; Fernandez-Gómez, F.J.; Concannon, C.G.; Segura, M.F.; Comella, J.X.; Prehn, J.H.; Jordan, J. Reactive oxygen species and p38 mitogen-activated protein kinase activate Bax to induce mitochondrial cytochrome c release and apoptosis in response to malonate. *Mol. Pharmacol.* **2007**, *71*, 736–743. [CrossRef]
199. Tsuruta, F.; Sunayama, J.; Mori, Y.; Hattori, S.; Shimizu, S.; Tsujimoto, Y.; Yoshioka, K.; Masuyama, N.; Gotoh, Y. JNK promotes Bax translocation to mitochondria through phosphorylation of 14-3-3 proteins. *EMBO J.* **2004**, *23*, 1889–1899. [CrossRef]
200. Alge, C.S.; Priglinger, S.G.; Neubauer, A.S.; Kampik, A.; Zillig, M.; Bloemendal, H.; Welge-Lussen, U. Retinal pigment epithelium is protected against apoptosis by alphaB-crystallin. *Investig. Ophthalmol. Vis. Sci.* **2002**, *43*, 3575–3582.
201. Li, D.W.; Liu, J.P.; Mao, Y.W.; Xiang, H.; Wang, J.; Ma, W.Y.; Dong, Z.; Pike, H.M.; Brown, R.E.; Reed, J.C. Calcium-activated RAF/MEK/ERK signaling pathway mediates p53-dependent apoptosis and is abrogated by alpha B-crystallin through inhibition of RAS activation. *Mol. Biol Cell* **2005**, *16*, 4437–4453. [CrossRef]
202. Zhou, P.; Kannan, R.; Spee, C.; Sreekumar, P.G.; Dou, G.; Hinton, D.R. Protection of retina by αB crystallin in sodium iodate induced retinal degeneration. *PLoS ONE* **2014**, *9*, e98275. [CrossRef]
203. Yaung, J.; Kannan, R.; Wawrousek, E.F.; Spee, C.; Sreekumar, P.G.; Hinton, D.R. Exacerbation of retinal degeneration in the absence of alpha crystallins in an in vivo model of chemically induced hypoxia. *Exp. Eye Res.* **2008**, *86*, 355–365. [CrossRef] [PubMed]
204. Watanabe, G.; Kato, S.; Nakata, H.; Ishida, T.; Ohuchi, N.; Ishioka, C. alphaB-crystallin: A novel p53-target gene required for p53-dependent apoptosis. *Cancer Sci.* **2009**, *100*, 2368–2375. [CrossRef] [PubMed]
205. Liu, S.; Li, J.; Tao, Y.; Xiao, X. Small heat shock protein alphaB-crystallin binds to p53 to sequester its translocation to mitochondria during hydrogen peroxide-induced apoptosis. *Biochem. Biophys. Res. Commun.* **2007**, *354*, 109–114. [CrossRef] [PubMed]
206. Mao, Y.W.; Liu, J.P.; Xiang, H.; Li, D.W. Human alphaA- and alphaB-crystallins bind to Bax and Bcl-X(S) to sequester their translocation during staurosporine-induced apoptosis. *Cell Death Differ.* **2004**, *11*, 512–526. [CrossRef]
207. Sreekumar, P.G.; Kannan, R.; Kitamura, M.; Spee, C.; Barron, E.; Ryan, S.J.; Hinton, D.R. αB crystallin is apically secreted within exosomes by polarized human retinal pigment epithelium and provides neuroprotection to adjacent cells. *PLoS ONE* **2010**, *5*, e12578. [CrossRef] [PubMed]
208. Sreekumar, P.G.; Chothe, P.; Sharma, K.K.; Baid, R.; Kompella, U.; Spee, C.; Kannan, N.; Manh, C.; Ryan, S.J.; Ganapathy, V.; et al. Antiapoptotic properties of α-crystallin-derived peptide chaperones and characterization of their uptake transporters in human RPE cells. *Investig. Ophthalmol. Vis. Sci.* **2013**, *54*, 2787–2798. [CrossRef] [PubMed]
209. O'Reilly, A.M.; Currie, R.W.; Clarke, D.B. HspB1 (Hsp 27) expression and neuroprotection in the retina. *Mol. Neurobiol.* **2010**, *42*, 124–132. [CrossRef] [PubMed]
210. Ryhänen, T.; Hyttinen, J.M.; Kopitz, J.; Rilla, K.; Kuusisto, E.; Mannermaa, E.; Viiri, J.; Holmberg, C.I.; Immonen, I.; Meri, S.; et al. Crosstalk between Hsp70 molecular chaperone, lysosomes and proteasomes in autophagy-mediated proteolysis in human retinal pigment epithelial cells. *J. Cell. Mol. Med.* **2009**, *13*, 3616–3631. [CrossRef]
211. Beere, H.M.; Wolf, B.B.; Cain, K.; Mosser, D.D.; Mahboubi, A.; Kuwana, T.; Tailor, P.; Morimoto, R.I.; Cohen, G.M.; Green, D.R. Heat-shock protein 70 inhibits apoptosis by preventing recruitment of procaspase-9 to the Apaf-1 apoptosome. *Nat. Cell Biol.* **2000**, *2*, 469–475. [CrossRef]
212. Bian, Z.M.; Elner, S.G.; Elner, V.M. Regulation of VEGF mRNA expression and protein secretion by TGF-beta2 in human retinal pigment epithelial cells. *Exp. Eye Res.* **2007**, *84*, 812–822. [CrossRef]
213. Halapin, N.A.; Bazan, N.G. NPD1 induction of retinal pigment epithelial cell survival involves PI3K/Akt phosphorylation signaling. *Neurochem. Res.* **2010**, *35*, 1944–1947. [CrossRef]
214. Tang, B.; Cai, J.; Sun, L.; Li, Y.; Qu, J.; Snider, B.J.; Wu, S. Proteasome inhibitors activate autophagy involving inhibition of PI3K-Akt-mTOR pathway as an anti-oxidation defense in human RPE cells. *PLoS ONE* **2014**, *9*, e103364. [CrossRef] [PubMed]
215. Gureev, A.P.; Popov, V.N.; Starkov, A.A. Crosstalk between the mTOR and Nrf2/ARE signaling pathways as a target in the improvement of long-term potentiation. *Exp. Neurol.* **2020**, *328*, 113285. [CrossRef] [PubMed]
216. Li, Z.; Dong, X.; Liu, H.; Chen, X.; Shi, H.; Fan, Y.; Hou, D.; Zhang, X. Astaxanthin protects ARPE-19 cells from oxidative stress via upregulation of Nrf2-regulated phase II enzymes through activation of PI3K/Akt. *Mol. Vis.* **2013**, *19*, 1656–1666. [PubMed]
217. Whitlock, N.A.; Lindsey, K.; Agarwal, N.; Crosson, C.E.; Ma, J.X. Heat shock protein 27 delays Ca2+-induced cell death in a caspase-dependent and -independent manner in rat retinal ganglion cells. *Investig. Ophthalmol. Vis. Sci.* **2005**, *46*, 1085–1091. [CrossRef]
218. Nahomi, R.B.; Palmer, A.; Green, K.M.; Fort, P.E.; Nagaraj, R.H. Pro-inflammatory cytokines downregulate Hsp27 and cause apoptosis of human retinal capillary endothelial cells. *Biochim. Biophys. Acta* **2014**, *1842*, 164–174. [CrossRef] [PubMed]
219. Kabbage, M.; Dickman, M.B. The BAG proteins: A ubiquitous family of chaperone regulators. *Cell. Mol. Life Sci.* **2008**, *65*, 1390–1402. [CrossRef] [PubMed]

220. Zoubeidi, A.; Chi, K.; Gleave, M. Targeting the cytoprotective chaperone, clusterin, for treatment of advanced cancer. *Clin. Cancer Res.* **2010**, *16*, 1088–1093. [CrossRef]
221. Alberti, S.; Esser, C.; Höhfeld, J. BAG-1--a nucleotide exchange factor of Hsc70 with multiple cellular functions. *Cell Stress Chaperones* **2003**, *8*, 225–231. [CrossRef]
222. Hanus, J.; Zhang, H.; Wang, Z.; Liu, Q.; Zhou, Q.; Wang, S. Induction of necrotic cell death by oxidative stress in retinal pigment epithelial cells. *Cell Death Dis.* **2013**, *4*, e965. [CrossRef] [PubMed]
223. Li, G.Y.; Fan, B.; Zheng, Y.C. Calcium overload is a critical step in programmed necrosis of ARPE-19 cells induced by high-concentration H_2O_2. *Biomed. Environ. Sci.* **2010**, *23*, 371–377. [CrossRef] [PubMed]
224. Fritsch, M.; Günther, S.D.; Schwarzer, R.; Albert, M.C.; Schorn, F.; Werthenbach, J.P.; Schiffmann, L.M.; Stair, N.; Stocks, H.; Seeger, J.M.; et al. Caspase-8 is the molecular switch for apoptosis, necroptosis and pyroptosis. *Nature* **2019**, *575*, 683–687. [CrossRef]
225. Tummers, B.; Mari, L.; Guy, C.S.; Heckmann, B.L.; Rodriguez, D.A.; Rühl, S.; Moretti, J.; Crawford, J.C.; Fitzgerald, P.; Kanneganti, T.D.; et al. Caspase-8-Dependent Inflammatory Responses Are Controlled by Its Adaptor, FADD, and Necroptosis. *Immunity* **2020**, *52*, 994–1006.e8. [CrossRef]
226. Trichonas, G.; Murakami, Y.; Thanos, A.; Morizane, Y.; Kayama, M.; Debouck, C.M.; Hisatomi, T.; Miller, J.W.; Vavvas, D.G. Receptor interacting protein kinases mediate retinal detachment-induced photoreceptor necrosis and compensate for inhibition of apoptosis. *Proc. Natl. Acad. Sci. USA* **2010**, *107*, 21695–21700. [CrossRef]
227. Qin, S.; Ni, M.; Wang, X.; Maurier-Mahé, F.; Shurland, D.L.; Rodrigues, G.A. Inhibition of RPE cell sterile inflammatory responses and endotoxin-induced uveitis by a cell-impermeable HSP90 inhibitor. *Exp. Eye Res.* **2011**, *93*, 889–897. [CrossRef]
228. Wu, Z.; Geng, Y.; Lu, X.; Shi, Y.; Wu, G.; Zhang, M.; Shan, B.; Pan, H.; Yuan, J. Chaperone-mediated autophagy is involved in the execution of ferroptosis. *Proc. Natl. Acad. Sci. USA* **2019**, *116*, 2996–3005. [CrossRef]
229. Hu, Y.; Wang, B.; Li, S.; Yang, S. Pyroptosis, and its Role in Central Nervous System Disease. *J. Mol. Biol.* **2022**, *434*, 167379. [CrossRef] [PubMed]
230. Ho, D.T.; Bardwell, A.J.; Grewal, S.; Iverson, C.; Bardwell, L. Interacting JNK-docking sites in MKK7 promote binding and activation of JNK mitogen-activated protein kinases. *J. Biol. Chem.* **2006**, *281*, 13169–13179. [CrossRef] [PubMed]
231. Gao, J.; Cui, J.Z.; To, E.; Cao, S.; Matsubara, J.A. Evidence for the activation of pyroptotic and apoptotic pathways in RPE cells associated with NLRP3 inflammasome in the rodent eye. *J. Neuroinflamm.* **2018**, *15*, 15. [CrossRef]
232. Tseng, W.A.; Thein, T.; Kinnunen, K.; Lashkari, K.; Gregory, M.S.; D'Amore, P.A.; Ksander, B.R. NLRP3 inflammasome activation in retinal pigment epithelial cells by lysosomal destabilization: Implications for age-related macular degeneration. *Investig. Ophthalmol. Vis. Sci.* **2013**, *54*, 110–120. [CrossRef] [PubMed]
233. Liao, Y.; Zhang, H.; He, D.; Wang, Y.; Cai, B.; Chen, J.; Ma, J.; Liu, Z.; Wu, Y. Retinal Pigment Epithelium Cell Death Is Associated With NLRP3 Inflammasome Activation by All-trans Retinal. *Investig. Ophthalmol. Vis. Sci.* **2019**, *60*, 3034–3045. [CrossRef]
234. Sun, Y.; Li, F.; Liu, Y.; Qiao, D.; Yao, X.; Liu, G.S.; Li, D.; Xiao, C.; Wang, T.; Chi, W. Targeting inflammasomes and pyroptosis in retinal diseases-molecular mechanisms and future perspectives. *Prog. Retin. Eye Res.* **2024**, *101*, 101263. [CrossRef] [PubMed]
235. Hytti, M.; Piippo, N.; Salminen, A.; Honkakoski, P.; Kaarniranta, K.; Kauppinen, A. Quercetin alleviates 4-hydroxynonenal-induced cytotoxicity and inflammation in ARPE-19 cells. *Exp. Eye Res.* **2015**, *132*, 208–215. [CrossRef] [PubMed]
236. Yang, H.J.; Hu, R.; Sun, H.; Chen, B.; Li, X.; Chen, J.B. 4-HNE induces proinflammatory cytokines of human retinal pigment epithelial cells by promoting extracellular efflux of HSP70. *Exp. Eye Res.* **2019**, *188*, 107792. [CrossRef]
237. Martine, P.; Chevriaux, A.; Derangère, V.; Apetoh, L.; Garrido, C.; Ghiringhelli, F.; Rébé, C. HSP70 is a negative regulator of NLRP3 inflammasome activation. *Cell Death Dis.* **2019**, *10*, 256. [CrossRef] [PubMed]
238. Mayor, A.; Martinon, F.; De Smedt, T.; Pétrilli, V.; Tschopp, J. A crucial function of SGT1 and HSP90 in inflammasome activity links mammalian and plant innate immune responses. *Nat. Immunol.* **2007**, *8*, 497–503. [CrossRef]
239. Ranta-Aho, S.; Piippo, N.; Korhonen, E.; Kaarniranta, K.; Hytti, M.; Kauppinen, A. TAS-116, a Well-Tolerated Hsp90 Inhibitor, Prevents the Activation of the NLRP3 Inflammasome in Human Retinal Pigment Epithelial Cells. *Int. J. Mol. Sci.* **2021**, *22*, 4875. [CrossRef]
240. Zuo, Y.; Wang, J.; Liao, F.; Yan, X.; Li, J.; Huang, L.; Liu, F. Inhibition of Heat Shock Protein 90 by 17-AAG Reduces Inflammation via P2X7 Receptor/NLRP3 Inflammasome Pathway and Increases Neurogenesis After Subarachnoid Hemorrhage in Mice. *Front. Mol. Neurosci.* **2018**, *11*, 401. [CrossRef] [PubMed]
241. Piippo, N.; Korhonen, E.; Hytti, M.; Skottman, H.; Kinnunen, K.; Josifovska, N.; Petrovski, G.; Kaarniranta, K.; Kauppinen, A. Hsp90 inhibition as a means to inhibit activation of the NLRP3 inflammasome. *Sci. Rep.* **2018**, *8*, 6720. [CrossRef]
242. Ma, J.; Liu, J.; Chen, S.; Zhang, W.; Wang, T.; Cao, M.; Yang, Y.; Du, Y.; Cui, G.; Du, Z. Understanding the Mechanism of Ferroptosis in Neurodegenerative Diseases. *Front. Biosci.* **2024**, *29*, 291. [CrossRef] [PubMed]
243. Gao, M.; Monian, P.; Pan, Q.; Zhang, W.; Xiang, J.; Jiang, X. Ferroptosis is an autophagic cell death process. *Cell Res.* **2016**, *26*, 1021–1032. [CrossRef] [PubMed]

244. Malaviya, P.; Kumar, J.; Kowluru, R.A. Role of ferroptosis in mitochondrial damage in diabetic retinopathy. *Free Radic. Biol. Med.* **2024**, *225*, 821–832. [CrossRef]
245. Wei, S.; Li, J.; Zhang, Y.; Li, Y.; Wang, Y. Ferroptosis in eye diseases: A systematic review. *Eye* **2024**, *39*, 18–27. [CrossRef]
246. Yang, M.; So, K.F.; Lam, W.C.; Lo, A.C.Y. Cell Ferroptosis: New Mechanism and New Hope for Retinitis Pigmentosa. *Cells* **2021**, *10*, 2153. [CrossRef] [PubMed]
247. Liu, Y.; Zhou, L.; Xu, Y.; Li, K.; Zhao, Y.; Qiao, H.; Xu, Q.; Zhao, J. Heat Shock Proteins and Ferroptosis. *Front. Cell Dev. Biol.* **2022**, *10*, 864635. [CrossRef]
248. Zhou, Y.; Liao, J.; Mei, Z.; Liu, X.; Ge, J. Insight into Crosstalk between Ferroptosis and Necroptosis: Novel Therapeutics in Ischemic Stroke. *Oxid. Med. Cell. Longev.* **2021**, *2021*, 9991001. [CrossRef]
249. Chen, C.; Wang, D.; Yu, Y.; Zhao, T.; Min, N.; Wu, Y.; Kang, L.; Zhao, Y.; Du, L.; Zhang, M.; et al. Legumain promotes tubular ferroptosis by facilitating chaperone-mediated autophagy of GPX4 in AKI. *Cell Death Dis.* **2021**, *12*, 65. [CrossRef] [PubMed]
250. Zhu, S.; Zhang, Q.; Sun, X.; Zeh, H.J.; Lotze, M.T.; Kang, R.; Tang, D. HSPA5 Regulates Ferroptotic Cell Death in Cancer Cells. *Cancer Res.* **2017**, *77*, 2064–2077. [CrossRef]
251. Chen, X.; Kang, R.; Tang, D. Ferroptosis by Lipid Peroxidation: The Tip of the Iceberg? *Front. Cell Dev. Biol.* **2021**, *9*, 646890. [CrossRef]
252. Zhang, Z.; Zhou, H.; Gu, W.; Wei, Y.; Mou, S.; Wang, Y.; Zhang, J.; Zhong, Q. CGI1746 targets σ1R to modulate ferroptosis through mitochondria-associated membranes. *Nat. Chem. Biol.* **2024**, *20*, 699–709. [CrossRef] [PubMed]
253. Jiang, B.; Zhao, Y.; Shi, M.; Song, L.; Wang, Q.; Qin, Q.; Song, X.; Wu, S.; Fang, Z.; Liu, X. DNAJB6 Promotes Ferroptosis in Esophageal Squamous Cell Carcinoma. *Dig. Dis. Sci.* **2020**, *65*, 1999–2008. [CrossRef] [PubMed]
254. Adedoyin, O.; Boddu, R.; Traylor, A.; Lever, J.M.; Bolisetty, S.; George, J.F.; Agarwal, A. Heme oxygenase-1 mitigates ferroptosis in renal proximal tubule cells. *Am. J. Physiol. Renal Physiol.* **2018**, *314*, F702–F714. [CrossRef] [PubMed]
255. Chen, H.; Zheng, C.; Zhang, Y.; Chang, Y.Z.; Qian, Z.M.; Shen, X. Heat shock protein 27 downregulates the transferrin receptor 1-mediated iron uptake. *Int. J. Biochem. Cell Biol.* **2006**, *38*, 1402–1416. [CrossRef] [PubMed]
256. Welsh, J.A.; Goberdhan, D.C.I.; O'Driscoll, L.; Buzas, E.I.; Blenkiron, C.; Bussolati, B.; Cai, H.; Di Vizio, D.; Driedonks, T.A.P.; Erdbrügger, U.; et al. Minimal information for studies of extracellular vesicles (MISEV2023): From basic to advanced approaches. *J. Extracell. Vesicles* **2024**, *13*, e12404. [CrossRef]
257. Gangalum, R.K.; Bhat, A.M.; Kohan, S.A.; Bhat, S.P. Inhibition of the Expression of the Small Heat Shock Protein αB-Crystallin Inhibits Exosome Secretion in Human Retinal Pigment Epithelial Cells in Culture. *J. Biol. Chem.* **2016**, *291*, 12930–12942. [CrossRef] [PubMed]
258. Kurzawa-Akanbi, M.; Whitfield, P.; Burté, F.; Bertelli, P.M.; Pathak, V.; Doherty, M.; Hilgen, B.; Gliaudelytė, L.; Platt, M.; Queen, R.; et al. Retinal pigment epithelium extracellular vesicles are potent inducers of age-related macular degeneration disease phenotype in the outer retina. *J. Extracell. Vesicles* **2022**, *11*, e12295. [CrossRef]
259. Atienzar-Aroca, S.; Flores-Bellver, M.; Serrano-Heras, G.; Martinez-Gil, N.; Barcia, J.M.; Aparicio, S.; Perez-Cremades, D.; Garcia-Verdugo, J.M.; Diaz-Llopis, M.; Romero, F.J.; et al. Oxidative stress in retinal pigment epithelium cells increases exosome secretion and promotes angiogenesis in endothelial cells. *J. Cell. Mol. Med.* **2016**, *20*, 1457–1466. [CrossRef]
260. Knickelbein, J.E.; Liu, B.; Arakelyan, A.; Zicari, S.; Hannes, S.; Chen, P.; Li, Z.; Grivel, J.C.; Chaigne-Delalande, B.; Sen, H.N.; et al. Modulation of Immune Responses by Extracellular Vesicles From Retinal Pigment Epithelium. *Investig. Ophthalmol. Vis. Sci.* **2016**, *57*, 4101–4107. [CrossRef]
261. Martins, B.; Pires, M.; Ambrósio, A.F.; Girão, H.; Fernandes, R. Contribution of extracellular vesicles for the pathogenesis of retinal diseases: Shedding light on blood-retinal barrier dysfunction. *J. Biomed. Sci.* **2024**, *31*, 48. [CrossRef]
262. Nishimura, Y.; Hara, H.; Kondo, M.; Hong, S.; Matsugi, T. Oxidative Stress in Retinal Diseases. *Oxidat. Med. Cell. Longev.* **2017**, *2017*, 4076518. [CrossRef] [PubMed]
263. Whitcup, S.M.; Nussenblatt, R.B.; Lightman, S.L.; Hollander, D.A. Inflammation in retinal disease. *Int. J. Inflam.* **2013**, *2013*, 724648. [CrossRef] [PubMed]
264. Viiri, J.; Amadio, M.; Marchesi, N.; Hyttinen, J.M.T.; Kivinen, N.; Sironen, R.; Rilla, K.; Akhtar, S.; Provenzani, A.; D'Agostino, V.G.; et al. Autophagy Activation Clears ELAVL1/HuR-Mediated Accumulation of SQSTM1/p62 during Proteasomal Inhibition in Human Retinal Pigment Epithelial Cells. *PLoS ONE* **2013**, *8*, e69563. [CrossRef] [PubMed]
265. Fernández-Albarral, J.A.; de Julián-López, E.; Soler-Domínguez, C.; de Hoz, R.; López-Cuenca, I.; Salobrar-García, E.; Ramírez, J.M.; Pinazo-Durán, M.D.; Salazar, J.J.; Ramírez, A.I. The Role of Autophagy in Eye Diseases. *Life* **2021**, *11*, 189. [CrossRef] [PubMed]
266. Levine, B.; Packer, M.; Codogno, P. Development of autophagy inducers in clinical medicine. *J. Clin. Investig.* **2015**, *125*, 14–24. [CrossRef] [PubMed]

International Journal of
Molecular Sciences

MDPI

Review

The Role of Selected Elements in Oxidative Stress Protection: Key to Healthy Fertility and Reproduction

Marcin Wróblewski [1,*], Weronika Wróblewska [2] and Marta Sobiesiak [3]

1 Department of Medical Biology and Biochemistry, Ludwik Rydygier Collegium Medicum in Bydgoszcz, Nicolaus Copernicus University in Torun, 87-100 Torun, Poland
2 Student Research Club of Medical Biology and Biochemistry, Department of Medical Biology and Biochemistry, Faculty of Medicine, Ludwik Rydygier Collegium Medicum in Bydgoszcz, Nicolaus Copernicus University in Torun, 87-100 Torun, Poland; 316714@stud.umk.pl
3 Department of Inorganic and Analytical Chemistry, Faculty of Pharmacy, Ludwik Rydygier Collegium Medicum in Bydgoszcz, Nicolaus Copernicus University in Torun, 87-100 Torun, Poland; marta.sobiesiak@cm.umk.pl
* Correspondence: marcin.wroblewski@cm.umk.pl

Abstract: Oxidative stress and its relationship to fertility and reproduction is a topic of interest in medicine, especially in the context of the effects of trace elements and micronutrients. Oxidative stress occurs when there is an excess of free radicals in the body, which can lead to cell and tissue damage. Free radicals are reactive oxygen species (ROS) that can be formed as a result of normal metabolic processes, as well as under the influence of external factors such as environmental pollution, UV radiation, and diet. Oxidative stress has a significant impact on fertility. In men, it can lead to DNA damage in sperm, which can result in reduced semen quality, reduced sperm motility and increased numbers of defective sperm, and free radical damage to sperm cell membranes causing a reduction in the number of available sperm. In women, oxidative stress can affect the quality of female reproductive cells, which can lead to problems with their maturation and with embryo implantation in the uterus and can also affect ovarian function and disrupt hormonal regulation of the menstrual cycle. A proper balance of trace elements and micronutrients is key to protecting against oxidative stress and maintaining reproductive health. Supplementation with appropriate elements such as zinc, selenium, copper, manganese, chromium, and iron can help reduce oxidative stress and improve fertility. This work discusses the effects of selected elements on oxidative stress parameters specifically in terms of fertility and reproduction.

Keywords: fertility; oxidative stress; reproduction; trace elements

1. Introduction

Oxidative stress (OS) occurs when the production of free radicals surpasses the body's antioxidant defenses. Reactive oxygen species (ROS) such as hydroxyl radicals ($^{\bullet}OH$), superoxide anion radicals ($O_2^{\bullet -}$), and hydrogen peroxide (H_2O_2) are highly reactive molecules that damage cellular components to balance their electron shortage [1,2]. The excessive production of ROS can coincide with an increased formation of reactive nitrogen species (RNS), resulting in oxidative/nitrosative stress, a condition commonly observed in a range of human diseases [3]. To counteract surplus ROS, cells have developed a comprehensive antioxidant defense system comprising both enzymatic antioxidants, such as superoxide dismutase (SOD), catalase (CAT), and glutathione peroxidases (GPx), and non-enzymatic antioxidants [3]. Antioxidants can also be categorized by their origin, including those synthesized endogenously, such as enzymes and small molecules, and those acquired from the diet, such as phenolic compounds, flavonoids, phenolic acids, carotenoids, vitamins, and trace elements (TE) [4]. TE, such as selenium (Se), manganese (Mn), and chromium (Cr), and micro-minerals, such as zinc (Zn), iron (Fe), and copper

(Cu), exhibit significant effects on reproductive health. Additionally, selenium, zinc, iron, copper, and manganese can be integral to the function of antioxidant enzymes, playing a vital role in sustaining enzymatic activity, and are also cofactors of many enzymes involved in redox reactions [5–8]. Infertility has become a growing global health concern over the last twenty years. Supplementing with specific trace elements can help mitigate oxidative damage and improve the chances of successful conception and a healthy pregnancy [9]. A healthy antioxidant system is essential for neutralizing ROS, protecting reproductive cells and tissues, and supporting normal fertility. Despite the known link between oxidative stress and unexplained infertility, effective treatments remain elusive. In this review, we briefly discuss the links between oxidative stress, trace elements, and infertility.

2. Methods

We conducted an extensive literature search using the ISI Web of Science/PubMed/Science Direct/Google Scholar databases containing information on the impact of trace elements on oxidative stress in the aspect of reproduction and fertility. The following keywords were used in the data search: ("fertility" and "trace elements", "selenium", "zinc", "copper", "iron", "manganese", "chromium"); ("oxidative stress" and "reproduction" and "selenium", "zinc", "copper", "iron", "manganese", "chromium"); ("oxidative stress" and "selenium", "zinc", "copper", "iron", "manganese", "chromium"); ("reactive oxygen species" and "fertility" or "reproduction").There were no restrictions in collecting the data. No language restrictions were applied during the analysis. Rather, we tried to select articles from the last 20 years. After searching, we further examined the full texts of the literature to determine eligibility for inclusion in this review. Editorials, conference abstracts, and studies with incomplete or unavailable data were excluded.

3. Reactive Oxygen Species in Reproductive Health

The balance between oxidants and antioxidants is crucial for women's reproductive health. Elevated levels of ROS are crucial in several stages of the ovulation process, such as cumulus expansion, progesterone synthesis, gene expression before ovulation, and the initiation of ovulatory signals [10,11]. A balanced ROS level in follicular fluid not only reflects healthy follicular metabolism but may also serve as a predictive marker for the success of in vitro fertilization–embryo transfer [10]. Additionally, nitric oxide ($^{\bullet}$NO) plays a vital role in maintaining various reproductive functions, including pregnancy, at normal physiological levels [11]. However, excessive ROS levels can damage cellular structures, impacting the environment of follicular, hydrosalpingeal, and peritoneal fluids, thus affecting egg quality, fertilization, implantation, and early embryo development, leading to oxidative stress-induced infertility [10–12]. Fertility declines with age, linked to a decrease in ovarian function, to mitochondrial issues, and to ROS accumulation [12,13]. Furthermore, endometriosis and polycystic ovary syndrome are among the most common diseases associated with infertility, and it has been suggested that they are linked to increased levels of oxidative stress [11,14].

A substantial proportion of infertile men, ranging from 30% to 80%, exhibit elevated levels of seminal ROS, highlighting the significance of assessing seminal oxidative stress in male reproductive health evaluations [7,8]. It is crucial to note that molecules such as $O_2^{\bullet -}$, H_2O_2, and nitric oxide ($^{\bullet}$NO) play pivotal roles at the onset of capacitation, serving as key regulators to promote processes such as cholesterol efflux, cAMP production, and protein tyrosine phosphorylation [15]. ROS contribute to increased membrane fluidity of spermatozoa and enhance the rates of sperm–oocyte fusion, which are critical during the capacitation and acrosome reaction stages [8]. The impact of ROS on capacitation is dose-dependent [15]. While moderate levels are beneficial, excessive ROS activity in sperm may lead to infertility by hindering the sperm's ability to fuse with oocytes [8]. High levels of ROS can result in sperm dysfunction due to DNA damage, lipid peroxidation, loss of membrane integrity, cell death, diminished antioxidant levels, and ultimately, compromised sperm quality [1]. Sperm cells are particularly susceptible to oxidative stress because of their high polyunsaturated fatty acids content in the membrane and their scarcity of cytoplasmic antioxidant enzymes [1,7]. Mitochondria are a primary source of ROS in cells, with $O_2^{\bullet-}$ being the principal mitochondrial ROS produced in sperm [16]. The generation of mitochondrial ROS is largely governed by the redox state of the respiratory chain complexes, including proton pumps regulated by the mitochondrial membrane potential (MMP). Maintaining a high MMP is essential for preserving chromatin structure, ensuring normal sperm morphology, promoting high motility, and inducing the acrosome reaction [16]. Excessive ROS production is linked to an increase in sperm subpopulations with low MMP among infertile men, leading to DNA fragmentation, reduced sperm viability, and impaired mitochondrial function, which decreases motility [1,15–17]. Another issue affecting male fertility is semen hyperviscosity (SHV). SHV affects male fertility through various mechanisms, including impaired sperm movement, inflammatory responses, oxidative damage, and alteration in levels of trace elements, and poses challenges in assisted reproductive techniques [1]. RNS, a subset of ROS, includes molecules such as peroxynitrite anions and $^{\bullet}NO$ [16]. At normal levels, RNS regulate various sperm functions such as capacitation, acrosomal reaction, and binding to the zona pellucida. Elevated RNS levels can induce nitrosative stress, affecting mitochondrial and sperm function by inhibiting electron transport chain components, altering protein functions through tyrosine nitration and S-glutathionylation, and disrupting energy production, which may lead to apoptosis and necrosis [9,11,16].

4. Impact of Trace Elements on Oxidative Stress

Selenium plays a critical role in the body's antioxidant defense system. In its inorganic form, selenium primarily occurs as Se^{4+} and Se^{6+} [18].Within living organisms, selenium is encountered as selenoproteins (SePs) and possesses antioxidant capabilities. It is integral to the structure of two essential amino acids, selenomethionine and selenocysteine. Selenocysteine is notably present in the catalytic centers of enzymes that exhibit antioxidant functions, such as GPx and thioredoxin reductases [19–22]. GPx protects cells from oxidative damage by reducing peroxides, harmful by-products of cellular metabolism that can damage cell components such as lipids, proteins, and DNA [23]. A selenium-deficient diet results in reduced GPx activity [24]. Thioredoxin reductase, a significant selenoenzyme, operates alongside NADPH and thioredoxin within the thioredoxin system, offering antioxidant benefits. The thioredoxin system plays a crucial role in mitigating oxidative stress by aiding in the reduction of peroxiredoxin proteins, which eliminate hydrogen peroxide and are oxidized in the process [25]. Selenoprotein P (SePP) functions as a redox protein and affects the redox balance utilizing its thioredoxin domain to distribute selenium necessary for the production of intracellular GPx [26].

Zinc is an essential micronutrient that plays a crucial role in the functioning of approximately 300 metalloenzymes, including oxidoreductases, hydrolases, and ligases [27]. It acts as a cofactor for these enzymes, and a deficiency in zinc can lead to reduced activity of enzymes that depend on it [28]. Zinc also has antioxidant capabilities, effectively blocking the formation of harmful radicals such as hydroxyl radicals ($^{\bullet}OH$) and superoxide radicals ($O_2^{\bullet-}$) by competing with pro-oxidative metals such as iron (Fe^{2+}) and copper (Cu^{2+}) [29]. Zinc is also a key component of superoxide dismutase, an essential enzyme that transforms singlet oxygen radicals into hydrogen peroxide [30]. Moreover, it inhibits NADPH oxidases, which are responsible for generating singlet oxygen radicals from oxygen using NADPH [30,31]. Zinc stimulates the production of metallothionein, which neutralizes ROS, and possesses potent anti-inflammatory and antioxidant properties, including inhibition of tumor necrosis factor-alpha (TNF-α)-induced nuclear factor kappa-light-chain-enhancer of activated B cells (NF-κB) activation [30,32].

Iron that is not attached to proteins poses a significant risk to cells by promoting oxidative stress [33]. Iron is an abundant transition metal in food, and in a healthy state, the vast majority of iron within the body is safely incorporated into proteins such as hemoglobin, myoglobin, transferrin, ferritin, and hemosiderin [6,34]. Iron is fundamentally important in hematopoiesis, primarily for its role in hemoglobin synthesis [6]. In biological systems, Fe^{2+} ions and H_2O_2 participate in the Fenton reaction, resulting in the formation of highly reactive hydroxyl radicals ($^\bullet OH$) and ferric ions (Fe^{3+}) [25,35]. Free iron acts as a source of ROS, contrasting with Fe bound to proteins which has antioxidant properties. The heme group, containing Fe^{3+}, is the catalytic center of all four polypeptide chains of CAT [36]. CAT is an enzyme belonging to the oxidoreductases class that facilitates the conversion of hydrogen peroxide into water and oxygen [37].

Copper, existing in both Cu^+ and Cu^{2+} forms, serves as a vital cofactor for numerous enzymes that play a role in electron transfer processes within metabolic pathways [38]. It is known for both its pro-oxidant and antioxidant characteristics [39,40]. Copper participates in a Fenton-like redox reaction that leads to the generation of ROS [6,41]. Exposure to elevated levels of copper significantly decreases glutathione level which is an important cellular antioxidant acting against this trace element toxicity [6]. Copper, alongside zinc, is integral to the functioning of antioxidant enzymes such as SOD-1 and SOD-3 [42,43]. A significant portion of copper found in blood plasma is incorporated into ceruloplasmin, a protein known for its antioxidant capabilities [44]. Classified as an oxidoreductase, ceruloplasmin's activity is dependent on copper ions [45]. This protein plays a crucial role in maintaining iron balance within the body by catalyzing the oxidation of Fe^{2+} to Fe^{3+} and reduces the generation of $^\bullet OH$ [46].

Manganese, particularly as Mn^{2+} ions, is a crucial component of the active sites in numerous enzymes [47]. It acts as a cofactor for various enzymes, such as manganese superoxide dismutase (MnSOD), arginase, glutamine synthetase, phosphoenolpyruvate decarboxylase, and pyruvate carboxylase [48]. MnSOD is found in mitochondria, where it forms the primary defense against oxidative stress produced by mitochondrial respiration [43,49]. This enzyme transforms superoxide anion radicals into hydrogen peroxide and oxygen within the mitochondria. Moreover, manganese can indirectly affect the redox state of iron, copper, and other transition metals, enhancing the production of mitochondrial peroxide (H_2O_2) even at physiological levels [48].

Chromium exists mainly in trivalent (Cr^{3+}) and hexavalent (Cr^{6+}) forms, with trivalent chromium being significantly less toxic than its hexavalent counterpart. However, at high concentrations or depending on its chemical ligands, Cr^{3+} can still be toxic [50]. Inside cells, chromium (VI) reduces to the more stable chromium (III), generating genotoxic intermediates responsible for its mutagenic and carcinogenic effects [8]. The reduction of Cr(VI) also produces various reactive oxygen species, such as hydroxyl radicals, singlet oxygen, and superoxide anions, leading to different types of DNA damage and oxidative stress, which can result in lipid peroxidation, altered cellular redox states, and apoptosis [6,51]. Table 1 summarizes the mechanism of action of selected elements on oxidative stress parameters.

Table 1. Effects of selected elements on oxidative stress parameters.

Element	Impact on Oxidative Stress	Mechanism of Action	Examples of Parameters/ Stressors	Reference
Selenium	Decreases oxidative stress	It is part of the enzyme glutathione peroxidase, which neutralizes lipid peroxides and hydrogen peroxide	Decreased MDA levels, increased GSH levels	[19–26,52]
Zinc	Decreases oxidative stress	Supports the function of antioxidant enzymes such as superoxide dismutase (SOD)	Reduced DNA damage, reduced levels of ROS	[27–32]
Copper	Decreases oxidative stress	It is a cofactor of superoxide dismutase (SOD), which neutralizes superoxide anions	Increased SOD activity, decreased in ROS levels	[39–44]
Manganese	Decreases oxidative stress	Manganese-dependent superoxide dismutase (MnSOD) cofactor	Decreased MDA levels, increased MnSOD activity	[47–49]
Iron	Increases oxidative stress	Participates in the Fenton reaction, which generates hydroxyl radicals	Increased MDA levels, DNA and protein damage	[25,33–36]

Table 1. *Cont.*

Element	Impact on Oxidative Stress	Mechanism of Action	Examples of Parameters/ Stressors	Reference
Chromium	Decreases oxidative stress	Trivalent chromium (Cr(III)) is an essential micronutrient that supports glucose and lipid metabolism, which can indirectly reduce oxidative stress	Reduced ROS levels, improved GSH levels	[8,50,51,53]
	Increases oxidative stress	Hexavalent chromium (Cr(VI)) is toxic and can generate ROS through redox reactions, leading to cellular damage	Increased levels of ROS, damage to DNA, lipids, and proteins, increased levels of MDA	

Abbreviations: GSH: glutathione; MDA: malondialdehyde; MnSOD: manganese-dependent superoxide dismutase; ROS: reactive oxygen species.

5. Trace Elements in Reproductive Health

Selenium is a trace element that is naturally present in Brazil nuts, green vegetables, shiitake and button mushrooms, and various seeds such as young barley seedlings. The Recommended Dietary Allowance for selenium is established at 70 µg/day for men and 55 µg/day for women to maintain optimal health. Nonetheless, these recommended levels are viewed as insufficient by some studies, which propose a minimum daily requirement of 90 µg for adults. The World Health Organization sets the maximum safe intake level at 400 µg/day or 5.1 µmol/day for adults 19 years and older, with intake above this threshold potentially being harmful [54]. Selenium is essential for the activity of selenoproteins that protect against oxidative damage [55]. This micronutrient, known for its antimutagenic properties, helps prevent the transformation of normal cells into cancerous ones. Its protective benefits are largely linked to its incorporation into GPx and thioredoxin reductase, enzymes that are essential for safeguarding DNA and cellular components from oxidative damage [12,56,57]. Functional selenium levels are crucial for thyroid hormone production and DNA formation [57]. These processes significantly affect fertility and reproductive outcomes. Despite limited research, evidence suggests selenium's critical role in ovarian follicle growth and maturity. Sodium selenite, a form of inorganic selenium, not only enhances the growth of oocytes but also boosts the proliferation rate of theca and granulosa cells [55]. In human studies, low levels of selenium in plasma, follicular fluid, amniotic fluid, or tissues, as well as reduced concentrations or activity of GPx in tissues, have been linked with unexplained infertility, miscarriage, preterm birth, and gestational diabetes [58]. Lower selenium levels in maternal blood are correlated with prolonged attempts to conceive and a 46% higher chance of subfertility, even when accounting for various maternal and paternal factors [57]. Only a limited number of studies have shown a link between female fertility, selenium status, and the activity of selenium-dependent GPx. Women undergoing in vitro fertilization (IVF) or experiencing unexplained infertility had lower selenium levels in their serum and follicular fluid. A reduction in selenium-dependent GPx activity within the follicular fluid has been linked to the lack of oocyte fertilization in women undergoing IVF treatments [59]. In a study exploring the effects of sodium selenite, Abedelahi et al. [49] found that selenium supplementation (with 5 or 10 ng/mL sodium selenite added to the culture medium) significantly reduced ROS-induced oxidative stress and enhanced the total antioxidant capacity (TAC) and GPx activity in derivedfromvitrified and non-vitrified ovarian tissue. The growth rates of follicles, oocytes, and embryos were notably higher in the selenium-supplemented groups. An interesting observation is that conditions such as endometriosis and polycystic ovary syndrome, characterized by an excessive production of ROS, may lead to a significant depletion of selenium due to its antioxidant properties [55].The beneficial effects of selenium are crucial for the growth of testicles, the production of sperm, the movement of spermatozoa, and testosterone secretion [5,54,60]. Increased dietary selenium intake has been correlated with higher sperm counts in the semen of infertile men. Studies on selenium supplementation in infertile men have shown enhancements in testicular antioxidant capability, semen selenium levels, sperm count, sperm shape and movement, and overall fertility [58]. Selenium may offer protection against oxidative DNA damage in human sperm cells [56]. Research has shown that men with normal sperm quality have significantly higher levels of glutathione (GSH) compared to those with sperm defects, with a notable correlation between seminal GSH levels and sperm motility [56]. GSH is a major intracellular antioxidant and has a protective role against cytotoxic aldehydes from lipid peroxidation [56,60]. Additionally, the antioxidant function of selenium may be a key factor in explaining the observed increase in lipid peroxidation among infertile patients. Lower selenium levels in the seminal fluid of infertile men have been associated with higher concentrations of malondialdehyde (MDA) in the same individuals, indicating oxidative stress [56]. MDA, the final product of the oxidation of polyunsaturated fatty acids, is a reliable and widely used biomarker for the assessment of lipid peroxidation [2]. Chiti et al. [61] discovered that dietary selenium supplementation led to a 3.7-fold increase in seminal total antioxidant capacity (TAC) levels and a 2.4-fold reduction in sperm with non-progressive motility. The study also noted that selenium, zinc, iron, copper, manganese, and chromium significantly affected the level of MDA in semen. Yin et al. [62] studied the associations between oxidative stress indicators (MDA, SOD, and GSH) in seminal plasma and the risk of idiopathic oligoasthenoteratozoospermia (OAT) among 148 participants, including 75 cases of idiopathic OAT and 73 controls. They also examined how these associations might be influenced by levels of essential trace elements such as selenium, copper, and iron. There was a negative association between GSH levels and the risk of idiopathic OAT in the high Se group, with no significant association in the low Se group. Subgroup analyses showed a positive association between MDA level and the risk of idiopathic OAT in the high Cu group, with no significant association in the low Cu group. The researchers concluded that there is an association between these markers of OS and OAT risk that may be influenced by Cu and Se levels in seminal plasma. The available literature also includes studies in which no correlation was found between Se content, semen parameters, and oxidative status [60].

Diet is a factor determining the level of zinc in the body, and the main sources of this element are legumes, cereals, and meat products. The recommended daily requirement for zinc is 2–3 mg in adults [63]. Recently, zinc has been identified as a crucial element necessary for the completion of meiosis and egg activation in vitro, as well as for follicle rupture and the completion of meiosis in vivo [64]. Zinc plays an antioxidant role in yak oocytes by reducing levels of ROS. Nonetheless, a substantial decrease in ROS levels, resulting from treatment with high concentrations of Zn, could negatively impact the development of bovine embryos and the viability of human cultured cells. In a laboratory setting, excessively high concentrations of zinc that are not physiological can negatively impact the development of embryos before implantation [64]. Zinc is involved in neutralizing free peroxide radicals as part of superoxide dismutase (Zn-SOD), preventing the oxidation of unsaturated fatty acids, regulating plasma vitamin A levels, and counteracting the effects of harmful metals such as cadmium and lead in the ovaries [12]. Zn-SOD is present in oocytes and pre-implantation embryos throughout their passage in the oviduct [64]. Anchordoquy et al. [65] discovered that the activity of SOD significantly increased following zinc supplementation, a change that was associated with improved embryonic development. According to their in vitro research, zinc supplementation during oocyte maturation enhanced SOD activity in cumulus cells, reducing DNA damage and apoptosis in these cells. As a result, the increased SOD activity helped protect the embryo from oxidative stress through to the blastocyst stage. Xiong et al. [64] investigated the impact of zinc supplementation on the maturation and developmental potential of yak (*Bos grunniens*) oocytes during in vitro maturation. The researchers concluded that zinc supplementation during the in vitro maturation process enhanced the maturation and subsequent development of yak oocytes by increasing GSH and SOD levels and reducing ROS levels in the oocytes, demonstrating a beneficial effect of zinc on oocyte quality and developmental capacity. Polycystic ovary syndrome (PCOS) is a prevalent condition that impedes fertility. Women with PCOS are prone to various metabolic issues, including hormonal imbalances and elevated indicators of inflammation and oxidative stress. Research indicates that taking magnesium and zinc supplements for 12 weeks, as compared to a placebo, can positively impact serum high-sensitivity C-reactive protein (hs-CRP), plasma MDA, TAC, and gene expression of interleukin-1 (IL-1) and TNF-α in women with PCOS. However, this supplementation does not affect •NO, GSH, MDA levels, or expression of interleukin-8 (IL-8), transforming growth factor-beta (TGF-β), and vascular endothelial growth factor (VEGF) [66]. Zinc is involved in production of luteinizing, follicular, and testicular hormones, with Zn levels significantly differing between fertile and infertile men [7,60]. Chiti et al. [61] showed that zinc and iron intake positively impacted sperm viability. Zinc is also known for its antioxidant properties and its role in stabilizing sperm DNA and membranes. An elevated zinc level, in the absence of a specific seminal vesicle zinc-binding agent, may hinder the process of sperm chromatin decondensation, potentially leading to chromatin instability. Poor chromatin packaging or integrity is linked to sperm DNA damage, elevating the risk of infertility and adverse pregnancy outcomes [1]. The role of zinc in supporting antioxidant defense suggests that its reduction could make sperm more susceptible to oxidative damage [60]. Nenkova et al. [60] detected lower levels of zinc in men exhibiting poor sperm quality than in controls (men with normozoospermia). Furthermore, these groups showed higher levels of MDA and lower levels of GSH, indicating the presence of oxidative stress. Atgi et al. [56] conducted a study to analyze the antioxidant content in the seminal plasma of men with varying fertility potential, focusing on GSH, zinc, and MDA. The study included semen samples from fertile men (normozoospermics) and infertile patients divided into three groups based on their infertility issues (asthenozoospermics, oligozoospermics, and teratozoospermics). The concentrations of Zn were higher in normozoospermics compared to the infertile groups. Zn was positively correlated with sperm motility and count. Total GSH (GSHt), reduced GSH (GSHr), and GSSG levels were significantly elevated in normozoospermics compared to the infertile groups, with significant correlations between GSH levels and sperm motility and count. MDA levels were markedly higher in the infertile groups, with negative correlations to sperm motility and concentration and a positive correlation with abnormal sperm morphology. The study concluded that lower levels of semen GSH and deficiencies in trace elements are linked to reduced semen quality, suggesting these may serve as important indirect biomarkers for idiopathic male infertility. The measurement of reduced and oxidized thiol levels, as well as the enzymes associated with them, can serve as a valuable approach for assessing the oxidative/antioxidative balance in clinical and epidemiological studies, particularly in relation to male fertility issues such as asthenospermia. Alsalman et al. [67] explored the impact of zinc supplementation on semen qualities and certain biochemical markers in men with asthenospermia. The asthenospermic men received 440 mg of zinc sulfate daily for three months. Semen analyses were conducted before and after the supplementation period, focusing on the levels of reduced and oxidized thiols, the thiol oxidation-reduction index, and the activity of thiol-dependent enzymes in both the sperm and seminal plasma. The findings indicated that before treatment, men with asthenospermia exhibited significantly higher levels of oxidized thiols and significantly lower levels of reduced thiols, sulfhydryl oxidase, and glutathione peroxidase activities compared to the fertile control group. After zinc supplementation, the levels of oxidized and reduced thiols, along with the activities of thiol-derived enzymes in the asthenospermic patients, normalized, suggesting an improvement in the oxidant/antioxidant balance. However, the specific levels of reduced and oxidized thiols within the sperm did not show significant changes post-treatment.

Iron, another critical micronutrient, is crucial for the synthesis of nucleic acids and proteins, electron transportation, cellular breathing, and the growth and specialization of cells, all of which are essential for the development of sperm and their metabolic functions. However, being a transition metal, iron can readily lose an electron during oxidation processes, becoming a source for the production of ROS [5]. Both a deficiency and an excess of iron can affect fertility. Establishing iron intake recommendations poses challenges due to issues in gauging iron consumption. The World Health Organization classifies iron deficiency as ferritin levels below 15 μg/L. However, there is a proposition to raise this threshold to below 30 μg/L to enhance the sensitivity of this measurement across populations, both with and without diseases [68]. Ferritin levels below 30 μg/L are linked to unexplained infertility and an increased likelihood of recurrent pregnancy loss [68]. Huang et al. [69] found higher levels of iron in men with asthenospermia compared to normal controls, as well as increased levels of MDA, but showed no significant associations between this micronutrient and MDA. They suggested that MDA was not formed in the seminal plasma itself, but it may be produced by the spermatozoa, prostate, or other accessory organs. However, during incubation of human sperm with Fe^{2+} it was demonstrated that elevated iron levels had a negative impact on sperm motility linked to lipid peroxidation [70]. Nenkova et al. [60] observed that in groups exhibiting poor sperm quality, iron levels were found to be elevated compared to the group with normal sperm (normozoospermia). Furthermore, these groups showed higher levels of malondialdehyde and lower levels of glutathione, indicating the presence of oxidative stress. The high MDA and the low tGSH (total glutathione) levels contributed to decreased sperm motility and increased abnormal spermatozoa. In a study involving sixty-five fertile male volunteers from the southern region of Poland, participants were categorized into two groups based on the iron levels in their seminal plasma: low iron levels (Fe-L) and high iron levels (Fe-H). The findings revealed that the Fe-H group had a significantly higher percentage of non-motile sperm compared to the Fe-L group. Additionally, activities of SOD and manganese-SOD, as well as MDA levels, were notably lower in the Fe-H group than in the Fe-L group. Conversely, total oxidant status (TOS) and oxidative stress index (OSI) values were significantly elevated in the Fe-H group. The research concluded that in fertile men, high iron levels may adversely affect sperm motility and exacerbate oxidative stress, as well as alter the levels of certain cytokines in human semen [71]. Iron overload leads to oxidative stress, which damages fats, proteins, and DNA, negatively impacting the production of sperm and their metabolic activities. Elevated iron levels are linked to reduced sperm movement and higher TOS values in the seminal fluid of men with normal fertility. Iron's role is critical not only in situations of infertility or reduced fertility but also in the general population of men with normal fertility [5,71].

Both severe copper deficiency and excess have been linked to increased production of ROS and cell death stemming from mitochondrial malfunction [43]. Furthermore, minor variations in cellular copper levels, which are not toxic, can affect cell growth or differentiation by altering mitochondrial metabolism. This change can regulate the balance between glycolysis and oxidative phosphorylation, as well as ROS production, thereby creating an oxidative cellular environment [72]. A lack of copper markedly impacts depletion or complete elimination of copper-reliant enzymes in the body, subsequently impeding cellular life functions. Conversely, due to its high reactivity, free copper can initiate the generation of an abundance of free radicals, resulting in significant damage to proteins and DNA [60]. The total amount of copper in normal adults is 50–150 mg; this metal concentration is tissue-dependent, varying between 3 mg (kidneys) and 46 mg (skeleton) (average adult $\approx$ 70 kg) [72,73]. Foods rich in copper, such as seafood (especially shellfish), organ meats, nuts, seeds, whole grains, lemons, raisins, coconuts, papaya, apples, and drinking water can help maintain adequate levels of this micro-mineral [72,74]. Copper is crucial for processes such as iron metabolism and enzyme reactions in nervous system function, as well as in hematopoietic, bone, and other systems [73]. Its role in energy production, iron metabolism, and cardiovascular health ensures optimal functioning of the body's systems, creating a favorable environment for conception and a healthy pregnancy [43,72]. Some researchers suggest that a mother's inability to maintain high Cu and Mn levels may contribute, at least in part, to reproductive health problems [75]. A study by Grieger et al. [57] involving 45 women participating in IVF found that elevated levels of copper in urine correlated with a greater total count of oocytes collected and improved embryo quality. Furthermore, an increase in both urinary copper and zinc levels was linked to the overall number of embryos produced. Current research highlights a noticeable increase in serum copper levels during pregnancy, while copper deficiency is often associated with pregnancy complications and miscarriage. An increase in copper concentration in the blood during pregnancy may suggest an increased demand for this metal due to its key role in embryogenesis and fetal development [75]. The existing literature does not contain any documented associations between oxidative stress, copper, and infertility in human females. In men, maintaining proper copper levels is linked to enhanced sperm quality, underscoring its critical function in male reproductive health. Copper is vital for the creation of male reproductive cells, supporting the structural robustness of sperm and boosting their mobility. This improvement in movement facilitates the sperm's journey through the female reproductive tract towards egg fertilization [76]. Both an excess and a deficiency of copper can markedly diminish male fertility, affecting a broad range of factors from sperm abnormalities, gonadal health, and hormone production to the distribution of micronutrients such as zinc and iron [76]. Enzymes reliant on copper, including ceruloplasmin, SOD (SOD-1 and SOD-3), metallothioneins, and cytochrome c oxidase, are active throughout all phases of gamete formation. They are also present in the somatic cells within the testes and the somatic cells of the epididymis [76]. Abdul-Rasheed [2] highlighted the direct relationship between copper levels in seminal fluid and SOD activity, noting a significant reduction in copper levels in patients with azoospermia, which may lead to reduced SOD activity. He also noticed that malondialdehyde (MDA) levels showed a significant increase over normal control values in oligozoospermic and azoospermic seminal plasma samples.

Typically, mammalian tissues hold between 0.3 and 3.0 μg of manganese per gram of wet tissue weight. Dietary consumption of food and drinking water usually fulfills the body's manganese nutritional needs. The estimated safe and adequate daily dietary intakes (ESADDI) of manganese necessary to preserve body manganese stores are 2 to 5 mg/day for adults and 1.5 to 2.0 mg/day for children aged 4 to 6 years [48]. Few studies have examined the effects of Mn exposure on male reproductive health. Chiti et al. [61] showed that higher intake of manganese was associated with a 7.8-fold increase in sperm count, and Lapointe et al. [77] found Mn^{2+} was a potent stimulator of sperm motility through stimulation of adenylate cyclase activity. Bansal and Kaur [78] observed that oxidative stress conditions in human sperm, induced by ferrous ascorbate, nicotine, or a combination of both, led to an increase in TSH and GSSG content alongside a decrease in GSH levels and the redox coefficient. The addition of 0.1 mM Mn^{2+} enhanced the TSH, GSH, and redox ratios, while also lowering GSSG levels. The researchers deduced that Mn^{2+} supplementation helps preserve thiol levels by mitigating oxidative stress.

Chromium exists in all animal tissues, typically ranging from a few to several tens of μg/kg and seldom surpassing 100 micrograms μg/kg [79]. Chromium(III) is involved in the metabolism of carbohydrates, lipids, and proteins [79]. Dietary sources of trivalent chromium are whole grains, fish and shellfish, dry edible beans, and potatoes [80]. Trivalent chromium enhances insulin's action, but it requires a suitable ligand, such as nicotinic acid, to function biologically [50]. This trace element is often included in dietary supplements [79]. Chromium supplements such as chromium picolinate and niacin-bound chromium(III) are marketed for their potential benefits on blood sugar and lipid levels, body composition, and muscle strength [50].

In addition, Amiri Siavashani et al. [81] hypothesized that chromium supplementation might be beneficial in women with PCOS and candidates for IVF who suffer from different metabolic abnormalities. In a randomized, placebo-controlled clinical trial, Jamilian et al. [82] found that administration of chromium (200 μg/day) for PCOS women for 8 weeks resulted in decreased MDA, along with a significant rise in plasma TAC concentrations, in comparison to the placebo. However, chromium supplementation did not significantly impact plasma concentrations of NO and GSH. In another study, a noticeable increase in TAC and a significant reduction in MDA levels, indicating a reduction in oxidative stress, were observed after chromium (200 μg/day) supplementation for 8 weeks in infertile women with PCOS preparing for in vitro fertilization [83]. Trivalent chromium influences male reproductive health by altering levels of ROS, testosterone, and sperm characteristics such as motility and the count of abnormal sperm. The compound can damage Leydig cells, which are crucial for male reproductive health, by blocking enzymes essential for steroid hormone production and triggering mutagenesis and cell death [84]. However, the impact of chromium(III) on oxidative stress in male reproductive organs is still uncertain, with mixed findings on its effects on lipid peroxidation in seminal plasma from numerous studies. Biswas et al. [85] observed a rise in MDA levels in the seminal plasma of male turkeys that received a diet supplemented with chromium picolinate. This increase in MDA levels may indicate heightened lipid peroxidation in the seminal plasma of the turkeys treated with chromium. Conversely, Shanmugan et al. [86] indicated that chromium supplementation did not influence lipid peroxidation of seminal fluid in Dahlem Red peripubertal roosters by measuring levels of MDA. Type 2 diabetes has a considerable impact on male fertility, negatively influencing semen quality by reducing its volume, sperm count, concentration, and motility [84]. In the study conducted by Imanparast et al. [87], 92 patients with type 2 diabetes mellitus received supplements of either chromium picolinate or trivalent chromium along with vitamin D. The research revealed a decrease in MDA levels in the serum after four months, indicating reduced OS. The supplementation notably increased total thiol groups, serving as innate antioxidants, and the combined administration of chromium picolinate and vitamin D significantly boosted the TAC, a critical marker of antioxidant status. This finding supports the antioxidant properties of chromium(III) within cellular systems. Reduced levels of testosterone are a significant factor contributing to male infertility since this hormone plays a vital role in numerous functions across the male reproductive system. Testosterone is synthesized by Leydig cells in the testes through a process known as testicular steroidogenesis. Any disruption in the functioning of Leydig cells can impair steroid production and, consequently, fertility. Endocrine-disrupting chemicals (EDCs), which interfere with hormonal signals, are particularly concerning for their impact on steroidogenesis. Chromium is identified as one of these EDCs [84]. Therefore, although trivalent chromium has been noted for its potential health benefits, its safety remains a contentious issue. However, exposure to chromium(VI) can adversely affect male fertility by causing oxidative stress in the testis, leading to spermatogenesis disruption, structural damage to seminiferous tubules, and decreased antioxidant defenses [8]. The selected elements discussed affect the fertility of both men and women and thus have important implications for reproduction (Figure 1).

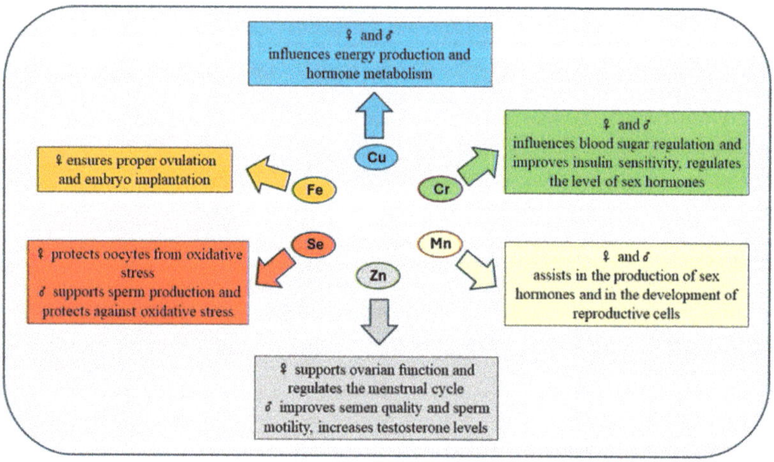

Figure 1. Importance of selected elements on male and female fertility and reproduction (♀—female, ♂—male).

6. Conclusions

Metal ions such as zinc, copper, selenium, and iron are important in female reproductive health. They are involved in various processes, including follicular development, ovulation, and early embryonic development. The seminal plasma contains various metal ions, including zinc, selenium, copper, iron, manganese, and chromium which play crucial roles in maintaining sperm health and function. Abnormal levels of these metal ions can lead to reproductive disorders, affecting fertility and pregnancy outcomes. Certainly, further research determining the correlations between the content of various components, including trace elements and micronutrients, is needed to get a more complete picture of the effects on many conditions in terms of oxidative stress, including fertility and reproductive problems.

Author Contributions: Design and methodology, M.W., W.W. and M.S.; writing—original draft preparation, M.W., W.W. and M.S.; writing—review and editing, M.W.; visualization W.W. and M.W.; supervision, M.W. All authors have read and agreed to the published version of the manuscript.

Funding: This research received no external funding.

Institutional Review Board Statement: Not applicable.

Informed Consent Statement: Not applicable.

Data Availability Statement: No new data were created or analyzed in this study. Data sharing is not applicable to this article.

Conflicts of Interest: The authors declare no conflicts of interest.

References

1. Harchegani, A.B.; Rahmani, H.; Tahmasbpour, E.; Shahriary, A. Hyperviscous Semen Causes Poor Sperm Quality and Male Infertility through Induction of Oxidative Stress. *Curr. Urol.* **2019**, *13*, 1–6. [CrossRef] [PubMed]
2. Abdul-Rasheed, O. Association Between Seminal Plasma Copper and Magnesium Levels with Oxidative Stress in Iraqi Infertile Men. *Oman Med. J.* **2010**, *25*, 168–172. [CrossRef] [PubMed]
3. Amorini, A.M.; Listorti, I.; Bilotta, G.; Pallisco, R.; Saab, M.W.; Mangione, R.; Manca, B.; Lazzarino, G.; Tavazzi, B.; Lazzarino, G.; et al. Antioxidant-Based Therapies in Male Infertility: Do We Have Sufficient Evidence Supporting Their Effectiveness? *Antioxidants* **2021**, *10*, 220. [CrossRef] [PubMed]
4. Flieger, J.; Flieger, W.; Baj, J.; Maciejewski, R. Antioxidants: Classification, Natural Sources, Activity/Capacity Measurements, and Usefulness for the Synthesis of Nanoparticles. *Materials* **2021**, *14*, 4135. [CrossRef]
5. Dobrakowski, M.; Kaletka, Z.; Machoń-Grecka, A.; Kasperczyk, S.; Horak, S.; Birkner, E.; Zalejska-Fiolka, J.; Kasperczyk, A. The Role of Oxidative Stress, Selected Metals, and Parameters of the Immune System in Male Fertility. *Oxidative Med. Cell. Longev.* **2018**, *2018*, 6249536. [CrossRef] [PubMed]
6. Jomova, K.; Valko, M. Advances in metal-induced oxidative stress and human disease. *Toxicology* **2011**, *283*, 65–87. [CrossRef]
7. Takalani, N.B.; Monageng, E.M.; Mohlala, K.; Monsees, T.K.; Henkel, R.; Opuwari, C.S. Role of oxidative stress in male infertility. *Reprod. Fertil.* **2023**, *4*, e230024. [CrossRef]
8. Pereira, S.C.; Oliveira, P.F.; Oliveira, S.R.; Pereira, M.d.L.; Alves, M.G. Impact of Environmental and Lifestyle Use of Chromium on Male Fertility: Focus on Antioxidant Activity and Oxidative Stress. *Antioxidants* **2021**, *10*, 1365. [CrossRef]
9. O'flaherty, C.; Scarlata, E. Oxidative stress and reproductive function: The protection of mammalian spermatozoa against oxidative stress. *Reproduction* **2022**, *164*, F67–F78. [CrossRef]
10. Yan, F.; Zhao, Q.; Li, Y.; Zheng, Z.; Kong, X.; Shu, C.; Liu, Y.; Shi, Y. The role of oxidative stress in ovarian aging: A review. *J. Ovarian Res.* **2022**, *15*, 100. [CrossRef]
11. Didziokaite, G.; Biliute, G.; Gudaite, J.; Kvedariene, V. Oxidative Stress as a Potential Underlying Cause of Minimal and Mild Endometriosis-Related Infertility. *Int. J. Mol. Sci.* **2023**, *24*, 3809. [CrossRef] [PubMed]
12. Wang, L.; Tang, J.; Wang, L.; Tan, F.; Song, H.; Zhou, J.; Li, F. Oxidative stress in oocyte aging and female reproduction. *J. Cell. Physiol.* **2021**, *236*, 7966–7983. [CrossRef]
13. Sasaki, H.; Hamatani, T.; Kamijo, S.; Iwai, M.; Kobanawa, M.; Ogawa, S.; Miyado, K.; Tanaka, M. Impact of Oxidative Stress on Age-Associated Decline in Oocyte Developmental Competence. *Front. Endocrinol.* **2019**, *10*, 811. [CrossRef]
14. Kicińska, A.M.; Maksym, R.B.; Zabielska-Kaczorowska, M.A.; Stachowska, A.; Babińska, A. Immunological and Metabolic Causes of Infertility in Polycystic Ovary Syndrome. *Biomedicines* **2023**, *11*, 1567. [CrossRef] [PubMed]
15. Jin, S.-K.; Yang, W.-X. Factors and pathways involved in capacitation: How are they regulated? *Oncotarget* **2016**, *8*, 3600–3627. [CrossRef] [PubMed]

16. Tiwari, S.; Dewry, R.; Srivastava, R.; Nath, S.; Mohanty, T. Targeted antioxidant delivery modulates mitochondrial functions, ameliorates oxidative stress and preserve sperm quality during cryopreservation. *Theriogenology* **2021**, *179*, 22–31. [CrossRef]
17. Esteves, S.C.; Zini, A.; Coward, R.M.; Evenson, D.P.; Gosálvez, J.; Lewis, S.E.M.; Sharma, R.; Humaidan, P. Sperm DNA fragmentation testing: Summary evidence and clinical practice recommendations. *Andrologia* **2020**, *53*, e13874. [CrossRef]
18. Hadrup, N.; Ravn-Haren, G. Acute human toxicity and mortality after selenium ingestion: A review. *J. Trace Elem. Med. Biol.* **2019**, *58*, 126435. [CrossRef]
19. Al-Mubarak, A.A.; Al-Mubarak, A.A.; van der Meer, P.; van der Meer, P.; Bomer, N.; Bomer, N. Selenium, Selenoproteins, and Heart Failure: Current Knowledge and Future Perspective. *Curr. Heart Fail. Rep.* **2021**, *18*, 122–131. [CrossRef]
20. Yang, F.; Ma, H.; Butler, M.R.; Ding, X.-Q. Deficiency of type 2 iodo- thyronine deiodinase reduces necroptosis activity and oxidative stress responses in retinas of Leber congenital amaurosis model mice. *FASEB J.* **2018**, *32*, 6316–6329. [CrossRef]
21. Lu, J.; Holmgren, A. The thioredoxin antioxidant system. *Free Radic. Biol. Med.* **2014**, *66*, 75–87. [CrossRef]
22. Andrade, I.G.A.; Suano-Souza, F.I.; Fonseca, F.L.A.; Lago, C.S.A.; Sarni, R.O.S. Selenium levels and glutathione peroxidase activity in patients with ataxia-telangiectasia: Association with oxidative stress and lipid status biomarkers. *Orphanet J. Rare Dis.* **2021**, *16*, 83. [CrossRef]
23. Ighodaro, O.M.; Akinloye, O.A. First line defence antioxidants-superoxide dismutase (SOD), catalase (CAT) and glutathione peroxidase (GPX): Their fundamental role in the entire antioxidant defence grid. *Alex. J. Med.* **2018**, *54*, 287–293. [CrossRef]
24. Hu, Y.; Xiao, T.; Wang, Q.; Liang, B.; Zhang, A. Effects of Essential Trace Elements and Oxidative Stress on Endemic Arsenism Caused by Coal Burning in PR China. *Biol. Trace Elem. Res.* **2020**, *198*, 25–36. [CrossRef]
25. Wróblewski, M.; Wróblewska, J.; Nuszkiewicz, J.; Pawłowska, M.; Wesołowski, R.; Woźniak, A. The Role of Selected Trace Elements in Oxidoreductive Homeostasis in Patients with Thyroid Diseases. *Int. J. Mol. Sci.* **2023**, *24*, 4840. [CrossRef] [PubMed]
26. Takamura, T. Hepatokine Selenoprotein P-Mediated Reductive Stress Causes Resistance to Intracellular Signal Transduction. *Antioxid. Redox Signal.* **2020**, *33*, 517–524. [CrossRef] [PubMed]
27. Livingstone, C. Zinc: Physiology, deficiency, and parenteral nutrition. *Nutr. Clin. Pract.* **2015**, *30*, 371–382. [CrossRef]
28. Lin, P.-H.; Sermersheim, M.; Li, H.; Lee, P.H.U.; Steinberg, S.M.; Ma, J. Zinc in Wound Healing Modulation. *Nutrients* **2017**, *10*, 16. [CrossRef]
29. Osredkar, J.; Sustar, N. Copper and Zinc, Biological Role and Significance of Copper/Zinc Imbalance. *J. Clin. Toxicol.* **2011**, *3*, 001. [CrossRef]
30. He, L.; He, T.; Farrar, S.; Ji, L.; Liu, T.; Ma, X. Antioxidants Maintain Cellular Redox Homeostasis by Elimination of Reactive Oxygen Species. *Cell. Physiol. Biochem.* **2017**, *44*, 532–553. [CrossRef]
31. Marreiro, D.D.N.; Cruz, K.J.C.; Morais, J.B.S.; Beserra, J.B.; Severo, J.S.; De Oliveira, A.R.S. Zinc and oxidative stress: Current mechanisms. *Antioxidants* **2017**, *6*, 24. [CrossRef] [PubMed]
32. Si, M.; Lang, J. The roles of metallothioneins in carcinogenesis. *J. Hematol. Oncol.* **2018**, *11*, 107. [CrossRef]
33. Wan, J.; Ren, H.; Wang, J. Iron toxicity, lipid peroxidation and ferroptosis after intracerebral haemorrhage. *Stroke Vasc. Neurol.* **2019**, *4*, 93–95. [CrossRef]
34. Isidori, A.; Loscocco, F.; Visani, G.; Chiarucci, M.; Musto, P.; Kubasch, A.-S.; Platzbecker, U.; Vinchi, F. Iron Toxicity and Chelation Therapy in Hematopoietic Stem Cell Transplant. *Biol. Blood Marrow Transplant.* **2020**, *27*, 371–379. [CrossRef]
35. Abe, C.; Miyazawa, T.; Miyazawa, T. Current Use of Fenton Reaction in Drugs and Food. *Molecules* **2022**, *27*, 5451. [CrossRef]
36. Glorieux, C.; Calderon, P.B. Catalase, a remarkable enzyme: Targeting the oldest antioxidant enzyme to find a new cancer treatment approach. *Biol. Chem.* **2017**, *398*, 1095–1108. [CrossRef] [PubMed]
37. Nandi, A.; Yan, L.-J.; Jana, C.K.; Das, N. Role of Catalase in Oxidative Stress- and Age-Associated Degenerative Diseases. *Oxidative Med. Cell. Longev.* **2019**, *2019*, 9613090. [CrossRef] [PubMed]
38. Cobine, P.A.; Moore, S.A.; Leary, S.C. Getting out what you put in: Copper in mitochondria and its impacts on human disease. *Biochim. Biophys. Acta (BBA)—Mol. Cell Res.* **2020**, *1868*, 118867. [CrossRef]
39. Yin, J.-J.; Fu, P.P.; Lutterodt, H.; Zhou, Y.-T.; Antholine, W.E.; Wamer, W. Dual Role of Selected Antioxidants Found in Dietary Supplements: Crossover between Anti- and Pro-Oxidant Activities in the Presence of Copper. *J. Agric. Food Chem.* **2012**, *60*, 2554–2561. [CrossRef]
40. Suntres, Z.E.; Lui, E.M. Prooxidative effect of copper–metallothionein in the acute cytotoxicity of hydrogen peroxide in Ehrlich ascites tumour cells. *Toxicology* **2006**, *217*, 155–168. [CrossRef]
41. Pham, A.N.; Xing, G.; Miller, C.J.; Waite, T.D. Fenton-like copper redox chemistry revisited: Hydrogen peroxide and superoxide mediation of copper-catalyzed oxidant production. *J. Catal.* **2013**, *301*, 54–64. [CrossRef]
42. Altobelli, G.G.; Van Noorden, S.; Balato, A.; Cimini, V. Copper/Zinc Superoxide Dismutase in Human Skin: Current Knowledge. *Front. Med.* **2020**, *7*, 183. [CrossRef]
43. Liu, Y.; Miao, J. An Emerging Role of Defective Copper Metabolism in Heart Disease. *Nutrients* **2022**, *14*, 700. [CrossRef] [PubMed]
44. Inoue, K.; Sakano, N.; Ogino, K.; Sato, Y.; Wang, D.-H.; Kubo, M.; Takahashi, H.; Kanbara, S.; Miyatake, N. Relationship between ceruloplasmin and oxidative biomarkers including ferritin among healthy Japanese. *J. Clin. Biochem. Nutr.* **2013**, *52*, 160–166. [CrossRef]
45. Chen, J.; Jiang, Y.; Shi, H.; Peng, Y.; Fan, X.; Li, C. The molecular mechanisms of copper metabolism and its roles in human diseases. *Pflügers Arch. Eur. J. Physiol.* **2020**, *472*, 1415–1429. [CrossRef]

46. Wang, B.; Wang, X.-P. Does Ceruloplasmin Defend Against Neurodegenerative Diseases? *Curr. Neuropharmacol.* **2019**, *17*, 539–549. [CrossRef]
47. Chen, P.; Bornhorst, J.; Aschner, M. Manganese metabolism in humans. *Front. Biosci.* **2018**, *23*, 1655–1679. [CrossRef] [PubMed]
48. Dorman, D.C. The Role of Oxidative Stress in Manganese Neurotoxicity: A Literature Review Focused on Contributions Made by Professor Michael Aschner. *Biomolecules* **2023**, *13*, 1176. [CrossRef]
49. Abedelahi, A.; Salehnia, M.; Allameh, A.; Davoodi, D. Sodium selenite improves the in vitro follicular development by reducing the reactive oxygen species level and increasing the total antioxidant capacity and glutathione peroxide activity. *Hum. Reprod.* **2010**, *25*, 977–985. [CrossRef]
50. Bagchi, D.; Stohs, S.J.; Downs, B.W.; Bagchi, M.; Preuss, H.G. Cytotoxicity and oxidative mechanisms of different forms of chromium. *Toxicology* **2002**, *180*, 5–22. [CrossRef]
51. Singh, V.; Singh, N.; Verma, M.; Kamal, R.; Tiwari, R.; Chivate, M.S.; Rai, S.N.; Kumar, A.; Singh, A.; Singh, M.P.; et al. Hexavalent-Chromium-Induced Oxidative Stress and the Protective Role of Antioxidants against Cellular Toxicity. *Antioxidants* **2022**, *11*, 2375. [CrossRef]
52. Zakeri, N.; Kelishadi, M.R.; Asbaghi, O.; Naeini, F.; Afsharfar, M.; Mirzadeh, E.; Naserizadeh, S.K. Selenium supplementation and oxidative stress: A review. *PharmaNutrition* **2021**, *17*, 100263. [CrossRef]
53. Morvaridzadeh, M.; Estêvão, M.D.; Qorbani, M.; Heydari, H.; Hosseini, A.S.; Fazelian, S.; Belančić, A.; Persad, E.; Rezamand, G.; Heshmati, J. The effect of chromium intake on oxidative stress parameters: A systematic review and meta-analysis. *J. Trace Elem. Med. Biol.* **2021**, *69*, 126879. [CrossRef]
54. Shreenath, A.P.; Hashmi, M.F.; Dooley, J. *Selenium Deficiency*; StatPearls Publishing: Treasure Island, FL, USA, 2024.
55. Qazi, I.H.; Angel, C.; Yang, H.; Pan, B.; Zoidis, E.; Zeng, C.-J.; Han, H.; Zhou, G.-B. Selenium, Selenoproteins, and Female Reproduction: A Review. *Molecules* **2018**, *23*, 3053. [CrossRef] [PubMed]
56. Atig, F.; Raffa, M.; Habib, B.-A.; Kerkeni, A.; Saad, A.; Ajina, M. Impact of seminal trace element and glutathione levels on semen quality of Tunisian infertile men. *BMC Urol.* **2012**, *12*, 6. [CrossRef] [PubMed]
57. Grieger, J.A.; Grzeskowiak, L.E.; Wilson, R.L.; Bianco-Miotto, T.; Leemaqz, S.Y.; Jankovic-Karasoulos, T.; Perkins, A.V.; Norman, R.J.; Dekker, G.A.; Roberts, C.T. Maternal Selenium, Copper and Zinc Concentrations in Early Pregnancy, and the Association with Fertility. *Nutrients* **2019**, *11*, 1609. [CrossRef] [PubMed]
58. Dahlen, C.R.; Reynolds, L.P.; Caton, J.S. Selenium supplementation and pregnancy outcomes. *Front. Nutr.* **2022**, *9*, 1011850. [CrossRef]
59. Paszkowski, T.; Traub, A.; Robinson, S.; McMaster, D. Selenium dependent glutathione peroxidase activity in human follicular fluid. *Clin. Chim. Acta* **1995**, *236*, 173–180. [CrossRef]
60. Nenkova, G.; Petrov, L.; Alexandrova, A. Role of Trace Elements for Oxidative Status and Quality of Human Sperm. *Balk. Med. J.* **2017**, *34*, 343–348. [CrossRef]
61. Chiti, H.; Hosseini, E.; Ebrahimi, V.; Mousavi, S.N. Correlation of Dietary Macro- and Micro-Mineral Intake with Seminal Plasma Quality/Quantity and Oxidant/Antioxidant Status in Infertile Compared to the Normal Men: A Case-Control Study. *Biol. Trace Elem. Res.* **2023**, *202*, 1991–1997. [CrossRef]
62. Yin, T.; Yue, X.; Li, Q.; Zhou, X.; Dong, R.; Chen, J.; Zhang, R.; Wang, X.; He, S.; Jiang, T.; et al. The Association Between the Levels of Oxidative Stress Indicators (MDA, SOD, and GSH) in Seminal Plasma and the Risk of Idiopathic Oligo-asthenoteratozoospermia: Does Cu or Se Level Alter the Association? *Biol. Trace Elem. Res.* **2023**, *202*, 2941–2953. [CrossRef] [PubMed]
63. Maret, W.; Sandstead, H.H. Zinc requirements and the risks and benefits of zinc supplementation. *J. Trace Elem. Med. Biol.* **2006**, *20*, 3–18. [CrossRef]
64. Xiong, X.; Lan, D.; Li, J.; Lin, Y.; Zi, X. Effects of Zinc Supplementation During In Vitro Maturation on Meiotic Maturation of Oocytes and Developmental Capacity in Yak. *Biol. Trace Elem. Res.* **2018**, *185*, 89–97. [CrossRef]
65. Anchordoquy, J.; Sirini, M.; Picco, S.; Peral-García, P.; Furnus, C. The Importance of Having Zinc During In Vitro Maturation of Cattle Cumulus–Oocyte Complex: Role of Cumulus Cells. *Reprod. Domest. Anim.* **2014**, *49*, 865–874. [CrossRef] [PubMed]
66. Ebrahimi, F.A.; Foroozanfard, F.; Aghadavod, E.; Bahmani, F.; Asemi, Z. The Effects of Magnesium and Zinc Co-Supplementation on Biomarkers of Inflammation and Oxidative Stress, and Gene Expression Related to Inflammation in Polycystic Ovary Syndrome: A Randomized Controlled Clinical Trial. *Biol. Trace Elem. Res.* **2017**, *184*, 300–307. [CrossRef] [PubMed]
67. Alsalman, A.R.S.; Almashhedy, L.A.; Hadwan, M.H. Effect of Oral Zinc Supplementation on the Thiol Oxido-Reductive Index and Thiol-Related Enzymes in Seminal Plasma and Spermatozoa of Iraqi Asthenospermic Patients. *Biol. Trace Elem. Res.* **2017**, *184*, 340–349. [CrossRef]
68. Holzer, I.; Ott, J.; Beitl, K.; Mayrhofer, D.; Heinzl, F.; Ebenbauer, J.; Parry, J.P. Iron status in women with infertility and controls: A case-control study. *Front. Endocrinol.* **2023**, *14*, 1173100. [CrossRef]
69. Huang, Y.-L.; Tseng, W.-C.; Cheng, S.-Y.; Lin, T.-H. Trace Elements and Lipid Peroxidation in Human Seminal Plasma. *Biol. Trace Elem. Res.* **2000**, *76*, 207–216. [CrossRef]
70. Huang, Y.-L.; Tseng, W.-C.; Lin, T.-H. In Vitro Effects of Metal Ions (Fe^{2+}, Mn^{2+}, Pb^{2+}) on Sperm Motility and Lipid Peroxidation in Human Semen. *J. Toxicol. Environ. Health Part A* **2001**, *62*, 259–267. [CrossRef]
71. Kasperczyk, A.; Dobrakowski, M.; Czuba, Z.; Kapka-Skrzypczak, L.; Kasperczyk, S. Influence of iron on sperm motility and selected oxidative stress parameters in fertile males—A pilot study. *Ann. Agric. Environ. Med.* **2016**, *23*, 292–296. [CrossRef]

72. Ruiz, L.M.; Libedinsky, A.; Elorza, A.A. Role of Copper on Mitochondrial Function and Metabolism. *Front. Mol. Biosci.* **2021**, *8*, 711227. [CrossRef]
73. Wang, P.; Yuan, Y.; Xu, K.; Zhong, H.; Yang, Y.; Jin, S.; Yang, K.; Qi, X. Biological applications of copper-containing materials. *Bioact. Mater.* **2020**, *6*, 916–927. [CrossRef] [PubMed]
74. Squitti, R.; Siotto, M.; Polimanti, R. Low-copper diet as a preventive strategy for Alzheimer's disease. *Neurobiol. Aging* **2014**, *35*, S40–S50. [CrossRef] [PubMed]
75. Skalnaya, M.G.; Tinkov, A.A.; Lobanova, Y.N.; Chang, J.-S.; Skalny, A.V. Serum levels of copper, iron, and manganese in women with pregnancy, miscarriage, and primary infertility. *J. Trace Elem. Med. Biol.* **2019**, *56*, 124–130. [CrossRef]
76. Ogórek, M.; Gąsior, Ł.; Pierzchała, O.; Daszkiewicz, R.; Lenartowicz, M. Role of copper in the process of spermatogenesis. *Postep. Hig. I Med. Dosw.* **2017**, *71*, 663–683. [CrossRef]
77. Lapointe, S.; Ahmad, I.; Buhr, M.; Sirard, M.-A. Modulation of Postthaw Motility, Survival, Calcium Uptake, and Fertility of Bovine Sperm by Magnesium and Manganese. *J. Dairy Sci.* **1996**, *79*, 2163–2169. [CrossRef]
78. Bansal, A.K.; Kaur, A.R.J. Cooperative functions of manganese and thiol redox system against oxidative stress in human spermatozoa. *J. Hum. Reprod. Sci.* **2009**, *2*, 76–80. [CrossRef] [PubMed]
79. Lewicki, S.; Zdanowski, R.; Krzyżowska, M.; Lewicka, A.; Dębski, B.; Niemcewicz, M.; Goniewicz, M. The role of Chromium III in the organism and its possible use in diabetes and obesity treatment. *Ann. Agric. Environ. Med.* **2014**, *21*, 331–335. [CrossRef]
80. Thor, M.Y.; Harnack, L.; King, D.; Jasthi, B.; Pettit, J. Evaluation of the comprehensiveness and reliability of the chromium composition of foods in the literature. *J. Food Compos. Anal.* **2011**, *24*, 1147–1152. [CrossRef]
81. Siavashani, M.A.; Modarres, S.Z.; Mirhosseini, N.; Aghadavod, E.; Salehpour, S.; Asemi, Z. The Effects of Chromium Supplementation on Gene Expression of Insulin, Lipid, and Inflammatory Markers in Infertile Women With Polycystic Ovary Syndrome Candidate for in vitro Fertilization: A Randomized, Double-Blinded, Placebo-Controlled Trial. *Front. Endocrinol.* **2018**, *9*, 726. [CrossRef]
82. Jamilian, M.; Bahmani, F.; Siavashani, M.A.; Mazloomi, M.; Asemi, Z.; Esmaillzadeh, A. The Effects of Chromium Supplementation on Endocrine Profiles, Biomarkers of Inflammation, and Oxidative Stress in Women with Polycystic Ovary Syndrome: A Randomized, Double-Blind, Placebo-Controlled Trial. *Biol. Trace Elem. Res.* **2015**, *172*, 72–78. [CrossRef] [PubMed]
83. Jamilian, M.; Modarres, S.Z.; Siavashani, M.A.; Karimi, M.; Mafi, A.; Ostadmohammadi, V.; Asemi, Z. The Influences of Chromium Supplementation on Glycemic Control, Markers of Cardio-Metabolic Risk, and Oxidative Stress in Infertile Polycystic ovary Syndrome Women Candidate for In vitro Fertilization: A Randomized, Double-Blind, Placebo-Controlled Trial. *Biol. Trace Elem. Res.* **2018**, *185*, 48–55. [CrossRef] [PubMed]
84. Moreira, R.; Martins, A.D.; Alves, M.G.; Pereira, M.d.L.; Oliveira, P.F. A Comprehensive Review of the Impact of Chromium Picolinate on Testicular Steroidogenesis and Antioxidant Balance. *Antioxidants* **2023**, *12*, 1572. [CrossRef] [PubMed]
85. Biswas, A.; Divya, S.; Mandal, A.; Majumdar, S.; Singh, R. Effects of dietary supplementation of organic chromium (picolinate) on physical and biochemical characteristics of semen and carcass traits of male turkeys. *Anim. Reprod. Sci.* **2014**, *151*, 237–243. [CrossRef]
86. Shanmugam, M.; Prakash, B.; Panda, A. Effect of dietary organic zinc and chromium supplementation on semen quality in layer breeders. *Indian. J. Poult. Sci.* **2020**, *55*, 133–138. [CrossRef]
87. Imanparast, F.; Mashayekhi, F.J.; Kamankesh, F.; Rafiei, F.; Mohaghegh, P.; Alimoradian, A. Improving the endothelial dysfunction in type 2 diabetes with chromium and vitamin D3 byreducing homocysteine and oxidative stress: A randomized placebo-controlled trial. *J. Trace Elem. Med. Biol.* **2020**, *62*, 126639. [CrossRef]

International Journal of
Molecular Sciences

MDPI

Review

Environmental and Genetic Determinants of Ankylosing Spondylitis

Rafał Bilski [1,*], Piotr Kamiński [2,3], Daria Kupczyk [1], Sławomir Jeka [4], Jędrzej Baszyński [2], Halina Tkaczenko [5] and Natalia Kurhaluk [5]

1 Department of Medical Biology and Biochemistry, Collegium Medicum in Bydgoszcz, Nicholaus Copernicus University, M. Karłowicz St. 24, 85-092 Bydgoszcz, Poland
2 Department of Medical Biology and Biochemistry, Division of Ecology and Environmental Protection, Collegium Medicum in Bydgoszcz, Nicolaus Copernicus University in Toruń, M. Skłodowska-Curie St. 9, 85-094 Bydgoszcz, Poland
3 Department of Biotechnology, Institute of Biological Sciences, Faculty of Biological Sciences, University of Zielona Góra, Prof. Z. Szafran St. 1, 65-516 Zielona Góra, Poland
4 Department of Rheumatology and Connective Tissue Diseases, Collegium Medicum, Nicolaus Copernicus University, University Hospital No. 2, Ujejski St. 75, 85-168 Bydgoszcz, Poland
5 Institute of Biology, Pomeranian University in Słupsk, Arciszewski St. 22 B, 76-200 Słupsk, Poland
* Correspondence: rafal.bilski@cm.umk.pl; Tel.: +48-52-585-37-37; Fax: +48-52-585-37-42

Abstract: Exposure to heavy metals and lifestyle factors like smoking contribute to the production of free oxygen radicals. This fact, combined with a lowered total antioxidant status, can induce even more damage in the development of ankylosing spondylitis (AS). Despite the fact that some researchers are looking for more genetic factors underlying AS, most studies focus on polymorphisms within the genes encoding the human leukocyte antigen (HLA) system. The biggest challenge is finding the effective treatment of the disease. Genetic factors and the influence of oxidative stress, mineral metabolism disorders, microbiota, and tobacco smoking seem to be of great importance for the development of AS. The data contained in this review constitute valuable information and encourage the initiation and development of research in this area, showing connections between inflammatory disorders leading to the pathogenesis of AS and selected environmental and genetic factors.

Keywords: ankylosing spondylitis; oxidative stress; pathophysiological state; reactive oxygen species; HLA; IL-23; IL-17

1. Introduction

Ankylosing spondylitis (AS) is a chronic disease of an unknown etiology, belonging to the group of spondyloarthropathy together with rheumatoid arthritis (RA) and psoriatic arthritis (PsA) [1,2]. The reasons for its occurrence are believed to be genetic and environmental factors [3,4]. Due to its autoimmune nature, AS is also difficult to diagnose before pain symptoms occur. Researchers have proven the presence of genetic links between people affected by the described disease [4,5]. However, modern works indicate an interaction of genetic and environmental factors in the pathogenesis of AS [3,4]. The essence of the disease is inflammatory changes in the synovium of the sacroiliac joints, consisting mainly of the infiltration of lymphocytes and plasma cells. Over time, the spaces in these joints slowly heal and ossify. As the disease progresses, it leads to similar changes in the apophysial joints of the spine. The next stage of changes is inflammatory ossifications in the perivertebral tissue and in the outer layer of the fibrous ring. Vertebrae that are adjacent to each other are connected by syndesmophytes (bone bridges) [6–8].

The pathogenesis of AS is not fully explained and the mechanisms underlying it are influenced by many factors. However, their basis is the inflammatory process resulting from the autoimmune response. The role of the IL-23/Il-17 pathway in autoimmune diseases has been proven by many researchers [9,10]. IL-23 is a heterodimeric cytokine secreted by monocytes, macrophages, and dendritic cells, often in response to information on threats [Figure 1]. IL-23 is crucial for the differentiation of helper T cells to the Th17 phenotype, as it polarizes and stabilizes this pro-inflammatory phenotype [11]. IL-17A is a pro-inflammatory cytokine also involved in maintaining mucosal immunity, especially in the skin and intestines. It is produced by various cell types, including CD4 cells (TH17), CD8+ T cells, and innate lymphoid cells [12]. The Il-23/Il-17 pathway is important in AS, as well as inflammatory bowel disease [13]. The concentration of IL-23 and IL-17 cytokines in these diseases is significantly increased, and it has also been shown that AS can be significantly alleviated by blocking the IL-23/IL-17 pathway, which further indicates the involvement of the discussed pathway in the pathogenesis of AS [14]. Pathogen-associated molecular patterns (PAMPs) activate TNF-α through the NF-κB pathway, which contributes to the expression of IL-17 [15]. Additionally, intracellular Nod-like receptors (NLRs) are able to recognize PAMPs. NLRs can then cause pro-inflammatory cytokines, such as IL-18, to be activated [16]. Modern scientists are searching for evidence of NLRP3 (NLR Family Pyrin Domain Containing 3), YAP (Yes1 Associated Transcriptional Regulator), and AMPK (AMP-activated protein kinase) pathway participation in the pathogenesis of AS. There is evidence that NLRP3 plays a significant role in human immunity and autoimmune disease development. Research conducted on mice proved the influence of the AMPK/YAP/NLRP3 signaling pathway in the pathogenesis of AS [16–18].

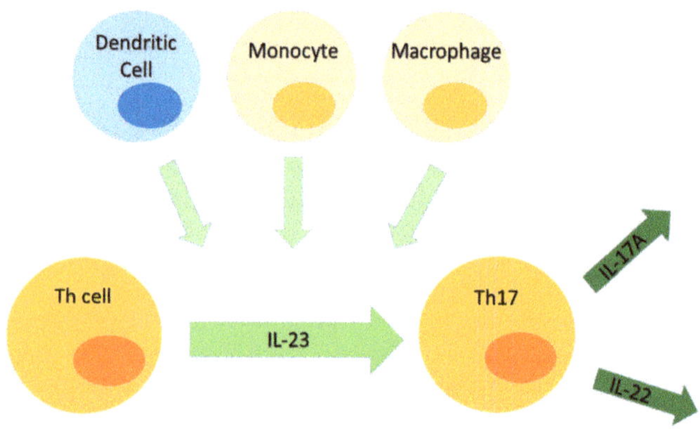

Figure 1. The schematic image of the IL-23/IL-17 pathway.

The hypothesis of the activation of bone morphogenetic protein signals and the reduction in sclerostin and Dickkopf-related protein 1 (DKK1), which are endogenous Wnt pathway inhibitors, is somewhat supported by the available data [19–21]. According to the modified Stoke Ankylosing Spondylitis Spine Score, radiographic damage may be correlated over time with disease activity [22].

Angiogenesis is an important stage in the development of rheumatic diseases. Patients with AS also have an increased atherosclerosis risk, therefore leading to an increased mortality from cardiovascular diseases [23]. Vascular endothelial growth factor (VEGF) is a key regulator of angiogenesis, with several isoforms and roles in different physiological processes. Mounting data indicate a correlation between the VEGF system and rheumatic diseases with anti-VEGF and VEGF receptor (VEGFR) [24]. Angiogenesis is a hypothesis that explains the process of new bone creation in AS. New bone growth, sacroiliitis, and enthesitis, which are different symptoms of AS, necessitate angiogenesis. VEGF plays a key role in regulating this process, according to Yamamoto's research in 2013 [25]. Fearon et al. noted that VEGF is present in early inflammatory arthritis and has a strong connection to angiopoietins [26]. It may play a role in regulating these processes in the development of early AS. VEGF can directly control the development of osteoblasts from synovial fibroblasts, which aids in the production of new bone in AS [27]. Furthermore, according to Kawashima's research in 2009, it can activate the COX2 pathway [28]. According to Drouart, VEGF is also involved in inflammation in AS [29]. Multiple investigations have confirmed that VEGF is associated with AS susceptibility, indicating that it is involved in the development of AS [30].

In this review article, we are trying to find the connection between the inflammatory processes underlying the pathogenesis of AS with selected genetic and environmental factors such as oxidative stress, the concentrations of selected elements, lifestyle, gut microbiota, infections, and genetic polymorphisms. Although the etiology of this disease is not fully known, environmental and lifestyle factors, which are modifiable factors, are implicated. Analysis of the environment in the form of pollutants and toxic substances can provide information about changes occurring at the cellular level, helping to better understand the pathomechanism of AS.

2. The Participation of Oxidative Stress in Pathogenesis of AS

Oxidative stress refers to an imbalance that favors oxidants over antioxidants, which can cause molecular damage and/or the disturbance of redox signaling and control [31]. More specifically, oxidative stress is defined as a disturbance in the balance between the production of reactive oxygen species (free radicals and non-free radicals) and antioxidant defenses [32]. Researchers' reports suggest that, in the chronic inflammatory process, significant DNA damage is involved. Oxidative stress is a phenomenon harmful to DNA, as evidenced by its numerous damages in the course of action of reactive forms of oxygen [33]. The production of inflammatory cytokines in the course of disease development will induce the production of oxidative stress markers [34]. Therefore, despite the fact that the etiology of AS is still not fully known, we cannot ignore the significant share of oxidative stress underlying AS, which has been proven by researchers [35].

The stimulation of the cell membrane during phagocytosis activates NADP oxidase, which catalyzes the formation of the superoxide anion O^{2-}. Thanks to the presence of superoxide dismutase (SOD), this radical is converted into hydrogen peroxide and singlet oxygen [36,37]. The reaction between the superoxide radical and H_2O_2 produces a hydroxyl radical. These three compounds are called reactive oxygen species (ROS) and play a significant role in inflammatory processes [36]. In the context of changes in the joints, the hydroxyl radical has the greatest share, having the ability to combine with glycoproteins, phospholipids of cell membranes, hyaluronic acid, DNA, and collagen, leading to the destruction of their structure [38,39]. The body has a number of defense mechanisms aimed at neutralizing the risk of changes caused by free oxygen radicals. The body's natural antioxidant barrier consists of a number of chemical compounds that help to reduce and inactivate ROS. Antioxidant defense mechanisms can be divided into enzymatic and non-enzymatic. The first group includes enzymes that catalyze ROS scavenging reactions—superoxide dismutase, catalase CAT, peroxidase (GPx), Glutathione S-transferase (GST), and glutathione reductase (GR) [36,37]. The second group includes compounds obtained from external sources, such as vitamins C, E, flavonoids, and β-carotene, but also those synthesized by the body, such as melatonin and glutathione [40]. Oxidative stress can trigger a large number of proteins such as heat shock proteins and turn on transcription factors such as AP-1 and NF-κB [41–43].

In inflammatory diseases, there is an increased production of superoxide anions at the site of inflammation [44]. This problem is important for the pathomechanism of many diseases, including AS discussed in this paper, RA, and PsA [37,45–48].

An important mechanism of action of oxidative stress is the disruption of redox signaling, causing molecular damage that inevitably affects angiogenesis, inflammation, and the function/activation of dendritic cells, lymphocytes, and keratinocytes [49]. This is also important for cytochrome C, which floats in the peripheral mitochondrial membrane and mediates the transfer of electrons in the respiratory chain [50–52]. The formation of reactive oxygen species induces the release of cytochrome c into the cytosol [53], where it binds to the apoptosis protease activation factor-1 (APAf-1), forming an apoptosome from ATP or dATP; this complex activates procaspase 9, which triggers an enzymatic cascade leading to cell apoptosis [54–56]. In this way, oxidative stress causes uncontrolled cell apoptosis, among others within the joint tissue, leading to its damage and inflammation [57]. Several clinical studies on diseases characterized by cell death, either due to apoptosis or necrosis, have reported the release of cytochrome c from the mitochondria into the extracellular space and ultimately into the circulation. Elevated levels of cytochrome c in serum have been found in chronic and acute diseases, including arthritis, myocardial infarction and stroke, and liver diseases [58–60]. Its role as an inducer of skin and joint inflammation has been proposed. In particular, an association between cytochrome c, skin inflammation, and keratinocyte proliferation during ROS production has been demonstrated [61]. The proliferation of keratinocytes is the basis of the pathomechanism of diseases such as psoriasis and PsA [62,63]. Dysregulations in ROS and cytochrome c release have been associated with psoriasis and been reported to cause associated skin inflammation [64]. This is one of the mechanisms that facilitates their growth and, at the joint level, the destruction of articular cartilage and the erosion and/or proliferation of bones [65]. Based on the important role played by ROS in some malignancies and rheumatic diseases, it is also possible that cells are involved in the activation of AS [66]. In patients with AS with inflammatory bone disease located in the lower part of the spine, an increase in the total oxidative status, as well as the advanced oxidation of protein products and lipid peroxidation, was demonstrated [35,45,67,68].

A significant role in the development of this disease is believed to be attributed to the disturbance of the body's oxidative balance [45]. Over recent years, the importance of disturbing the balance between the production of ROS and the ability to inactivate them via the antioxidant system in disturbances of body homeostasis has been increasingly discussed. There are studies that have confirmed the significant role of oxidative stress in the pathogenesis of chronic inflammatory autoimmune diseases [69]. AS may reduce the body's condition because it disturbs the balance between the interactions of individual antioxidant enzymes. This is related to changes in their activity compared to healthy people in whom no pathological changes in the osteoarticular system are observed [35]. Moreover, excess free radicals may contribute to the damage of biological structures due to the oxidation processes of the compounds present in them [70]. Free oxygen radicals cause the oxidation of proteins, lipids, and DNA, and, thus, may contribute to tissue damage. ROS sources can be divided into endogenous and exogenous. Endogenous sources include numerous biochemical processes taking place in physiological conditions, and the most efficient source is the respiratory chain. In turn, external factors include, among others, environmental pollution, tobacco smoke, ionizing radiation, ultraviolet radiation, and ultrasound [71]. As a result of the oxidation reaction, toxic products are formed, which, by their action on the cell, have a cytostatic effect and lead to damage to cell membranes and the activation of apoptosis mechanisms [72]. The body has a natural defense system that protects cells against excessive amounts and harmful effects of ROS. The pro- and antioxidant balance is ensured by antioxidant enzymes, including SOD, CAT, GPx, and GR and a number of non-enzymatic substances, including glutathione, vitamin E or vitamin C, and ceruloplasmin (CP). These are compounds that, through their action, enable the removal of excess reactive oxygen species from the cell. Small-molecule antioxidants are mostly compounds of exogenous origin that are supplied to the body through the diet. They act as responses to non-specific reactions that result in the inactivation of free radicals [73]. ROS can have a significant impact on the development and progression of inflammatory conditions in the osteoarticular system, including ankylosing spondylitis [74]. Moreover, the increased production of ROS, which intensifies joint inflammation, may be related to a low supply of fish in the diet of AS patients. This fact is then explained by the low content of antioxidant vitamins and omega 3 acids, which may have a protective effect and counteract the effects of oxidative stress. As a result of the ongoing inflammatory process and excess ROS, the body's oxidative reserves are depleted, and damage caused by ROS accumulates in the joints [75]. Stanek et al. [76] showed the reduced effectiveness of the natural antioxidant barrier in patients with AS. Karakoc et al. [45] described the total potential of antioxidant parameters in patients with AS compared to a group of healthy people. Their research also showed a reduced antioxidant potential in patients with AS. In turn, Ozgocmen et al. [67], in their study, did not show statistically significant differences in SOD activity between groups of AS patients and healthy people. However, in the study by Pishgahi et al. [77], oxidative stress parameters were assessed in AS patients with metabolic syndrome compared to AS patients without comorbidities. It turned out that, in patients with the above-mentioned syndrome, the levels of SOD and CAT were increased compared to patients with AS without concomitant diseases. In the study by Danaii et al. [78], a statistically significant variability was obtained regarding the parameters of oxidative stress in people with AS. The study involved 35 people with diagnosed disease and 40 healthy people who constituted the control group. Significantly increased levels of SOD and CAT were observed in the group of people suffering from AS. The total antioxidant status (TAS) was proven to be lower in patients with ankylosing spondylitis compared to healthy individuals [45,76,79].

An antioxidant that also belongs to the group of acute-phase proteins is ceruloplasmin. The concentration of this protein increases in chronic inflammation. Also, in the research conducted by Jayson et al. [80], significantly higher concentrations of ceruloplasmin were found in patients with AS. Similar results were shown in the study by Aiginger et al. [81], which found an increased level of ceruloplasmin among patients with AS compared to a group of healthy people. Similarly, in an experiment conducted by Li et al. [82], a higher ceruloplasmin expression was shown in AS patients compared to the control group. A high level of free oxygen radicals, combined with a deficiency in the activity of antioxidant enzymes, may result in damage to the body at the molecular level [72,75].

Research conducted by Salmón et al. [83] confirmed that antioxidant defense mechanisms adapt to greater amounts of pollution in the urban environment, but not enough to fully neutralize the damage caused by greater exposure to oxidative stress.

The oxidation reactions of unsaturated fatty acids proceed in stages and lead to the formation of end products harmful to the body, including malondialdehyde (MDA) as an indicator of lipid peroxidation and advanced oxidation protein products (AOPP) [84]. These are compounds that can inhibit the processes of replication, transcription, and DNA strand breaks. They can also inhibit the activity of selected enzymes or change the antigenic properties of proteins. MDA and AOPP, respectively, can function as secondary mediators to exacerbate damage and accelerate the course of the disease by stimulating the TNF-α and NF-κB pathways, which are critical for the inflammatory process associated with AS [85]. In the study by Kozaci et al. [34], a higher level of MDA was found among patients with AS compared to the control group. The results of Ozgocmen et al. [67] and Ayala et al. [86] are similar.

The oxidative biomarkers in the serum that were higher in AS patients and the part that oxidative stress plays in the pathophysiology of AS may be explained by the following theories. First off, oxidative stress can harm proteins, lipids, nucleic acids, cell membranes, and the extracellular matrix, which raises the concentration of oxidative injury products [87]. When leukocytes become more activated at the site of inflammation in AS, ROS and enzymes like Myeloperoxidase (MPO) are released, which can cause hypochlorous acid to form. By reacting with the components of cartilage, hypochlorous acid causes tissue damage and protein oxidation. [68,85]. Second, by reacting with cell membranes and triggering phospholipase A2, lipid peroxidation affects the fluidity and receptor functions of cell membranes. Thus, it is possible to induce T cells to release more interleukins, which will set off a vicious cycle by increasing lipid peroxidation and inflammation [88]. The research conducted by Feng et al. showed promising results in mice by applying Punicalagin, which effectively decreased oxidative stress and regulated the NF-kβ pathway [89]. Kiranatlioglu-Firat et al. proved, in their research, the participation of oxidative stress, oxidative DNA damage, and chromosomal DNA damage in patients with AS [33]. Exogenous ROS stimulate the induction of VEGF expression in various cell types, such as endothelial cells, smooth muscle cells, and macrophages, where VEGF induces endothelial cell migration and proliferation through an increase in intracellular ROS [90–93]. Thus, the reciprocity between oxidative stress and angiogenesis is centered on the VEGF signaling pathway. A number of studies have demonstrated this positive interrelationship between ROS and angiogenesis. Hydrogen peroxide induces VEGF expression in vascular smooth muscle cells, as well as endothelial cells, thereby promoting angiogenic responses [90,92]. Oxidative stress enhances metabolism toward glycolysis, possibly leading to increased inflammatory processes and impaired angiogenesis in RA [94,95].

The overproduction of free radicals and weakening of mechanisms for their neutralization are two of the causes of the development of the inflammatory process in ankylosing spondylitis. The toxic accumulation of metabolites first leads to functional impairment and then the destruction of the structure of the cell membrane in cells, resulting in inflammation. As a result, a number of inflammatory mediators are secreted within the tissue. This process leads to uncontrolled apoptosis and tissue damage. In the pathogenesis of AS, a significant role is attributed to disorders in the pro- and antioxidant systems. Malfunctioning defense systems do not protect the body against the accumulation of free radicals; therefore, they are unable to protect against the development of inflammation and the destruction of individual structures. From this point of view, it seems important, among others, to supply exogenous antioxidants in the diet or avoid stimulants.

In recent years, the importance of disturbing the balance between the production of ROS and the ability of the antioxidant system to neutralize them has been increasingly discussed. The significant role of ROS in the pathogenesis of chronic autoimmune diseases has been confirmed. An excess of ROS can lead to damage to structures at the cell level due to the oxidation of the chemical compounds contained therein. Additionally, researchers wonder whether the increased synthesis of free radicals is a causative factor or whether it appears secondary to the disease. Therefore, conducting research on the involvement of ROS in the etiopathogenesis of AS is justified. The ongoing oxidation reactions may initiate a chronic inflammatory process and contribute to tissue destruction in the course of AS. From this point of view, antioxidant treatment seems to be beneficial, which may have a significant impact on the direction of AS pharmacotherapy. Moreover, the generation of significant amounts of free oxygen radicals is caused by an excessive influx of neutrophils, which leads to a disturbance of the pro- and antioxidant balance in cells and, consequently, to the uncontrolled and progressive destruction of structural elements of the joints [94–96]. Therefore, eliminating the phenomenon of oxidative stress in places affected by inflammation may significantly influence the inhibition of the development of the disease, improving the quality of life of patients. The levels of the antioxidant enzymes SOD, CAT, GPx, and GR were found to be elevated or reduced in certain research. One possible explanation is that the levels of antioxidant enzymes may not be reliable predictors of high or low oxidative stress. Instead, the evaluation of oxidative stress is mostly based on the harm it causes to important biological molecules such as proteins, lipids, and DNA. MDA serves as a marker for this assessment, as it is one of the byproducts of lipid peroxidation. Furthermore, TAS is an unreliable indicator of overall antioxidant status due to its use of manufactured free radicals as one of its reagents, which are not naturally produced in human bodies. Elevated levels of MDA and total oxidative stress (TOS) provide evidence of the influence of this phenomenon in individuals with AS. A summary of oxidative stress indicators is shown below (Table 1).

Table 1. The changes in oxidative stress parameters in AS patients in comparison to healthy controls.

Parameter	Result in AS Patients	Reference
SOD	Increased	[76,78]
	Decreased	[76,97]
	No significant change	[67]
CAT	Increased	[76,78]
	Decreased	[76,97]
GPx	Increased	[76]
	Decreased	[76,97]
GST	Decreased	[76]
GR	Increased	[98]
	Decreased	[76]
TAS	Decreased	[45,76,79]
TOS	Increased	[45,98]
MDA	Increased	[34,67,97,98]
AOPP	Increased	[97]
CP	Increased	[80–82]

3. The Participation of Elements as Environmental Factors in AS Development

The importance of environmental factors in the etiopathogenesis of diseases of the osteoarticular system has been noted. The osteoarticular system plays a special role in assessing the risk of exposure to xenobiotics. Bone reconstruction and mineralization take a very long time, usually the entire life of the organism, so they may be a determinant of the long-term accumulation of bioelements and toxic metals. Unfavorable changes in the skeletal and joint systems are most often the result of the long-term impact of heavy metals on the body. It is suggested that elements easily absorbed by bone tissue, e.g., cadmium, lead, aluminum, and strontium, may be involved in the development of diseases of the osteoarticular system (osteoarthritis, rheumatoid arthritis, and osteoporosis). The proper functioning of the osteoarticular system is determined by various factors, depending on systemic and multi-organ changes [99–101]. Although bones or joint elements could be a very valuable material for assessing the state of elements in humans, except in special situations (orthopedic surgeries, e.g., arthroplasty), such material is difficult to obtain intravitally. Hence, most studies are usually based on material derived from serum. Metabolic diseases of the osteoarticular system are clinical syndromes developing as a result of the interaction of many factors, some of which are the impact of the external environment, including exposure to chemicals or elements that potentially act as triggers of autoimmune reactions [102]. Other environmental factors that play a significant role in the pathomechanism of AS include an abnormal element metabolism in the body and exposure to heavy metals and other pollutants [103].

Calcium plays a role in inflammatory processes as a secondary messenger. Its abnormal deposition in the extracellular matrix, occurring during dystrophic or repair processes, is the basis for ectopic calcifications in bone and extraskeletal spaces. Ankylosing spondylitis is a rheumatic disease characterized by the inflammation and final calcification of tendon attachments, which are an important site for tendons, joint capsules, and ligaments [104]. Since AS is often accompanied by non-specific intestinal inflammation and disorders in its microflora, it is believed that it is important for calcium absorption. As a result of the above-mentioned comorbidities, the expression of Ca transporters such as calbindin D9K is inhibited, and, therefore, the absorption of calcium in the intestine is impaired [105]. Calcium can influence the immune response and repair mechanisms in several ways. Intracellular calcium can form complexes with calmodulin and activate calcium/calmodulin-dependent kinases or calcineurin, which, respectively, phosphorylate κB kinase inhibitor (IKK2) and dephosphorylated nuclear factor of activated T-cells (NFAT) [Figure 2].

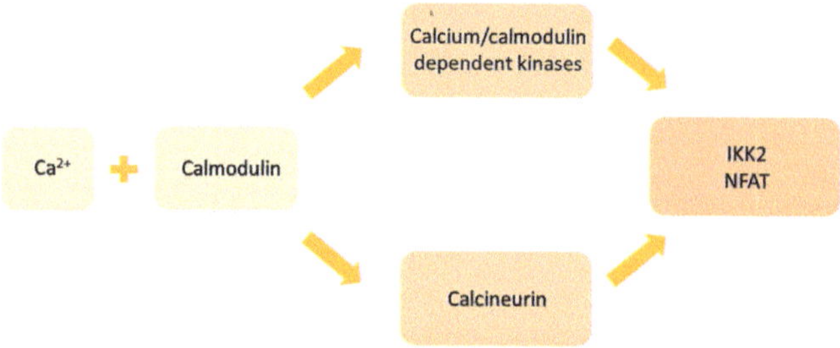

Figure 2. The schematic image of calcium activated immune response.

This leads to the activation and translocation of either nuclear factor kappa light chain enhancer of activated B cells (NF-κB) or NFAT to the nucleus, where they can control the transcription of various genes related to proliferation and inflammation [106]. Additionally, the importance of calcium in inflammation has been revealed by the role played by calcium-sensing receptors (CaSR) in immune cells. In fact, hypocalcemia often occurs in sepsis or other systemic inflammatory conditions due to the inhibition of parathyroid hormone secretion. This may depend on the abnormal expression of calcium receptors on peripheral mononuclear cells, which use calcium to secrete cytokines and activate the inflammasome. Therefore, cytokines such as IL-6, TNF-α, and IL-1 may indirectly influence the secretion of parathyroid hormone (PTH) by increasing the expression and activation of calcium receptors on immune cells [107]. Additionally, intracellular calcium has been shown to be increased in neutrophils from AS patients, where it may contribute to lipid peroxidation, apoptosis, and the activation of caspases 3 and 9. Of note, the anti-TNF monoclonal antibody infliximab, used in the treatment of many rheumatic diseases, including AS, may prevent the influx of calcium into neutrophils, suggesting that the opening of calcium channels may be a result of the response to cytokines [108]. Similarly, a blockade of lymphocyte potassium channels may have anti-inflammatory properties in patients with AS and rheumatoid arthritis, preventing the differentiation of CD4+ and CD8+ T cells [109]. The calcium levels in the body can be influenced by vitamin D deficiency. In the case of this deficiency, calcium is reabsorbed from the bones to maintain the correct serum levels [110]. Chen et al., in their metanalysis, indicated a strong correlation between elevated levels of vitamin D in the blood and the ability to manage the disease and enhance the overall well-being of individuals with AS. Furthermore, the overall blood 25(OH)D levels and the inverse correlation between serum vitamin D levels and Bass Ankylosing Spondylitis Disease Activity Index in patients with AS may be considerably impacted by differences in continents and ethnicities. Thus, individuals diagnosed with AS are recommended to augment their vitamin D intake through either exposure to sunlight or the use of dietary supplements. This is enacted to reduce the severity of clinical symptoms and improve overall quality of life [111]. A metanalysis by Diao et al. showed similar results, proving that patients with higher levels of vitamin D had a lower C-reactive protein and erythrocyte sedimentation rate, although there were no significant changes in the serum calcium levels between patients with AS and healthy controls [112]. On the other hand, a two-sample randomized Mendelian study by Jiang et al. showed no strong evidence to connect vitamin D levels and AS. [113]. The vitamin D levels in the body are strongly connected to sun exposure and ethnicity. Although there is no evidence of differences in the serum calcium levels in patients with AS and healthy people, the real determinant of calcium–phosphate metabolism disorders would be an analysis of bone composition due to the involvement of factors such as vitamin D or parathromon. The level of vitamin D is significantly related to exposure to sunlight, which depends on the geographical location and ethnic origin of the patients.

Another important element that is a component of nervous tissue proteins is magnesium. The role of this element cannot be ignored due to its involvement in the bone mineralization process and its impact on the progression of AS [114]. It is responsible for the proper absorption of calcium in the bones, regulates the transport of this element, activates the ossification process, stimulates bone-forming cells to incorporate calcium into the bone structure, and also prevents pathological calcium deposition in soft tissues [115]. Moreover, by reducing the level of cytokines, i.e., factors that are crucial for pathological changes in the joints in the course of AS, it influences the body's immune response. This element reduces the concentration of inflammatory factors, including TNF-α, interleukins IL-1β, IL-6, and NF-κB [116]. Tumor necrosis factor α stimulates NF-κB to bind to TNF receptors, leading to the activation of a kinase cascade. The above processes lead to the activation of pathological responses triggered in the course of AS [117]. Increased phosphorus concentrations lead to the release of calcium from the bones, thus increasing its content in the blood. This not only disturbs bone mineralization processes, but also leads to the induction of inflammatory reactions resulting from hypercalcemia. Researchers have shown reduced magnesium concentrations in the blood of people with AS [118].

Selenium, on the other hand, optimizes the action of glutathione peroxidase, one of the main enzymatic antioxidants responsible for the transformation of glutathione and combating reactive oxygen species. In case of its deficiency, the body's antioxidant defense decreases, which may result in the development of a number of abnormalities. Its action is based on the metabolism of selenoglutathione. The two-step reaction of selenium reduction to hydride takes place in both the liver and erythrocytes. In the case of red blood cells, this is important because it protects hemoglobin against oxidation, which reduces its affinity for oxygen. Selenium takes part in the control of lipid peroxidation and nucleic acid synthesis processes. Thanks to its ability to penetrate cell membranes, it can induce the production of antibodies [119]. Selenoproteins containing selenocysteine have antioxidant activity, protecting the components of cell membranes and cytosol against reactive oxygen species. Selenium reduces the expression of inflammatory genes and may reduce the ability of matrix metalloproteinases (MMPs) to break down cartilage by inducing tissue inhibitors of MMPs (TIMPs). Selenoprotein P scavenges peroxynitrite, a potent inflammatory factor produced in inflamed joints. Selenium can inhibit neovascularization, which is crucial for the development and perpetuation of rheumatoid synovitis [119,120]. An incorrect concentration of selenium in the blood is an important factor for the osteoarticular system. Deficiency may lead to autoimmune changes in the joints in the course of diseases such as AS or RA [121]. Reduced selenium concentrations in the blood have been observed in patients with autoimmune diseases compared to healthy people. However, the supply of selenium in patients suffering from autoimmune diseases helps to alleviate their course [122]. The reasons for such improvement are believed to be the antioxidant properties of this element. In diseases such as RA, the administration of selenium helps to reduce ailments such as joint pain and swelling [123]. A recent meta-analysis [124] showed a significant difference in serum Se between RA patients and healthy controls. A meta-regression analysis determined that age, gender, and disease duration did not significantly affect serum selenium concentrations. However, such an effect has been demonstrated with the administration of steroids. Their use was positively associated with serum Se concentration. Significant differences in the Se concentration in the serum of RA patients on different continents were found. Hence, there is a suggestion that the relationship between RA and Se results from an insufficient intake of this element [124,125]. Another aspect is the influence of some factors on the bioavailability and activity of Se, such as geographical region, drugs, and genetic polymorphisms. Although clinical trials have been conducted to analyze the effect of Se on patients with RA, the expected results have not been obtained [126–129]. On the other hand, preclinical studies examining the effects of new selenium nanoparticles have yielded favorable results [130–132]. It seems necessary to conduct studies assessing both the concentration of Se in the diet and its concentration in the serum of this population in order to more comprehensively determine whether there is a relationship between the deficiency of this element and the progression of disease activity.

Zinc is one of the most important elements for the natural antioxidant barrier. It is necessary for the formation of superoxide dismutase. Zinc, as a cofactor in over 3000 human proteins and a signal ion, affects many pathways important for AW. It is known that this is not only a matter of zinc status, but also of mutations in many genes encoding proteins that maintain cellular zinc homeostasis, such as metallothioneins and zinc transporters from the ZIP (Zrt/Irt-like protein) and ZnT families. There are also many pathways that change the expression of these proteins [133]. Zinc also has regulatory influence on glutathione peroxidase and metallothionein [134,135]. A deficiency of this element can lead to a weakening of antioxidant defense and the induction of proinflammatory processes. In the case of arthritis, reduced Zn assimilation causes an increase in inflammatory status, immune system changes, extracellular matrix breakdown, Zn-mediated disruption of the Th1/Th2 balance with an increase in Th17, and a loss in PMN phagocytic activity. The mechanisms underlying this are related to an increased secretion of IL-1β, IL-8, and TNF-α and a reduced secretion of IFN-γ and IL-2 [136–138].

There are studies showing no connection between the concentrations of elements and the occurrence of AS. This study provides insufficient evidence to indicate a causal relationship between calcium, zinc, copper, and selenium and AS, however, the authors suggested that more extensive research should be conducted on this matter [139]. According to the literature data, an assessment of the zinc content in the hair of non-smokers and smokers with RA showed significantly reduced zinc concentrations in both groups [140]. Similarly, the total level of zinc in the serum of people with RA was low and negatively correlated with the levels of secreted pro-inflammatory markers [124,141]. A comprehensive review of the pathogenesis of RA from a medical point of view did not mention Zn at all [142]. As has been shown in many studies, zinc is involved in arthritis because it plays an essential role in the innate and adaptive parts of the immune system, in the regulation of various aspects of the inflammatory response, and in bone growth and regeneration [143–145]. The assessment of the zinc content in the hair of non-smokers and smokers with RA showed significantly reduced zinc concentrations in both groups [140]. A large meta-analysis of 16 studies showed a similar correlation for zinc and selenium and an inverse correlation for copper (increased Cu concentration vs. control group) [124].

Cadmium disturbs the functions of zinc, and zinc deficiency causes its increased absorption. Exposure to cadmium through inhalation is considered to be a trigger for the activation of macrophages in a pro-inflammatory state. It is suggested that this initiates and liberates a specific form of RA through the formation of pulmonary nodules [133,146]. Significantly increased cadmium concentrations were observed, among others, in smokers suffering from RA [140]. There is evidence to suggest a link between the development of moderate and severe forms of osteoarthritis and the use of tobacco and the associated higher Cd concentration in the blood serum. Cd may trigger oxidative stress and inflammation, promoting cartilage loss [147]. Research conducted by Shiue [148] suggested a relationship between high antimony, tungsten, cadmium, uranium, and trimethylarsine oxide concentrations in the urine and the severity of pathological changes in the cervical spine of patients with AS. Abnormal levels of these elements and compounds can induce oxidative stress in patients with AS. Cadmium exposure was also proven to be a factor in induction and/or exacerbation mechanisms in the development of autoimmune joint diseases. These mechanisms include oxidative stress, disruptions in biogenic element metabolisms such as zinc and iron, and inflammation [149].

Vanadium also affects the NF-κB signaling pathway, leading to its activation [150]. The above studies confirm the involvement of vanadium in the potential pathogenesis of autoimmune processes underlying diseases such as RA, psoriatic arthritis (PsA), and AS [151].

A summary of elements' influence on the pathomechanisms of AS is shown in the schematic below [Figure 3].

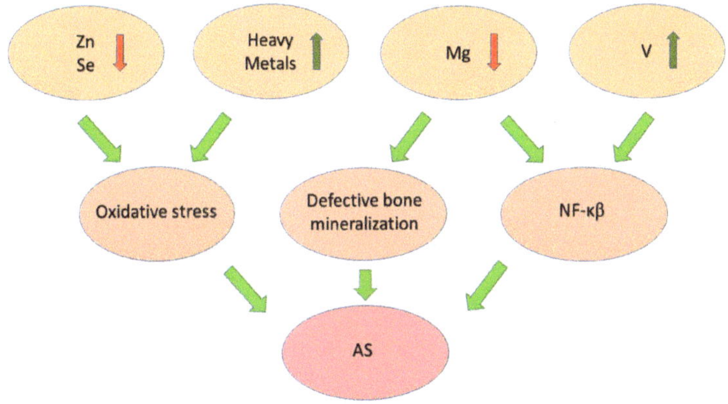

Figure 3. The schematic shows abnormalities in element concentration and their influence on the pathomechanisms underlying AS.

Destabilization in bone mineralization processes in the body results in remodeling of the extracellular matrix, inflammation, and calcification. Hence, calcium, magnesium, and phosphorus levels play significant roles in the pathogenesis of AS. In turn, selenium and zinc are closely related to the functioning of the antioxidant barrier, and, therefore, the neutralization of free radicals, which accumulate in the described disease.

4. Lifestyle as an Environmental Factor in AS Development

Smoking is one of most common addictions related to pathogenesis of diseases. In addition to contributing to the pathophysiology of a number of rheumatic diseases, smoking increases the difficulty of therapy for patients when compared to non-smoking patients [152–154]. According to certain research, smoking can cause rheumatoid arthritis, which carries a higher risk of functional disability due to elevated risk factors, more severe disease activity, and considerable arthrosis damage [155,156]. Cigarette smoking has been linked to the advancement of spinal radiography in patients with early axial spondyloarthritis, according to Poddubnyy et al. [7]. An epidemiological study with 75 AS patients in Taiwan also revealed that smoking AS patients had significantly worse physical mobility parameters than non-smoking AS patients, including modified Schober's index, cervical rotation, later lumbar flexion, chest expansion, and occiput-to-wall distances. Smoking AS patients also had significantly higher systemic inflammation parameters, such as ESR [157]. The study performed by Zhang et al. confirmed that smoking negatively impacts the physical functioning and disease activity of patients with AS [158]. Zhao et al. [159], in a large cross-sectional study, showed unequivocal evidence that both past and current smoking are associated with incrementally worse disease across a wide range of severity measures using a large and well-characterized national cohort of patients with axial spondyloarthropathies. Moreover, it was found that current smoking was associated with lower odds of acute anterior uveitis (AAU) and higher odds of psoriasis than either ex- or never smokers. A recent study showed that smoking can influence the development of AS, leading to particular pathological changes in the hip joints. Exposure to fewer than 10 pack-years of smoke can also increase the prevalence of hip involvement in AS [160]. The summary of smoking influence on AS patients is shown in Table 2.

Table 2. The influence of smoking in AS patients.

Factor	Participation in AS	Population	References
Smoking	Decreased physical mobility parameters including modified Schober's index, cervical rotation, later lumbar flexion, chest expansion, and occiput-to-wall distances. Increased systemic inflammation parameters	Chinese	[157]
	Exposure of no more than ten pack-years may contribute to a higher prevalence of hip involvement in AS	Chinese	[160]
	Increased risk of psoriasis, decreased risk of acute anterior uveitis	United Kingdom	[159]
	Smoking is associated with spinal radiographic progression in patients in first 10 years of the disease	German	[7]

The association between alcohol consumption and AS is currently underreported. A 1998 study based on national sickness insurance data in Finland established that excessive alcohol consumption was a significant factor contributing to the excess number of accidents and violent deaths among Finnish patients with AS. The average lifespan of patients with AS was 6–8 years shorter than the average lifespan of the Finnish population [161]. Zhang et al. reported no statistically significant variation in the analysis of the relationship between alcohol consumption and disease activity [158]. According to Zhao et al.'s findings, drinking alcohol is linked to lower measurements of the severity of axSpA disease compared to abstainers, although the author suggested a wider study on that matter due to the limitations of their study [162]. Research conducted by Min et al. found a substantial correlation between alcohol consumption and the advancement of spinal structural damage in individuals with axial spondyloarthropathies (axSpA). These findings suggest that alcohol consumption may have a deleterious impact on the development of spinal structural damage in individuals with axSpA [163].

Du et al. conducted research regarding the influence of work exposure on ankylosing spondylitis. The study compared the influence of three types of work: working in shifts, heavy lifting work, and mainly standing or walking work. It showed that there is no direct link between AS and the type of work exposure [164]. On the other hand, there are studies showing a link between the disease and physical activity. The two-sample Mendelian randomization proved the casual association between AS and the duration of physical exercise, which has been shown to be protective against its risk [165].

The other important factor that may affect AS patients is the quality of their sleep. Batmaz et al. proved that patients with AS suffer from pain, depression, and a low quality of life, which are correlated with a poor sleep quality [166]. The research conducted by Yüce et al. proved that poor sleep, depressive symptoms, and a poor quality of life might have a negative impact on the diseases [167]. Therefore, sleep quality analysis may be a good indicator to evaluate AS patients [168].

In patients with AS, worse disease outcomes have been linked to long-term exposure to air pollution and a high-fat diet [103]. Diet as an important factor in AS was also proven by Vergne-Salle et al. The study confirmed that an insufficient consumption of omega-3 polyunsaturated fatty acids and fiber was linked to elevated spondyloarthritis (SpA) activity. Additionally, a higher intake of ultra-transformed foods was found to be associated with increased SpA activity in a more comprehensive analysis. These findings support the idea that omega-3 polyunsaturated fatty acids and fiber play a role in decreasing immuno-inflammatory responses and preserving the integrity of the intestinal barrier. However, additional research is needed to fully understand the underlying mechanism of ultra-transformed foods [169]. The researchers Kao et al. stated that exposure to ambient

CO can be an aggravating factor, but exposure to ambient NO_2 can be a protective factor regarding the use of biologics indicated for high AS activity [170].

Research shows that lifestyle has a significant impact on the development of a given disease. For direct health behaviors, lifestyle factors include low physical activity, smoking cigarettes, drinking large amounts of alcohol, or a diet rich in highly processed fats and sugars. Appropriate lifestyle modification is necessary to minimize the risk occurrence of lifestyle diseases, including AS.

5. Infection in AS

Numerous studies have explored AS-related infections, including bacterial [171–174], viral [172,175–177], fungal [178], and microorganisms with sizes between bacteria and viruses [173,179,180]. Infected sites include the respiratory [173,179–181], immunological [176,177,179–182], digestive [171,173,177,179–183], and genitourinary systems [178]. However, there is no unanimity in the link between infections and the risk of AS.

Case–control and cohort investigations have proven that infections are found to increase the risk of AS. These findings are consistent across the majority of investigations, including four Asian studies [174,175,179,182], two European studies [172,183], and one North American study [176]. However, Bartels et al. found the opposite [171], indicating that past *Helicobacter pylori* (*H. pylori*) infection may lessen the likelihood of acquiring this disease. Another study discovered that H. pylori infection was eradicated in more than 80% of cases in the same cohort as Bartels et al. [184], implying that, when H. pylori is eradicated, it leaves a protective potential for the development of AS later in life [171]. Furthermore, the microbiota in the gastrointestinal tract change during H. pylori eradication, which may influence AS development [185,186]. Regarding infection types, we found no evidence that bacterial infections increase the incidence of AS. Only Keller et al.'s case–control study found a relationship between AS and a prior diagnosis of chronic periodontitis, which is defined by an oral bacterial infection [174]. This could be because rheumatic illnesses and chronic periodontitis have similar pathogenic features, such as inflammatory mechanism malfunction and an imbalance of proinflammatory and anti-inflammatory cytokines [174,187–189]. In cohort studies, Bartels et al. found that H. pylori may be a protective factor for AS [171]. Nielsen et al. found that bacterial infections are linked to the development of AS in the general population [172]. AS caused by different forms of infection is not well understood. For example, one study proposed that exposure to Candida albicans could induce AS via a T-cell-driven model of Th17 responses [175]. Another study found that Mycoplasma pneumoniae had a considerable impact on immune cells and the host's immune system, including the polyclonal stimulation of T and B cells and the release of associated cytokines [179], resulting in a breakdown in immunological tolerance. In addition, subgroup analysis revealed that viruses play a significant role in the risk of AS in cohort studies. According to one study, viruses (such as human papillomavirus) may cause inflammatory or immune-mediated disease by activating the pathogenic IL-23/IL-17 axis, which results in higher serum levels of Th17 cells, IL-17, and IL-23, as well as an imbalance of IL-17A/IL-23 cytokines [178]. In a subgroup analysis of infection sites, we discovered that immune system infections were strongly related to the incidence of AS in case–control studies. Some immunological organs, such as the tonsils, play a role in allergen tolerance by producing allergen-specific FOXP3+ regulatory T cells, implying that they are essential in the establishment of immune tolerance [190]. Some investigations have hypothesized that the modification of immunological tolerance in tonsillitis patients can lead to inflammatory problems in autoimmune arthritis, including AS; hence, tonsillitis could be exacerbated by spondylitis, leading to the diagnosis of AS [181,182]. Furthermore, the increased incidence of AS among infected individuals may be explained by HIV-induced antigen-driven immune responses [191], T cell imbalance [192], and molecular mimicry between HIV proteins and self-antigens [193]. In cohort studies, infections at other sites were strongly related to the incidence of AS, indicating that respiratory tract and genitourinary system infections may be the cause. The pathophysiology of AS caused by genitourinary infection is complex.

In one study, it was suggested that human papillomavirus in the genitourinary system may cause AS by activating the IL-23/IL-17 axis [178]. Klebsiella pneumoniae in the respiratory system may cause a decrease in the number of particular T cells, indicating an insufficient host defense against Klebsiella, allowing AS to be impacted by bacterial antigens that reach the joint [173].

In conclusion, this article demonstrates that infections are associated with an elevated risk of AS, despite the substantial heterogeneity of the included studies. While the mechanism of infection and the effect of bacterial and viral infections on this disease have yet to be determined, more studies, particularly higher-quality prospective cohort studies and case–control studies, are needed to confirm a true cause-and-effect relationship between infections and the risk of developing AS.

The available data indicate that, the HLA system, and more specifically its subtypes involved in the pathogenesis of this disease unit, have a large share in AS. CD8+ T lymphocytes are presented with arthretogenic peptides from intestinal microbes in order to induce an immune response. Proteins with an abnormal folding structure then accumulate and increase the production of pro-inflammatory cytokines. While the etiopathogenesis of this unit is not fully known, it indicates a large proportion of infectious agents.

6. Microbiota in AS

Recent research has indicated that the presence and progression of AS are connected to alterations in the variety and composition of the gut microbiome. Zhou et al. [194] discovered, using metagenomic shotgun sequencing, that *Bacteroides coprophilus, Parabacteroides distasonis, Eubacterium siraeum, Acidaminococcus fermentans,* and *Prevotella copri* were more abundant in AS. On the other hand, *Enterococcus faecium* E980 and TX0133a01 were shown to be less abundant. In their study, Liu et al. [195] conducted 16S rRNA gene sequencing on stool samples obtained from both AS patients and healthy controls (HCs). They discovered that the relative abundance of *Bacteroidetes* was lower in AS cases compared to HCs. Conversely, the levels of *Firmicutes* and *Verrucobacterium* were higher in AS cases. In addition, particular gut microorganisms were linked to the disease activity of patients with this disease. Costello et al. [196] discovered higher levels of *Lachnospiriaceae, Ruminococcus, Rikenellaceae, Porphyromonadaceae,* and *Bacteroidaceae* but lower levels of *Veillonellaceae* and *Prevotellaceae* in the terminal ileum of patients with AS. These findings indicate that certain species rich in AS may potentially initiate autoimmunity. The ratio of *Firmicutes* to *Bacteroidetes* (F/B ratio) is commonly linked to the maintenance of a normal intestinal balance, and an elevated presence of certain species of Firmicutes leads to AS [195,197]. Furthermore, the researchers also noted alterations in metabolites both within and outside the gastrointestinal tract in patients with AS. Alterations in intestinal metabolites are intricately linked to the metabolism of intestinal microbes, whereas modifications in extra-intestinal metabolites may also be somewhat associated with the transmission of intestinal microbial metabolites. Several studies have also demonstrated the crucial functions of certain bacteria in the development of AS. The prevalence of Intestinal *Klebsiella* was strongly associated with disease activity in AS [198]. There is a strong correlation between *Klebsiella* antibodies and intestinal inflammation in patients with axial AS [199]. The discovery by Wen et al. showed that *Actinobacteria* are more abundant in patients with AS and may have a role in regulating the ubiquitination of IkB-a. Consequently, this triggers NF-kB signaling and facilitates the accumulation of proinflammatory cytokines in patients with AS [200]. Furthermore, the contact between fungi and bacteria, known as cross-kingdom interactions, may have a role in the advancement of AS. SpA has shown the presence of heightened antibodies against mannan, a component of the fungal cell wall [197].

Genetic variations that affect the functioning of genes, including CARD9 and interleukin (IL)23R, have been discovered. These genes have a role in controlling the body's natural immunological response to fungi [201]. Multiple studies support the notion that the gut microbiota can heighten the likelihood of AS by interacting with HLA-B27. HLA-B27 transgenic rats exhibited SpA-like disease without any external intervention. This disease is dependent on the gut microbiota triggering the IL-23/IL-17 pathway in the inflamed colon and joints [202,203]. Curiously, rats that were genetically modified to have the HLA-B27 gene and grown in a sterile environment did not show any obvious signs of inflammation in their intestines and joints [204]. Conversely, the presence of typical luminal bacteria consistently and uniformly led to persistent inflammation in the colon, stomach, and joints of B27 transgenic rats [205]. This indicates that the participation of bacterial flora is essential for the development of HLA-B27-related diseases.

The IL-23/IL-17 axis plays a crucial role in the development of AS by affecting the immune system. IL-23 primarily induces the secretion of DCs and macrophages, which, in turn, stabilizes the characteristics of T helper 17 (Th17) cells. It also plays a crucial role in the differentiation of different subsets of cells that secrete IL-17 [206,207]. The Th17 cell was formerly believed to be the main producer of IL-17. Recent research has discovered that various types of lymphocytes, including CD8+ cytotoxic T cells, Tc17 cells, γδ T cells, MAIT cells, NK cells, and ILCs cells, are capable of producing significant amounts of IL-17 [208–210]. IL-17 enhances the activation of T cells and induces the secretion of pro-inflammatory cytokines and chemokines by fibroblasts, epithelial cells, endothelial cells, and immune cells such as macrophages [211].

Intestinal inflammation in HLA-B27/β2 microglobulin (β2m) transgenic rats is accompanied by an elevated production of IL-23 and IL-17 in the colon tissue [212]. The interaction between cell surface ligands and particular bacteria triggers the development of cells that secrete IL-17 and IL-22, therefore determining their ability to cause disease [213,214]. Dysbiosis, which is the imbalance of microbial communities, can lead to an overabundance of Prevotella species at mucosal sites. This can have a direct or indirect impact on type 17 immune responses by changing the levels of microbial metabolites or affecting barrier function. As a result, immune tolerance can be disrupted and there can be an increase in pro-inflammatory cytokines like IL-23, which can trigger ankylosing spondylitis in susceptible populations. Prevotella primarily stimulates TLR2, leading to the production of Th17-polarizing cytokines, such as IL-1 and IL-23, by antigen-presenting cells. Prevotella further induces the secretion of IL-6, IL-8, and CCL20 by epithelial cells, thus facilitating the recruitment of neutrophils to the mucosal area. This mechanism results in the widespread spread of bacteria, bacterial metabolites, and inflammatory mediators, which, in turn, affects the outcomes of systemic diseases [215]. These findings indicate that an imbalance of microorganisms in the body leads to alterations in the balance of the immune system in both the intestines and joints, resulting in inflammation. This process is mediated by the IL-23/IL-17 pathway in AS. The bacterial diversity of the gut microbiota affects the balance between Th17 and Treg cells in the lamina propria (LP), which, in turn, can impact gut immunity, tolerance, and susceptibility to AS. Segmented filamentous bacteria (SFB) stimulate the synthesis of serum amyloid A (SAA) in epithelial cells, which, in turn, triggers the production of interleukin-6 (IL-6) and interleukin-23 (IL-23) by dendritic cells (DCs). This ultimately leads to the development of Th17 cells [216]. SFB induces aberrant T cell differentiation in the mouse gut via TLR5 [217]. The differentiation of Th17 cells is also associated with the presence of *Cytophaga-Flavobacterium-Bacteroides* (CFB) bacteria in the gut. This process is not influenced by Toll-like receptors (TLRs), interleukin-21 (IL-21), or interleukin-23 (IL-23) signaling, but it does depend on the appropriate activation of transforming growth factor beta (TGF-β). The absence of Th17 cell stimulation by bacteria is associated with an elevation in Foxp3+ Treg in the lamina propria [218]. Tregs were predominantly found in the colonic mucosa of mice. The guts of AS patients exhibited a prevalence of active Treg responses characterized by the production of IL-10. Additionally, there was a significant decrease in the number of Treg cells in the lamina propria of germ-free mice. However, the lost bacteria can potentially be restored with the addition of specific bacteria, such as *Bacteroides fragilis*, *Clostridium consortium* (particularly Cluster IV and XIVa), and the "altered Scheidler flora" (a combination of eight recognized symbiotic bacteria) [219–221]. Polysaccharide A (PSA), the chemical in *B. fragilis* which affects the immune system, stimulates the conversion of CD4+ T cells into Foxp3 + Treg cells. These cells play an active role in maintaining tolerance in the mucosal lining by generating IL-10 in the presence of normal bacteria [222]. It may be deduced that the immune system's cells detect and are influenced by the metabolic byproducts, which, in turn, impact the equilibrium between pro-inflammatory and anti-inflammatory cells. For example, it has been documented that *C. Consortium* and *B. Fragilis* can stimulate the development of Treg cells by generating short-chain fatty acids (SCFAs) from carbohydrates found in the diet [223]. Recent data suggest that gut microorganisms play a role in the development of AS. The gut microbiome undergoes continuous and fluctuating changes. Hence, specific microorganisms can serve as markers for monitoring disease activity and assessing the efficacy of treatment. To accurately diagnose and treat patients with AS in the future, clinicians may need to perform rapid and thorough analyses of the gut flora [224].

Therefore, there are many studies that indicate that a relationship between the composition of the microbiome and the development of autoimmune diseases, including AS. The bacteria that are found in the intestines can act in two ways: to silence or intensify the immune response. This way, modified cells can migrate from the intestines to the joints, contributing, for example, to their reactive inflammation. From this point of view, lifestyle factors, in particular, a proper diet and the appropriate contribution of micro- and macro-elements, are important.

7. Genetic Polymorphism in HLA System

With 231 protein subtypes, the HLA-B27 family exhibits a high degree of genetic polymorphism. These subtypes differ from one another by only a small number of amino acids, which may change the molecule's ability to bind peptides [3,225].

Human MHC, called the HLA complex, belongs to cell surface proteins operating in the process of acquired immunity. There are three subgroups in the MHC gene family: class I, II, and III. MHC class I encodes HLA-A, HLA-B, and HLA-C and is present on all human nucleated cells and platelets, presenting epitopes to T cell receptors (TCR) on the surface of cytotoxic T lymphocytes [3]. The heterodimeric subgroup of MHC class I consists of a polymorphic heavy chain. The chain contains three domains, i.e., $\alpha1$, $\alpha2$, and $\alpha3$. The $\alpha1$ domain binds non-covalently to the non-MHC $\beta2m$ molecule, while $\alpha3$ spans the cell membrane and interacts with the T cell CD8 coreceptor [226–228]. The MHC class I complex can connect to peptides 8–10 amino acids long through one cleavage separated from both $\alpha1$ and $\alpha2$, which leads to the initiation and propagation of the immune response [229]. A stable MHC molecule must be properly packaged and then folded in the endoplasmic reticulum of cell organelles (ER) under the guidance of chaperone proteins (calreticulin and tapasin) [226]. Although the classical MHC I class contains one heavy chain, there are three distinct MHC-I structures, including cell surface HLA-B27 homodimers and intracellular and exosomal MHC-I dimers [230]. These components may function in various pathophysiological processes. The exact mechanism binding the HLA-B27 antigen and AS has not yet been identified, but it is assumed that the intracellular process of HLA-B27 antigen formation is the subject of further research in this area [231]. Studies have shown that 85% of AS patients are characterized by the presence of the classic HLA-B27 allele. This does not mean, however, that this allele is responsible for causing the disease, because only 1–5% of carriers develop AS [232,233]. The arthritis-causing peptide hypothesis assumes that structurally mutually exclusive peptide–MHC complexes can directly initiate the specific HLA-B27 autoimmune response, relying on the primary structure of antigenic peptides [234]. Some microbial peptides are similar to self-peptides in body tissues and can activate the response of some HLA-B27-specific CD8+ T cells. T cells react with HLA-B27 peptide complexes, leading to autoreactivity and autoimmune disease [235].

HLA-B27 tends to misfold within the endoplasmic reticulum, leading to the formation of dimers and multimeters. Without the proper folding, HLA-B27 would be produced and transported to the cell surface as homodimers containing only heavy chains. The presence of such HLA-B27 variants correlates with the occurrence of AS [236]. Such misfolded proteins accumulate in the endoplasmic reticulum, activating autophagy and the IL-23/IL-17 pathway. They can also affect the functioning of the endoplasmic reticulum, causing a pro-inflammatory response to the presence of misfolded proteins, which, in turn, leads to the activation of the above-mentioned pathway [237]. Another hypothesis suggests that HLA-B27 homodimers are associated with the receptors on NK cells, myelomonocytes and lymphocytes. Binding occurs through the killer cell immunoglobulin-like receptor (KIR) and the leukocyte immunoglobulin-like receptor (LILR). This leads to the release of pro-inflammatory cytokines such as IL-17, TNF-α, and IFN-γ [238]. Although attempts have been made to demonstrate associations between HLAs other than B-27, particularly with HLA-B40 [239–242] and HLA-A02 [243], most of them have not been confirmed or repeated in independent studies. The only confirmed case of an antigen that may be important for the pathomechanism of AS is HLA-B14. The reports do not explain how this antigen contributes to the development of AS, but confirm its occurrence in a population of patients in whom HLA-B27 has not been detected [244,245].

Some researchers look for connections between AS and other genetic polymorphisms. In the study by Remans et al. [246], it was found that the CTLA4 gene has a protective effect against the phenomenon of oxidative stress, which may suggest that the presence of the discussed polymorphism actually reduces the antioxidant potential, but there are no reports in the literature that the polymorphism of the discussed gene has been tested in this aspect. To confirm the conclusions described, studies should be carried out on a larger population of patients with the described polymorphism. The research conducted by Dahmani et al. established a link between ankylosing spondylitis and the CT60/rs3087243 polymorphism of the CTLA4 gene. The study found that the HLA-B27 antigen and variations in the CTLA4 3′UTR region played an essential role in ankylosing spondylitis susceptibility in a west Algerian population. The genetic difference observed between the B27+ and B27− groups explains the disease's variability [247]. STAT4 gene polymorphism is also important for the functioning of antioxidant enzymes in the bodies of patients with RA, PsA, and AS. In research conducted by Liu et al. [248] and Ebrahimiyan et al. [249], the involvement of the T allele of the STAT4 gene polymorphism in the occurrence of rheumatoid diseases such as RA and AS was confirmed. Research by García-Ruiz et al. [250] stated, however, that it is the high level of MDA that stimulates the expression of the COL1A1 gene. However, there is a lack of research in the literature presenting the above phenomena on a broader scale.

In recent years, researchers' interest has focused on attempts to study genes outside the major histocompatibility complex. This seems important, because the etiopathogenesis of this disease is still not fully understood; therefore, it seems that the relationship between genes and AS remains one of the strongest and can be used to determine susceptibility to AS. Examples of such genes have been collected and presented in the Table 3.

Table 3. New genetic polymorphisms' influences on AS.

Polymorphism	Genotype	Influence on AS	Population	References
IL-1A-889 (*rs1800587*)	various	increases the risk of AS in studied populations	English and Tunisian	[251]
IL1F7 exon 2 (*rs3811047*)	G allele	is negatively correlated with susceptibility to AS	Canadian and Chinese	[251]
IL5 (*rs2069812*)	A allele	associated with lower CRP and VAS values	Polish	[252]
IL9 (*rs2069885*)	A allele	associated with lower CRP and VAS values	Polish	[252]
IL17F (*rs763780*)	G allele	should be considered as a promising biomarker of disease activity and anti-TNF treatment outcome	Polish	[253]

Table 3. *Cont.*

Polymorphism	Genotype	Influence on AS	Population	References
IL17RA (*rs48419554*)	G allele	may serve as a potential marker of disease severity in Polish AS patients	Polish	[253]
TLR2 (*rs5743708*)	A allele	haplotypes appear to be involved in the development of clinical forms of SpA and can be a possible therapeutic target for the spondyloarthritis.	Brasil	[254]
TLR9 (*rs187084_rs5743836*)	T/C allele	haplotypes appear to be involved in the development of clinical forms of SpA and can be a possible therapeutic target for the spondyloarthritis.	Brasil	[254]
TIMP3 (*rs11547635*)		may be associated with susceptibility to AS	Chinese	[255]
RUNX3 (*rs760805*)	T allele	can contribute to AS incidence	Chinese	[256]
TNFα (*rs1799724*)	C vs T allele	reduced risk of AS	Asians	[257]
TNFα (*rs1800629*)	G vs A allele	significantly increased the risk of AS in Caucasians and decreased the risk of AS in mixed populations.	Asians, Caucasian	[257]
TNFα (*rs361525*)	G vs A allele, GG vs GA genotype	linked to an elevated AS susceptibility	Asians	[257]
TNFα (*rs1800630*)	C allele	linked to an elevated AS susceptibility	Asians	[257]
TNFAIP3 (*rs10499194*)	T allele, CT genotype	may be associated with a reduced risk of AS.	Chinese Han	[258]

8. Conclusions

Comprehensive medical care for a patient with AS should also include a detailed analysis of lifestyle and factors that may increase oxidative stress, such as smoking or excessive alcohol consumption.

The review carried out in this study indicates the involvement of many factors underlying AS. Factors related to lifestyle should be mentioned here, in which smoking cigarettes comes to the fore and, consequently, the production of free radicals. This is important, because the disruption of the pro- and antioxidant balance triggers a cascade of reactions that produce toxic products. In turn, the inflammatory process that accompanies this disease results in the release of a number of mediators and molecules that change the functions performed by cells, leading to the activation of pathological reactions. Moreover, proteins are secreted, the presence of which indicates an ongoing chronic inflammatory process. Another factor that plays a role in the pathomechanism of AS is exposure to heavy metals and environmental pollutants. The role of genetic factors is also important. There is a relationship between the occurrence of specific polymorphisms and, for example, a reduction in antioxidant potential, which may lead to the development of a condition called oxidative stress, which is related to the pathomechanism of AS. The data contained in this review constitute valuable information and encourage the initiation and development of research in this area.

Author Contributions: Conceptualization, R.B. and P.K.; methodology, R.B. and J.B.; software, R.B.; formal analysis, P.K.; investigation, R.B., D.K. and J.B.; resources, S.J.; data curation, R.B.; writing—original draft preparation, R.B.; writing—review and editing, D.K.; visualization, R.B.; supervision, H.T. and N.K.; project administration, S.J.; funding acquisition, P.K. All authors have read and agreed to the published version of the manuscript.

Funding: This research received no external funding.

Institutional Review Board Statement: Not applicable.

Informed Consent Statement: Not applicable.

Data Availability Statement: Not applicable.

Acknowledgments: Figure created in the Mind the Graph platform: https://mindthegraph.com (accesed on 13 April 2024).

Conflicts of Interest: The authors declare no conflicts of interest.

References

1. Sharip, A.; Kunz, J. Understanding the Pathogenesis of Spondyloarthritis. *Biomolecules* **2020**, *10*, 1461. [CrossRef]
2. Akhondi, H.; Varacallo, M. *Rheumatoid Arthritis And Ankylosing Spondylitis*; StatPearls Publishing: Treasure Island, FL, USA, 2023.
3. Hwang, M.C.; Ridley, L.; Reveille, J.D. Ankylosing spondylitis risk factors: A systematic literature review. *Clin. Rheumatol* **2021**, *40*, 3079–3093. [CrossRef]
4. Liao, H.-T.; Tsai, C.-Y.; Lai, C.-C.; Hsieh, S.-C.; Sun, Y.-S.; Li, K.-J.; Shen, C.-Y.; Wu, C.-H.; Lu, C.-H.; Kuo, Y.-M.; et al. The Potential Role of Genetics, Environmental Factors, and Gut Dysbiosis in the Aberrant Non-Coding RNA Expression to Mediate Inflammation and Osteoclastogenic/Osteogenic Differentiation in Ankylosing Spondylitis. *Front. Cell Dev. Biol.* **2022**, *9*, 748063. [CrossRef]
5. Wordsworth, B.P.; Cohen, C.J.; Davidson, C.; Vecellio, M. Perspectives on the Genetic Associations of Ankylosing Spondylitis. *Front. Immunol.* **2021**, *12*, 603726. [CrossRef]
6. Voruganti, A.; Bowness, P. New developments in our understanding of ankylosing spondylitis pathogenesis. *Immunology* **2020**, *161*, 94–102. [CrossRef]
7. Poddubnyy, D.; Haibel, H.; Listing, J.; Märker-Hermann, E.; Zeidler, H.; Braun, J.; Sieper, J.; Rudwaleit, M. Baseline radiographic damage, elevated acute-phase reactant levels, and cigarette smoking status predict spinal radiographic progression in early axial spondylarthritis. *Arthritis Rheum.* **2012**, *64*, 1388–1398. [CrossRef]
8. de Bruin, F.; de Koning, A.; van den Berg, R.; Baraliakos, X.; Braun, J.; Ramiro, S.; van Gaalen, F.A.; Reijnierse, M.; van der Heijde, D. Development of the CT Syndesmophyte Score (CTSS) in patients with ankylosing spondylitis: Data from the SIAS cohort. *Ann. Rheum. Dis.* **2018**, *77*, 371–377. [CrossRef]
9. Tsukazaki, H.; Kaito, T. The Role of the IL-23/IL-17 Pathway in the Pathogenesis of Spondyloarthritis. *Int. J. Mol. Sci.* **2020**, *21*, 6401. [CrossRef]
10. Baeten, D.; Adamopoulos, I.E. IL-23 Inhibition in Ankylosing Spondylitis: Where Did It Go Wrong? *Front. Immunol.* **2021**, *11*, 623874. [CrossRef]
11. Gravallese, E.M.; Schett, G. Effects of the IL-23-IL-17 pathway on bone in spondyloarthritis. *Nat. Rev. Rheumatol.* **2018**, *14*, 631–640. [CrossRef]
12. Tang, M.; Lu, L.; Yu, X. Interleukin-17A Interweaves the Skeletal and Immune Systems. *Front. Immunol.* **2021**, *11*, 625034. [CrossRef]
13. Vebr, M.; Pomahačová, R.; Sýkora, J.; Schwarz, J. A Narrative Review of Cytokine Networks: Pathophysiological and Therapeutic Implications for Inflammatory Bowel Disease Pathogenesis. *Biomedicines* **2023**, *11*, 3229. [CrossRef]
14. Klasen, C.; Meyer, A.; Wittekind, P.S.; Waqué, I.; Nabhani, S.; Kofler, D.M. Prostaglandin receptor EP4 expression by Th17 cells is associated with high disease activity in ankylosing spondylitis. *Arthritis Res. Ther.* **2019**, *21*, 159. [CrossRef]
15. Verstrepen, L.; Bekaert, T.; Chau, T.L.; Tavernier, J.; Chariot, A.; Beyaert, R. TLR-4, IL-1R and TNF-R signaling to NF-κB: Variations on a common theme. *Cell. Mol. Life Sci.* **2008**, *65*, 2964–2978. [CrossRef]
16. Li, Z.; Guo, J.; Bi, L. Role of the NLRP3 inflammasome in autoimmune diseases. *Biomed. Pharmacother.* **2020**, *130*, 110542. [CrossRef]
17. Coll, R.C.; Schroder, K.; Pelegrín, P. NLRP3 and pyroptosis blockers for treating inflammatory diseases. *Trends Pharmacol. Sci.* **2022**, *43*, 653–668. [CrossRef]
18. Fu, R.; Guo, X.; Pan, Z.; Wang, Y.; Xu, J.; Zhang, L.; Li, J. Molecular mechanisms of AMPK/YAP/NLRP3 signaling pathway affecting the occurrence and development of ankylosing spondylitis. *J. Orthop. Surg. Res.* **2023**, *18*, 831. [CrossRef]
19. Kwon, S.R.; Lim, M.J.; Suh, C.H.; Park, S.G.; Hong, Y.S.; Yoon, B.Y.; Kim, H.A.; Choi, H.J.; Park, W. Dickkopf-1 level is lower in patients with ankylosing spondylitis than in healthy people and is not influenced by anti-tumor necrosis factor therapy. *Rheumatol. Int.* **2012**, *32*, 2523–2527. [CrossRef]

20. Qin, X.; Zhu, B.; Jiang, T.; Tan, J.; Wu, Z.; Yuan, Z.; Zheng, L.; Zhao, J. miR-17-5p Regulates Heterotopic Ossification by Targeting ANKH in Ankylosing Spondylitis. *Mol. Ther. Nucleic Acids* **2019**, *18*, 696–707. [CrossRef]
21. Solmaz, D.; Uslu, S.; Kozacı, D.; Karaca, N.; Bulbul, H.; Tarhan, E.F.; Ozmen, M.; Can, G.; Akar, S. Evaluation of periostin and factors associated with new bone formation in ankylosing spondylitis: Periostin may be associated with the Wnt pathway. *Int. J. Rheum. Dis.* **2018**, *21*, 502–509. [CrossRef]
22. Ramiro, S.; van der Heijde, D.; van Tubergen, A.; Stolwijk, C.; Dougados, M.; van den Bosch, F.; Landewé, R. Higher disease activity leads to more structural damage in the spine in ankylosing spondylitis: 12-year longitudinal data from the OASIS cohort. *Ann. Rheum. Dis.* **2014**, *73*, 1455–1461. [CrossRef]
23. Łosińska, K.; Korkosz, M.; Kwaśny-Krochin, B. Endothelial Dysfunction in Patients with Ankylosing Spondylitis. *Rheumatology* **2019**, *57*, 100–105. [CrossRef]
24. Le, T.H.V.; Kwon, S.-M. Vascular Endothelial Growth Factor Biology and Its Potential as a Therapeutic Target in Rheumatic Diseases. *Int. J. Mol. Sci.* **2021**, *22*, 5387. [CrossRef]
25. Yamamoto, T. Angiogenic and Inflammatory Properties of Psoriatic Arthritis. *Int. Sch. Res. Not.* **2013**, *2013*, 630620. [CrossRef]
26. Fearon, U.; Griosios, K.; Fraser, A.; Reece, R.; Emery, P.; Jones, P.F.; Veale, D.J. Angiopoietins, growth factors, and vascular morphology in early arthritis. *J. Rheumatol.* **2003**, *30*, 260–268.
27. Liu, K.G.; He, Q.H.; Tan, J.W.; Liao, G.J. Expression of TNF-α, VEGF, and MMP-3 mRNAs in Synovial Tissues and Their Roles in Fibroblast-Mediated Osteogenesis in Ankylosing Spondylitis. *Genet. Mol. Res.* **2015**, *14*, 6852–6858. [CrossRef]
28. Kawashima, M.; Fujikawa, Y.; Itonaga, I.; Takita, C.; Tsumura, H. The Effect of Selective Cyclooxygenase-2 Inhibitor on Human Osteoclast Precursors to Influence Osteoclastogenesis In Vitro. *Mod. Rheumatol.* **2009**, *19*, 192–198. [CrossRef]
29. Drouart, M.; Saas, P.; Billot, M.; Cedoz, J.-P.; Tiberghien, P.; Wendling, D.; Toussirot, É. High Serum Vascular Endothelial Growth Factor Correlates with Disease Activity of Spondylarthropathies. *Clin. Exp. Immunol.* **2003**, *132*, 158–162. [CrossRef]
30. Liu, Z.; Cai, M.; Ke, H.; Deng, H.; Ye, W.; Wang, T.; Chen, Q.; Cen, S. Fibroblast Insights into the Pathogenesis of An-kylosing Spondylitis. *J. Inflamm. Res.* **2023**, *16*, 6301–6317. [CrossRef]
31. Sies, H. Oxidative Stress: Concept and Some Practical Aspects. *Antioxidants* **2020**, *9*, 852. [CrossRef]
32. Betteridge, D.J. What Is Oxidative Stress? *Metab. Clin. Exp.* **2000**, *49*, 3–8. [CrossRef] [PubMed]
33. Kiranatlioglu-Firat, F.; Demir, H.; Cuce, I.; Altın-Celik, P.; Eciroglu, H.; Bayram, F.; Donmez-Altuntas, H. Increased Oxidative and Chromosomal DNA Damage in Patients with Ankylosing Spondylitis: Its Role in Pathogenesis. *Clin. Exp. Med.* **2023**, *23*, 1721–1728. [CrossRef] [PubMed]
34. Kozaci, L.D.; Sari, I.; Alacacioglu, A.; Akar, S.; Akkoc, N. Evaluation of inflammation and oxidative stress in ankylosing spondylitis: A role for macrophage migration inhibitory factor. *Mod. Rheumatol.* **2010**, *20*, 34–39. [CrossRef] [PubMed]
35. Solmaz, D.; Kozacı, D.; Sarı, İ.; Taylan, A.; Önen, F.; Akkoç, N.; Akar, S. Oxidative stress and related factors in patients with ankylosing spondylitis. *Eur. J. Rheumatol.* **2016**, *3*, 20–24. [CrossRef] [PubMed]
36. Alfadda, A.A.; Sallam, R.M. Reactive oxygen species in health and disease. *J. Biotechnol. Biomed.* **2012**, *2012*, 936486. [CrossRef] [PubMed]
37. Belenguer-Varea, Á.; Tarazona-Santabalbina, F.J.; Avellana-Zaragoza, J.A.; Martínez-Reig, M.; Mas-Bargues, C.; Inglés, M. Oxidative stress and exceptional human longevity: Systematic review. *Free Radic. Biol. Med.* **2020**, *149*, 51–63. [CrossRef] [PubMed]
38. Ezzat, S.; Louka, M.L.; Zakaria, Z.M.; Nagaty, M.M.; Metwaly, R.G. Autophagy in osteoporosis: Relation to oxidative stress. *J. Cell. Biochem.* **2018**, *120*, 2560–2568. [CrossRef] [PubMed]
39. Ansari, M.Y.; Ahmad, N.; Haqqi, T.M. Oxidative stress and inflammation in osteoarthritis pathogenesis: Role of polyphenols. *Biomed. Pharmacother.* **2020**, *129*, 110452. [CrossRef] [PubMed]
40. Veen, L.; Hantikainen, E.; Bellocco, R.; Ye, W.; Serafini, M.; Ponzano, M.; Grotta, A.; Trolle Lagerros, Y. Dietary antioxidants, non-enzymatic antioxidant capacity and the risk of osteoarthritis in the Swedish National March Cohort. *Eur. J. Nutr.* **2021**, *60*, 169–178. [CrossRef]
41. Wang, Y.N.; Miao, H.; Yu, X.Y.; Guo, Y.; Su, W.; Liu, F.; Cao, G.; Zhao, Y.Y. Oxidative stress and inflammation are mediated via aryl hydrocarbon receptor signalling in idiopathic membranous nephropathy. *Free Radic. Biol. Med.* **2023**, *207*, 89–106. [CrossRef]
42. Mussbacher, M.; Salzmann, M.; Brostjan, C.; Hoesel, B.; Schoergenhofer, C.; Datler, H.; Hohensinner, P.; Basílio, J.; Petzelbauer, P.; Assinger, A.; et al. Cell Type-Specific Roles of NF-κB Linking Inflammation and Thrombosis. *Front. Immunol.* **2019**, *10*, 85. [CrossRef]
43. Vatner, S.F.; Zhang, J.; Oydanich, M.; Berkman, T.; Naftalovich, R.; Vatner, D.E. Healthful aging mediated by inhibition of oxidative stress. *Ageing Res. Rev.* **2020**, *64*, 101194. [CrossRef] [PubMed]
44. Henrotin, Y.; Kurz, B.; Aigner, T. Oxygen and reactive oxygen species in cartilage degradation: Friends or foes? *Osteoarthr. Cartil.* **2005**, *13*, 643–654. [CrossRef]
45. Karakoc, M.; Altindag, O.; Keles, H.; Soran, N.; Selek, S. Serum oxidative-antioxidative status in patients with ankylosing spondilitis. *Rheumatol. Int.* **2007**, *27*, 1131–1134. [CrossRef]
46. Łuszczewski, A.; Matyska-Piekarska, E.; Trefler, J.; Wawer, I.; Łącki, J.; Śliwińska-Stańczyk, P. Review article Reactive oxygen species—Physiological and pathological function in the human body. *Rheumatology* **2007**, *45*, 284–289.
47. Kabuyama, Y.; Kitamura, T.; Yamaki, J.; Homma, M.K.; Kikuchi, S.; Homma, Y. Involvement of thioredoxin reductase 1 in the regulation of redox balance and viability of rheumatoid synovial cells. *Biochem. Biophys. Res.* **2008**, *367*, 491–496. [CrossRef]

48. Farsi, F.; Heshmati, J.; Keshtkar, A.; Irandoost, P.; Alamdari, N.M.; Akbari, A.; Janani, L.; Morshedzadeh, N.; Vafa, M. Can coenzyme Q10 supplementation effectively reduce human tumor necrosis factor-α and interleukin-6 levels in chronic inflammatory diseases? A systematic review and meta-analysis of randomized controlled trials. *Pharmacol. Res.* **2019**, *148*, 104290. [CrossRef] [PubMed]
49. Bushnell, G.W.; Louie, G.V.; Brayer, G.D. High-resolution three-dimensional structure of horse heart cytochrome c. *J. Mol. Biol.* **1990**, *214*, 585–595. [CrossRef] [PubMed]
50. Patriarca, A.; Eliseo, T.; Sinibaldi, F.; Piro, M.C.; Melis, R.; Paci, M.; Cicero, D.O.; Polticelli, F.; Santucci, R.; Fiorucci, L. ATP acts as a regulatory effector in modulating structural transitions of cytochrome c: Implications for apoptotic activity. *Biochemistry* **2009**, *48*, 3279–3287. [CrossRef]
51. Yadav, N.; Gogada, R.; O'Malley, J.; Gundampati, R.K.; Jayanthi, S.; Hashmi, S.; Lella, R.; Zhang, D.; Wang, J.; Kumar, R.; et al. Molecular insights on cytochrome c and nucleotide regulation of apoptosome function and its implication in cancer. *Biochim. Biophys. Acta (BBA)-Mol. Cell Res.* **2020**, *1867*, 118573. [CrossRef]
52. Hannibal, L.; Tomasina, F.; Capdevila, D.A.; Demicheli, V.; Tórtora, V.; Alvarez-Paggi, D.; Jemmerson, R.; Murgida, D.H.; Radi, R. Alternative Conformations of Cytochrome c: Structure, Function, and Detection. *Biochemistry* **2016**, *55*, 407–428. [CrossRef]
53. Santucci, R.; Sinibaldi, F.; Polticelli, F.; Fiorucci, L. Role of cardiolipin in mitochondrial diseases and apoptosis. *Curr. Med. Chem.* **2014**, *21*, 2702–2714. [CrossRef] [PubMed]
54. Jemmerson, R.; Staskus, K.; Higgins, L.; Conklin, K.; Kelekar, A. Intracellular leucine-rich alpha-2-glycoprotein-1 competes with Apaf-1 for binding cytochrome c in protecting MCF-7 breast cancer cells from apoptosis. *Apoptosis* **2021**, *26*, 71–82. [CrossRef]
55. Tang, M.; Li, G.; Chen, L.; Tu, J. Application of Cytochrome C-Related Genes in Prognosis and Treatment Prediction of Lung Adenocarcinoma. *Dis. Markers* **2022**, *2022*, 8809956. [CrossRef]
56. Patriarca, A.; Polticelli, F.; Piro, M.C.; Sinibaldi, F.; Mei, G.; Bari, M.; Santucci, R.; Fiorucci, L. Conversion of cytochrome c into a peroxidase: Inhibitory mechanisms and implication for neurodegenerative diseases. *Arch. Biochem. Biophys.* **2012**, *522*, 62–69. [CrossRef]
57. Hamanaka, R.B.; Chandel, N.S. Mitochondrial reactive oxygen species regulate hypoxic signaling. *Curr. Opin. Cell Biol.* **2009**, *21*, 894–899. [CrossRef]
58. Kretz-Rommel, A.; Rubin, R.L. Early cellular events in systemic autoimmunity driven by chromatin-reactive T cells. *Cell. Immunol.* **2001**, *208*, 125–136. [CrossRef] [PubMed]
59. Omori, E.; Morioka, S.; Matsumoto, K.; Ninomiya-Tsuji, J. TAK1 regulates reactive oxygen species and cell death in keratinocytes, which is essential for skin integrity. *J. Biol. Chem.* **2008**, *283*, 26161–26168. [CrossRef]
60. Shi, Q.; Yang, X.; Mattes, W.B.; Mendrick, D.L.; Harrill, A.H.; Beger, R.D. Circulating mitochondrial biomarkers for drug-induced liver injury. *Biomark. Med.* **2015**, *9*, 1215–1223. [CrossRef] [PubMed]
61. Rashmi, R.; Rao, K.S.; Basavaraj, K.H. A comprehensive review of biomarkers in psoriasis. *Clin. Exp. Dermatol.* **2009**, *34*, 658–663. [CrossRef]
62. Gladman, D.D.; Antoni, C.; Mease, P.; Clegg, D.O.; Nash, P. Psoriatic arthritis: Epidemiology, clinical features, course, and outcome. *Ann. Rheum. Dis.* **2005**, *64*, ii14–ii17. [CrossRef] [PubMed]
63. Fitzgerald, O.; Dougados, M. Psoriatic arthritis: One or more diseases? Best practice & research. *Clin. Rheumatol.* **2006**, *20*, 435–450.
64. Sommerhoff, C.P.; Bode, W.; Pereira, P.J.; Stubbs, M.T.; Stürzebecher, J.; Piechottka, G.P.; Matschiner, G.; Bergner, A. The structure of the human betaII-tryptase tetramer: Fo(u)r better or worse. *Proc. Natl. Acad. Sci. USA* **1999**, *96*, 10984–10991. [CrossRef] [PubMed]
65. Chimenti, M.S.; Triggianese, P.; Botti, E.; Narcisi, A.; Conigliaro, P.; Giunta, A.; Teoli, M.; Perricone, R.; Costanzo, A. S100A8/A9 in psoriatic plaques from patients with psoriatic arthritis. *Int. J. Med. Res.* **2016**, *44*, 33–37. [CrossRef]
66. Couto, D.; Ribeiro, D.; Freitas, M.; Gomes, A.; Lima, J.L.; Fernandes, E. Scavenging of reactive oxygen and nitrogen species by the prodrug sulfasalazine and its metabolites 5-aminosalicylic acid and sulfapyridine. *Redox Rep.* **2010**, *15*, 259–267. [CrossRef]
67. Ozgocmen, S.; Sogut, S.; Ardicoglu, O.; Fadillioglu, E.; Pekkutucu, I.; Akyol, O. Serum nitric oxide, catalase, superoxide dismutase, and malondialdehyde status in patients with ankylosing spondylitis. *Rheumatol. Int.* **2004**, *24*, 80–83. [CrossRef]
68. Yazici, C.; Köse, K.; Calis, M.; Kuzugüden, S.; Kirnap, M. Protein oxidation status in patients with ankylosing spondylitis. *Rheumatology* **2004**, *43*, 1235–1239. [CrossRef] [PubMed]
69. Hassan, S.Z.; Gheita, T.A.; Kenawy, S.A.; Fahim, A.T.; El-Sorougy, I.M.; Abdou, M.S. Oxidative stress in systemic lupus erythematosus and rheumatoid arthritis patients: Relationship to disease manifestations and activity. *Int. J. Rheum. Dis.* **2011**, *14*, 325–331. [CrossRef] [PubMed]
70. Ye, G.; Xie, Z.; Zeng, H.; Wang, P.; Li, J.; Zheng, G.; Wang, S.; Cao, Q.; Li, M.; Liu, W.; et al. Oxidative stress-mediated mitochondrial dysfunction facilitates mesenchymal stem cell senescence in ankylosing spondylitis. *Cell Death Dis.* **2020**, *11*, 775. [CrossRef]
71. Phaniendra, A.; Jestadi, D.B.; Periyasamy, L. Free radicals: Properties, sources, targets, and their implication in various diseases. *Indian J. Clin. Biochem.* **2015**, *30*, 11–26. [CrossRef]
72. Ryter, S.W.; Kim, H.P.; Hoetzel, A.; Park, J.W.; Nakahira, K.; Wang, X.; Choi, A.M. Mechanisms of cell death in oxidative stress. *Antioxid. Redox Signal.* **2007**, *9*, 49–89. [CrossRef] [PubMed]
73. Sardesai, V.M. Role of antioxidants in health maintenance. *Nutr. Clin. Pract.* **1995**, *10*, 19–25. [CrossRef] [PubMed]

74. Li, J.; Liu, S.; Cui, Y. Oxidative and Antioxidative Stress Linked Biomarkers in Ankylosing Spondylitis: A Systematic Review and Meta-analysis. *Oxidative Med. Cell. Longev.* **2020**, *2020*, 4759451. [CrossRef]
75. Surh, Y.J.; Kundu, J.K.; Na, H.K.; Lee, J.S. Redox-sensitive transcription factors as prime targets for chemoprevention with anti-inflammatory and antioxidative phytochemicals. *J. Nutr.* **2005**, *135*, 2993–3001. [CrossRef] [PubMed]
76. Stanek, A.; Cieślar, G.; Romuk, E.; Kasperczyk, S.; Sieroń-Stołtny, K.; Birkner, E.; Sieroń, A. Decrease in antioxidant status of plasma and erythrocytes from patients with ankylosing spondylitis. *Clin. Biochem.* **2010**, *43*, 566–570. [CrossRef] [PubMed]
77. Pishgahi, A.; Abolhasan, R.; Danaii, S.; Amanifar, B.; Soltani-Zangbar, M.S.; Zamani, M.; Kamrani, A.; Ghorbani, F.; Mehdizadeh, A.; Kafil, H.S.; et al. Immunological and oxidative stress biomarkers in Ankylosing Spondylitis patients with or without metabolic syndrome. *Cytokine* **2020**, *128*, 155002. [CrossRef] [PubMed]
78. Danaii, S.; Abolhasani, R.; Soltani-Zangbar, M.S.; Zamani, M.; Mehdizadeh, A.; Amanifar, B.; Yousefi, B.; Nazari, M.; Pourlak, T.; Hajialiloo, M.; et al. Oxidative stress and immunological biomarkers in Ankylosing spondylitis patients. *Gene Rep.* **2020**, *18*, 100574. [CrossRef]
79. Adıgüzel, T.K.; Yurdakul, F.G.; Kürklü, N.S.; Yaşar, E.; Bodur, H. Relationship between diet, oxidative stress, and inflammation in ankylosing spondylitis. *Arch. Rheumatol.* **2021**, *37*, 1–10. [CrossRef]
80. Jayson, M.I.; Davis, P.; Whicher, J.T.; Walters, G. Serum copper and caeruloplasmin in ankylosing spondylitis, systemic sclerosis, and morphea. *Ann. Rheum. Dis.* **1975**, *35*, 443–445. [CrossRef]
81. Aiginger, P.; Kolarz, G.; Willvonseder, R. Copper in ankylosing spondylitis and rheumatoid arthritis. *Scand. J. Rheumatol.* **1978**, *7*, 75–78. [CrossRef]
82. Li, T.; Huang, Z.; Zheng, B.; Liao, Z.; Zhao, L.; Gu, J. Serum disease-associated proteins of ankylosing spondylitis: Results of a preliminary study by comparative proteomics. *Clin. Exp. Rheumatol.* **2010**, *28*, 201–207.
83. Salmón, P.; Watson, H.; Nord, A.; Isaksson, C. Effects of the Urban Environment on Oxidative Stress in Early Life: Insights from a Cross-fostering Experiment. *Integr. Comp. Biol.* **2018**, *58*, 986–994. [CrossRef] [PubMed]
84. Damavandi, D.R.; Mousavi, S.N.; Shidfar, F.; Mohammadi, V.; Rajab, A.; Hosseini, S.; Heshmati, J. Effects of Daily Consumption of Cashews on Oxidative Stress and Atherogenic Indices in Patients with Type 2 Diabetes: A Randomized, Controlled-Feeding Trial. *Int. J. Endocrinol. Metab.* **2019**, *17*, e70744. [CrossRef] [PubMed]
85. Vanaki, N.; Aslani, S.; Jamshidi, A.; Mahmoudi, M. Role of innate immune system in the pathogenesis of ankylosing spondylitis. *Biomed. Pharmacother.* **2018**, *105*, 130–143. [CrossRef] [PubMed]
86. Ayala, M.; Muroz, F.; Arguelles, S. Lipid peroxidation: Production, metabolism, and signaling mechanisms of malondialdehyde and 4-hydroxy-2-nonenal. *Oxidative Med. Cell. Longev.* **2014**, *2014*, 360438. [CrossRef]
87. Türkön, H.; Gökmen, F.; Çakir, D.Ü.; Sehitoğlu, M.H.; Reşorlu, H.; Döner, D.; Eşsizoğlu, E.; Akbal, A.; Kizilkaya, B.; Uysal, S. Increased Levels of Serum Ischemia Modified Albumin in Patients with Ankylosing Spondylitis. *Clin. Lab.* **2016**, *62*, 645–649. [CrossRef]
88. Nazıroğlu, M.; Akkuş, S.; Celik, H. Levels of lipid peroxidation and antioxidant vitamins in plasma and erythrocytes of patients with ankylosing spondylitis. *Clin. Biochem.* **2011**, *44*, 1412–1415. [CrossRef]
89. Feng, X.; Yang, Q.; Wang, C.; Tong, W.; Xu, W. Punicalagin Exerts Protective Effects against Ankylosing Spondylitis by Regulating NF-κB-TH17/JAK2/STAT3 Signaling and Oxidative Stress. *BioMed Res.* **2020**, *2020*, 4918239. [CrossRef]
90. Ruef, J.; Hu, Z.Y.; Yin, L.-Y.; Wu, Y.; Hanson, S.R.; Kelly, A.B.; Harker, L.A.; Rao, G.N.; Runge, M.S.; Patterson, C. Induction of Vascular Endothelial Growth Factor in Balloon-Injured Baboon Arteries. *Circ. Res.* **1997**, *81*, 24–33. [CrossRef]
91. Knighton, D.R.; Hunt, T.K.; Scheuenstuhl, H.; Halliday, B.J.; Werb, Z.; Banda, M.J. Oxygen Tension Regulates the Expression of Angiogenesis Factor by Macrophages. *Science* **1983**, *221*, 1283–1285. [CrossRef]
92. Chua, C.C.; Hamdy, R.C.; Chua, B.H.L. Upregulation of Vascular Endothelial Growth Factor by H_2O_2 in Rat Heart Endothelial Cells. *Free Radic. Biol. Med.* **1998**, *25*, 891–897. [CrossRef] [PubMed]
93. Wang, Y.; Zang, Q.S.; Liu, Z.; Wu, Q.; Maass, D.; Dulan, G.; Shaul, P.W.; Melito, L.; Frantz, D.E.; Kilgore, J.A.; et al. Regulation of VEGF-Induced Endothelial Cell Migration by Mitochondrial Reactive Oxygen Species. *Am. J. Physiol. Cell Physiol.* **2011**, *301*, C695–C704. [CrossRef] [PubMed]
94. Balbir-Gurman, A.; Fuhrman, B.; Braun-Moscovici, Y.; Markovits, D.; Aviram, M. Consumption of pomegranate decreases serum oxidative stress and reduces disease activity in patients with active rheumatoid arthritis: A pilot study. *Isr. Med. Assoc. J.* **2011**, *13*, 474–479. [PubMed]
95. Balogh, E.; Veale, D.J.; McGarry, T.; Orr, C.; Szekanecz, Z.; Ng, C.-T.; Fearon, U.; Biniecka, M. Oxidative Stress Impairs Energy Metabolism in Primary Cells and Synovial Tissue of Patients with Rheumatoid Arthritis. *Arthritis Res. Ther.* **2018**, *20*, 95. [CrossRef] [PubMed]
96. Khan, S.; Yousaf, M.J.; Rashid, A.; Majeed, A.; Javed, A. Comparison of oxidative stress, lipid peroxidation and inflammatory markers between rheumatoid arthritis and ankylosing spondylitis patients. *JPMA. J. Pak. Med. Assoc.* **2024**, *74*, 886–890. [CrossRef] [PubMed]
97. Wang, L.; Gao, L.; Jin, D.; Wang, P.; Yang, B.; Deng, W.; Xie, Z.; Tang, Y.; Wu, Y.; Shen, H. The Relationship of Bone Mineral Density to Oxidant/Antioxidant Status and Inflammatory and Bone Turnover Markers in a Multicenter Cross-Sectional Study of Young Men with Ankylosing Spondylitis. *Calcif. Tissue Int.* **2015**, *97*, 12–22. [CrossRef] [PubMed]

98. Stanek, A.; Cholewka, A.; Wielkoszyński, T.; Romuk, E.; Sieroń, K.; Sieroń, A. Increased Levels of Oxidative Stress Markers, Soluble CD40 Ligand, and Carotid Intima-Media Thickness Reflect Acceleration of Atherosclerosis in Male Patients with Ankylosing Spondylitis in Active Phase and without the Classical Cardiovascular Risk Factors. *Oxidative Med. Cell. Longev.* **2017**, *2017*, 9712536. [CrossRef] [PubMed]
99. Jeka, S.; Murawska, A. Ultrasonography of synovium in rheumatological diseases. *Reumatologia* **2009**, *47*, 339–343.
100. Jeka, S. Soft tissue rheumatism. *Reumatologia* **2012**, *50*, 166–176.
101. Kamiński, P.; Wiśniewska, A.; Jerzak, L.; Żuchowski, P.; Jeka, S.; Kavanagh, B.P.; Kozera, W.; Woźniak, A. Ecophysiological Determinants of the Human Skeletal System. *Med. Res. J.* **2018**, *3*, 47–54. [CrossRef]
102. Somers, E.C.; Richardson, B.C. Environmental exposures, epigenetic changes and the risk of lupus. *Lupus* **2014**, *23*, 568–576. [CrossRef] [PubMed]
103. Soleimanifar, N.; Nicknam, M.H.; Bidad, K.; Jamshidi, A.R.; Mahmoudi, M.; Mostafaei, S.; Hosseini-Khah, Z.; Nikbin, B. Effect of food intake and ambient air pollution exposure on ankylosing spondylitis disease activity. *Adv. Rheumatol.* **2019**, *59*, 9. [CrossRef] [PubMed]
104. Wang, M.; Fu, W.; Meng, L.; Liu, J.; Wu, L.; Peng, Y.; Li, Z. SWE and SMI ultrasound techniques for monitoring needling treatment of ankylosing spondylitis: Study protocol for a single-blinded randomized controlled trial. *Trials* **2021**, *22*, 385. [CrossRef] [PubMed]
105. Huybers, S.; Apostolaki, M.; van der Eerden, B.C.; Kollias, G.; Naber, T.H.; Bindels, R.J.; Hoenderop, J.G. Murine $TNF^{\Delta ARE}$ Crohn's disease model displays diminished expression of intestinal Ca^{2+} transporters. *Inflamm. Bowel Dis.* **2008**, *14*, 803–811. [CrossRef] [PubMed]
106. Pinto, M.C.; Kihara, A.H.; Goulart, V.A.; Tonelli, F.M.; Gomes, K.N.; Ulrich, H.; Resende, R.R. Calcium signaling and cell proliferation. *Cell. Signal.* **2015**, *27*, 2139–2149. [CrossRef] [PubMed]
107. Hendy, G.N.; Canaff, L. Calcium-sensing receptor, proinflammatory cytokines and calcium homeostasis. *Semin. Cell Dev. Biol.* **2016**, *49*, 37–43. [CrossRef] [PubMed]
108. Ugan, Y.; Nazıroğlu, M.; Şahin, M.; Aykur, M. Anti-tumor Necrosis Factor Alpha (Infliximab) Attenuates Apoptosis, Oxidative Stress, and Calcium Ion Entry Through Modulation of Cation Channels in Neutrophils of Patients with Ankylosing Spondylitis. *J. Membr. Biol.* **2016**, *249*, 437–447. [CrossRef] [PubMed]
109. Toldi, G.; Munoz, L.; Herrmann, M.; Schett, G.; Balog, A. The effects of Kv1.3 and IKCa1 channel inhibition on cytokine production and calcium influx of T lymphocytes in rheumatoid arthritis and ankylosing spondylitis. *Immunol. Res.* **2016**, *64*, 627–631. [CrossRef] [PubMed]
110. Staud, R. Vitamin D: More than Just Affecting Calcium and Bone. *Curr. Rheumatol. Rep.* **2005**, *7*, 356–364. [CrossRef]
111. Chen, M.; Li, W.; Li, L.; Chai, Y.; Yang, Y.; Pu, X. Ankylosing Spondylitis Disease Activity and Serum Vitamin D Levels: A Systematic Review and Meta-Analysis. *Medicine* **2022**, *101*, e31764. [CrossRef]
112. Diao, M.; Peng, J.; Wang, D.; Wang, H. Peripheral Vitamin D Levels in Ankylosing Spondylitis: A Systematic Review and Meta-Analysis. *Front. Med.* **2022**, *9*, 972586. [CrossRef] [PubMed]
113. Jiang, J.; Shao, M.; Wu, X. Vitamin D and Risk of Ankylosing Spondylitis: A Two-Sample Mendelian Randomization Study. *Hum. Immunol.* **2022**, *83*, 81–85. [CrossRef] [PubMed]
114. Lee, J.S.; Lee, E.J.; Lee, J.H.; Hong, S.C.; Lee, C.K.; Yoo, B.; Oh, J.S.; Lee, S.H.; Kim, T.J.; Lee, S.H.; et al. Autoantibodies against Protein Phosphatase Magnesium-Dependent 1A as a Biomarker for Predicting Radiographic Progression in Ankylosing Spondylitis Treated with Anti-Tumor Necrosis Factor Agents. *J. Clin. Med.* **2020**, *9*, 3968. [CrossRef] [PubMed]
115. Zhang, J.; Tang, L.; Qi, H.; Zhao, Q.; Liu, Y.; Zhang, Y. Dual Function of Magnesium in Bone Biomineralization. *Adv. Healthc. Mater.* **2019**, *8*, e1901030. [CrossRef] [PubMed]
116. Brenner, M.; Laragione, T.; Gulko, P.S. Short-term low-magnesium diet reduces autoimmune arthritis severity and synovial tissue gene expression. *Physiol. Genom.* **2017**, *49*, 238–242. [CrossRef] [PubMed]
117. Sode, J.; Bank, S.; Vogel, U.; Andersen, P.S.; Sørensen, S.B.; Bojesen, A.B.; Andersen, M.R.; Brandslund, I.; Dessau, R.B.; Hoffmann, H.J.; et al. Genetically determined high activities of the TNF-alpha, IL23/IL17, and NFkB pathways were associated with increased risk of ankylosing spondylitis. *BMC Med. Genet.* **2018**, *19*, 165. [CrossRef] [PubMed]
118. Gönüllü, E.; Bilge, N.Ş.Y.; Cansu, D.U.; Bekmez, M.; Musmul, A.; Akçar, N.; Kaşifoğlu, T.; Korkmaz, C. Risk factors for urolithiasis in patients with ankylosing spondylitis: A prospective case-control study. *Urolithiasis* **2017**, *45*, 353–357. [CrossRef] [PubMed]
119. Kabata-Pendias, A.; Szteke, B. *Trace Elements in Abiotic and Biotic Environments*, 1st ed.; CRC Press, Taylor & Francis LLC: Boca Raton, FL, USA, 2015; pp. 281–293.
120. Rayman, M.P.; Pattison, D.J. Dietary Manipulation in Musculoskeletal Conditions. *Best Pract. Res. Clin. Rheumatol.* **2008**, *22*, 535–561. [CrossRef]
121. Deyab, G.; Hokstad, I.; Aaseth, J.; Småstuen, M.C.; Whist, J.E.; Agewall, S.; Lyberg, T.; Tveiten, D.; Hjeltnes, G.; Zibara, K.; et al. Effect of anti-rheumatic treatment on selenium levels in inflammatory arthritis. *J. Trace Elem. Med. Biol.* **2018**, *49*, 91–97. [CrossRef]
122. Sahebari, M.; Rezaieyazdi, Z.; Khodashahi, M. Selenium and Autoimmune Diseases: A Review Article. *Curr. Rheumatol. Rev.* **2019**, *15*, 123–134. [CrossRef]
123. Qamar, N.; John, P.; Bhatti, A. Emerging role of selenium in treatment of rheumatoid arthritis: An insight on its antioxidant properties. *J. Trace Elem. Med. Biol.* **2021**, *66*, 126737. [CrossRef] [PubMed]

124. Ma, Y.; Zhang, X.; Fan, D.; Xia, Q.; Wang, M.; Pan, F. Common Trace Metals in Rheumatoid Arthritis: A Systematic Review and Meta-Analysis. *J. Trace Elem. Med. Biol.* **2019**, *56*, 81–89. [CrossRef] [PubMed]
125. Turrubiates-Hernández, F.J.; Márquez-Sandoval, Y.F.; González-Estevez, G.; Reyes-Castillo, Z.; Muñoz-Valle, J.F. The Relevance of Selenium Status in Rheumatoid Arthritis. *Nutrients* **2020**, *12*, 3007. [CrossRef] [PubMed]
126. Tarp, U.; Overvad, K.; Thorling, E.B.; Graudal, H.; Hansen, J.C. Selenium Treatment in Rheumatoid Arthritis. *Scand. J. Rheumatol.* **1985**, *14*, 364–368. [CrossRef] [PubMed]
127. Jäntti, J.; Vapaatalo, H.; Seppala, E.; Ruutsalo, H.M.; Isomaki, H. Treatment of rheumatoid arthritis with fish oil, selenium, Vitamins A and E, and placebo. *Scand. J. Rheumatol.* **1991**, *20*, 225.
128. Heinlei, K.; Adam, A.; Gradl, M.; Wiseman, M.; Adam, O. Selenkonzentration in den Erythrozyten bei Patienten mit rheumatoider. *Arthritis Med. Klin.* **1997**, *92*, 29–31. [CrossRef] [PubMed]
129. Peretz, A.; Siderova, V.; Nève, J. Selenium supplementation in rheumatoid arthritis investigated in a double blind, placebo-controlled trial. Scand. *J. Rheumatol.* **2001**, *30*, 208–212.
130. Malhotrai, S.; Welling, M.N.; Mantri, S.B.; Desai, K. In vitro and in vivo antioxidant, cytotoxic, and anti-chronic inflammatory arthritic effect of selenium nanoparticles. *J. Biomed. Mater. Res. B Appl. Biomater.* **2016**, *104*, 993–1003. [CrossRef] [PubMed]
131. Liu, Y.; Ma, L.; Zhou, H.; Zhu, X.; Yu, Q.; Chen, X.; Zhao, Y.; Liu, J. Polypeptide Nano-Se Targeting Inflammation and Theranostic Rheumatoid Arthritis by Anti-Angiogenic and NO Activating AMPKα Signaling Pathway. *J. Mater. Chem. B* **2018**, *6*, 3497–3514. [CrossRef]
132. Ren, S.X.; Zhang, B.; Lin, Y.; Ma, D.S.; Yan, H. Selenium Nanoparticles Dispersed in Phytochemical Exert Anti-Inflammatory Activity by Modulating Catalase, GPx1, and COX-2 Gene Expression in a Rheumatoid Arthritis Rat Model. *Med. Sci. Monit.* **2019**, *25*, 991–1000. [CrossRef]
133. Frangos, T.; Maret, W. Zinc and Cadmium in the Aetiology and Pathogenesis of Osteoarthritis and Rheumatoid Arthritis. *Nutrients* **2021**, *13*, 53. [CrossRef] [PubMed]
134. Rodríguez-Menéndez, S.; García, M.; Fernández, B.; Álvarez, L.; Fernández-Vega-Cueto, A.; Coca-Prados, M.; Pereiro, R.; González-Iglesias, H. The Zinc-Metallothionein Redox System Reduces Oxidative Stress in Retinal Pigment Epithelial Cells. *Nutrients* **2018**, *10*, 1874. [CrossRef]
135. Eide, D.J. The oxidative stress of zinc deficiency. *Metallomics* **2011**, *3*, 1124–1129. [CrossRef]
136. Goggs, R.; Vaughan-Thomas, A.; Clegg, P.D.; Carter, S.D.; Innes, J.F.; Mobasheri, A.; Shakibaei, M.; Schwab, W.; Bondy, C.A. Nutraceutical therapies for degenerative joint diseases: A critical review. *Crit. Rev. Food Sci. Nutr.* **2005**, *45*, 145–164. [CrossRef] [PubMed]
137. Tsai, Y.L.; Ko, W.S.; Hsiao, J.L.; Pan, H.H.; Chiou, Y.L. Zinc sulfate improved the unbalanced T cell profiles in Der p-allergic asthma: An ex vivo study. *Clin. Respir. J.* **2018**, *12*, 563–571. [CrossRef]
138. Higashimura, Y.; Takagi, T.; Naito, Y.; Uchiyama, K.; Mizushima, K.; Tanaka, M.; Hamaguchi, M.; Itoh, Y. Zinc Deficiency Activates the IL-23/Th17 Axis to Aggravate Experimental Colitis in Mice. *J. Crohn's Colitis* **2020**, *14*, 856–866. [CrossRef] [PubMed]
139. Sun, X.; Deng, Y.; Ma, Y.; Shao, M.; Ni, M.; Zhang, T.; Wang, X.; Xu, S.; Chen, Y.; Xu, S.; et al. Common mineral nutrients in ankylosing spondylitis: A 2-sample Mendelian randomization study. *Int. J. Rheum. Dis.* **2022**, *25*, 1129–1136. [CrossRef]
140. Afridi, H.I.; Kazi, T.G.; Brabazon, D.; Naher, S. Association between essential trace and toxic elements in scalp hair samples of smokers' rheumatoid arthritis subjects. *Sci. Total Environ.* **2011**, *412–413*, 93–100. [CrossRef]
141. Mierzecki, A.; Strecker, D.; Radomska, K. A pilot study on zinc levels in patients with rheumatoid arthritis. *Biol. Trace Elem. Res.* **2011**, *143*, 854–862. [CrossRef]
142. Smolen, J.S.; Aletaha, D.; Bijlsma, J.W.J.; Breedveld, F.C.; Boumpas, D.; Burmester, G.; Combe, B.; Cutolo, M.; de Wit, M.; Dougados, M.; et al. Treating Rheumatoid Arthritis to Target: Recommendations of an International Task Force. *Ann. Rheum. Dis.* **2010**, *69*, 631–637. [CrossRef]
143. Hall, S.C.; Knoell, D.L.; Fukada, T.; Kambe, T. Zinc signals in inflammation. In *Zinc Signalling*; Springer: Singapore, 2019; Volume 2019, pp. 279–304.
144. Maywald, M.; Rink, L.; Fukada, T.; Kambe, T. Zinc signals in immunology. In *Zinc Signalling*; Springer: Singapore, 2019; pp. 243–277.
145. O'Connor, J.P.; Kanjilal, D.; Teitelbaum, M.; Lin, S.S.; Cottrell, J.A. Zinc as a therapeutic agent in bone regeneration. *Materials* **2020**, *13*, 2211. [CrossRef] [PubMed]
146. Murphy, D.; Sinha-Royle, E.; Bellis, K.; Harrington, C.; Hutchinson, D. Nodular Rheumatoid Arthritis (RA): A Distinct Disease Subtype, Initiated by Cadmium Inhalation Inducing Pulmonary Nodule Formation and Subsequent RA-Associated Autoantibody Generation. *Med. Hypotheses* **2019**, *122*, 48–55. [CrossRef] [PubMed]
147. Fernández-Torres, J.; Zamudio-Cuevas, Y.; Martínez-Nava, G.A.; Aztatzi-Aguilar, O.G.; Sierra-Vargas, M.P.; Lozada-Pérez, C.A.; Suárez-Ahedo, C.; Landa-Solís, C.; Olivos-Meza, A.; Del Razo, L.M.; et al. Impact of Cadmium Mediated by Tobacco Use in Musculoskeletal Diseases. *Biol. Trace Elem. Res.* **2021**, *200*, 2008–2015. [CrossRef] [PubMed]
148. Shiue, I. Relationship of environmental exposures and ankylosing spondylitis and spinal mobility: US NHAENS, 2009–2010. *Int. J. Environ. Res. Public Health* **2015**, *25*, 322–329. [CrossRef] [PubMed]
149. Markiewicz-Górka, I.; Chowaniec, M.; Martynowicz, H.; Wojakowska, A.; Jaremków, A.; Mazur, G.; Wiland, P.; Pawlas, K.; Poręba, R.; Gać, P. Cadmium Body Burden and Inflammatory Arthritis: A Pilot Study in Patients from Lower Silesia, Poland. *Int. J. Environ. Res. Public Health* **2022**, *19*, 3099. [CrossRef] [PubMed]

150. Chen, F.; Demers, L.M.; Vallyathan, V.; Ding, M.; Lu, Y.; Castranova, V.; Shi, X. Vanadate induction of NF-κB involves IκB kinase β and SAPK/ERK kinase 1 in macrophages. *J. Biol. Chem.* **1999**, *274*, 20307–20312. [CrossRef] [PubMed]
151. Pessoa, J.C. Thirty years through vanadium chemistry. *J. Inorg. Biochem.* **2015**, *147*, 4–24. [CrossRef] [PubMed]
152. Sparks, J.A.; Chen, C.Y.; Hiraki, L.T.; Malspeis, S.; Costenbader, K.H.; Karlson, E.W. Contributions of familial rheumatoid arthritis or lupus and environmental factors to risk of rheumatoid arthritis in women: A prospective cohort study. *Arthritis Care Res.* **2014**, *66*, 1438–1446. [CrossRef] [PubMed]
153. Pfohler, C.; Muller, C.S.; Vogt, T. Psoriasis vulgaris and psoriasis pustulosa—Epidemiology, quality of life, comorbidities and treatment. *Curr. Rheumatol. Rev.* **2013**, *9*, 2–7. [CrossRef]
154. Yazisiz, V.; Arslan, G.; Ozbudak, I.H.; Turker, S.; Erbasan, F.; Avci, A.B.; Ozbudak, O.; Terzioglu, E. Lung involvement in patients with primary Sjögren's syndrome: What are the predictors? *Rheumatol. Int.* **2010**, *30*, 1317–1324. [CrossRef]
155. Jiang, L.; Yin, J.; Ye, L.; Yang, J.; Hemani, G.; Liu, A.J.; Zou, H.; He, D.; Sun, L.; Zeng, X.; et al. Novel risk loci for rheumatoid arthritis in Han Chinese and congruence with risk variants in Europeans. *Arthritis Rheumatol.* **2014**, *66*, 1121–1132. [CrossRef] [PubMed]
156. Saevarsdottir, S.; Rezaei, H.; Geborek, P.; Petersson, I.; Ernestam, S.; Albertsson, K.; Forslind, K.; van Vollenhoven, R.F.; SWEFOT Study Group. Current smoking status is a strong predictor of radiographic progression in early rheumatoid arthritis: Results from the SWEFOT trial. *Ann. Rheum. Dis.* **2015**, *74*, 1509–1514. [CrossRef] [PubMed]
157. Chen, C.H.; Chen, H.A.; Lu, C.L.; Liao, H.T.; Liu, C.H.; Tsai, C.Y.; Chou, C.T. Association of cigarette smoking with Chinese ankylosing spondylitis patients in Taiwan: A poor disease outcome in systemic inflammation, functional ability, and physical mobility. *Clin. Rheumatol.* **2013**, *32*, 659–663. [CrossRef] [PubMed]
158. Zhang, S.; Li, Y.; Xu, X.; Feng, X.; Yang, D.; Lin, G. Effect of cigarette smoking and alcohol consumption on disease activity and physical functioning in ankylosing spondylitis: A cross-sectional study. *Int. J. Clin. Exp.* **2015**, *8*, 13919–13927.
159. Zhao, S.; Jones, G.T.; Macfarlane, G.J.; Hughes, D.M.; Dean, L.E.; Moots, R.J.; Goodson, N.J. Associations between smoking and extra-axial manifestations and disease severity in axial spondyloarthritis: Results from the BSR Biologics Register for Ankylosing Spondylitis (BSRBR-AS). *Rheumatology* **2019**, *58*, 811–819. [CrossRef] [PubMed]
160. Hu, Z.; Li, Y.; Hu, L.; Ji, X.; Wang, L.; Li, K.; Zhu, J.; Zhang, J.; Huang, F. Cigarette Smoking Increases the Prevalence of Hip Joint Involvement in Ankylosing Spondylitis: A Real-World Case-Control Study. *J. Rheumatol.* **2023**, *50*, 901–906. [CrossRef] [PubMed]
161. Myllykangas-Luosujärvi, R.; Aho, K.; Lehtinen, K.; Kautiainen, H.; Hakala, M. Increased incidence of alcohol-related deaths from accidents and violence in subjects with ankylosing spondylitis. *Br. J. Rheumatol.* **1998**, *37*, 688–690. [CrossRef] [PubMed]
162. Zhao, S.; Thong, D.; Duffield, S.J.; Hughes, D.; Goodson, N.J. Alcohol and disease activity in axial spondyloarthritis: A cross-sectional study. *Rheumatol. Int.* **2018**, *38*, 375–381. [CrossRef] [PubMed]
163. Min, H.K.; Lee, J.; Ju, J.H.; Park, S.-H.; Kwok, S.-K. Alcohol consumption as a predictor of the progression of spinal structural damage in axial spondyloarthritis: Data from the Catholic Axial Spondyloarthritis COhort (CASCO). *Arthritis Res. Ther.* **2019**, *21*, 187. [CrossRef]
164. Du, K.; Zhang, C.Y.; Li, A.; Hu, J.Z.; Guo, R.; Li, S.M. Causality of occupational exposure on rheumatoid arthritis and ankylosing spondylitis: A two-sample mendelian randomization study. *Front. Immunol.* **2023**, *14*, 1223810. [CrossRef]
165. Hu, S.; Xing, H.; Wang, X.; Zhang, N.; Xu, Q. Causal Relationships Between Total Physical Activity and Ankylosing Spondylitis: A Mendelian Randomization Study. *Front Immunol.* **2022**, *13*, 887326. [CrossRef] [PubMed]
166. Batmaz, İ.; Sarıyıldız, M.A.; Dilek, B.; Bez, Y.; Karakoç, M.; Çevik, R. Sleep Quality and Associated Factors in Ankylosing Spondylitis: Relationship with Disease Parameters, Psychological Status and Quality of Life. *Rheumatol. Int.* **2012**, *33*, 1039–1045. [CrossRef]
167. Yüce, E.; Şentürk, E.; Sağaltıcı, E.; Şentürk, İ.A.; Aytekin, E. Sleep quality and depression in patients with ankylosing spondylitis and their associations with clinical parameters: A cross-sectional, case-control study. Ankilozan spondilitli hastalarda uyku kalitesi, depresyon ve bunların klinik parametrelerle ilişkisi: Kesitsel, vaka-kontrol çalışması. *AGRI* **2023**, *35*, 1–9. [PubMed]
168. Aydin, E.; Bayraktar, K.; Turan, Y.; Omurlu, I.; Tastaban, E.; Sendur, O.F. Qualidade Do Sono Em Pacientes Com Espondilite Anquilosante. *Rev. Bras. Reumatol.* **2015**, *55*, 340–345. [CrossRef] [PubMed]
169. Vergne-Salle, P.; Salle, L.; Fressinaud-Marie, A.C.; Descamps-Deplas, A.; Montestruc, F.; Bonnet, C.; Bertin, P. Diet and Disease Activity in Patients with Axial Spondyloarthritis: SpondyloArthritis and NUTrition Study (SANUT). *Nutrients* **2022**, *14*, 4730. [CrossRef] [PubMed]
170. Kao, C.M.; Chen, Y.M.; Huang, W.N.; Chen, Y.H.; Chen, H.H. Association between air pollutants and initiation of biological therapy in patients with ankylosing spondylitis: A nationwide, population-based, nested case-control study. *Arthritis Res. Ther.* **2023**, *25*, 75. [CrossRef] [PubMed]
171. Bartels, L.E.; Pedersen, A.B.; Kristensen, N.R.; Vilstrup, H.; Stengaard-Pedersen, K.; Dahlerup, J.F. A positive Helicobacter pylori test is associated with low spondylarthritis incidence in a Danish historical cohort study. *Rheumatol. Int.* **2020**, *40*, 359–366. [CrossRef] [PubMed]
172. Nielsen, P.R.; Kragstrup, T.W.; Deleuran, B.W.; Benros, M.E. Infections as risk factor for autoimmune diseases—A nationwide study. *J. Autoimmun.* **2016**, *74*, 176–181. [CrossRef] [PubMed]
173. Stone, M.A.; Payne, U.; Schentag, C.; Rahman, P.; Pacheco-Tena, C.; Inman, R.D. Comparative immune responses to candidate arthritogenic bacteria do not confirm a dominant role for Klebsiella pneumonia in the pathogenesis of familial ankylosing spondylitis. *Rheumatology* **2004**, *43*, 148–155. [CrossRef]

174. Keller, J.J.; Kang, J.H.; Lin, H.C. Association between ankylosing spondylitis and chronic periodontitis: A population-based study. *Arthritis Rheumatol.* **2013**, *65*, 167–173. [CrossRef]
175. Wei, C.Y.; Lin, J.Y.; Wang, Y.T.; Huang, J.Y.; Wei, J.C.; Chiou, J.Y. Risk of ankylosing spondylitis following human papillomavirus infection: A nationwide, population-based, cohort study. *J. Autoimmun.* **2020**, *113*, 102482. [CrossRef]
176. Damba, J.J.; Laskine, M.; Jin, Y.; Sinyavskaya, L.; Durand, M. Incidence of autoimmune diseases in people living with HIV compared to a matched population: A cohort study. *Clin. Rheumatol.* **2021**, *40*, 2439–2445. [CrossRef] [PubMed]
177. Yen, Y.F.; Chuang, P.H.; Jen, I.A.; Chen, M.; Lan, Y.C.; Liu, Y.L.; Lee, Y.; Chen, Y.H.; Chen, Y.A. Incidence of autoimmune diseases in a nationwide HIV/AIDS patient cohort in Taiwan, 2000–2012. *Ann. Rheum. Dis.* **2017**, *76*, 661–665. [CrossRef] [PubMed]
178. Wei, J.C.; Chou, M.C.; Huang, J.Y.; Chang, R.; Hung, Y.M. The association between Candida infection and ankylosing spondylitis: A population-based matched cohort study. *Curr. Med. Res. Opin.* **2020**, *36*, 2063–2069. [CrossRef] [PubMed]
179. Chu, K.A.; Chen, W.; Hung, Y.M.; Wei, J.C. Increased risk of ankylosing spondylitis after Mycoplasma pneumonia: A Nationwide population-based study. *Medicine* **2019**, *98*, e15596. [CrossRef] [PubMed]
180. Feng, X.G.; Xu, X.J.; Ye, S.; Lin, Y.Y.; Chen, P.; Zhang, X.J.; Lin, G.Y.; Lin, X.Q. Recent Chlamydia pneumoniae infection is highly associated with active ankylosing spondylitis in a Chinese cohort. Scand. *J. Rheumatol.* **2011**, *40*, 289–291.
181. Lindström, U.; Exarchou, S.; Lie, E.; Dehlin, M.; Forsblad-d'Elia, H.; Askling, J.; Jacobsson, L. Childhood hospitalisation with infections and later development of ankylosing spondylitis: A national case-control study. *Arthritis Res. Ther.* **2016**, *18*, 240. [CrossRef] [PubMed]
182. Chao, W.C.; Lin, C.H.; Chen, Y.M.; Jiang, R.S.; Chen, H.H. Association between tonsillitis and newly diagnosed ankylosing spondylitis: A nationwide, population-based, case-control study. *PLoS ONE* **2019**, *14*, e0220721. [CrossRef] [PubMed]
183. Abbood, H.M.; Pathan, E.; Cherukara, G.P. The link between ankylosing spondylitis and oral health conditions: Two nested case-control studies using data of the UK Biobank. *J. Appl. Oral Sci.* **2018**, *27*, e20180207. [CrossRef]
184. Bartels, L.E.; Jepsen, P.; Christensen, L.A.; Gerdes, L.U.; Vilstrup, H.; Dahlerup, J.F. Diagnosis of Helicobacter Pylori Infection is Associated with Lower Prevalence and Subsequent Incidence of Crohn's Disease. *J. Crohn's Colitis* **2016**, *10*, 443–448. [CrossRef]
185. Yap, T.W.; Gan, H.M.; Lee, Y.P.; Leow, A.H.; Azmi, A.N.; Francois, F.; Perez-Perez, G.I.; Loke, M.F.; Goh, K.L.; Vadivelu, J. Helicobacter pylori Eradication Causes Perturbation of the Human Gut Microbiome in Young Adults. *PLoS ONE* **2016**, *11*, e0151893. [CrossRef] [PubMed]
186. Yang, L.; Wang, L.; Wang, X.; Xian, C.J.; Lu, H. A Possible Role of Intestinal Microbiota in the Pathogenesis of Ankylosing Spondylitis. *Int. J. Mol. Sci.* **2016**, *17*, 2126. [CrossRef]
187. Al-Katma, M.K.; Bissada, N.F.; Bordeaux, J.M.; Sue, J.; Askari, A.D. Control of periodontal infection reduces the severity of active rheumatoid arthritis. *J. Clin. Rheumatol.* **2007**, *13*, 134–137. [CrossRef]
188. de Pablo, P.; Dietrich, T.; McAlindon, T.E. Association of periodontal disease and tooth loss with rheumatoid arthritis in the US population. *J. Rheumatol.* **2008**, *35*, 70–76. [PubMed]
189. Pischon, N.; Pischon, T.; Kröger, J.; Gülmez, E.; Kleber, B.M.; Bernimoulin, J.P.; Landau, H.; Brinkmann, P.G.; Schlattmann, P.; Zernicke, J.; et al. Association among rheumatoid arthritis, oral hygiene, and periodontitis. *J. Periodontol.* **2008**, *79*, 979–986. [CrossRef]
190. Palomares, O.; Rückert, B.; Jartti, T.; Kücüksezer, U.C.; Puhakka, T.; Gomez, E.; Fahrner, H.B.; Speiser, A.; Jung, A.; Kwok, W.W.; et al. Induction and maintenance of allergen-specific FOXP3+ Treg cells in human tonsils as potential first-line organs of oral tolerance. *J. Allergy Clin. Immunol.* **2012**, *129*, 510–520.e9. [CrossRef] [PubMed]
191. Itescu, S.; Dalton, J.; Zhang, H.Z.; Winchester, R. Tissue infiltration in a CD8 lymphocytosis syndrome associated with human immunodeficiency virus-1 infection has the phenotypic appearance of an antigenically driven response. *J. Clin. Investig.* **1993**, *91*, 2216–2225. [CrossRef]
192. Fife, D.J.; Waller, J.M.; Jeffes, E.W.; Koo, J.Y. Unraveling the paradoxes of HIV-associated psoriasis: A review of T-cell subsets and cytokine profiles. *Dermatol. Online J.* **2007**, *13*, 4. [CrossRef]
193. Russo, S.; Lopalco, L. Is autoimmunity a component of natural immunity to HIV? *Curr. HIV Res.* **2006**, *4*, 177–190. [CrossRef]
194. Zhou, C.; Zhao, H.; Xiao, X.Y.; Chen, B.D.; Guo, R.J.; Wang, Q.; Chen, H.; Zhao, L.D.; Zhang, C.C.; Jiao, Y.H.; et al. Metagenomic profiling of the pro-inflammatory gut microbiota in ankylosing spondylitis. *J. Autoimmun.* **2020**, *107*, 102360. [CrossRef]
195. Liu, G.; Hao, Y.; Yang, Q.; Deng, S. The Association of Fecal Microbiota in Ankylosing Spondylitis Cases with C-Reactive Protein and Erythrocyte Sedimentation Rate. *Mediat. Inflamm.* **2020**, *2020*, 8884324. [CrossRef] [PubMed]
196. Costello, M.E.; Ciccia, F.; Willner, D.; Warrington, N.; Robinson, P.C.; Gardiner, B.; Marshall, M.; Kenna, T.J.; Triolo, G.; Brown, M.A. Brief Report: Intestinal Dysbiosis in Ankylosing Spondylitis. *Arthritis Rheumatol.* **2015**, *67*, 686–691. [CrossRef] [PubMed]
197. Li, M.; Dai, B.; Tang, Y.; Lei, L.; Li, N.; Liu, C.; Ge, T.; Zhang, L.; Xu, Y.; Hu, Y.; et al. Altered Bacterial-Fungal Interkingdom Networks in the Guts of Ankylosing Spondylitis Patients. *mSystems* **2019**, *4*, e00176-18. [CrossRef] [PubMed]
198. Ebringer, R.W.; Cawdell, D.R.; Cowling, P.; Ebringer, A. Sequential studies in ankylosing spondylitis. association of klebsiella pneumoniae with active disease. *Ann. Rheum. Dis.* **1978**, *37*, 146–151. [CrossRef] [PubMed]
199. Mäki-Ikola, O.; Leirisalo-Repo, M.; Turunen, U.; Granfors, K. Association of gut inflammation with increased serum IgA class Klebsiella antibody concentrations in patients with axial ankylosing spondylitis (AS): Implication for different aetiopathogenetic mechanisms for axial and peripheral AS? *Ann. Rheum. Dis.* **1997**, *56*, 180–183. [CrossRef] [PubMed]
200. Wen, C.; Zheng, Z.; Shao, T.; Liu, L.; Xie, Z.; Le Chatelier, E.; He, Z.; Zhong, W.; Fan, Y.; Zhang, L.; et al. Quantitative metagenomics reveals unique gut microbiome biomarkers in ankylosing spondylitis. *Genome Biol.* **2017**, *18*, 142.

201. Laurence, M.; Asquith, M.; Rosenbaum, J.T. Spondyloarthritis, Acute Anterior Uveitis, and Fungi: Updating the Catterall-King Hypothesis. *Front. Med.* **2018**, *5*, 80. [CrossRef]
202. Hammer, R.E.; Maika, S.D.; Richardson, J.A.; Tang, J.P.; Taurog, J.D. Spontaneous inflammatory disease in transgenic rats expressing HLA-B27 and human beta 2m: An animal model of HLA-B27-associated human disorders. *Cell* **1990**, *63*, 1099–1112. [CrossRef] [PubMed]
203. Glatigny, S.; Fert, I.; Blaton, M.A.; Lories, R.J.; Araujo, L.M.; Chiocchia, G.; Breban, M. Proinflammatory Th17 cells are expanded and induced by dendritic cells in spondylarthritis-prone HLA-B27-transgenic rats. *Arthritis Rheum.* **2012**, *64*, 110–120. [CrossRef]
204. Taurog, J.D.; Richardson, J.A.; Croft, J.T.; Simmons, W.A.; Zhou, M.; Fernández-Sueiro, J.L.; Balish, E.; Hammer, R.E. The germfree state prevents development of gut and joint inflammatory disease in HLA-B27 transgenic rats. *J. Exp. Med.* **1994**, *180*, 2359–2364. [CrossRef]
205. Rath, H.C.; Herfarth, H.H.; Ikeda, J.S.; Grenther, W.B.; Hamm, T.E.; Balish, E., Jr.; Taurog, J.D.; Hammer, R.E.; Wilson, K.H.; Sartor, R.B. Normal luminal bacteria, especially Bacteroides species, mediate chronic colitis, gastritis, and arthritis in HLA-B27/human beta2 microglobulin transgenic rats. *J. Clin. Investig.* **1996**, *98*, 945–953. [CrossRef] [PubMed]
206. Schlosstein, L.; Terasaki, P.I.; Bluestone, R.; Pearson, C.M. High association of an HL-A antigen, W27, with ankylosing spondylitis. *N. Engl. J. Med.* **1973**, *288*, 704–706. [CrossRef] [PubMed]
207. Taurog, J.D.; Chhabra, A.; Colbert, R.A. Ankylosing Spondylitis and Axial Spondyloarthritis. *N. Engl. J. Med.* **2016**, *374*, 2563–2574. [CrossRef] [PubMed]
208. Rudwaleit, M.; van der Heijde, D.; Landewé, R.; Listing, J.; Akkoc, N.; Brandt, J.; Braun, J.; Chou, C.T.; Collantes-Estevez, E.; Dougados, M.; et al. The development of Assessment of SpondyloArthritis international Society classification criteria for axial spondyloarthritis (part II): Validation and final selection. *Ann. Rheum. Dis.* **2009**, *68*, 777–783. [CrossRef] [PubMed]
209. Rudwaleit, M.; van der Heijde, D.; Landewé, R.; Akkoc, N.; Brandt, J.; Chou, C.T.; Dougados, M.; Huang, F.; Gu, J.; Kirazli, Y.; et al. The Assessment of SpondyloArthritis International Society classification criteria for peripheral spondyloarthritis and for spondyloarthritis in general. *Ann. Rheum. Dis.* **2011**, *70*, 25–31. [CrossRef] [PubMed]
210. Li, J.; Casanova, J.L.; Puel, A. Mucocutaneous IL-17 immunity in mice and humans: Host defense vs. excessive inflammation. *Mucosal Immunol.* **2018**, *11*, 581–589. [CrossRef] [PubMed]
211. Fragoulis, G.E.; Liava, C.; Daoussis, D.; Akriviadis, E.; Garyfallos, A.; Dimitroulas, T. Inflammatory bowel diseases and spondyloarthropathies: From pathogenesis to treatment. *World J. Gastroenterol.* **2019**, *25*, 2162–2176. [CrossRef] [PubMed]
212. DeLay, M.L.; Turner, M.J.; Klenk, E.I.; Smith, J.A.; Sowders, D.P.; Colbert, R.A. HLA-B27 misfolding and the unfolded protein response augment interleukin-23 production and are associated with Th17 activation in transgenic rats. *Arthritis Rheumatol.* **2009**, *60*, 2633–2643. [CrossRef] [PubMed]
213. Yang, Y.; Torchinsky, M.B.; Gobert, M.; Xiong, H.; Xu, M.; Linehan, J.L.; Alonzo, F.; Ng, C.; Chen, A.; Lin, X.; et al. Focused specificity of intestinal TH17 cells towards commensal bacterial antigens. *Nature* **2014**, *510*, 152–156. [CrossRef]
214. Zielinski, C.E.; Mele, F.; Aschenbrenner, D.; Jarrossay, D.; Ronchi, F.; Gattorno, M.; Monticelli, S.; Lanzavecchia, A.; Sallusto, F. Pathogen-induced human TH17 cells produce IFN-γ or IL-10 and are regulated by IL-1β. *Nature* **2012**, *484*, 514–518. [CrossRef]
215. Larsen, J.M. The immune response to prevotella bacteria in chronic inflammatory disease. *Immunology* **2017**, *151*, 363–374. [CrossRef]
216. Goto, Y.; Panea, C.; Nakato, G.; Cebula, A.; Lee, C.; Diez, M.G.; Laufer, T.M.; Ignatowicz, L.; Ivanov, I.I. Segmented filamentous bacteria antigens presented by intestinal dendritic cells drive mucosal Th17 cell differentiation. *Immunity* **2014**, *40*, 594–607. [CrossRef]
217. Rogier, R.; Koenders, M.I.; Abdollahi-Roodsaz, S. Toll-like receptor mediated modulation of T cell response by commensal intestinal microbiota as a trigger for autoimmune arthritis. *J. Immunol. Res.* **2015**, *2015*, 527696. [CrossRef]
218. Ivanov, I.I.; Frutos, R. deL.; Manel, N.; Yoshinaga, K.; Rifkin, D.B.; Sartor, R.B.; Finlay, B.B.; Littman, D.R. Specific microbiota direct the differentiation of IL-17-producing T-helper cells in the mucosa of the small intestine. *Cell Host Microbe* **2008**, *4*, 337–349. [CrossRef]
219. Chambers, R.E.; MacFarlane, D.G.; Whicher, J.T.; Dieppe, P.A. Serum amyloid-A protein concentration in rheumatoid arthritis and its role in monitoring disease activity. *Ann. Rheum. Dis.* **1983**, *42*, 665–667. [CrossRef]
220. Geuking, M.B.; Cahenzli, J.; Lawson, M.A.; Ng, D.C.; Slack, E.; Hapfelmeier, S.; McCoy, K.D.; Macpherson, A.J. Intestinal bacterial colonization induces mutualistic regulatory T cell responses. *Immunity* **2010**, *34*, 794–806. [CrossRef]
221. Round, J.L.; Lee, S.M.; Li, J.; Tran, G.; Jabri, B.; Chatila, T.A.; Mazmanian, S.K. The Toll-like receptor 2 pathway establishes colonization by a commensal of the human microbiota. *Science* **2011**, *332*, 974–977. [CrossRef]
222. Round, J.L.; Mazmanian, S.K. Inducible Foxp3+ regulatory T-cell development by a commensal bacterium of the intestinal microbiota. *Proc. Natl. Acad. Sci. USA* **2010**, *107*, 12204–12209. [CrossRef]
223. Smith, P.M.; Howitt, M.R.; Panikov, N.; Michaud, M.; Gallini, C.A.; Bohlooly-Y, M.; Glickman, J.N.; Garrett, W.S. The microbial metabolites, short-chain fatty acids, regulate colonic Treg cell homeostasis. *Science* **2013**, *341*, 569–573. [CrossRef] [PubMed]
224. Song, Z.Y.; Yuan, D.; Zhang, S.X. Role of the microbiome and its metabolites in ankylosing spondylitis. *Front. Immunol.* **2022**, *13*, 1010572. [CrossRef] [PubMed]
225. Galocha, B.; López de Castro, J.A. Mutational analysis reveals a complex interplay of peptide binding and multiple biological features of HLA-B27. *J. Biol. Chem.* **2010**, *285*, 39180–39190. [CrossRef] [PubMed]

226. Madden, D.R. The three-dimensional structure of peptide-MHC complexes. *Annu. Rev. Immunol.* **1995**, *13*, 587–622. [CrossRef] [PubMed]
227. Toh, H.; Savoie, C.J.; Kamikawaji, N.; Muta, S.; Sasazuki, T.; Kuhara, S. Changes at the floor of the peptide-binding groove induce a strong preference for proline at position 3 of the bound peptide: Molecular dynamics simulations of HLA-A*0217. *Biopolymers* **2000**, *54*, 318–327. [CrossRef] [PubMed]
228. Nguyen, T.T.; Chang, S.C.; Evnouchidou, I.; York, I.A.; Zikos, C.; Rock, K.L.; Goldberg, A.L.; Stratikos, E.; Stern, L.J. Structural basis for antigenic peptide precursor processing by the endoplasmic reticulum aminopeptidase ERAP1. *Nat. Struct. Mol. Biol.* **2011**, *18*, 604–613. [CrossRef] [PubMed]
229. Yewdell, J.W. DRiPs solidify: Progress in understanding endogenous MHC class I antigen processing. *Trends Immunol.* **2011**, *32*, 548–558. [CrossRef] [PubMed]
230. Alvarez-Navarro, C.; López de Castro, J.A. ERAP1 structure, function and pathogenetic role in ankylosing spondylitis and other MHC-associated diseases. *Mol. Immunol.* **2014**, *57*, 12–21. [CrossRef] [PubMed]
231. Schittenhelm, R.B.; Sian, T.C.; Wilmann, P.G.; Dudek, N.L.; Purcell, A.W. Revisiting the arthritogenic peptide theory: Quantitative not qualitative changes in the peptide repertoire of HLA-B27 allotypes. *Arthritis Rheumatol.* **2015**, *67*, 702–713. [CrossRef] [PubMed]
232. Australo-Anglo-American Spondyloarthritis Consortium (TASC); Reveille, J.D.; Sims, A.M.; Danoy, P.; Evans, D.M.; Leo, P.; Pointon, J.J.; Jin, R.; Zhou, X.; Bradbury, L.A.; et al. Genome-wide association study of ankylosing spondylitis identifies non-MHC susceptibility loci. *Nat. Genet.* **2010**, *42*, 123–127. [CrossRef] [PubMed]
233. Evans, D.M.; Spencer, C.C.; Pointon, J.J.; Su, Z.; Harvey, D.; Kochan, G.; Oppermann, U.; Dilthey, A.; Pirinen, M.; Stone, M.A.; et al. Interaction between ERAP1 and HLA-B27 in ankylosing spondylitis implicates peptide handling in the mechanism for HLA-B27 in disease susceptibility. *Nat. Genet.* **2011**, *43*, 761–767. [CrossRef]
234. Chatzikyriakidou, A.; Voulgari, P.V.; Drosos, A.A. What is the role of HLA-B27 in spondyloarthropathies? *Autoimmun. Rev.* **2011**, *10*, 464–468. [CrossRef]
235. Faham, M.; Carlton, V.; Moorhead, M.; Zheng, J.; Klinger, M.; Pepin, F.; Asbury, T.; Vignali, M.; Emerson, R.O.; Robins, H.S.; et al. Discovery of T Cell Receptor β Motifs Specific to HLA-B27-Positive Ankylosing Spondylitis by Deep Repertoire Sequence Analysis. *Arthritis Rheumatol.* **2017**, *69*, 774–784. [CrossRef] [PubMed]
236. Antoniou, A.N.; Ford, S.; Taurog, J.D.; Butcher, G.W.; Powis, S.J. Formation of HLA-B27 homodimers and their relationship to assembly kinetics. *J. Biol. Chem.* **2004**, *279*, 8895–8902. [CrossRef] [PubMed]
237. Colbert, R.A.; DeLay, M.L.; Layh-Schmitt, G.; Sowders, D.P. HLA-B27 misfolding and spondyloarthropathies. *Prion* **2009**, *3*, 15–26. [CrossRef] [PubMed]
238. Bowness, P.; Ridley, A.; Shaw, J.; Chan, A.T.; Wong-Baeza, I.; Fleming, M.; Cummings, F.; McMichael, A.; Kollnberger, S. Th17 cells expressing KIR3DL2+ and responsive to HLA-B27 homodimers are increased in ankylosing spondylitis. *J. Immunol.* **2011**, *186*, 2672–2680. [CrossRef] [PubMed]
239. Wu, X.; Wu, J.; Li, X.; Wei, Q.; Lv, Q.; Zhang, P.; Zheng, X.; Chen, Z.; Cao, S.; Tu, L.; et al. The Clinical Characteristics of Other HLA-B Types in Chinese Ankylosing Spondylitis Patients. *Front. Med.* **2021**, *7*, 568790. [CrossRef] [PubMed]
240. Brown, M.A.; Pile, K.D.; Kennedy, L.G.; Calin, A.; Darke, C.; Bell, J.; Wordsworth, B.P.; Cornélis, F. HLA class I associations of ankylosing spondylitis in the white population in the United Kingdom. *Ann. Rheum. Dis.* **1996**, *55*, 268–270. [CrossRef]
241. Brown, M.A.; Kennedy, L.G.; MacGregor, A.J.; Darke, C.; Duncan, E.; Shatford, J.L.; Taylor, A.; Calin, A.; Wordsworth, P. Susceptibility to ankylosing spondylitis in twins: The role of genes, HLA, and the environment. *Arthritis Rheum.* **1997**, *40*, 1823–1828. [CrossRef] [PubMed]
242. Wei, J.C.; Tsai, W.C.; Lin, H.S.; Tsai, C.Y.; Chou, C.T. HLA-B60 and B61 are strongly associated with ankylosing spondylitis in HLA-B27-negative Taiwan Chinese patients. *Rheumatology* **2004**, *43*, 839–842. [CrossRef]
243. International Genetics of Ankylosing Spondylitis Consortium (IGAS). Identification of multiple risk variants for ankylosing spondylitis through high-density genotyping of immune-related loci. *Nat. Genet.* **2013**, *45*, 730–738. [CrossRef]
244. López-Larrea, C.; Mijiyawa, M.; González, S.; Fernández-Morera, J.L.; Blanco-Gelaz, M.A.; Martínez-Borra, J.; López-Vázquez, A. Association of ankylosing spondylitis with HLA-B*1403 in a West African population. *Arthritis Rheum.* **2002**, *46*, 2968–2971. [CrossRef]
245. Díaz-Peña, R.; Blanco-Gelaz, M.A.; Njobvu, P.; López-Vazquez, A.; Suárez-Alvarez, B.; López-Larrea, C. Influence of HLA-B*5703 and HLA-B*1403 on susceptibility to spondyloarthropathies in the Zambian population. *J. Rheumatol.* **2008**, *35*, 2236–2240. [CrossRef] [PubMed]
246. Remans, P.H.; Wijbrandts, C.A.; Sanders, M.E.; Toes, R.E.; Breedveld, F.C.; Tak, P.P.; van Laar, J.M.; Reedquist, K.A. CTLA-4IG suppresses reactive oxygen species by preventing synovial adherent cell-induced inactivation of Rap1, a Ras family GTPASE mediator of oxidative stress in rheumatoid arthritis T cells. *Arthritis Rheum.* **2006**, *54*, 3135–3143. [CrossRef] [PubMed]
247. Dahmani, C.A.; Benzaoui, A.; Amroun, H.; Mecabih, F.; Sediki, F.Z.; Zemani-Fodil, F.; Fodil, M.; Boughrara, W.; Mecheti, B.; Attal, N.; et al. Association of the HLA-B27 antigen and the CTLA4 gene CT60/rs3087243 polymorphism with ankylosing spondylitis in Algerian population: A case-control study. *Int. J. Immunogenet.* **2018**, *45*, 109–117. [CrossRef] [PubMed]
248. Liu, Z.; Zhang, P.; Dong, J. Genetic variants of STAT4 are associated with ankylosing spondylitis susceptibility and severity in a Chinese Han population. *Int. J. Clin. Exp. Med.* **2014**, *7*, 5877–5881. [PubMed]

249. Ebrahimiyan, H.; Mostafaei, S.; Aslani, S.; Jamshidi, A.; Mahmoudi, M. Studying the Association between STAT4 Gene Polymorphism and Susceptibility to Rheumatoid Arthritis Disease: An Updated Meta-Analysis. *Iran. J. Immunol.* **2019**, *16*, 71–83. [PubMed]
250. García-Ruiz, I.; de la Torre, P.; Díaz, T.; Esteban, E.; Fernández, I.; Muñoz-Yagüe, T.; Solís-Herruzo, J.A. Sp1 and Sp3 transcription factors mediate malondialdehyde-induced collagen alpha 1(I) gene expression in cultured hepatic stellate cells. *J. Biol. Chem.* **2002**, *277*, 30551–30558. [CrossRef] [PubMed]
251. Gao, M.; Li, T.; Song, Z.; Wang, X.; Zhang, X.; Liu, W. Relationship Between IL1 Gene Polymorphism and Susceptibility to Ankylosing Spondylitis: An Updated and Supplemented Meta-Analysis. *Biochem. Genet.* **2022**, *60*, 1025–1038. [CrossRef]
252. Biały, S.; Iwaszko, M.; Świerkot, J.; Bugaj, B.; Kolossa, K.; Jeka, S.; Bogunia-Kubik, K. Th2 Cytokines (Interleukin-5 and -9) Polymorphism Affects the Response to Anti-TNF Treatment in Polish Patients with Ankylosing Spondylitis. *Int. J. Mol. Sci.* **2022**, *23*, 13177. [CrossRef]
253. Wielińska, J.; Świerkot, J.; Kolossa, K.; Bugaj, B.; Chaszczewska-Markowska, M.; Jeka, S.; Bogunia-Kubik, K. Polymorphisms within Genes Coding for IL-17A and F and Their Receptor as Clinical Hallmarks in Ankylosing Spondylitis. *Mediat. Inflamm.* **2021**, *2021*, 3125922. [CrossRef] [PubMed]
254. Oliveira-Toré, C.F.; Moraes, A.G.; Martinez, G.F.; Neves, J.S.F.; Macedo, L.C.; Rocha-Loures, M.A.; Quirino, M.G.; Alves, H.V.; Sell, A.M.; Visentainer, J.E.L. Genetic Polymorphisms of Toll-like receptors 2 and 9 as Susceptibility Factors for the Development of Ankylosing Spondylitis and Psoriatic Arthritis. *J. Immunol. Res.* **2019**, *2019*, 1492092. [CrossRef]
255. Wu, J.; Xiong, Z.; Sun, Y.; Song, J.; Niu, F.; Yan, M.; Jin, T. TIMP3 gene polymorphisms and relation to Ankylosing spondylitis susceptibility in Chinese Han population. *Int. J. Immunogenet.* **2019**, *46*, 472–478. [CrossRef] [PubMed]
256. Liu, H.; Fu, L.; He, D.; Deng, J.; Zhu, J.; Xu, K.; Hu, D.; Li, J.; Wang, Y.; Hu, W.; et al. RUNX3 Polymorphisms Affect the Risk of Ankylosing Spondylitis. *Med. Sci. Monit.* **2020**, *26*, e919528. [CrossRef] [PubMed]
257. Gao, S.; Liang, W.; Xu, T.; Xun, C.; Cao, R.; Deng, Q.; Zhang, J.; Sheng, W. Associations of Tumor Necrosis Factor Alpha Gene Polymorphisms and Ankylosing Spondylitis Susceptibility: A Meta-analysis Based on 35 Case-control Studies. *Immunol. Investig.* **2022**, *51*, 859–882. [CrossRef] [PubMed]
258. Yang, J.; Hu, X.; Wu, M.; Ma, Y.; Zhang, X.; Chen, M.; Yuan, Y.; Han, R.; Liu, R.; Guan, S.; et al. TNFAIP3 genetic polymorphisms reduce ankylosing spondylitis risk in Eastern Chinese Han population. *Sci. Rep.* **2019**, *9*, 10209. [CrossRef] [PubMed]

International Journal of
Molecular Sciences

MDPI

Review

The Current State of Knowledge Regarding the Genetic Predisposition to Sports and Its Health Implications in the Context of the Redox Balance, Especially Antioxidant Capacity

Paweł Sutkowy [1], Martyna Modrzejewska [1], Marta Porzych [2] and Alina Woźniak [1,*]

[1] Department of Medical Biology and Biochemistry, Ludwik Rydygier Collegium Medicum in Bydgoszcz, Nicolaus Copernicus University in Toruń, 85-092 Bydgoszcz, Poland; p.sutkowy@cm.umk.pl (P.S.); martyna.modrzejewska@cm.umk.pl (M.M.)

[2] Student Research Club of Medical Biology and Biochemistry, Department of Medical Biology and Biochemistry, Faculty of Medicine, Ludwik Rydygier Collegium Medicum in Bydgoszcz, Nicolaus Copernicus University in Toruń, 24 Karłowicza St., 85-092 Bydgoszcz, Poland; m.porzych@o2.pl

* Correspondence: al1103@cm.umk.pl

Abstract: The significance of physical activity in sports is self-evident. However, its importance is becoming increasingly apparent in the context of public health. The constant desire to improve health and performance suggests looking at genetic predispositions. The knowledge of genes related to physical performance can be utilized initially in the training of athletes to assign them to the appropriate sport. In the field of medicine, this knowledge may be more effectively utilized in the prevention and treatment of cardiometabolic diseases. Physical exertion engages the entire organism, and at a basic physiological level, the organism's responses are primarily related to oxidant and antioxidant reactions due to intensified cellular respiration. Therefore, the modifications involve the body adjusting to the stresses, especially oxidative stress. The consequence of regular exercise is primarily an increase in antioxidant capacity. Among the genes considered, those that promote oxidative processes dominate, as they are associated with energy production during exercise. What is missing, however, is a look at the other side of the coin, which, in this case, is antioxidant processes and the genes associated with them. It has been demonstrated that antioxidant genes associated with increased physical performance do not always result in increased antioxidant capacity. Nevertheless, it seems that maintaining the oxidant–antioxidant balance is the most important thing in this regard.

Keywords: gene; polymorphism; exercise; antioxidant; oxidant

1. Introduction

1.1. Exercise—Health Implications

Physical exercise/effort (PE) is currently an important issue in the context of public health. A sedentary lifestyle is considered a significant pathological factor of many civilization diseases, i.e., cardiometabolic [1] and neurodegenerative diseases, as well as mental disorders (e.g., diabetes, heart attack, Alzheimer's disease, depression) [2]. PE improves mood and physical performance, delays aging, and ameliorates cognitive function. Physically, PE is associated with increased oxygen consumption (VO_2), which strictly affects the oxidant–antioxidant (redox) balance. Intensified cellular respiration in mitochondria increases the production of reactive oxygen species (ROSs), escalating oxidation reactions. Appropriately intensive PE may lead to a state of oxidative stress, the risk of cell damage, and, thus, inflammation. The effect of exercise depends on the intensity and duration of the effort. "Small" negative post-exercise changes can contribute to positive adaptive effects, such as increased antioxidant capacity (primarily higher activity of antioxidant enzymes) [2].

The effects of exercise on an organism are being increasingly utilized in the rehabilitation of heart diseases [3]. Moreover, PE is regarded as the most significant predictor of cardiometabolic diseases. The absence of PE is the most significant risk factor for the development of such diseases. Regular exercise is the most effective means of enhancing cardiorespiratory and metabolic health. The entire organism—all organs and physiological systems—is involved in PE [4]. Therefore, all possible kinds of genes are activated in response to the effort. This covers particularly protein genes, such as those of structural, enzymatic, and signaling proteins (muscle proteins, enzymes, receptors, cytokines, and hormones). The fundamental differences result only from the type of exercise bout (ratio of aerobic versus anaerobic efforts, endurance vs. resistance) [5]. There are parameters that connect the fields of sports and healthcare. For instance, the maximal fat oxidation rate is a predictor of metabolic flexibility, body weight loss, and endurance performance. It can be evaluated and defined not only on the basis of physical and biochemical parameters but also on the basis of genetic parameters. This is because, as is well known, everything has its origin in the genes [6].

1.2. Sports Genomics and the Redox Balance

It appears that the pinnacle of the human body's capabilities has been reached in many sports. Nevertheless, the pursuit of improvement in top results remains constant. The level of professionalism in sports is on the rise, accompanied by a concurrent growth in knowledge. This has always included technique and technology, while a relatively new field of study is athletic genetics, which has significant potential for development. This is an increasingly prominent area of sports theory, representing one of many avenues in the pursuit of enhanced sports performance. For example, the use of athletic genetics could facilitate the assignment of a novice athlete to a sport discipline that is most suited to their abilities at an early stage of training. This would obviate the necessity for the athlete to undergo the arduous process of searching for a suitable sport through the conventional means of sports testing, thus allowing the athlete to be trained more expeditiously and effectively.

A study that involved monozygotic and dizygotic twins revealed that DNA may be responsible for approximately 66% of the interindividual variance in terms of sports predisposition (microsatellite variations and single nucleotide polymorphisms, SNPs) [7], and over 200 genes may even be closely related to sports performance [8]. A recent review of the literature indicates that 128 genetic markers, distributed virtually across the entire genome, may be associated with professional athletes. These include 41 polymorphisms related to endurance performance, 45 polymorphisms related to power, and 42 polymorphisms related to strength [9].

The first papers that appeared at the end of the twentieth century focused on only one gene that potentially strongly determines physical performance. One of the first ones was the gene for angiotensin-converting enzyme (ACE) [10]. Presently, a genome-wide association study (GWAS) is a method for evaluating genetic predisposition for sports. This is based on determining several hundred thousand to 5 million loci in one DNA sample (no assumptions regarding loci that are potentially related to the trait of interest) [11]. Sports genomics are primarily concerned with the following genes: ACE, α-actininin-3 (ACTN3), peroxisome proliferator-activated receptors (PPARs) $\alpha/\gamma/\delta$, hypoxia-inducible factor-1α (HIF1α), and endothelial nitric oxide synthase (eNOS) [5,12,13]. ACE is an indirect prooxidant enzyme, since angiotensin II, which ACE produces, increases superoxide anion radical ($O_2^{\cdot-}$) production through the activation of nicotinamide adenine dinucleotide phosphate (NADPH) oxidase (NOX) [14]. ACTN3 is a protein filament of muscle fiber involved in contraction; thus, it is not associated with the redox balance [5]. PPARs, in turn, may indirectly promote antioxidant signaling through transcriptional or post-translational activities [15]. HIF1α is a crucial subunit for transcriptional hypoxia-inducible factor, while hypoxia is a state associated with oxidative stress [16]. Likewise, eNOS, which is a source of nitric oxide ($^{\cdot}NO$), is a kind of reactive nitrogen species (RNS) with a potent vasodilatory effect that produces peroxynitrite ($ONOO^-$) in contact with $O_2^{\cdot-}$. In turn, $ONOO^-$ is an ROS with a greatly strong oxidizing and vasoconstricting effect [17]. Table 1 presents a list of loci for which there is reliable evidence of a beneficial effect on physical performance.

Table 1. Alleles that have been demonstrated to enhance physical performance [9].

Gene	Locus	Alleles	Type of Benefit
AMPD1 (Adenosine Monophosphate Deaminase 1)	1p13	rs17602729 C/T	endurance/power (C allele)
CDKN1A (Cyclin-Dependent Kinase Inhibitor 1A)	6p21.2	rs236448 A/C	endurance (A allele)/power (C allele)
HFE (Homeostatic Iron Regulator)	6p21.3	rs1799945 C/G	endurance (G allele)
MYBPC3 (Myosin Binding Protein C3)	11p11.2	rs1052373 A/G	endurance (G allele)
NFIA-AS2 (NFIA antisense RNA 2)	1p31.3	rs1572312 C/A	endurance (C allele)
PPARA (Peroxisome Proliferator Activated Receptor A)	22q13.31	rs4253778 G/C	endurance (G allele)
PPARGC1A (Peroxisome Proliferative Activated Receptor Γ coactivator 1 A)	4p15.1	rs8192678 G/A	endurance (G allele)
ACTN3 (Actinin A 3)	11q13.1	rs1815739 C/T	power/strength (C allele)
CPNE5 (Copine V)	6p21.2	rs3213537 G/A	power (G allele)
GALNTL6 (Polypeptide N-acetylgalactosaminyltransferase Like 6)	4q34.1	rs558129 T/C	power (T allele)
IGF2 (Insulin-Like Growth Factor 2)	11p15.5	rs680 A/G	power (G allele)
IGSF3 (Immunoglobulin Superfamily Member 3)	1p13.1	rs699785 G/A	power (A allele)
NOS3 (Nitric Oxide Synthase 3)	7q36	rs2070744 T/C	power (T allele)
TRHR (Thyrotropin-Releasing Hormone Receptor	8q23.1	rs7832552 C/T	power (T allele)
AR (Androgen Receptor)	Xq12	CAG repeats	strength (allele of ≥21 CAG repeats)
LRPPRC (Leucine-Rich Pentatricopeptide Repeat Cassette)	2p21	rs10186876 A/G	strength (A allele)
MMS22L (Methyl Methanesulfonate-Sensitivity Protein 22-Like)	6q16.1	rs9320823 T/C	strength (T allele)
PHACTR1 (Phosphate and Actin Regulator 1)	6p24.1	rs6905419 C/T	strength (C allele)
PPARG (Peroxisome Proliferator Activated Receptor Γ)	3p25.2	rs1801282 G/C	strength (G allele)

The existing literature indicates a growing interest in the field of sports genetics. However, genes that have a direct impact on physical performance are still poorly understood. Antioxidant genes are an especially new approach to this issue, while antioxidant capacity is an essential part of post-exercise adaptation changes. Therefore, we decided to review the knowledge about genes responsible for antioxidant defense in the context of physical performance.

2. Body Functions during Physical Exercise

It is important to know the general physiological variations in the human organism during PE for this article. The body adjusts its physiological functions to accommodate the increased demand for energy and oxygen during physical activity. These changes include increased body temperature, respiration rate, heart rate, and blood pressure, while at a molecular level, hydrolysis and recovery of adenosine triphosphate (ATP) are intensified [18].

2.1. Energy Metabolism

ATP hydrolysis and recovery are essential to maintain body functions during physical activity. The energy required for muscle contractions, enabling movement, is produced through ATP hydrolysis in muscle mitochondria. For aerobic exertion, this includes the following metabolic pathways of cellular respiration: glycolysis, pyruvate oxidation, the tricarboxylic acid cycle, and oxidative phosphorylation [18]. Anaerobic bouts of exercise involve oxygen-free mechanisms such as anaerobic glycolysis and the phosphagen system (phosphocreatine utilization) based on muscle stores. The anaerobic activity involves the very rapid production of ATP, faster than under aerobic conditions, but ATP is quickly depleted because the stores of glucose and phosphocreatine in muscles are rapidly used up. Hence, anaerobic efforts allow for the generation of high power output; however, they do not last longer than 6 s. The consequence of anaerobic glycolysis is lactate production. Very intense PE with about 100% maximal VO_2, lasting 15–30 s, involves the rapid use of glycogen stores in skeletal muscles and maximally intensified oxidative phosphorylation. The availability of O_2 during endurance exercise and the lower intensity of the exercise allow one to continue the effort over a much longer time. Up to approximately 2 h, apart from muscle glycogen reserves, plasma glucose, plasma free fatty acids, and muscle triglycerides are used for energy production, while in addition to muscles, many other organs and body systems are involved in this process. Endurance exercise bouts of longer than 2 h are associated with a metabolic shift from carbohydrate oxidation to lipid oxidation [19].

2.2. Energy Resources

Physical exertion may be referred to as a stressor because it causes a reduction in fatty acid oxidation and increases glucose metabolism to cover energy demand. In contrast, fatty acid oxidation is the primary source of energy flow when resting. An instant source of energy is produced when glycogenolysis converts glycogen stored in muscle and liver cells into glucose. When the glucose concentration is too low (consumption of glycogen stores and lack of current supply of simple sugars), ATP is produced using the β-oxidation of fatty acids [18]. However, specific alterations must be considered in relation to exercise intensity and diet. Moderate-intensity exercise (40–55% VO_{2max}) results in the oxidation of both lipids and carbohydrates. Hepatic glycogen stores are then mainly used, but they can also be recovered if glucose is available from digested food/drink (small intestine). An increase in the intensity of exercise results in greater involvement of skeletal muscles. PE above 75% VO_{2max} is primarily related to the consumption of muscle glycogen, which is used up very quickly during maximal and supramaximal efforts [19]. A diet rich in carbohydrates, particularly polysaccharides, can result in a high glycogen concentration within 24 to 36 h. However, a continuous supply of monosaccharides, disaccharides, and oligosaccharides is essential for prolonged endurance exercise. Interestingly, a low-carbohydrate high-fat diet increases glycogen stores but impairs aerobic metabolism and, thus, endurance exercise performance [20].

Glycogen is a glucose polymer and the main carbohydrate used for energy production. The conversion of glycogen into glucose is catalyzed by glycogen phosphorylase, which completely decomposes the glucose polymer in conjunction with a debranching enzyme [19]. To produce energy, glycogen cooperates with specific cellular components in skeletal muscles, e.g., mitochondria, myofilaments, and the sarcoplasmic reticulum. The complex of glycogen and the sarcoplasmic reticulum has been well described in fast-twitch skeletal muscle. It is believed that phosphorylase-mediated glycogen decomposition provides the availability of glucose-1-phosphate for glycolytic ATP production at this complex. Moreover, glycolytic ATP supply may facilitate Ca^{2+} reuptake into the sarcoplasmic reticulum, ensuring muscle relaxation during the contractile cycle. This is possible thanks to the interaction of glycolytic enzymes with calcium ATPases in the reticulum [21].

Depending on exercise intensity, duration, and nutritional status, adipose-derived fatty acids can be a significant or dominant origin of energy during PE. Fatty acid absorption within the cell initiates the process, while the absorption may depend on particular muscle proteins. The first step is activating fatty acid molecules with acyl-CoA synthetase found in the endoplasmic reticulum and on the surface of the outer mitochondrial membrane. The enzyme uses coenzyme A to perform ATP-dependent thioesterification of fatty acid molecules. Fatty acid combined with coenzyme A (long-chain acyl-CoA) additionally connects with a molecule of carnitine and, in this form, enters the mitochondrion. Eventually, the fatty acid undergoes oxidation inside the mitochondrion, which involves the deletion of two carbon atoms called beta and gamma (detachment of acetyl-CoA at the carboxyl end—β-oxidation of fatty acid) [22].

3. The Effect of Physical Exercise on the Redox Balance

The other important part of this article's topic is the redox balance in humans, as it is a crucial part of human homeostasis during exercise. As mentioned in the introduction, changes in an organism caused by exercise result mainly from the impact on this part of the organism's functions.

3.1. Mitochondrial Respiratory Chain and ROS Production

The solubility of molecular oxygen (O_2) is up to eight-fold higher in organic solvents than in water [23]. The oxygen concentration within a cell may locally differ. Generally, the highest O_2 concentration occurs near cellular membranes, which have physical properties similar to those of organic solvents. Mitochondria are characterized by the lowest O_2 concentration compared to other cellular organelles, as they are central oxygen consumers. Molecular oxygen reacts with organic compounds and oxidizes them by acquiring their electrons. The complete reduction of O_2 to water is problematic because it requires the double acquisition of two electrons (and two protons; four electrons and four protons in total) from the molecule being oxidized, and the overwhelming majority of them have paired electrons. Therefore, oxygen reacts with many compounds in the one-electron way to form $O_2^{\cdot-}$ [24]. In aerobic cells, a complete reduction of oxygen occurs in the mitochondrial respiratory chain. However, 1–4% of the oxygen consumed by mitochondria is converted into $O_2^{\cdot-}$ [24,25]. In vivo, $O_2^{\cdot-}$ reacts predominantly with iron–sulfur centers and transition metal ions. The protonated form of $O_2^{\cdot-}$, i.e., hydroperoxyl radical ($HO^{\cdot}{}_2$), which is devoid of an electric charge, penetrates cellular membranes more easily and stays longer in their hydrophobic interior (inaccessible to $O_2^{\cdot-}$), where it can initiate a reaction of lipid peroxidation. It should be noted that lipid peroxidation is an element of normal cell metabolism. Increased exposure to ROSs enhances it but does not initiate it. The attachment of another electron (and two protons) to $O_2^{\cdot-}$ produces hydrogen peroxide (H_2O_2). This weak oxidant reacts mainly with the thiol group of protein cysteine residues (e.g., protein tyrosine phosphatases, G proteins, some ion channels, and some transcription factors) and transition metal ions (Fe^{2+}/Cu^{+}). The latter group of reactions leads to the formation of the most reactive oxygen species in biological systems: hydroxyl radical ($^{\cdot}OH$), the product of the attachment of three electrons to molecular oxygen. In the body, $^{\cdot}OH$ reacts non-specifically with biomolecules of all major classes (low-molecular-mass compounds, proteins, lipids, carbohydrates, nucleic acids). In contrast, $O_2^{\cdot-}$ and H_2O_2 mainly react with enzymes that specifically break them down [26]. The enzymes catalase (CAT) and superoxide dismutase (SOD) catalyze the dismutation reaction of H_2O_2 and $O_2^{\cdot-}$, respectively. Peroxidases consume H_2O_2 to oxidize their substrates. The activity of these enzymes determines the protection of eukaryotic cells against $O_2^{\cdot-}$ and H_2O_2. The protection against $^{\cdot}OH$ is mainly based on preventing its formation [24].

3.2. The Components of Oxidant–Antioxidant Balance

The emergence of one ROS entails the generation of others. As a result of the decomposition of $O_2^{\cdot-}$, H_2O_2 is formed. That, in turn, in the presence of transition metal ions, creates the possibility of the formation of $^{\cdot}OH$ in the Haber–Weiss reaction ($H_2O_2 + O_2^{\cdot-} + Fe^{2+}/Fe^{3+} \rightarrow {}^{\cdot}OH + OH^- + O_2$) or other ROSs. Physical factors such as ionizing radiation, ultraviolet radiation, photochemical reactions, and ultrasound are sources of ROSs with marginal biological significance. Intracellular sources of ROSs are much more important; they result primarily from the one-electron oxidation of reduced forms of many compounds (such as cysteine, glutathione (GSH), glucose, flavins, quinones, and nucleotides—$FMNH_2$, $FADH_2$, and catecholamines) by O_2. Another source of ROSs in cells is specific enzymatic reactions. Two enzymes generating $O_2^{\cdot-}$ are particularly important here: xanthine oxidase (peroxisomes) and NOX, which are present in the plasma membrane of phagocytic cells. $O_2^{\cdot-}$ is also generated by the smooth endoplasmic reticulum and lysosomal redox chain. The reaction of $O_2^{\cdot-}$ with $^{\cdot}NO$ produces $ONOO^-$, a strong oxidant with a bactericidal effect. However, the most important cellular source of ROSs is the mitochondrial respiratory chain composed of large protein complexes: NADH-ubiquinone oxidoreductase (complex I), ubiquinone-cytochrome c oxidoreductase (complex III), and cytochrome c oxidase (complex IV). Succinate dehydrogenase (complex II) does not pump protons but provides reduced ubiquinone [24,27–29]. Electrons enter the chain from NADH via complex I and from succinate via complex II. Complexes I, III, and IV pump protons across the inner mitochondrial membrane, which flow back into the matrix via the ATP synthase, driving the rotor to produce ATP [29].

In living cells, oxidative damage caused by ROSs generated in the environment and during aerobic metabolism may lead to DNA mutations, protein inactivation, and cell death. On the other hand, ROSs may exert beneficial effects for proper cellular development and proliferation. For instance, they take part in signaling pathways, imitate and enhance the action of growth factors, or have mitogenic effects [26]. Thus, cells should have a dynamic balance between the rate of ROS formation and the decomposition rate (Figure 1). To maintain this redox balance, aerobic cells evolved defense mechanisms, which include prevention (counteracting ROS reactions with biologically important compounds), intervention (terminating free radical chain reactions), and elimination or repair (removing the products of ROS reactions with biomolecules). These mechanisms may complement and replace each other. For instance, mammalian tissues contain glutathione peroxidases (GPXs), which catalyze the reaction between GSH and H_2O_2, forming the oxidized form of GSH—glutathione disulfide. This disulfide may ultimately lead to protein inactivation by oxidizing the thiol groups in proteins and forming disulfide bridges (forming mixed disulfides with proteins containing thiol groups). Fortunately, GPX enzymes cooperate with glutathione reductase (GR), which recreates the reduced form of glutathione at the expense of NADPH oxidation. NADPH is regenerated by, for example, glucose-6-phosphate dehydrogenase or isocitrate dehydrogenase [24].

A heme-containing enzyme, cytochrome c peroxidase, located in the intermembrane space of mitochondria, decomposes H_2O_2 generated by mitochondrial SOD (MnSOD) with the use of the reduced form of cytochrome c. Under certain conditions, other hemoproteins (hemoglobin, myoglobin) may also protect cells against ROSs through their pseudoperoxidase activity in cooperation with low-molecular-mass antioxidants (e.g., ascorbate) [30,31]. Unfortunately, reactive ferryl forms of hemoproteins may then be formed. Flavins seem to be safer substrates for hemoproteins because their oxidized forms can be reduced by NADPH-dependent methemoglobin reductase (i.e., flavin reductase) [32].

Iron ions do not occur in free form inside or outside cells. They remain firmly bound to specific proteins (ferritin of the intestinal mucosa, transferrin—a plasma protein transporting iron, lactoferrin—a neutrophil protein) and are maintained in oxidized form—ferric ions, which cannot initiate the Fenton reaction. Haptoglobin (Hp) strongly binds free hemoglobin, while hemopexin binds heme, thus preventing lipid peroxidation catalyzed by hemoglobin and heme [24].

Copper ions in blood plasma are bound to a specific protein—ceruloplasmin—and in small amounts by albumin, transcuprein, and amino acids [33]. Intracellular copper ions are bound by metallothionein and GSH [34]. Ceruloplasmin demonstrates low SOD and ferroxidase activities, oxidizing Fe^{2+} ions to Fe^{3+} ions with the complete reduction of oxygen.

The primary role of metallothioneins is to bind and detoxify heavy metal ions (cadmium, mercury). Still, they are also involved in the homeostasis of metals necessary for organisms (zinc and copper). A typical mammalian metallothionein protein contains approximately 20 cysteinyl residues out of 61 amino acid residues [35]. They can, therefore, react quickly with $O_2^{\cdot-}$ and $^{\cdot}OH$. In addition to their primary functions, some highly concentrated proteins may also act as antioxidants. For example, albumin in blood plasma binds fatty acids (potential peroxidation substrates), copper, and heme and reacts with ROSs, thereby protecting more critical macromolecules of an organism. Damaged albumin molecules undergo proteolysis and are replaced by new ones. Similarly, GSH serves as a protective target for H_2O_2 in cells. GPX directs the attack of H_2O_2 on GSH. Thus, it protects thiol groups of enzymes and prevents H_2O_2 from participating in the Fenton reaction.

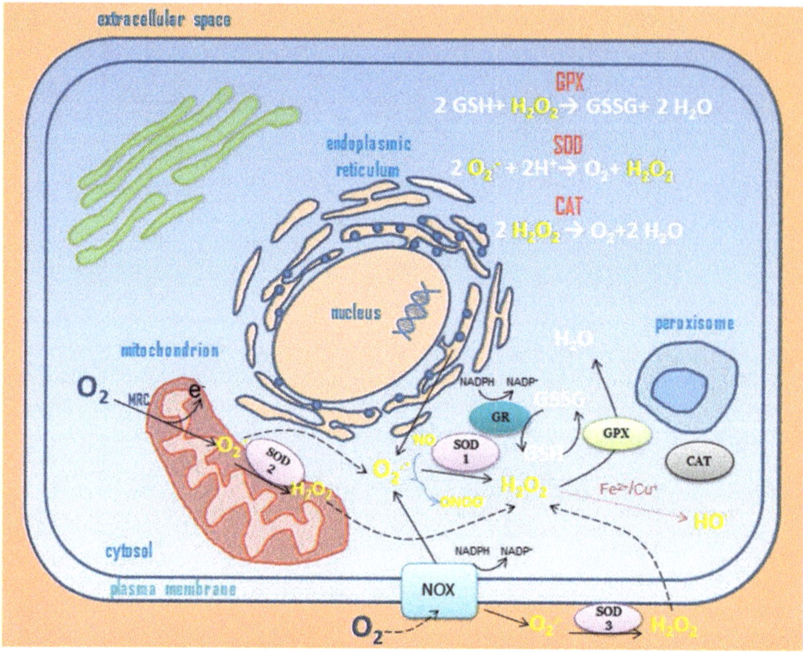

Figure 1. Maintaining a dynamic equilibrium between the formation and elimination of reactive oxygen species in eukaryotic cells. CAT: catalase; GPX: glutathione peroxidase; GR: glutathione reductase: GSH/GSSG: reduced/oxidized form of glutathione; H_2O_2: hydrogen peroxide; MRC: mitochondrial respiratory chain; NADPH/NADP$^+$: reduced/oxidized form of nicotinamide adenine dinucleotide phosphate; $^{\cdot}$NO: nitric oxide; NOX: NADPH oxidase; O_2: molecular oxygen; $O_2^{\cdot-}$: superoxide anion radical; $^{\cdot}$OH, hydroxyl radical; $ONOO^-$: peroxynitrite; SOD 1–3: superoxide dismutases 1–3.

The fundamental differences in the physical properties of polar and non-polar cellular environments entail the existence of water-soluble antioxidants, i.e., ascorbate, GSH, uric acid, bilirubin, glucose, and pyruvate, as well as lipid-soluble antioxidants, i.e., α, γ-tocopherol, α, β-carotene, lycopene, lutein, reduced coenzyme Q, and retinol. The reactions of low-molecular-mass antioxidants with ROSs and free organic radicals are less specific than enzymatic reactions, which is not a disadvantage. These compounds are more universal defenders of the body and can play different roles. They can react with the fraction of $O_2^{\cdot-}$ and H_2O_2 molecules that escaped the action of antioxidant enzymes, reducing the chance of hydroxyl radical formation [24,26].

3.3. Exercise-Induced Changes in the Redox Balance

PE can induce changes in the concentration/activity of oxidative stress markers resulting from the increased generation of ROSs. Increased ROS generation during exercise results from enhanced oxygen demand, mainly in skeletal muscles [2,36]. Large amounts of ROSs can also be released by phagocytes that infiltrate tissue at sites of muscle damage [37]. Exercise-induced ROS generation has been shown to have both positive and negative physiological effects [37,38]. High levels of ROSs can lead to damage to cellular structures and cell death. On the other hand, ROSs play an essential role in cell signaling, for example, during muscles' adaptation to exercise [39]. It is common knowledge that physical training improves physical fitness. Myokines, among others, are involved in this process; these are cytokines produced and released from skeletal muscles during exercise that affect metabolic and cellular processes in various tissues and organs. Some of them can affect the oxidant–antioxidant balance [40]. It has also been demonstrated that ROS-linked pathways can participate in myokine induction [39].

Numerous studies confirm the effects of exercise on the redox balance. However, the effects of this influence vary depending on the nature of changes in individual markers of the balance. ROS overproduction depends mainly on the intensity or volume of exercise [41]. For example, cycling with intensities of 50%, 60%, and 70% of the maximal VO_2 for 10, 20, and 30 min has enhanced oxidative stress and antioxidant activities in men who lead sedentary lifestyles. However, the changes in individual oxidative stress markers observed by the researchers varied and depended on the duration of PE. SOD activity in erythrocytes, for example, increased immediately after exercise of all intensities and durations, except for an intensity of 70% of the maximal VO_2 for 30 min, when SOD activity decreased. In the case of GPX, the only statistically significant changes consisted of a decrease in the activity of this enzyme after 20 and 30 min at all intensity levels [42]. Other studies, in turn, showed no statistically significant changes in the serum malondialdehyde (MDA) and GPX4 levels of 20 healthy young adults (men and women) and a lack of regular exercise habits after aerobic exercise at moderate intensity on a power bike (five times per week for 4 weeks). The SOD levels, on the other hand, decreased in a statistically significant manner [43]. After a 12 week maximum fat oxidation intensity exercise, however, there was a decrease in CAT activity and MDA levels and an increase in serum SOD activity in obese patients (men and women) with nonalcoholic fatty liver disease (all junior students without physical education courses). The GPX activity and total antioxidant capacity did not change in a statistically significant manner [44]. There were no changes in MnSOD, CAT, GPX, and hemeoxygenase-1 in muscle biopsy material after a 4 week training intervention consisting of high-intensity interval training in healthy volunteers [45]. A specific training program was shown to improve elite karate athletes' oxidant–antioxidant balance. After 3 months of karate training, the authors observed an increase in SOD and CAT activity, while the MDA concentrations decreased [46].

Studies confirm that suitable exercise maintains physiological levels of ROSs, which allows skeletal muscle to function properly and facilitates adaptation to exercise [46,47]. Among other things, regular PE induces adaptive changes in endogenous antioxidant mechanisms [17,37,48,49]. These changes can involve both enzymatic and non-enzymatic antioxidants. It has been shown that oxidants can affect gene expression. Changes in gene expression involving ROSs have been demonstrated at the level of transcription, mRNA stability, and signal transduction. Under the influence of exercise, changes occur in SOD activity, among other things [50]. Comparing the mRNA levels of antioxidant enzymes before and after physical training, the authors showed, for example, an increase in the mRNA levels of SOD1 in peripheral blood mononuclear cells after completing a 2 week training period (jogging for 30 min 5 days per week) in healthy subjects. The mRNA levels of SOD1 and SOD2 were also found to be higher in a statistically significant manner two weeks after the completion of the training program. Before exercise, the SOD2 mRNA levels were higher in those with exercise habits compared to those without [51]. It is considered that regular exercise induces so-called "oxidative eustress" [52]. This type of oxidative stress, in contrast with "oxidative distress", helps maintain redox homeostasis. This is possible, among other means, through the activation of transcription factors including erythroid-related nuclear factor 2 (NRF2) [52], which is considered a key regulator of antioxidant defense [53,54]. This factor regulates the expression of GR, SOD, GPX, thioredoxin reductase, hemeoxygenase-1, peroxiredoxin, metallothionein, and thioredoxin, among others. Numerous studies confirm NRF2 activation in response to exercise [55,56].

4. Genes Responsible for the Production of Antioxidant Capacity in the Context of Athletic Performance

4.1. Polymorphisms of Antioxidant Genes and Physical Performance

There are not many publicly available papers on research about the direct impact of antioxidant genes on physical performance (Table 2). Interestingly, the investigated antioxidant gene polymorphisms are not associated with improved antioxidant capacity. This covers changes leading to a weakening of the antioxidant barrier. Several SNPs of antioxidant enzyme genes have been noted to cause a decrease in the levels or activities of antioxidant enzymes. They include, among others, a valine/alanine substitution at position 9 of the amino acid chain in the mitochondrial targeting sequence (MTS) of MnSOD (Val-9Ala: NCBI, refSNP ID: rs1799725), adenine/thymine substitution at position 21 of the promoter region of the CAT gene (21A/T: NCBI, refSNP ID: rs7943316), and proline/leucine substitution at position 198 of the amino acid chain in GPX1 (Pro-198Leu: NCBI, refSNP ID: rs1050450) [57].

Table 2. Polymorphisms of antioxidant parameters assessed for their relationships with physical performance.

Parameter	Genotypes (Single-Nucleotide Polymorphisms)	Relationship with Physical Performance	References
Manganese superoxide dismutase (SOD 2, EC 1.15.1.1)	Val/Val, Ala/Val, Ala/Ala	Yes	[57–59]
Catalase (CAT, EC 1.11.1.6)	AA, AT, TT	No	[57]
Glutathione peroxidase 1 (GPX 1, EC 1.11.1.9)	Pro/Pro, Pro/Leu, Leu/Leu	No	[57]
Haptoglobin (Hp)	1F-1F, 1F-1S, 1S-1S, 1F-2, 1S-2, 2-2	Yes	[57]
Glutathione S-transferase (GST, EC 2.5.1.18)	GSTM1-, GSTM1+, GSTT1-, GSTT1+, GSTM1-T1-, GSTM1+T1-, GSTM-T+, GSTM+T+	No	[57]

Val: valine; Ala: alanine; A: adenine; T: thymine; Pro: proline; Leu: leucine.

Ben-Zaken et al. [58] proved that the Ala allele in the MnSOD gene polymorphism occurs more often than the Val allele in professional athletes (N = 195) in both endurance and resistance in comparison with healthy controls (N = 240). The frequency of this allele compared to the Val allele also increases with the athlete's training level (higher in athletes at the Olympic level than in athletes at the national level) but is not influenced by the essential type of PE (aerobic/endurance or anaerobic/resistance). MTS directs the primary translation product of MnSOD to the mitochondrion; therefore, the replacement of valine with alanine causes a change in the spatial conformation of MTS from an α-helix to a β-sheet, impairs the transport of the enzyme, and ultimately reduces the antioxidant capacity in mitochondria. The authors suggest that increased physical performance in this way results from angiogenesis, mitochondrial biosynthesis, and muscle hypertrophy, which are intensified due to higher ROS concentrations [58]. A point mutation in the MnSOD gene with a similar redox effect was also investigated by Ahmetov et al. [59]. Specifically, this is about the Ala16Val polymorphism (rs4880 C/T), which is concerned with a substitution of cytosine for thymine, resulting in a substitution of alanine for valine at position 16 of the amino acid chain in MTS. Consequently, a reduction in the mitochondrial antioxidant capacity is observed as a result of the inefficient transport of MnSOD across the mitochondrial membrane. In contrast to the findings of Ben-Zaken et al. [58], the authors of this study demonstrated that the genotype associated with reduced MnSOD antioxidant capacity is less prevalent among athletes. However, this refers only to strength athletes (N = 524), as no differences were observed between endurance athletes (N = 180) and controls (N = 917). The mentioned antioxidant gene polymorphisms were also evaluated in the context of physical performance by Akimoto et al. [57]. They examined long-distance runners (135 individuals in total, 82 men and 53 women aged 15–58 years) with the following gene polymorphisms: Val/Val, Ala/Val, and Ala/Ala genotypes of MnSOD, AA, AT, and TT genotypes of CAT, and Pro/Pro, Pro/Leu, and Leu/Leu genotypes of GPX1. The DD, ID, and II genotypes of ACE were also evaluated. The impact of a given genotype on endurance performance was assessed. This was done using the parameters of post-exercise damage in venous blood serum: aspartate aminotransferase, alanine aminotransferase, creatine kinase, thiobarbituric acid reactive substances (TBARSs, mainly MDA), as well as using a comet assay of nucleated blood cells (leukocytes). The runners were tested with a single run over a distance of 4–21 km. The study shows that only the MnSOD polymorphisms are related to aerobic capacity. The authors found statistically significant lower serum activity of creatine kinase in runners with the Ala/Ala MnSOD genotype compared to the enzyme activities in runners with the other MnSOD genotypes. The study by Ahmetov et al. suggests an opposing relationship. In their study, elevated creatine kinase activity (women) and creatinine concentration (men and women) were positively correlated in professional athletes with the MnSOD polymorphism, which results in a reduction in mitochondrial antioxidant capacity [59]. Akimoto et al. [57] were also the first researchers to examine glutathione S-transferase (GST) in the context of physical performance. They studied cytosolic GST-μ and GST-θ, which are encoded by the glutathione S-transferase M1 and T1 genes, respectively. The family of those isoenzymes (EC 2.5.1.18) plays a key role in redox neutralization (reduction) of potentially harmful metabolic products using GSH. These mainly include a wide spectrum of xenobiotics and oxidative stress products [60]. It has been found that polymorphisms of these genes, which are based on allelic deletion (null genotype), result in a higher prevalence of neoplasm [60,61], and these are the loci that the authors examined, but they found no relation between them and endurance performance [57]. Moreover, Akimoto et al. [57] revealed that haptoglobin, a serum glycoprotein that is a type of acute-phase protein, may also be involved in endurance performance. The serum TBARS concentration was lower in runners with the 1S-1S Hp genotype than in runners with the 1F-1S and 1S-2 genotypes (no statistically significant differences compared to the other genotypes: 1F-1F, 1F-2, and 2-2). Hp is responsible for binding free hemoglobin in the bloodstream, which prevents oxidative stress resulting from the Fenton reaction [62]. Hp dysfunctions are associated with a higher incidence and clinical stage of many inflammatory and autoimmune diseases [63]. For instance, the 2-2 genotype is associated with an increased risk of developing vascular complications in patients suffering from diabetes. This results from the fact that heme iron combined with this type of Hp is more susceptible to redox reactions than other Hp complexes with heme iron, which results in higher concentrations of oxidized lipids and, thus, increased levels of dysfunctional lipoproteins [64].

4.2. *Antioxidant Adaptation as a Result of Exercise*

As mentioned in the previous section, regular PE has been widely shown to increase the ability to alleviate inflammation and oxidative stress by inducing an adaptive response in the endogenous antioxidant system. This is manifested by, among other things, the positive impact of exercise on ${}^{\cdot}$NO bioavailability in the endothelium. It has been found that the ${}^{\cdot}$NO concentration increases as a result of regular physical activity in both young and elderly individuals [65], which may be accompanied at the same time by the maintained redox balance (correspondingly higher plasma antioxidant capacity) and eventually proper cutaneous microcirculatory functions in older people [66]. Another research study also points to this; 36 older (47–74 years) professional long-distance runners had better regulation of blood circulation (endothelium function) and higher plasma antioxidant capacity than 36 controls (sex-matched, untrained people aged 46–77 years, with no cardiovascular disease or any risk factors of the disease). Blood flow in the hands and feet was examined in the participants under conditions of increased temperature (44 °C), as well as ischemia and hyperemia (3 min of brachial artery occlusion), using a laser Doppler flowmetry and biochemical parameters in venous blood (left cephalic vein). Higher plasma concentrations of ˙NO in the form of its metabolites (nitrites and nitrates), mRNA of PGC-1α (PPARγ coactivator-1 α), total antioxidant capacity, and sirtuin 1, as well as higher miR29 concentrations in mononuclear cells of the blood, were found in the athletes [67]. The increase in the eNOS activity and phosphorylation of serine at the 1177 position of its amino acid chain is most probably an adaptive and vasoprotective effect of PE. The molecular changes also include increased levels of integrins (cell membrane proteins) and H_2O_2 in the endothelium. Exercise-induced eNOS activation is transient and reversible and is regulated in redox reactions, including the upregulation of SOD (SOD1 and SOD3) and downregulation of NOX [68].

The same can be said of myokines, which, as previously stated, play a role in the physiological adaptations that occur during exercise. The induction and release of myokines are, in part, mediated by the muscles' production of ROSs [39]. At the same time, some myokines exhibit considerable antioxidant potential and the capacity to regulate the redox balance. For example, under ischemia/reperfusion-induced oxidative stress in the heart, brain-derived neurotrophic factor (BDNF) has been shown to reduce the concentration of H_2O_2, while leukemia inhibitory factor (LIF) and fibroblast growth factor 21 (FGF-21) have been demonstrated to result in a lower concentration of $O_2^{\cdot-}$. A similar effect regarding $O_2^{\cdot-}$ has been found for interleukin-6 (IL-6) in the brain and regarding $ONOO^-$ for fibroblast growth factor 2 (FGF-2) in the kidneys. Concurrently, the same or other myokines contribute to the higher activities of antioxidant enzymes (SOD, GPX) and/or oxidant enzymes (NOS, NOX) [40]. In general, myokines are signaling molecules that act in a number of ways—inside muscle cells ("in situ"), including in an autocrine manner, as well as in paracrine and endocrine manners on other tissues throughout the body. They may protect against oxidative stress while also acting as modulators of metabolism via redox reactions in both physiological and pathological conditions (e.g., PE and aging, as well as cardiometabolic diseases and cancer, respectively). For example, irisin, a recently identified myokine, appears to possess the ability to mitigate and even repair oxidative damage to muscle tissue caused by the aging process in individuals who engage in regular exercise. Myokines are also involved in the regulation of muscle tissue reconstruction. In satellite cells, the Pax genes are activated, and the sequential expression of myogenic regulatory factors occurs as follows: MyoD, Myf5, myogenin, and MRF4. The result is the proliferation, differentiation, and fusion of satellite cells into new multinucleated muscle cells, processes that are dependent on a transient increase in the concentration of ROSs [68]. Furthermore, numerous animal and in vitro studies have corroborated these findings [40,68]. For instance, cardiomyocyte culture and adult mice subjected to a pro-inflammatory agent (lipopolysaccharide) have been shown to enhance the cardiac concentration of FGF-21, which resulted in increased expression of antioxidant genes (SOD-2 and uncoupling protein 3, UCP-3) and capacity via an autocrine manner [69].

Finally, sometimes, a gene seemingly unrelated to the oxidant–antioxidant balance can affect it. An example is the MYBPC3 gene, in which SNPs are associated with either the phenotype of an elite athlete's heart or hypertrophy cardiomyopathy, depending on the gene allele. MYBPC3 encodes myosin-binding protein C3. Phosphorylation of this structural protein increases cardiac contraction, and its isoforms are also present in skeletal muscles. The polymorphisms of the gene present in professional athletes are probably related to theophylline, quinate, and decanoylcarnitine, metabolites whose level increases with aerobic capacity and that increase antioxidant capacity (theophylline and decanoylcarnitine). In turn, hypertrophy cardiomyopathy, a condition characterized by oxidative stress, is associated with ursodeoxycholate [70].

5. Conclusions and Future Perspectives

5.1. Conclusions

The importance of physical exertion in healthcare is increasingly being recognized, with evidence indicating that it is an effective means of improving metabolic and cardiorespiratory health. In this context, exercise is accorded particular attention with regard to the prevention and treatment of metabolic syndrome and its associated diseases [3,4]. In contrast, in the field of sports, novel approaches to enhancing performance are consistently being investigated. Therefore, investigating the origin of an organism's characteristics, specifically its genes, can facilitate the realization of health benefits that extend beyond the primary objective [6].

The literature in the field of sports genomics is mainly focused on genes that regulate muscle functions during physical exertion and genes of structural proteins in muscles, especially those responsible for contraction. Most of these genes are directly or indirectly related to oxidant processes, not antioxidant ones [5,9,12–17]. Another issue is inconsistent conclusions regarding the impact of exercise on the redox balance. On the one hand, the human body needs free radicals for effective PE, as they are signal conductors between cells/tissues. In any case, the consequence of this effort is the augmented generation of free radicals. ROSs and RNSs can also potentially threaten an organism in excessive concentrations. They may contribute to severe oxidative damage of cellular components and lead to pathological changes, starting with inflammation. However, this is not the case with exercise, as the effects of PE on an organism remain within the parameters of physiological responses. The consequences of free radicals, oxidative-stress-induced microdamage, and inflammation as a result of exercise are adaptive changes in the body, which are expressed in higher antioxidant capacity. This is a condition that an organism can cope with and ultimately benefit from [36–39]. The entire effect of exercise on the human redox equilibrium depends on the intensity, type, and duration of an exercise bout. The physical fitness of a given individual is also important, as well as whether an individual is a healthy or sick person [42–46]. In general, however, an organism's reactions tend to maintain the redox balance in each condition (keeping the system's self-regulation). This should be remembered when concluding research studies on antioxidant genes in the context of physical performance. On the one hand, it has been demonstrated that the frequency of prooxidant modification in the MnSOD gene (Ala allele) is strongly and positively correlated with the training level without an association with the type of exercise [57,58]. On the other hand, there are alleles of antioxidant genes that increase antioxidant capacity and, thereby, physical performance (e.g., 1S-1S allele of the Hp gene) [57]. Maintenance of the redox balance—the cooperation between oxidants and antioxidants in an organism to keep its proper functions—is well illustrated by the endothelium's already-mentioned regulation of blood pressure. For appropriate endothelium function, both oxidants ($^{\cdot}NO$, H_2O_2) and antioxidants (higher total plasma antioxidant capacity) are required [65–68]. A comparable redox effect is observed in the case of myokines [40,69].

5.2. Future Perspectives

It should be remembered that genes interact in many different ways. Thus, an examination of particular genes in one research experiment may provide different results in another, depending on the subjects, i.e., on the genotypes included in a study. The entire genotype shapes the phenotype. SNPs are not the only possible origin of differences. Many other genetic markers should also be considered (e.g., rare mutations), as well as epigenetic features. Future studies should also involve genes linked to other sport-related predispositions, in addition to those representing exercise physiology and anatomy. For instance, genes involved in shaping personality and mental traits should be included.

Author Contributions: Conceptualization: P.S. and A.W.; Writing—original draft: P.S., M.M., M.P. and A.W.; Writing—review and editing: P.S., M.M., M.P. and A.W. All authors have read and agreed to the published version of the manuscript.

Funding: This research received no funding. The authors declare no competing financial interests.

Conflicts of Interest: The authors declare no conflict of interest.

References

1. Świątkiewicz, I.; Wróblewski, M.; Nuszkiewicz, J.; Sutkowy, P.; Wróblewska, J.; Woźniak, A. The Role of Oxidative Stress Enhanced by Adiposity in Cardiometabolic Diseases. *Int. J. Mol. Sci.* **2023**, *24*, 6382. [CrossRef] [PubMed]
2. Sutkowy, P.; Woźniak, A.; Mila-Kierzenkowska, C.; Szewczyk-Golec, K.; Wesołowski, R.; Pawłowska, M.; Nuszkiewicz, J. Physical Activity vs. Redox Balance in the Brain: Brain Health, Aging and Diseases. *Antioxidants* **2021**, *11*, 95. [CrossRef] [PubMed]
3. Corbi, G.; Conti, V.; Troisi, J.; Colucci, A.; Manzo, V.; Di Pietro, P.; Calabrese, M.C.; Carrizzo, A.; Vecchione, C.; Ferrara, N.; et al. Cardiac Rehabilitation Increases SIRT1 Activity and β-Hydroxybutyrate Levels and Decreases Oxidative Stress in Patients with HF with Preserved Ejection Fraction. *Oxidative Med. Cell. Longev.* **2019**, *2019*, 7049237. [CrossRef] [PubMed]
4. Belanger, M.J.; Rao, P.; Robbins, J.M. Exercise, Physical Activity, and Cardiometabolic Health. *Cardiol. Rev.* **2021**, *30*, 134–144. [CrossRef] [PubMed]
5. Ahmetov, I.I.; Fedotovskaya, O.N. Sports genomics: Current state of knowledge and future directions. *Cell. Mol. Exerc. Physiol.* **2012**, *1*, e1. [CrossRef]
6. Chávez-Guevara, I.A.; Hernández-Torres, R.P.; González-Rodríguez, E.; Ramos-Jiménez, A.; Amaro-Gahete, F.J. Biomarkers and genetic polymorphisms associated with maximal fat oxidation during physical exercise: Implications for metabolic health and sports performance. *Eur. J. Appl. Physiol.* **2022**, *122*, 1773–1795. [CrossRef] [PubMed]
7. De Moor, M.H.M.; Spector, T.D.; Cherkas, L.F.; Falchi, M.; Hottenga, J.J.; Boomsma, D.I.; De Geus, E.J.C. Genome-Wide Linkage Scan for Athlete Status in 700 British Female DZ Twin Pairs. *Twin Res. Hum. Genet.* **2007**, *10*, 812–820. [CrossRef] [PubMed]
8. Bray, M.S.; Hagberg, J.M.; Pérusse, L.; Rankinen, T.; Roth, S.M.; Wolfarth, B.; Bouchard, C. The Human Gene Map for Performance and Health-Related Fitness Phenotypes. *Med. Sci. Sports Exerc.* **2009**, *41*, 34–72. [CrossRef] [PubMed]
9. Semenova, E.A.; Hall, E.C.R.; Ahmetov, I.I. Genes and Athletic Performance: The 2023 Update. *Genes* **2023**, *14*, 1235. [CrossRef]
10. Montgomery, H.E.; Marshall, R.; Hemingway, H.; Myerson, S.; Clarkson, P.; Dollery, C.; Hayward, M.; Holliman, D.E.; Jubb, M.; World, M.; et al. Human gene for physical performance. *Nature* **1998**, *393*, 221–222. [CrossRef]
11. Ahmetov, I.; Kulemin, N.; Popov, D.; Naumov, V.; Akimov, E.; Bravy, Y.; Egorova, E.; Galeeva, A.; Generozov, E.; Kostryukova, E.; et al. Genome-wide association study identifies three novel genetic markers associated with elite endurance performance. *Biol. Sport* **2014**, *32*, 3–9. [CrossRef] [PubMed]
12. Drozdovska, S.B.; Lysenko, O.M.; Dosenko, V.E.; Ilyin, V.N. Dependence of Aerobic Performance of Athletes on Polymorphism of Genes. *Centr. Eur. J. Sport Sci. Med.* **2015**, *9*, 65–73.
13. Végh, D.; Reichwalderová, K.; Slaninová, M.; Vavák, M. The Effect of Selected Polymorphisms of the *ACTN3*, *ACE*, *HIF1A* and *PPARA* Genes on the Immediate Supercompensation Training Effect of Elite Slovak Endurance Runners and Football Players. *Genes* **2022**, *13*, 1525. [CrossRef] [PubMed]
14. A Rabelo, L.; Alenina, N.; Bader, M. ACE2–angiotensin-(1–7)–Mas axis and oxidative stress in cardiovascular disease. *Hypertens. Res.* **2010**, *34*, 154–160. [CrossRef] [PubMed]
15. Kim, T. Peroxisome-proliferator-activated receptors regulate redox signaling in the cardiovascular system. *World J. Cardiol.* **2013**, *5*, 164–174. [CrossRef] [PubMed]
16. Kumar, H.; Choi, D.-K. Hypoxia Inducible Factor Pathway and Physiological Adaptation: A Cell Survival Pathway? *Mediat. Inflamm.* **2015**, *2015*, 584758. [CrossRef] [PubMed]
17. Sutkowy, P.; Wróblewska, J.; Wróblewski, M.; Nuszkiewicz, J.; Modrzejewska, M.; Woźniak, A. The Impact of Exercise on Redox Equilibrium in Cardiovascular Diseases. *J. Clin. Med.* **2022**, *11*, 4833. [CrossRef] [PubMed]
18. Yalçınkaya, B.H.; Genc, S.; Yılmaz, B.; Özilgen, M. How does exercise affect energy metabolism? An in silico approach for cardiac muscle. *Heliyon* **2023**, *9*, e17164. [CrossRef] [PubMed]
19. Ashcroft, S.P.; Stocks, B.; Egan, B.; Zierath, J.R. Exercise induces tissue-specific adaptations to enhance cardiometabolic health. *Cell Metab.* **2024**, *36*, 278–300. [CrossRef]
20. Burke, L.M.; Whitfield, J.; Heikura, I.A.; Ross, M.L.R.; Tee, N.; Forbes, S.F.; Hall, R.; McKay, A.K.A.; Wallett, A.M.; Sharma, A.P. Adaptation to a low carbohydrate high fat diet is rapid but impairs endurance exercise metabolism and performance despite enhanced glycogen availability. *J. Physiol.* **2021**, *599*, 771–790. [CrossRef]
21. Vigh-Larsen, J.F.; Ørtenblad, N.; Nielsen, J.; Andersen, O.E.; Overgaard, K.; Mohr, M. The Role of Muscle Glycogen Content and Localization in High-Intensity Exercise Performance: A Placebo-Controlled Trial. *Med. Sci. Sports Exerc.* **2022**, *54*, 2073–2086. [CrossRef] [PubMed]
22. Maunder, E.; Rothschild, J.A.; Fritzen, A.M.; Jordy, A.B.; Kiens, B.; Brick, M.J.; Leigh, W.B.; Chang, W.-L.; Kilding, A.E. Skeletal muscle proteins involved in fatty acid transport influence fatty acid oxidation rates observed during exercise. *Pflügers Arch. Eur. J. Physiol.* **2023**, *475*, 1061–1072. [CrossRef] [PubMed]
23. Sato, T.; Hamada, Y.; Sumikawa, M.; Araki, S.; Yamamoto, H. Solubility of Oxygen in Organic Solvents and Calculation of the Hansen Solubility Parameters of Oxygen. *Ind. Eng. Chem. Res.* **2014**, *53*, 19331–19337. [CrossRef]
24. Bartosz, G. *The Other Face of Oxygen. Free Radicals in Nature*; Polish Scientific Publishers: Warsaw, Poland, 2005.
25. Boveris, A. Mitochondrial Production of Superoxide Radical and Hydrogen Peroxide. In *Tissue Hypoxia and Ischemia*; Reivich, M., Coburn, R., Lahiri, S., Chance, B., Eds.; Springer: Boston, MA, USA, 1977; pp. 67–82.
26. Bartosz, G. Reactive oxygen species: Destroyers or messengers? *Biochem. Pharmacol.* **2009**, *77*, 1303–1315. [CrossRef] [PubMed]

27. Babiy, A.V.; Gebicki, S.; Gebicki, J.M. Protein Peroxides: Formation by Superoxide-Generating Systems and During Oxidation of Low Density Lipoprotein. In *Free Radicals from Basic Science to Medicine*; Springer: Berlin/Heidelberg, Germany, 1993; pp. 340–348. [CrossRef]
28. Gebicki, S.; Gebicki, J.M. Formation of peroxides in amino acids and proteins exposed to oxygen free radicals. *Biochem. J.* **1993**, *289*, 743–749. [CrossRef] [PubMed]
29. Sousa, J.S.; D'Imprima, E.; Vonck, J. Mitochondrial Respiratory Chain Complexes. In *Membrane Protein Complexes: Structure and Function*; Harris, J.R., Boekema, E.J., Eds.; Springer: Singapore, 2018; pp. 167–227.
30. Harel, S.; Kanner, J. Haemoglobin and Myoglobin as Inhibitors of Hydroxyl Radical Generation in a Model System of "Iron Redox" Cycle. *Free. Radic. Res. Commun.* **1989**, *6*, 1–10. [CrossRef] [PubMed]
31. Yang, W.; de Bono, D. Myoglobin protects against endothelial cell membrane damage associated with hydrogen peroxide or xanthine/xanthine oxidase. *FEBS Lett.* **1993**, *319*, 145–150. [CrossRef] [PubMed]
32. Hultquist, D.E.; Xu, F.; Quandt, K.S.; Shlafer, M.; Mack, C.P.; Till, G.O.; Seekamp, A.; Betz, A.L.; Ennis, S.R. Evidence that NADPH-dependent methemoglobin reductase and administered riboflavin protect tissues from oxidative injury. *Am. J. Hematol.* **1993**, *42*, 13–18. [CrossRef]
33. Linder, M.C. *Biochemistry of Copper*; Springer Science and Business Media LLC: Dordrecht, The Netherlands, 1991.
34. Freedman, J.H.; Ciriolo, M.R.; Peisach, J. The role of glutathione in copper metabolism and toxicity. *J. Biol. Chem.* **1989**, *264*, 5598–5605. [CrossRef]
35. Sato, M.; Bremner, I. Oxygen free radicals and metallothionein. *Free. Radic. Biol. Med.* **1993**, *14*, 325–337. [CrossRef]
36. Dimauro, I.; Paronetto, M.P.; Caporossi, D. Exercise, redox homeostasis and the epigenetic landscape. *Redox Biol.* **2020**, *35*, 101477. [CrossRef] [PubMed]
37. Kawamura, T.; Muraoka, I. Exercise-Induced Oxidative Stress and the Effects of Antioxidant Intake from a Physiological Viewpoint. *Antioxidants* **2018**, *7*, 119. [CrossRef] [PubMed]
38. Canale, R.E.; Farney, T.M.; McCarthy, C.G.; Bloomer, R.J. Influence of acute exercise of varying intensity and duration on postprandial oxidative stress. *Eur. J. Appl. Physiol.* **2014**, *114*, 1913–1924. [CrossRef] [PubMed]
39. Ost, M.; Coleman, V.; Kasch, J.; Klaus, S. Regulation of myokine expression: Role of exercise and cellular stress. *Free. Radic. Biol. Med.* **2016**, *98*, 78–89. [CrossRef] [PubMed]
40. Szabó, M.R.; Pipicz, M.; Csont, T.; Csonka, C. Modulatory Effect of Myokines on Reactive Oxygen Species in Ischemia/Reperfusion. *Int. J. Mol. Sci.* **2020**, *21*, 9382. [CrossRef] [PubMed]
41. Nobari, H.; Nejad, H.A.; Kargarfard, M.; Mohseni, S.; Suzuki, K.; Adsuar, J.C.; Pérez-Gómez, J. The Effect of Acute Intense Exercise on Activity of Antioxidant Enzymes in Smokers and Non-Smokers. *Biomolecules* **2021**, *11*, 171. [CrossRef] [PubMed]
42. Daud, D.M.A.; Ahmedy, F.; Baharuddin, D.M.P.; Zakaria, Z.A. Oxidative Stress and Antioxidant Enzymes Activity after Cycling at Different Intensity and Duration. *Appl. Sci.* **2022**, *12*, 9161. [CrossRef]
43. Ji, Y.; Ni, X.; Zheng, K.; Jiang, Y.; Ren, C.; Zhu, H.; Xiao, M.; Wang, T. Synergistic effects of aerobic exercise and transcranial direct current stimulation on executive function and biomarkers in healthy young adults. *Brain Res. Bull.* **2023**, *202*, 110747. [CrossRef] [PubMed]
44. Yang, R.; Wan, L.; Zhu, H.; Peng, Y. The effect of 12 week-maximum fat oxidation (FATmax) intensity exercise on microvascular function in obese patients with nonalcoholic fatty liver disease and its mechanism. *Gen. Physiol. Biophys.* **2023**, *42*, 251–262. [CrossRef]
45. Flockhart, M.; Nilsson, L.C.; Tais, S.; Ekblom, B.; Apró, W.; Larsen, F.J. Excessive exercise training causes mitochondrial functional impairment and decreases glucose tolerance in healthy volunteers. *Cell Metab.* **2021**, *33*, 957–970.e6. [CrossRef]
46. Jemili, H.; Mejri, M.; Bouhlel, E.; Amri, M. Biochemical status, oxidative and antioxidant responses after 3-month specific training in elite karate athletes. *Imaging* **2017**, *104*, 344–354. [CrossRef] [PubMed]
47. Wang, F.; Wang, X.; Liu, Y.; Zhang, Z. Effects of Exercise-Induced ROS on the Pathophysiological Functions of Skeletal Muscle. *Oxidative Med. Cell. Longev.* **2021**, *2021*, 3846122–5. [CrossRef] [PubMed]
48. Thirupathi, A.; Pinho, R.A. Effects of reactive oxygen species and interplay of antioxidants during physical exercise in skeletal muscles. *J. Physiol. Biochem.* **2018**, *74*, 359–367. [CrossRef] [PubMed]
49. Quan, H.; Koltai, E.; Suzuki, K.; Aguiar, A.S.; Pinho, R.; Boldogh, I.; Berkes, I.; Radak, Z. Exercise, redox system and neurodegenerative diseases. *Biochim. et Biophys. Acta (BBA)—Mol. Basis Dis.* **2020**, *1866*, 165778. [CrossRef] [PubMed]
50. Rahal, A.; Kumar, A.; Singh, V.; Yadav, B.; Tiwari, R.; Chakraborty, S.; Dhama, K. Oxidative Stress, Prooxidants, and Antioxidants: The Interplay. *BioMed Res. Int.* **2014**, *2014*, 761264. [CrossRef]
51. Ogino, S.; Ogino, N.; Tomizuka, K.; Eitoku, M.; Okada, Y.; Tanaka, Y.; Suganuma, N.; Ogino, K. SOD2 mRNA as a potential biomarker for exercise: Interventional and cross-sectional research in healthy subjects. *J. Clin. Biochem. Nutr.* **2021**, *69*, 137–144. [CrossRef] [PubMed]
52. Lisi, V.; Senesi, G.; Bertola, N.; Pecoraro, M.; Bolis, S.; Gualerzi, A.; Picciolini, S.; Raimondi, A.; Fantini, C.; Moretti, E.; et al. Plasma-derived extracellular vesicles released after endurance exercise exert cardioprotective activity through the activation of antioxidant pathways. *Redox Biol.* **2023**, *63*, 102737. [CrossRef]
53. Ngo, V.; Duennwald, M.L. Nrf2 and Oxidative Stress: A General Overview of Mechanisms and Implications in Human Disease. *Antioxidants* **2022**, *11*, 2345. [CrossRef]

54. Kasai, S.; Shimizu, S.; Tatara, Y.; Mimura, J.; Itoh, K. Regulation of Nrf2 by Mitochondrial Reactive Oxygen Species in Physiology and Pathology. *Biomolecules* **2020**, *10*, 320. [CrossRef]
55. Souza, J.; da Silva, R.A.; Scheffer, D.d.L.; Penteado, R.; Solano, A.; Barros, L.; Budde, H.; Trostchansky, A.; Latini, A. Physical-Exercise-Induced Antioxidant Effects on the Brain and Skeletal Muscle. *Antioxidants* **2022**, *11*, 826. [CrossRef]
56. Gao, L.; Wang, H.-J.; Tian, C.; Zucker, I.H. Skeletal Muscle Nrf2 Contributes to Exercise-Evoked Systemic Antioxidant Defense Via Extracellular Vesicular Communication. *Exerc. Sport Sci. Rev.* **2021**, *49*, 213–222. [CrossRef]
57. Akimoto, A.K.; Miranda-Vilela, A.L.; Alves, P.C.Z.; Pereira, L.C.d.S.; Lordelo, G.S.; Hiragi, C.d.O.; da Silva, I.C.R.; Grisolia, C.K.; Klautau-Guimarães, M.d.N. Evaluation of gene polymorphisms in exercise-induced oxidative stress and damage. *Free Radic. Res.* **2010**, *44*, 322–331. [CrossRef]
58. Ben-Zaken, S.; Eliakim, A.; Nemet, D.; Kassem, E.; Meckel, Y. Increased prevalence of MnSOD genetic polymorphism in endurance and power athletes. *Free Radic. Res.* **2013**, *47*, 1002–1008. [CrossRef]
59. Ahmetov, I.I.; Naumov, V.A.; Donnikov, A.E.; Maciejewska-Karłowska, A.; Kostryukova, E.S.; Larin, A.K.; Maykova, E.V.; Alexeev, D.G.; Fedotovskaya, O.N.; Generozov, E.V.; et al. *SOD2* gene polymorphism and muscle damage markers in elite athletes. *Free Radic. Res.* **2014**, *48*, 948–955. [CrossRef]
60. Cotton, S.C.; Sharp, L.; Little, J.; Brockton, N. Glutathione S-Transferase Polymorphisms and Colorectal Cancer: A HuGE Review. *Am. J. Epidemiol.* **2000**, *151*, 7–32. [CrossRef]
61. Cho, H.-J.; Lee, S.-Y.; Ki, C.-S.; Kim, J.-W. GSTM1, GSTT1 and GSTP1 Polymorphisms in the Korean Population. *J. Korean Med. Sci.* **2005**, *20*, 1089–1092. [CrossRef]
62. Guéye, P.M.; Glasser, N.; Férard, G.; Lessinger, J.-M. Influence of human haptoglobin polymorphism on oxidative stress induced by free hemoglobin on red blood cells. *Clin. Chem. Lab. Med.* **2006**, *44*, 542–547. [CrossRef]
63. Langlois, M.R.; Delanghe, J.R. Biological and clinical significance of haptoglobin polymorphism in humans. *Clin. Chem.* **1996**, *42*, 1589–1600. [CrossRef]
64. Goldenstein, H.; Levy, N.S.; Levy, A.P. Haptoglobin genotype and its role in determining heme-iron mediated vascular disease. *Pharmacol. Res.* **2012**, *66*, 1–6. [CrossRef]
65. Franzoni, F.; Ghiadoni, L.; Galetta, F.; Plantinga, Y.; Lubrano, V.; Huang, Y.; Salvetti, G.; Regoli, F.; Taddei, S.; Santoro, G.; et al. Physical activity, plasma antioxidant capacity, and endothelium-dependent vasodilation in young and older men. *Am. J. Hypertens.* **2005**, *18*, 510–516. [CrossRef]
66. Franzoni, F.; Plantinga, Y.; Femia, F.R.; Bartolomucci, F.; Gaudio, C.; Regoli, F.; Carpi, A.; Santoro, G.; Galetta, F. Plasma antioxidant activity and cutaneous microvascular endothelial function in athletes and sedentary controls. *Biomed. Pharmacother.* **2004**, *58*, 432–436. [CrossRef]
67. Scarfò, G.; Daniele, S.; Chelucci, E.; Rizza, A.; Fusi, J.; Freggia, G.; Costa, B.; Taliani, S.; Artini, P.; Martini, C.; et al. Regular exercise delays microvascular endothelial dysfunction by regulating antioxidant capacity and cellular metabolism. *Sci. Rep.* **2023**, *13*, 17671. [CrossRef]
68. Kojda, G.; Hambrecht, R. Molecular mechanisms of vascular adaptations to exercise. Physical activity as an effective antioxidant therapy? *Cardiovasc. Res.* **2005**, *67*, 187–197. [CrossRef]
69. Mancinelli, R.; Checcaglini, F.; Coscia, F.; Gigliotti, P.; Fulle, S.; Fanò-Illic, G. Biological Aspects of Selected Myokines in Skeletal Muscle: Focus on Aging. *Int. J. Mol. Sci.* **2021**, *22*, 8520. [CrossRef]
70. Riguene, E.; Theodoridou, M.; Barrak, L.; Elrayess, M.A.; Nomikos, M. The Relationship between Changes in *MYBPC3* Single-Nucleotide Polymorphism-Associated Metabolites and Elite Athletes' Adaptive Cardiac Function. *J. Cardiovasc. Dev. Dis.* **2023**, *10*, 400. [CrossRef]

International Journal of
Molecular Sciences

MDPI

Review

Immunogenetic and Environmental Factors in Age-Related Macular Disease

Sylwia Brodzka [1,2], Jędrzej Baszyński [1], Katarzyna Rektor [2], Karolina Hołderna-Bona [1], Emilia Stanek [1], Natalia Kurhaluk [3], Halina Tkaczenko [3], Grażyna Malukiewicz [4], Alina Woźniak [5] and Piotr Kamiński [1,2,*]

1 Division of Ecology and Environmental Protection, Department of Medical Biology and Biochemistry, Faculty of Medicine, Collegium Medicum in Bydgoszcz, Nicolaus Copernicus University in Toruń, M. Skłodowska-Curie St. 9, PL 85-094 Bydgoszcz, Poland; sylwia.brodzka@cm.umk.pl (S.B.); jedrzej.baszynski@cm.umk.pl (J.B.); karolina.holderna@cm.umk.pl (K.H.-B.); emilia.stanek@cm.umk.pl (E.S.)

2 Department of Biotechnology, Institute of Biological Sciences, Faculty of Biological Sciences, University of Zielona Góra, Prof. Z. Szafran St. 1, PL 65-516 Zielona Góra, Poland; k.rektor@wnb.uz.zgora.pl

3 Institute of Biology, Pomeranian University in Słupsk, Arciszewski St. 22 B, PL 76-200 Słupsk, Poland; natalia.kurhaluk@upsl.edu.pl (N.K.); halina.tkaczenko@upsl.edu.pl (H.T.)

4 Department of Eye Diseases, University Hospital No. 1, Faculty of Medicine, Collegium Medicum in Bydgoszcz, Nicolaus Copernicus University in Toruń, M. Skłodowska-Curie St. 9, PL 85-092 Bydgoszcz, Poland; g.malukiewicz@cm.umk.pl

5 Department of Medical Biology and Biochemistry, Faculty of Medicine, Collegium Medicum in Bydgoszcz, Nicolaus Copernicus University in Toruń, M. Karłowicz St. 24, PL 85-092 Bydgoszcz, Poland; al1103@cm.umk.pl

* Correspondence: piotr.kaminski@cm.umk.pl or p.kaminski@wnb.uz.zgora.pl

Abstract: Age-related macular degeneration (AMD) is a chronic disease, which often develops in older people, but this is not the rule. AMD pathogenesis changes include the anatomical and functional complex. As a result of damage, it occurs, in the retina and macula, among other areas. These changes may lead to partial or total loss of vision. This disease can occur in two clinical forms, i.e., dry (progression is slowly and gradually) and exudative (wet, progression is acute and severe), which usually started as dry form. A coexistence of both forms is possible. AMD etiology is not fully understood. Extensive genetic studies have shown that this disease is multifactorial and that genetic determinants, along with environmental and metabolic-functional factors, are important risk factors. This article reviews the impact of heavy metals, macro- and microelements, and genetic factors on the development of AMD. We present the current state of knowledge about the influence of environmental factors and genetic determinants on the progression of AMD in the confrontation with our own research conducted on the Polish population from Kuyavian-Pomeranian and Lubusz Regions. Our research is concentrated on showing how polluted environments of large agglomerations affects the development of AMD. In addition to confirming heavy metal accumulation, the growth of risk of acute phase factors and polymorphism in the genetic material in AMD development, it will also help in the detection of new markers of this disease. This will lead to a better understanding of the etiology of AMD and will help to establish prevention and early treatment.

Keywords: age-related macular degeneration; AMD; chemical elements impact; element-element interactions; oxidative stress; metallothioneins; lifestyle factors; genetic factors

1. Introduction

Age-related macular degeneration (AMD) is a chronic eye disease that usually appears in people after the age of 50, but in highly developed countries it can also occur in younger people. AMD is an important cause of blindness in the USA and accounts for 54% of blindness in Caucasian patients, 4.4% in black patients, and 14.3% in Spanish ones. Among white people aged 40 and older, AMD is the most common cause of visual impairment and blindness in the USA [1]. Due to the increasing life expectancy, it is expected that the frequency of AMD will increase [2–8].

Currently, AMD is considered to be the most prevalent retinal disease in the Western world. Furthermore, it is the most prevalent cause of blindness in developed nations. Considering ethnic factors, the problem is particularly remarkable among Caucasians compared to Hispanic, African, and Asian populations. A large-scale estimation revealed that in 2020, the worldwide population struggling with AMD encompassed approximately 196 million of individuals. The prognosis concerning a longer period of time (by year of 2040) is very alarming, and projects an increase to 288 million individuals. This phenomenon may largely arise from certain diet and lifestyle factors, as well as increasing lifespan [9]. Diverse causes and risk factors of AMD are summarized in Tables 1 and 2. Neovascular AMD is responsible for 90% of blindness connected with this disease. The mainstream treatment for slowing its progression is anti-vascular endothelial growth factor (anti-VEGF) therapy (aflibercept, ranibizumab). Nevertheless, this method is burdened with disadvantages such as high financial costs, requirement of frequent treatments and finally lack of entire treatment of the disease. Future approaches will be undoubtedly concentrated on the elaboration of new anti-VEGF drugs that conduce to increment in efficacy and prolongation of the treatment intervals, which ultimately make the therapy less expensive [9].

Oxidative stress plays an important role in the pathological events occurring in AMD. In this context, there are statements from clinical studies with post-mortem donor eyes that confirm presence of DNA, protein, and lipid biomarkers of oxidative stress in the course of AMD. The occurrence of oxidative DNA damage, which includes 8-hydroxy-2-deoxyguanosine (8-OHdG) is notable, particularly in dry form with RPE atrophy. Similarly, enhanced levels of carboxy-ethyl-pyrrole (CEP) protein adducts (which are indicative of docosahexaenoic acid-containing lipid peroxidation) in the Bruch's membrane of AMD donor eyes confirms augmented vulnerabilities toward oxidative stress (compared to non-AMD eyes) [12].

The dry form of disease may be connected with 60% elevation of CEP levels in comparison with control. A 60% increment in plasma CEP levels may be associated with increased complement factor H, ARMS2 or complement C3. Finally, augmented malondialdehyde (MDA) levels are also reported in serum from AMD individuals, which affirms the systemic elevation of oxidative stress [12]. There are certain observations confirming functionality of compensatory mechanisms of RPE cells to counteract oxidative stress, e.g., antioxidative agents such as catalase (CAT) and superoxide dismutase (SOD) tend to be upregulated in immunoblots of AMD donor eyes, especially in the early or intermediate stages of AMD, demonstrating the existence of a pro-survival mechanism in response to escalated oxidative stress. In cases of wet AMD, there are statements of increment of total oxidant status in serum of patients and decrease of total antioxidant status. Similarly, SOD, glutathione peroxidase (GPx), and glutathione reductase (GR) tend to be decreased in serum of wet AMD individuals compared to healthy individuals, while the activity of SOD, CAT, and GPx also tend to decrease as a consequence of wet AMD when analyzed in the erythrocytes of AMD patients in comparison with control. Therefore, augmented oxidative stress and diminished antioxidative defenses in wet AMD propel further AMD progression and underline the critical role of oxidative stress in the course of disease [12].

Table 1. Phenotypic, demographic, and environmental risk factors of AMD; based on Heesterbeek et al. 2020 [10].

Risk Factor	**Phenotypic Risk Factors**
Drusen	Predictive potential in context of AMD
	Deposits of extracellular debris are located between Bruch's membrane and retinal pigment epithelium (RPE). Different types of drusen and its location, volume, measured area, or total number of drusen may serve as predictive factors for AMD progression. Small drusen may be connected with a low probability (0.4%) of developing late AMD within five years in patients older than 55 years. Medium or large drusen with or without accompanying pigmentary abnormalities may inform about early or intermediate AMD stage. Individuals with medium drusen are burdened with probability between 2% and 20% of developing late AMD within five years, while large drusen increases such probability to between 13% and 47%.
	Different types of drusen and possible types of AMD
	Individuals with calcified drusen have a probability of 26% of developing geographic atrophy (GA, subtype of late AMD) within 5 years. The presence of reticular pseudo-drusen is highly prevalent in fellow-eyes of individuals with unilateral neovascular AMD. Progression rate in cases of neovascular AMD is estimated at the level of 31% in a period of two years. Finally, cuticular drusen may be connected with variants of *CFH* gene, and may also be helpful in the prediction of the development of nAMD and GA (progression probabilities of 8.7–12.5% and 25–28% in a period of 5 years).
Alternations of the retinal pigment epithelium	Pigmentary changes and hyperreflective foci in context of AMD risk
	The presence of pigmentary changes in addition to drusen is connected with dramatic increase of risk of developing late AMD (probability of 47% within 5 years in cases of large drusen and accompanying pigmentary changes). Hyperreflective foci may be connected with progression rates of both nAMD (47% of eyes with hyperreflective foci develop this form of AMD after two years) and GA (approximately 50% of eyes with hyperreflective foci develop GA within a period of 28 months).
	Types of pigment epithelial detachment and AMD
	Fibrovascular and serious types are both connected with nAMD. Drusenoid type may be engaged in the progression of different kinds of AMD, as well as being connected with the development of pigmentary alternations and calcified drusen. In cases of GA, features like lesion size, number, location (increased rates of GA growth in eyes with multifocal and extrafoveal lesions compared to unifocal and foveal lesions) and shape (lesions with irregular shape are connected with a faster growth rate compared to lesions with a circular shape) are important.
Vascular and other features	Choroid vascular structures alternations and flow anomalies in the context of AMD
	There is a possibility that choroid vascular structures deplete in eyes with early AMD, nAMD, and GA, as well as that irregular choroidal vessels may be a predictive factor for the development of nAMD and GA. Similarly, flow abnormalities in the choriocapillaris surrounding the atrophic lesions may be connected with the escalation in GA growth rate. Moreover, in cases of eyes with a quiescent choroidal neovascularization, there is considerably higher risk of becoming exudative (15–18) compared to eyes lacking a precursor quiescent choroidal neovascularization.
	Outer retinal tubulations and dynamics of AMD
	Outer retinal tubulations can be present in areas of fibrosis after the development of exudative nAMD or at the border of GA. In cases of individuals with GA, the localization of outer retinal tubulations at the border of GA lesion promote slower rate of GA growth compared to individuals without outer retinal tubulations.
Risk factor	**Demographic and Environmental Risk Factors**
Age	The prevalence of early AMD increases from 3.5% in individuals aged 55–59 years up to 17.6% in individuals 85 years and older. Furthermore, the prevalence rates increase from 0.1% to 9.8% respectively, as well as for late AMD.
Sex	There is a tendency of a higher progression rate to early AMD and late AMD (mainly nAMD) in cases of female sex.
Smoking	Toxic compounds contained in cigarette smoke may trigger the formation of retinal oxidative stress, vascular changes in the choroidal vessels as well as the inflammation in RPE cells. Thus, smoking is connected with a two- to four-fold increased risk for any form of AMD. Smoking may also be associated with escalation in GA growth. Quitting smoking generally mitigates the risk of developing AMD; however, risk probabilities appear to be comparable to that of non-smoking individuals after as long as 20 years cessation of smoking.
Body composition	Higher body mass index may be connected with greater probability of AMD developing. Especially obese individuals (body mass index BMI > 30) are in danger of developing late AMD compared with individuals with normal weight (BMI 20–25).

Table 1. *Cont.*

Risk factor	Demographic and Environmental Risk Factors
Diet	The number of vegetables and fruits in a diet is considered to be a protective factor in context of AMD progression, because these components of diet are rich in such antioxidants as carotenoids and vitamins. Similarly, fish consumption (rich in protective fatty acids) tends to mitigate the risk of AMD progression.
Physical activity	Physical activity tends to be connected with lower odds of both late and early AMD, because regular exercise is considered to be positively associated with activity of antioxidative enzymes.
Education	Higher education level generally acts as alleviating factor for early AMD and late AMD development (inverse relation between education level and development of AMD).
Sunlight exposure	Sunlight exposure escalates oxidative stress in the retina and triggers the development of AMD. This factor may be unreliable (difficulties in quantification of the total amount of sunlight exposure).
Ethnicity	There are evident differences between ethnic groups in context of prevalence rates of any form of AMD (5.4% Caucasians, 4.6% Chinese, 4.2% Hispanics, 2.4% Africans).
Comorbidity	Certain accompanying diseases may exert influence on AMD development and progression, e.g., cataract surgery may increase the incidence of AMD and intensify its progression because of augmented inflammatory processes during surgery and increased exposure to ultraviolet light afterward.

Table 2. Immunological, genetic, and oxidative stress driven causes of AMD; based on Shughoury et al. 2022 [11].

Causative Factor	Immune Dysregulation and the Complement System Factors
Innate immune system	Immune complex deposition is engaged in the formation and biomolecular makeup of drusen. AMD pathogenesis may be mediated by localized inflammation and microglial cell recruitment.
Complement components	Terminal fragment C5b is engaged in the uprising of a membrane attack complex (MAC) along with such complement factors as C6, C7, C8, C9. Finally, MAC is responsible for disruption of the lipid bilayer that ultimately leads to cell lysis. In this context, complement factor H (*CFH* gene) on chromosome 1 is recognized as a major susceptibility locus for AMD development. In vitro knockdown of this gene may be connected with escalated MAC deposition in choroidal endothelial cells. Common loss-of-function *CFH* variants *rs1061170 (Y402H)* and *rs1410996* constitute important genetic risk factors of AMD. Another *CFH rs121913059 (p.Arg1210Cys)* may confer approximately a 20-fold increment in AMD threat. *Complement Factor I* (*CFI*) gene on chromosome 4 and such variants as *p.Gly119Arg* and *p.Leu131Arg* may be implicated in a reduction of CFI concentration and activity subsequently leading to the development of AMD. The variant in the complement *C3* gene of chromosome 19, *rs2230199* (*p.Arg102Gly*), is connected with disrupted CFH binding, conferring resistance to CFH-mediated inactivation. This variant, as well as the associated polymorphism *rs1047286* (*p.Pro292Leu*), exist commonly in Caucasian populations and considerably augment the risk of AMD. Another candidate factor potentially connected with AMD may be complement component 5 because of its presence in drusen, as well as in the context of elevated serum C5a levels in course of disease. In turn, variants in the gene coding complement component 9 may be engaged in progression to more advanced stages of AMD. Especially *rs34882957* (*p.Pro167Ser*) is connected with elevated serum concentration of C9 and increased polymerization rates, which results in increased MAC formation and advances in course of AMD. Contrariwise, certain rare variants in the *complement component 2* and *complement factor B* genes on chromosome 6 may act as protective factors against the development of AMD. Variants mitigating the function of coded enzymes subsequently influence on the activity of certain complement cascades and exert this beneficial effect.
Extracellular Matrix/Metabolism of Lipids/Angiogenesis/Oxidative Stress/Multiple Pathways	
Extracellular Matrix Remodeling	The structure of Bruch's membrane depends on the balance between certain matrix metalloproteinases (MMPs) and their tissue inhibitors (TIMPs) (especially MMP-1, MMP-2, MMP-3, MMP-9, TIMP-1, TIMP-2, TIMP-3). TIMP-3 excess leads to impaired extracellular matrix turnover as well as pathologic thickening of Bruch's membrane. Decreased TIMP-3 activity may lead to escalated angiogenesis. Nine rare variants in *TIMP-3* are considered to be cumulatively connected with over 30-fold increased risk of AMD (*rs5754227, rs713685, rs743751, rs5749482 TIMP-3* intron variants constitute particularly important risk factors). In the matter of MMPs -gene *MMP-2* appear to be significant. There are admissions that *T* allele (*TT* and *CT* genotypes) of *rs243865* polymorphism may influences protectively against AMD, whereas homozygous *CC* genotype is possibly connected with hard drusen development in the course of disease. Variants of *MMP-9* (*rs142450006, rs3918241, rs3918241, rs3918242, rs4810482, rs17576, rs17577*) appear to be connected with threat of AMD and possible progression to the form of macular neovascularization. Also *MMP-9 CA* (13-27) microsatellite expansion variant may be related with threat of the progression to macular neovascularization. Among other genes coding for extracellular matrix components, gene variant *rs140647181*, near the *Collagen Type 8 α1* (*COL8A1*) gene is considered to be a risk factor for AMD. Presumptions about connections with AMD also apply to certain rare, protein-altering variants in the *COL8A1* gene itself.

Table 2. *Cont.*

Extracellular Matrix/Metabolism of Lipids/Angiogenesis/Oxidative Stress/Multiple Pathways	
Lipid metabolism	Lipids constitute important components of drusen making up over 40% of its volume. Pathologic accumulation of lipids results in RPE and photoreceptor loss in AMD while. Oxidation of lipoproteins participate in the progression of AMD. Therefore, variants in several genes coding for proteins engaged in lipid metabolism and cholesterol transport may be considered to be risk factors of AMD. Genes involved in functionality of high density lipoprotein (HDL) appear particularly important due to its fundamental transport and anti-inflammatory significance. The *apolipoprotein E* (*APOE*) gene is recognized as possibly connected with AMD in various contexts. E.g., allelic variant *ApoE4* may exert protective effect against the disease and confers up to 40% reduction in the threat of AMD developing. Oppositely, *ApoE2* is connected with a slightly escalated risk of AMD or pathogenesis of macular neovascularization. Important modulator is hepatic lipase (LIPC) involved in intra-retinal lipid transport and regulation of plasma HDL levels. Effects are diversified from protective (against AMD development) connected with such *LIPC* variants as *rs493258*, *rs10468017*, *rs9621532* or *rs11755724* to escalating the risk of disease associated with such *LIPC* variants as *rs13095226* and *rs3748391*. Moreover, cholesteryl ester transfer protein (CETP) may be engaged in AMD due to its role in the transport of cholesterol from peripheral tissue to the liver through transfer of cholesterol esters from low-density lipoproteins to HDLs. Mainly *CETP rs3764261* is connected with escalated risk of AMD, while two intronic variants of *CETP* (*rs17231506* and *rs5817082*) are associated respectively with slightly increased and reduced risk of AMD. Another factor is ATP-binding cassette transporter A1 (ABCA1) protein because its function in the elimination of excessive tissue cholesterol by triggering the formation of HDL. The *rs1883025 ABCA1* protein seems to exert the opposite influence due to its two alleles (*C* allele connected with increment of plasma HDL levels and escalated threat of AMD and *T* allele related to decreased HDL and lower risk of AMD).
Angiogenesis	Angiogenic processes may be engaged in the development of macular neovascularization. Vascular endothelial growth factor (VEGF) is considered to be a driving force of neovascularization. Among its isoforms elevated levels of VEGF-A are connected with ocular neovascular diseases. The *T* allele of *VEGF-A rs3025000* variant augments possibility of clinical response to anti-VEGF therapy in macular neovascularization. The *VEGF-A rs3025033* variant and haplotype of *rs1570360A-rs699947A-rs3025033G-rs2146323A* may be connected with decreased threat of neovascular AMD. Finally, fibulin 5 (FBLN5) dysfunction may pose important factor in AMD pathogenesis because this extracellular matrix protein is engaged in modulation of angiogenesis and antagonizing VEGF.
Oxidative stress	Certain genetic mutations heighten susceptibility to oxidative damages that leads to photoreceptor dysfunction and AMD pathogenesis. Rare mutations in *RAD51B* gene are connected with an augmented threat of disease and individuals with AMD may characterize themselves with abnormally decreased serum concentrations of RAD51B. Additionally, the tumor necrosis factor receptor superfamily member *10A* (*TNFRSF10A*) gene mutations may increase the risk of AMD, mainly in Asian populations. Animal models confirmed that decreased expression of oxidative stress is connected with reduced RPE cell viability and escalated apoptosis. Mutations in excision *repair cross complex 6* (*ERCC6*) gene may exert certain influence on the augmentation of AMD risk because individuals with AMD may demonstrate decreased retinal ERCC6 expression. Subsequently, in such patients ERCC6 functionality in context of transcription-coupled excision repair of DNA mutations undergoes perturbation.
Genes implicated in multiple pathways	The region of chromosome 10q26 spanning high-temperature requirement factor A serine peptidase 1 (*HTRA1*) gene promoter and the age-related maculopathy susceptibility 2 (*ARMS2*) gene coding region constitutes loci of particular significance in context of AMD susceptibility. In cases of Caucasians and East Asians, *ARMS2-HTRA1* region together with *CFH*, is responsible for over half of the genetic threat related to AMD. *ARMS2-HTRA1* variants are connected with more rapid course of AMD. Certain local *ARMS2* dysfunction may be associated with oxidative damages of the retina. In turn, *HTRA1* shows implication in extracellular matrix remodeling, TGF-β cytokine signaling as well as acts as regulator of local subretinal inflammatory processes and angiogenesis. Both gene variants *HTRA1 rs11200638* and *ARMS2 rs10490924* are connected with an escalated AMD threat and younger onset of the disease. *ARMS2 rs10490924* may implicate inflammatory response due to correlation with elevated C-reactive protein levels, while G allele is connected with augmented threat of AMD progression to advanced stages. However, this variant pose as more significant risk factor of the disease in Asian populations compared to Europeans (risk allele frequencies of 40% versus 20%). This fact also gains importance in the context of a slight connection between *ARMS2* dysfunction and progression to neovascular disease. Other *ARMS-HTRA1* variant *rs2284665* seems to be associated with development of macular neovascularization. Finally, *HTRA1* variants *rs1049331*, *rs2293870*, and *rs2284665* are reported as connected with AMD progression, possibly through disrupted HTRA1-mediated inhibition of cellular apoptosis and insulin-like growth factor 1.

A series of changes caused by aging of RPE cells can lead to AMD. The aging of retinal pigment epithelium cells (RPE) break the balance of enzymes in the extracellular matrix in macular area gathers on the Bruch membrane (BrM). Metabolites accumulate on the BrM, forming vitreous warts, damaging adjacent retinal tissues, and reducing the blood supply of retina. RPE cells senescence leads immune cells to produce vascular endothelial growth factor (VEGF). The calcification, rupture, and phagocytosis of BrM produce blood vessels which may lead to AMD (Figure 1).

A fundamental factor in the process of neovascularization is VEGF-A, therefore it is also considered to be the the main focus of anti-VEGF therapies. Pathological angiogenesis is connected with the improper regulation of VEGF, while this factor also promotes proliferation and tube formation, as well as the migration and vascular permeability of endothelial cells. The fact that RPE releases VEGF in the fetal development is necessary in the development of choriocapillaris. Furthermore, low ocular level of VEGF generally is considered natural under physiological conditions [13]. In turn, pathological conditions may be characterized by abnormally high levels of VEGF in affected zones, which may indicate advances in neovascularization. Another characteristic feature for AMD patients is improper regulation of para-inflammatory responses connected with the promotion of macular damages. These processes are additionally stimulated by certain behavioral risk factors, genetic predispositions, and advanced age [13].

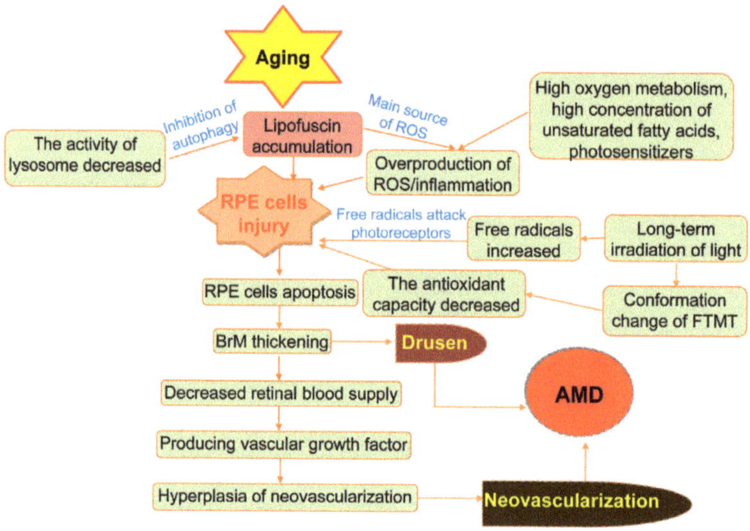

Figure 1. The formation of AMD during aging. Non-genetic mechanisms of AMD are induced by retinal pigment epithelium (RPE) cell senescence, oxidative stress, hemodynamics, and during aging (modified after Deng et al. (2022) [14].

The precise mechanism of this impact is still unknown. Previous studies have shown that, due to pathological changes in retinal pigment epithelium (RPE), the macula is damaged, which may lead to partial or complete loss of vision. The disease occurs in two clinical forms, dry and exudative (wet, neovascular), while the dry form tends to convert into a wet form. A dry form of AMD is a mild and gradual disease. However, it displays an acute and violent nature in the neovascular form. Both forms of AMD can occur simultaneously [3–8]. Environmental and genetic factors contribute significantly to the development of AMD. Studies have confirmed the role of heavy metal ions, chemicals, light, and many genes that increase the risk of disease [5,7,8] (Figure 2).

In context of immune surveillance, there is also a critical role of complement system because its overactivation may lead to AMD. Therefore, certain complement inhibitors are in advanced clinical trials for the treatment of AMD. Such therapies can be applied directly to the eye by intravitreal, subretinal, and suprachoroidal injections, as well as delivered systemically. However, high-resolution ophthalmic imaging revealed that the study of BrM appears fundamental for understanding of AMD pathogenesis, because BrM is responsible for key transport functions and constitutes structural support to RPE, while, in the process of aging, it undergoes detrimental structural changes including thickening, decreased elasticity, and reduced permeability. Thus, perception of transit of complement proteins and vectors across BrM and RPE is also pivotal for the efficacy of therapies based on application of complement inhibitors and regulators [15]. In cases of therapeutic use and possible clinical applications, certain biomaterials are praised for their potential in drug release control, redistribution, and the establishment of long-term delivery. In AMD therapy, certain port delivery systems (PDSs) have achieved long-term performance in delivering VEGF-A antibodies.

Additionally, the application of specifically fabricated micro-boxes may facilitate long-term delivery of various types of therapeutics. Finally, nanobiomaterials are able to control the intracellular inflammatory signals by acting as ROS scavengers and redox balance stabilizers. Such nanoplatforms may deliver anti-inflammatory agents to the disease regions, e.g., anti-VEGF-A antibody-loaded betasone phosphate-based hydrogel guarantees not only prolonged attenuation of choroidal neovascularization but also scavenges ROS to reduce inflammations [16].

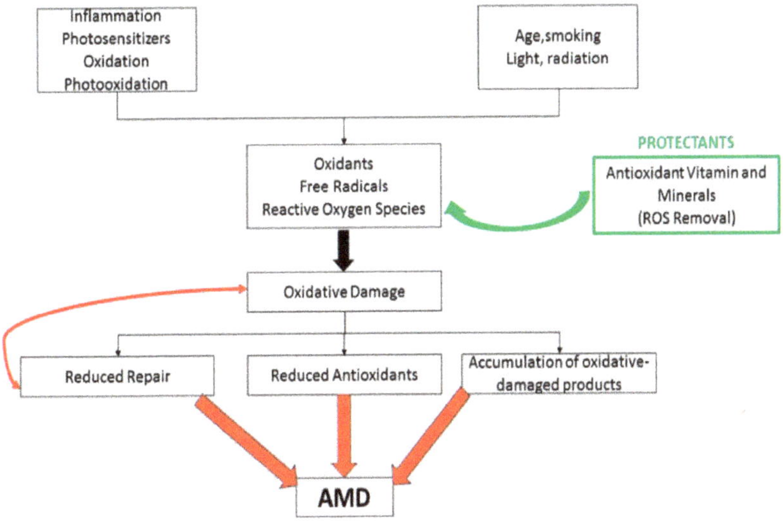

Figure 2. AMD development and protective mechanism of antioxidants (modified after Arslan et al., 2018 [17]. Arrows indicate mutual relations between parameters in rectangles.

Previous studies indicate that the impact of heavy metals leads to accumulation of reactive oxygen species (ROS). Unfortunately, there is no clear evidence that oxidative stress is a major factor in AMD development, but ROS continues to be an important factor in the course of disease [3–8,18]. ROS is a group of compounds of hydrogen peroxide (H_2O_2), singlet oxygen ($^1O_2$), superoxide radical anion ($O_2{}^{*-}$), and hydroxyl radicals (*OH) [3–8,19]—and the harmful activity of ROS damages lipid membranes, proteins, and nucleic acids [3–8,20–22]. The most exposed element for oxidative stress is the retina. This is due to its continuous exposure to radiation, high concentrations of O_2, high levels of polyunsaturated fatty acids in external photoreceptors, and the presence of many chromophores (lipofuscin, melanin, rhodopine, cytochrome C oxidase) in the retina and retinal pigment epithelial cells RPE generating ROS during phagocytosis of photoreceptor disks by RPE [3–8,19,20,22]. Molecules damaged by ROS called advanced glycation products AGE affect cell DNA and increase expression of genes that promote aging of RPE cells [23].

To counteract the effects of oxidative stress, a response is triggered in two stages. The first one is the reaction where cytochrome P450 monooxygenase system is involved.

The action involves the oxidation and reduction of dangerous compounds. In the next phase, these products are coupled with hydrophilic molecules. Antioxidants involved in the second phase are divided into "direct" and "indirect". The "direct" group includes SOD, glutathione GSH, and thioredoxin TR. The first two components oxidize dangerous compounds and quickly regenerate themselves. The "indirect" enzymes participate in the biosynthesis and regeneration of GSH and TR. They also participate in the removal of oxidized compounds [24]. The nuclear factor erythroid-2 related factor 2 NrF2 plays a major role in the antioxidant response. It activates transcription of the leucine slider by binding to the antioxidant response element ARE within the promoter of target genes. Its role is to maintain redox homeostasis in the cell [4,6,24–27]. NrF2 regulates both so-called "direct" and "indirect" enzymes [24,28]. The main task of this factor is to induce antioxidant response [4,24,27]. NrF2 precisely controls the activation of antioxidant genes in response to oxidative stress, which affects cell survival. Any defects in the antioxidant defense, controlled by NrF2, can contribute to the cell death by apoptosis. NrF2 signaling play an important role in the functioning of RPE cells [4,24,29–31].

Retinal cells are exposed to oxidative stress due to a large extent to light effects [3–8,19,20,22]. As a result of the increasing aging processes, cells do not respond to mitogenic factors and lose their ability to proliferate. Studies have confirmed the presence of AGE-exposed genes in RPE cells, as well as the occurrence of shorter telomeres [32]. AGE undergoes metabolism in a much more difficult manner, and as a result they are deposited in the lysosomes in the form of lipofuscin, which, in turn, is deposited in the form of drusen [33]. As a result, this disrupts the metabolic exchange between the choroid and retina [34]. Additionally, the work of photoreceptors is weakened [35]. Then, a low intensity inflammatory process begins, which increases the levels of acute-phase proteins, reducing total antioxidant status (TAS) [3–8,20,22,36–39]. The next stage is neovascularization, which occurs due to the secretion of proangiogenic factors during para-inflammatory process. New capillaries come from the choroid and are formed under the retina in the area of macula, where they should not occur [7]. As a result, photoreceptors are destroyed in the macula [3–8,20,22].

Thus, taking into account the above interrelationships, we consider the following aspects:

1. Presentation of the current state of knowledge about the impact of environmental and genetic factors on the development of AMD in relation to our research conducted on the Polish population in the Kuyavian–Pomeranian and Lubusz Regions (Central and SW Poland).
2. Analyzing how the polluted environment of large agglomerations affects the development of AMD, because this disease has only recently been a research subject and its etiology is still not well known.
3. Indication of new markers, because the review of the current literature from this field is very optimistic, which means that emphasis are changes in the protein and enzyme activity, metal concentration, gene polymorphism. This is the starting point for the analyses that we intend to perform within our studies. We investigated concentration of chemical elements Na, K, Ca, Mg, Be, V, Cr, Mn, Fe, Co, Ni, Cu, Zn, Al, Si, P, S, As, Se, Mo, Cd, activity of antioxidant enzymes (SOD, CAT, GPx, GR); intensity of lipoperoxidation (MDA), stress proteins (ceruloplasmin CP, haptoglobin HPT, ferritin FRT), non-enzymatic mechanisms (total antioxidant status TAS, uric acid, bilirubin).
4. Analysis of all the above relationships, which can bring us closer to a multi-level response to the question of the causes and development of pathophysiological changes that accompany AMD.

2. Role of Toxic Heavy Metals in AMD Development

As heavy metals are characterized by greater atomic mass than sodium atomic mass, with a density greater than 5 g·cm^{-3} and include elements of semi-metals (metaloids, e.g., As, Sb), they are classified as most important environmental and occupational poisons. As, Cr, Cd, Ni, Pb, Hg, Cr, Ni are highly allergenic [40]. High concentrations of heavy metals induce immunosuppressive effects on the immune system, while low concentrations cause immunostimulation. Metal ions cause oxidative stress in different ways, e.g., Fe, Cr, and Cu undergo redox reactions by increasing ROS leading to their over-production. In turn, Cd, Pb, and Ni decrease the antioxidative potential of cells, leading to ROS accumulation due to their reduced neutralization [41].

Metal ions have less molecules than organic chemicals, and the bonding they create is not a permanent covalent bond but a reversible one. In this case, the activation of immune cells by metal ions may take place in a manner different from that of classical hapten. Metal ions with high reduction and oxidation potential, such as Au (IV), can lead to irreversible denaturation of proteins by oxidation of sulfur-containing side chains in amino acids [42]. Metals primarily induce changes in protein synthesis, damage cell membranes and impair ATP production. Most react with sulfhydryl, carboxyl, and phosphate groups. The processes taking place in the cell depend on the amount of metals and their degree of affinity. The changes in metabolic processes in the body are revealed in the form of biochemical or clinical effects. Their occurrence is related to the critical concentration of elements in the organs [43].

Much research has been devoted to the impact of heavy metals on the nuclear factor erythroid-2 related with factor 2 (NrF2). Mice experiments have shown that, in response to light toxicity in vivo, thioredoxin is induced via thiophene [24,44]. In the experiments conducted by Gao and Talalay (2004) [45], human ARPE-19 cells were used, the sulforaphane was isolated by HPLC and concentration measured spectrophotometrically. The results indicated activation of NrF2 sulforaphane as a protective factor for RPE. One year later, Western Blot and immunohistochemically tested on human RPE K-1034 cells were assayed using murine cells and analyzed by flow cytometry and electrophoresis. The results confirmed activation via NrF2 and showed direct protection of RPE by the alpha-tocopherol and intermediate by inducing thioredoxin [24,44]. Glutathione synthesis is also induced by this factor by the activation of glutathione S-transferase and NADPH-quinone reductase.

Research conducted by Nelson et al. (2002) [46] included hRPE cells cultured at different concentrations of oltipraz and were subjected to a chemical oxidant. The level of glutathione in the mitochondria was determined by the HPLC method, and glutathione reductase activity was determined spectrophotometrically. The result was a significant increase in GST activity caused by the compound used. Research has speculated that oliprase may bind to glutathione and then to the nucleophilic center of transcriptional regulatory factor, and by interacting with NrF2, affecting the transcription of enzymes in the second phase of antioxidative response [24]. Conversely, studies by Ha et al. (2006) [47] tested the protective effect of zinc on RPE. The experiment was carried out on human ARPE-19 cell culture. Isolation of glutathione was made by HPLC. Gene expression was tested using RT-PCR. As a result, zinc has been shown to increase the amount of glutathione in RPE cells, and through NrF2 increases its de novo synthesis and protects RPE [24]. The impact of toxic metals on humans is presented in Table 3.

Table 3. Impact of toxic metals (As, Cr, Ni, Cd, Pb, Hg) on human health.

Chemical Element	Harmful Compounds	Mechanism of Action/Health Consequences of Exposition
Arsenic [As]	As_2S_3, AsS, AsO_6, FeAsS	Absorption
		The most common arsenic compounds are mainly absorbed by respiratory tract and accumulated in keratin-rich structures (hair, nails, skin), and in the placenta [48].
		Health effect and biochemical impact
		Arsenic and its compounds are carcinogenic for respiratory system, skin and other organs triggering tumors, and impaired metabolic processes of hepatocytes and kidneys. Arsenic blocks sulfhydryl groups of proteins and inhibits enzymes. It can react directly with thiols, especially glutathione, cysteine, and dithiides [48].
Chromium [Cr]	Valuable III active compounds	Consequences of reactivity
		Compounds react with substances that allow the diffusion of ions (pyrophosphates, methionine, glycine, leucine, lysine, proline). They form stable protein complexes and generate the mechanisms of harmful influence of chromium [49].
		Health effect
		Cr compounds damage the digestive system, cause skin lesions, mutagenic, embryotoxic, and teratogenic effects [49].
Nickel [Ni]	Ni^{2+} compounds and ions	Physiological significance
		Nickel in low concentrations is needed for normal functioning, whereas deficiency causes a reduction of oxygen consumption by the liver and an increase of fat accumulation. Physiologically, Ni activates certain enzymes, increases the activity of humoral response, stabilizes the nucleic acids, and plays a role in lipid metabolism [50].
		Absorption and health effect
		Nickel is a major causative agent of contact allergy [50]. A total of 75–80% of inhaled nickel remains in the lungs, when administered orally 90% is excreted. Nickel allergy is most often manifested as allergic contact dermatitis (Ni-ACD). Hand eczema is also observed, resulting from prolonged exposure to Ni contained in detergents or nickel-plated items. There are known cases of vesicular eczema following the consumption of foods containing Ni [50].
Cadmium [Cd]	CdO, $CdCl_2$, $CdSO_4$	Absorption
		When absorbed from food, cadmium effects undergo modulation in the range of digestive tract thanks to protein agents, Zn, Cu, Ca, and Fe contained in the diet. Their low content in the food cause increases in the absorption of Cd from the gastrointestinal tract and influence its accumulation in the body [51]. The greatest amount of Cd is absorbed in the duodenum. During absorption in the enterocytes, the transport of Cd occurs (non-specific divalent metal transporter (DMT1) bivalent conveyor and metal tolerance protein (MTP1) conveyor). Absorption can also occur through calcium channels of transporters responsible for Zn transport or can be absorbed from gastrointestinal tract in the combination with cysteine or glutathione-thiol groups [52].
		Accumulation and health effect
		Cd accumulates in the liver and kidneys, which are the target organ of toxicity. In the liver, Cd induces the synthesis of low molecular weight metallothionein (MT) proteins that bind Cd (II) ions to Cd-MT complexes. In the kidneys, Cd-MT complexes are easily filtered in glomeruli and resorbed in proximal tubules, where they degrade Cd ions. These structures are particularly exposed to metal toxicity and causing subjected to damages them and resulting that may result in resorption disorders [20–22,43,53,54].
		AMD risk assessment
		The results suggest that people with higher levels of Cd exposure are at increased risk for AMD. Concentration of Cd in the blood is a good reflection of recent exposure, especially in exposed workers. Additionally, nicotine is one of the main risk factors for AMD development and the main source of human exposure to Cd [53].

Table 3. *Cont.*

Chemical Element	Harmful Compounds	Mechanism of Action/Health Consequences of Exposition
Lead [Pb]	All Pb compounds and metallic Pb	Absorption and health effect
		Lead is mainly absorbed through respiratory and digestive tract [43,49,55]. In children exposed to Pb in the pre- or postnatal period, a number of deviations in mental development were demonstrated. The most commonly observed aggression seizures, emotional lability, memory disorders, and learning and reading difficulties [43,55]. The most common disease is saturnism. It is chronic poisoning with lead and its salts. It occurs in employees of printers, battery factories and lead paint factories [43,49]. Pb can cause high blood pressure. Through nephrotoxic activity lead may cause renal dysfunction and develop the renal hypertension. Pb can generate oxidative stress, which in turn results in dysfunction of sympathetic system. It also reduces baroreceptor sensitivity and vagus nerve tension. Pb significantly influences the conduction of stimuli in the parasympathetic system. This can result in ventricular arrhythmias or myocardial infarction [21,43,55]. Pb and Cd induce changes in vascular endothelium and thus contribute the development of atherosclerotic lesions. Lead affects the course of early and late inflammatory reactions, T and B lymphocytes, macrophages and cytotoxic cells. Bone marrow damage is a common complication of Pb poisoning. Hematopoietic disorders manifest as thrombocytopaenia, leukopenia or anemia. In advanced bone marrow failure, all marrow cell lines undergo perturbation. The higher levels of immunoglobulins, especially IgA, IgE and IgG, were also observed in patients with higher Pb. Neurological disorders after long exposure to Pb usually take the form of Pb neuropathy. It usually involves the motor neurons of the upper limbs, and in particular the radial nerves. Exposure of men to Pb results in an increased risk of spontaneous abortion in their partners. In men, gonadal dysfunction occurs in the form of decreased sperm quality [43,55].
		AMD risk assessment
		The present levels of Pb exposure in adult Americans are lower than in the past, and recent levels of exposure determined by blood lead measurements are not high enough to increase the risk of AMD. Such a hypothesis is probable, however, Pb accumulates in RPE may participate in ocular cancer as well as promote chronic disease by increasing chronic oxidative stress. As the level of antioxidant enzymes decreases with age, the retina is susceptible to stress and free radicals, which can lead to AMD development [56,57].
Mercury [Hg]	Inorganic compounds $HgCl_2$, $HgCn_2$, HgS	Absorption and consequence of exposure
		In people exposed to professional exposure to mercury, it penetrates through respiratory tract. A total of 80% is retained in the body. Sulfhydryl ligands (RSH) form complexes with Hg (reduced glutathione) and amino acids with cysteine and glycine. Directly high mercury poisoning occurs in cases of metallic form, from which 70% is removed to the outside of lungs. It partly penetrates into the brain [43,49].
		Accumulation and mechanism of toxicity
		Mercury remains in the blood and penetrates the blood–brain barrier and placental barrier. Consequently, it accumulates in the brain and fetal tissues. Hg^{2+} ions have the ability to form complexes with proteins and with compounds containing SH-groups. Fetal hemoglobin has a high affinity for Hg, and fetuses have a higher concentration of Hg than mothers. Due to the fact that mercury easily diffuses into lipids it is also present in breast milk of mothers. It is deposited in the hair, its concentration there is proportional to the concentration in the blood. The first mercury-invaded cell element is the cell membrane. This is due to the presence of SH-groups. Most proteins and enzymes in their composition contain amino acids of these groups, therefore Hg poisoning can interfere with enzymatic reactions [43,49]. Irrespective of the form of Hg, it is deposited in the form of a complex with metallothioneins (MT), because of the induction of MT production. Synthesis of these proteins play a key role in the detoxification process [20–22,43,49,54].

2.1. Element-Element Interactions in AMD

According to research focused on the impact of lead, it was positively associated with both early and late AMD form in all studied local and regional areas. Hg and Cd also had a positive relationship with late AMD in all studied areas, but did not affect early AMD. In contrast, Mn and Zn were the opposite of late AMD [58]. Mn and Zn are not associated with early AMD [59]. Pb and Cd can accumulate in human retinal tissues which may also be damaged in result of oxidative stress. They may therefore play a role in the development of AMD. AMD has a great significance as a cause of blindness in the USA (that accounts even for 54% of blindness in Caucasian patients, and is also considered an important risk factor for black patients, and people from Spain). Certainly, age and ethnic determinations play important role and among white people aged 40 and older, AMD is the predominant reason of visual impairment and loss of vision in the US [1]. Due to increasing life expectancy, the incidence of AMD is expected to increase [3–8].

Studies have shown that lead and cadmium can promote chronic disease, increasing chronic oxidative stress by damaging DNA, reducing antioxidant capacity of defense systems, stimulating the production of inflammatory cytokines [60], and ROS production, also in retinal pigmented epithelial cells (RPE) [3–8,21,57]. Cells involved in the transport of trace metals such as Pb and Cd are particularly susceptible to toxicity. RPE can interact with metals, including basic metals such as Zn, Cu, Fe, but also with toxic heavy metals, including Pb and Cd, which have a high affinity for melanin in RPE melanosomes [61,62]. Pb and Cd ions are similar in size to Zn, Cu, and Fe. Cd concentrations in human eyes increase with age and are higher in women than in men and in smokers compared to non-smokers [56,57]. Pb concentration in human retina also increases with age [56].

Problematic matter is detection of presence of multiple toxic metals within the intact retina. It is also difficult to separate primary metal toxicity from secondary uptake of metals in defective tissue. However, elemental bioimaging enables detection of toxic metals in various segments of eye [63]. Certain examinations use such methods as laser ablation-inductively coupled plasma mass spectrometry that make possible the detection of multiple elements in tissues as well as histochemical technique of auto-metallography that allow to detect the presence of inorganic Hg. Therefore, the presence of such toxic elements as Ni, Fe, Hg, and Al can be confirmed in retinal pigment epithelium, choriocapillaris or optic nerve head. Subsequently, this knowledge is fundamental for inferences about injuries to RPE from toxic metals, damages of neuroprotective functions or violations of outer neural retina [63]. Due to the fact that the degree of bioaccumulation of heavy metals in human body is high, there is a constant need to monitor their levels and influence. The development of biological methods of identification and quantification as well as usage of biomarkers create a possibility of establishment of reliable criteria and standards. It gains importance because the retinal epithelium is a chelating metal tissue, while melanin forms bonds with heavy metals that increase tissue affinity for these elements. Finally, wide variety of anthropogenic activities have led to the continuous release of toxic pollutants into the environment and made this problem so urgent [64].

2.2. Accumulation of Heavy Metal Ions in Eye Tissues

Research on human eyes confirmed the ability of the retina and choroid to accumulate heavy metals [65]. Analyses carried out on human eyes using X-ray showed that sometimes Al, Hg, and Se may be present in melanosomes in retinal pigment epithelium. In turn, lead spectra were observed in the choroid and retina. Accurate studies to determine the limit values of heavy metals for humans were carried out in 2005. The ability of Pb, Cd, Tl, and Hg metals to accumulate in the tissues of human eye has been confirmed. Studies were conducted using plasma-mass spectrometer and histologic examination. It has been shown the greatest affinity of these metals to the pigmental epithelium of retina and choroid. Pigmentary eye tissues, such as retina pigmentary epithelium, choroidal, iris, and ciliary body, have a high affinity for heavy metal ions. Additionally, melanin present in pigmented granules binds metal ions that bind to melanosomes according to atomic weight and volume [56,61]. It has also been proven that heavy metals, such as Pb and Cd, interfere with biochemical balance in cell and cause oxidative stress resulting from the production of ROS [21,56,66].

The occurrence of AMD correlated with cutaneous signs of Hg toxicity with skin lesions induced by toxic influence of this element [67]. Both histopathological studies of the skin biopsy, fundus, and determination of Hg concentration in the blood were performed.

Experiments were performed on the Grover's disease patients stated that skin lesions in the course of the Grover's disease may be an AMD marker. In addition, even low levels of mercury correlate with development of both diseases. However, the author also pointed out that if the results are confirmed in other studies, the elimination of Hg from the environment reduce the number of patients on both units [67].

Studies conducted in 2005 on the Caucasian race have demonstrated the presence of Pb and Cd in all examined eye tissues. Their concentration was higher in the retinal pigment epithelium/choroid than in the retina. The work involved histological examination and measurement of metal concentrations in tissues and blood by mass spectroscopy. As a result, it has been demonstrated that in the retinal pigment of retina and choroid, as well as in ciliary Cd and Pb of retina occur at the concentrations higher than in the blood and the fluid of the eye. The results confirmed the ability of Cd and Pb to accumulate in RPE and choroid. However, there are no known diseases that can be caused by this phenomenon [56].

Very important amphoteric chemical element, aluminum, was tested to correlate with AMD development. Studies have shown that Al salts activate the NALP3 inflammasome, causing lesions and rupture of lysosomes as well as migration of immune cells and inflammation. As a result, tissue damage and tumor growth occurred. Studies were conducted on mice using flow cytometry, confocal microscopy, as well as immunochemical methods and Western Blot [68]. Additionally, Cd and Pb concentrations in human eye tissue were studied. As a result, it has been shown that Pb accumulation is related to AMD's development and significantly affects homeostasis of the eye tissue. The neural cells of retina with AMD had increased Pb concentration. In turn, Cd concentrations in patients were not significantly higher than in healthy patients [69].

In 2008–2011, Korean team measured concentrations of Hg, Cd, Mn, Zn, and Pb in human peripheral blood in correlation with AMD. The results showed that high concentrations of lead were associated with development of both early and late AMD. In turn, Hg and Cd only correlated with pathogenesis of late AMD. The results for Mn and Zn were different; their high concentrations had protective effects in the late form of AMD [59]. Other studies have identified Zn as another potential protective factor for AMD, even in patients with polymorphisms that increase the risk of AMD development [5,70].

3. Role of Oxidative Stress

Cellular imbalance occurs as a result of oxidative stress. This is caused by an elevated oxidation reaction that produces ROS (superoxide radical anion (O_2^{*-}), hydroxyl radical (*OH), hydrogen peroxide (H_2O_2), and singlet oxygen ($^1O_2$). ROS can be formed in many ways as a product of respiratory chain in mitochondria, in photochemical or enzymatic reactions, as effect of exposure to UV light, ionizing radiation, or the influence of metal ions. Hydrogen peroxide is a low reactivity molecule, but it can easily penetrate cell membranes and generate the most reactive forms of oxygen, i.e., hydroxyl radical through Fenton's reaction [3–8,20–22].

ROS play an important role in the regulation of many physiological processes by participating in intracellular signaling [21,71], and causes serious damage of biomolecules by lipid peroxidation. This is a process of oxidation of polyunsaturated fatty acids due to the presence of several double bonds in the structure and involves the production of organic peroxides and reactive free radicals. These molecules react with other fatty acids, and initiate a chain reaction. ROS also attack structural and enzymatic proteins by oxidation of amino acid residues, a formation of transverse bonds and protein aggregates as well as proteolysis. Deactivation of key proteins can have serious consequences in important metabolic pathways. ROS can also react with nucleic acids (attacking nitrogenous bases and phosphate skeleton of sugar), and can cause damage to single- and double-stranded DNA. The inability of the cell to repair damage may lead to death, or alternatively, there may be mutations in the DNA, which leads to the development of cancer or neurodegenerative diseases [3–8,20–22,72].

Oxidative stress, causing damage to the eye, can occur in various forms and can be stimulated by various factors [3–8,20–22,73]. In human eyes, local exposure to light combined with locally high oxygen content, which is greater than in other tissues, leads to a high tendency for oxidative stress. In combination with systemic exposure to oxidative stressors, conditioned by lifestyle or other factors, relative oxidative stress can quickly become disproportionately high. Oxidative stress is associated with the promotion of inflammation and has a number of negative effects associated with AMD progression [3–8,20–22,54,73–77]. As a result of the recognition of system disorders, anomalies in DNA repair genes, inefficient repair of DNA damage, and acceleration of aging of the organism occur, leading to the dysfunction of cells and tissues. The aging is unavoidable because ROS formation is the result of normal, daily cellular metabolism. In effect, the cells have developed the complex of defense mechanisms to fight both the resulting of ROS and their effects [3–8,20–22,54]. Cells most susceptible to oxidative stress damage are non-proliferating postmitotic cells, including photoreceptors and RPE cells, because they do not have any DNA damage detection systems at the cell cycle checkpoints [3–8,20,22].

RPE is a retinal epithelium, a pigmented monolayer, located between the retina and the choroid. RPE is essential for the preservation and survival of photoreceptor cells because it performs a number of critical functions such as the formation of an external retinal barrier, transport, retinoid retention, phagocytosis, degradation of segmental photoreceptors, and protection against light and oxidative stress [78]. Contaminated environment promotes ROS production [3–8,20,22]. In the macula, the dominant photoreceptors are cones, which exhibit higher demand and energy production than rods, and therefore higher oxygen requirements [79–82]. Rod cells and cones differ in their susceptibility to oxidative stress, and cones show greater sensitivity to free radicals [83]. Macula is consistently exposed to a high metabolic activity and oxidative stress due to a high partial pressure of choriocapillaries and polyunsaturated fatty acids (PUFAs) from external segments of the retina [3–8,20,22,54,84]. This is considered in inducing drusen formation between RPE cells and Bruch's membrane.

Lipofuscin is a chromophore serving as the primary RPE photooxidation agent [85] which, after absorbing high energy photons, especially blue light, undergoes a series of photochemical reactions involving ROS formation, which in turn induces photochemical damage in the retina and RPE cells [3–8,20,22,54,86]. Autophagocytosis and RPE homeostasis pathways play an important role in the aging of cells and affect oxidative stress in AMD. Autophagy, cellular and molecular pathways were investigated for oxidative stress. The results indicated that acute exposure to oxidative stress increased autophagy activity, while chronic exposure to oxidative stress reduced autophagy. Based on murine and human models, it has been observed that autophagy is significantly elevated in early AMD, while its activity decreases in late AMD. Moreover, it has been observed that reduced autophagocytosis effect makes RPE more susceptible to oxidative stress and that autophagy system is enhanced to protect RPE from oxidative damage [78].

Oxidative stress in retina may be triggered by chronic UVA exposure, while blue light exposure also promotes this phenomenon in retinal pigment epithelium layer. In the unfavorable conditions results may be deepened by cigarette smoke or aging process. Another factor contributing to oxidative stress in RPE is phagocytosis of outer segments of the retina, which are abundant in polyunsaturated fatty acids. Subsequently their oxidative alteration cause formation of carboxy-ethyl-pyrrole adducts as well as advanced glycation end-products which may be detected within drusen. Ultimately, the accumulation of lipofuscin occurs within the lysosomal compartments and lysosomal degradation undergoes disturbance that results in amassment of protein and lipid extracellular deposits. Existence of certain non-invasive imaging techniques like optical coherence tomography (OCT), fundus autofluorescence or OCT angiography enable the insight into key retinal and choroidal anomalies. This makes possible the characterization of AMD features (drusen, neovascularization, GA) and investigation of disease advancements [87].

4. Role of Metallothioneins in AMD

Metallothioneins are low-molecular-weight (6–7 kDa) intracellular proteins, which were firstly isolated from horse kidneys and appreciated for high potential in binding cadmium [54,88,89]. These proteins are able to bind metals and act as free radical scavengers. This ability result in considering metallothioneins as one of important components of antioxidative defense (beside typical enzymatic mechanisms like dismutases or catalases, as well as nonenzymatic like glutathione, ascorbate, or carotenoids) [20–22,88]. These proteins have two domains, α and β. The α-domain (consists of 11 cysteine residues) is able to bind four Zn^{2+}, four Cd^{2+}, or six Cu^{+} ions. The β-domain (consists of nine cysteine residues) can binds three Cd^{2+}, three Zn^{2+}, or six Cu^{+} ions [21,22,88]. Metallothioneins are also stabilized by interactions with other metal ions such as Hg, Pb, Ni, or Co [54]. Single molecule of metallothioneine may contain 12 Cu atoms, as well as 7 Cd or Zn atoms [21].

Metallothioneins were identified in many cellular structures (nucleus, cytosol, lysosomes, mitochondrial intermembrane space). A total of four isoforms of metallothioneins (MT-1, MT-2, MT-3, MT-4) have been identified in the human body. Their location is presented in Table 4. The first two (MT-1 and MT-2) undergo expression in many tissues, while the expression of the third one (MT-3) is concentrated on central nervous system. The expression of all three isoforms was detected in the human brain and retina [20–22,54,88]. The construction of metallothioneine molecules (content of metals and thiol groups) enables them to make quick interactions with ROS (hydroxyl radical and superoxide anion radical) [21].

Table 4. Location of metallothioneines in the human body [88,89].

Isoform	Location of Expression
MT-1	brain (mainly astrocytes), liver, kidney, eye
MT-2	brain (mainly astrocytes), liver, kidney, eye
MT-3	brain (astrocytes, neurons, cortex, hippocampus), heart, kidney, eye
MT-4	squamous epithelia (upper gastrointestinal tract and skin), maternal deciduum

Through capturing of cations (like Fe) metallothioneins prevent from completing Fenton reaction, in which ROS may be generated [20]. In this reaction, hydrogen peroxide reacts with iron (II) ion creating hydroxyl radical (one of the most reactive oxidants). The high redox potential of this molecule enables it to interact with practically all substances present in the organism [21]. The ability of metallothioneins to counteract against oxidative stress as well as their presence in retina prompt to reflection about the role of these proteins in the process of AMD development, strictly connected with age-related progressive degenerative changes in retina [3,4,6–8]. A certain chance of counteraction against these destructive changes appeared to be possible with the introduction of antioxidants (vitamin C, E, β-carotene) to the patient's diet or zinc supplementation [4,5,88]. Integrated action of antioxidants and Zn is able to slow down AMD progression and metallothioneins by binding Zn participate in this process as well [5,88]. Metallothioneins are able to interact with retinol dehydrogenase through this participation in retinol reconstruction in the visual cycle [6]. Alvarez et al. (2012) [22] describe the close correlation between zinc and metallothioneins expression. According to these relations, the presence of Zn increases the expression of metallothioneine, simultaneously decreasing expression of inflammatory cytokines.

It was noted that two isoforms of metallothionein (MT-1, MT-2) come under particularly high expression in the front structures of eye in range of cornea and lens. High expression of metallothioneins in this area creates natural antioxidative barrier in the structures subject for light, UV, or other kind of radiation [88]. MT-3 is expressed at high levels in the neurosensory retina. This isoform is involved in CNV and vascular leakage from CNV. Moreover, MT-3 is able to protect the outer nuclear layer and photoreceptor cells related to dry AMD [89]. However, increasing the metallothineins expression cause enlargement in amount of bounded Zn [22]. In this mutually regulated mechanism, we can see the potential of metallothioneins to protect the retina against oxidative stress, particularly in the retinal pigment epithelium, neurons, and photoreceptors. The increment of metallothioneins secretion in these regions is a typical defense reaction against growing oxidative stress [20,22,88]. This way, the risk of extension of inflammation to internal structures of eye is reduced. Therefore, two roles of metallothioneins (antioxidative and anti-inflammatory) are replenished [22]. An additional trump of metallothioneins is the considerable speed of interaction with superoxide anion radical or hydroxyl radical. Thanks to this elimination of oxidative stress, precursors may occur effectively [21].

Swindell (2011) [54] consider metalthioneins to be "universal free radical scavengers" that take part in cellular response for diversified stress conditions. In his opinion antioxidative potential of metallothioneins confirm their significant role, especially in diseases which grow more intense in advanced age. Oxidative stress itself may be considered to be a group of factors accelerating biological aging of cells. Swindell (2011) [54] also enumerates some studies confirming a regulatory influence of metallothioneins on cellular respiration. Metallothioneins act on two levels, participating in the detoxification of free radicals, as well as not permitting to their rise [54]. There is evidence (via animal models) confirming that these proteins may positively influence the length of life, as well as that of the expression of metallothionein increases in response to calorific restriction (undernutrition), and in unfavorable or toxic conditions [54].

In addition to regulatory function for toxic metals, metallothioneins counteract against cellular apoptosis, alleviate cellular inflammatory processes, support growth, and sedate cell metabolism [54]. All mentioned functions of metallothioneins apply to protection of nervous system cells as well. This way, metallothioneins support the regenerative processes occurring in the nervous system in response to widely undrstood damages [20,22,54]. These statements may be referred to as AMD (as a degenerative disease, which aftereffects intensify in advanced age, leading even to complete loss of vision sense) because metallothioneins appear to support the survival of cells and the structures of organism through a counteraction against the aging processes stimulated by oxidative stress [3–8]. Thus, metallothioneins eliminate one of the important factors of oxidative stress stimulation by binding and the detoxification of heavy metals. The binding of Zn and Cu favors the maintenance of homeostasis of these elements. There is some mutually regulated mechanism between zinc and metallothioneins; they bind Zn, and moreover, exposure to Zn promotes the production of metallothioneins. The increase in the synthesis of metallothioneins (MT-1, MT-2, MT-3) also occurs in inflammatory conditions in response to the increased level of inflammatory cytokines [20–22,54,88,89].

5. Lifestyle and Environmental Risk Factors

5.1. Smoking

Different studies show that smoking is one of the most common causes of change leading to AMD. There are thousands of chemicals in the cigarettes, which have harmful influences on the organism. Numerous substances increase the level of oxidative stress, which lead to lipid peroxidation [10,90–94]. According to Velilla et al. (2013) [90], smoking affects the occurrence of molecular and pathological changes, which very strongly affect the macula and decide on the incidence of AMD. Studies showed that patients with AMD have a higher concentration of C-reactive protein (CRP). Unfortunately, a higher level of CRP is not specific to smoking; concentration of this protein increases during inflammation in the body. Research by Seddon et al. (2004) [95] and Molins et al. (2018) [96] showed that a higher level of CRP can increase the risk of AMD independently of smoking.

Smoking has multiple influence on the development of AMD. Nicotine from cigarettes stimulate the organism to produce vascular endothelial growth factor (VEGF). This factor has proangiogenic activity, which promotes angiogenesis. This process has an important role in the development of wet AMD. Smoking causes inflammation by decreasing concentration of complement factor H (CFH) in the blood and activating inflammatory mediators (e.g., complement C3) [90–95]. Environmental independent studies from Australia, Europe, and North America prove the role of smoking in the development of AMD [90]. Research by Thornton et al. (2005) [91] suggests a causal relationship between cigarette smoking and various types of AMD obtained by epidemiological evidence. The results show that the risk of AMD among smokers is three times greater in comparison to non-smokers. This difference is even greater in cases of exudative AMD, and is up to a four-fold increase in the risk of disease [91]. However, most patients and individuals do not show that they increase their probability of the development of AMD by smoking cigarettes, which may lead to total vision loss. Fortunately, for many smokers, the risk of blindness is the most mobilizing factor to quit smoking [91]. It has been confirmed that smoking cessation reduces the risk of developing AMD. Research proves that 20 years after the end of smoking, the risk from cigarette substances decreases to zero, as in people who had never smoked [10,90,97].

5.2. The Impact of Light

The light is form of electromagnetic energy, part of this spectrum, which interacts with the eye is referred to as optical radiation. This spectrum includes wavelengths from ultraviolet (UV, 100–400 nm), visible light (400–760 nm), and infrared (IR, 760–10,000 nm). In this optical radiation the Commission Internationale de l'Eclairage had defined several subgroups of wavelengths with similar photon energy. Today, UV radiation is classified into three subgroups: UVA (315–400 nm), UVB (260–315 nm), UVC (100–260 nm), and UVD (20–100 nm). Infrared light (IR) consists of three groups: IRA (700–1400 nm), IRB (1400–3000 nm), and IRC (3000–10,000 nm) [98,99].

The spectrum of light that reaches the eye is absorbed by its various tissues. The cornea absorbs almost all radiation below 295 nm. The lens absorbs wavelength range from 300 nm to 400 nm, which is part of UVB (315–400 nm) and total UVA [99–101]. The vitreous gel propagates wavelengths in the visual spectrum (400–700 nm) and IRA (700–1400 nm), but the UV, IRB, and IRC bands are almost entirely absorbed. The remainder of the optical radiation, which has a range of 400 nm (380 nm) to 1400 nm, reaches the retina of the eye and is referred to as the retinal hazard region [99,101]. The light can damage the retina by the photothermal, photomechanical, and photochemical pathways [99]. In cases of AMD development, the most important is the last one, in which oxidative stress plays a key role [74,102]. The retina's oxygen-rich and photosensitizer-rich environment combined with constant radiation exposure is perfect for the production of extremely lethal ROS [103].

Susceptibility to damage depends on the wavelength, which is inversely proportional to the energy it delivers. Studies show that the most important is high frequency and high energy visible light of 400–500 nm blue-violet band, referred to as blue light or high energy visible light (HEV). The ability to cause retinal damage as a result of HEV light is known as the blue light hazard (BLH) [100,101]. Green light, having a longer wavelength (495–570 nm) and lower energy, is 50–80 times less phototoxic than blue [104]. In studies by Braunstein and Sparrow (2005) [105], as well as Remé et al. (2000) [106], in vitro cell cultures, under the influence of green light, in contrast to those illuminated with blue light, were not damaged [105,106]. Having an even lower frequency, infrared radiation needs as much as 1000 times more power to produce similar phototoxic effects like HEV light in a given unit of time [107].

Epidemiological studies to assess exposure and associated risk of AMD are difficult to construct. The problem is a thorough estimation of exposure, as well as the potential impact of many blackout factors [108,109]. Nevertheless, Taylor et al. (1990) [110] studying of over 800 people working at the Chesapeake Bay found a significant correlation between exposure to blue light or the full range of visible light, and macular degeneration [110]. The Beaver Dam Eye Study published in 2004 also found a significant correlation between increased sunlight accumulation and a 10-year risk of early onset of macular degeneration [111]. European Eye Study (EUREYE), carried out for nearly 5000 participants aged 65 years or older demonstrated a positive correlation between AMD and exposure to blue light in people with low levels of antioxidants [102].

Conversely, in a large project of Pathologies Oculaires Lieesa l'Age (POLA), in Sète (S France), unlike the earlier mentioned, there was no significant relationship between exposure to sunlight and the development of AMD [112]. A meta-analysis published in 2013 showed that individuals with higher levels of sunlight exposure have significantly increased risk of AMD. In the same analysis, the latitude could be a covariate, which is negatively correlated with the strength of association, but the meta-regression had no sufficient evidence to support the thesis [108].

5.3. Diet and BMI

Type of diet appears important, e.g., the adherence to Mediterranean diet generally reduces the risk of developing late AMD (this diet characterizes itself with high content of antioxidants and very limited consumption of red meat). Conversely, consumption of a high-glycemic-index diet may pose a risk factor for the development of AMD [10]. AN unbalanced western diet may deliver additional dietary oxidants creating the risk of elevated ROS in RPE. Conversely, there are some statements that the association of early AMD prevalence tend to decrease with plant and sea-food-based diet. High-fat consumption certainly creates a risk of progression of AMD. Such a threat may be connected both with animal-based or plant-based high-fat diet and is associated with intensification of oxidative stress [113]. It has been confirmed that the role of nutritional factors in modulation of risk of developing AMD is significant.

Certain expectations are connected with administration of antioxidants because there were some promising effects of nutritional supplementation with vitamins C, E, β-carotene, Zn in delaying of AMD progression and alleviating oxidative stress. The usage of lutein and zeaxanthin replacing β-carotene appeared beneficial in context of intermediate-stage AMD. However, the effectiveness of such approaches is still controversial. Especially there are discussions about genetics–nutrient interactions and the usage of supplements in different ethnic groups [113]. The products containing β-carotene and omega-3 acids are known for their protective role against the development of AMD [10,114–117], because of their antioxidant, anti-inflammatory and anti-angiogenic effects [114]. It is also known that taking dietary supplements containing lutein, zeaxanthin, antioxidants (vitamin A, C, E and others) or minerals factors protect against the development of wet AMD [10,74,114,117,118]. Therefore, a diet low in antioxidants is one of the risk factors for AMD development [10,119]. However, supplementation of antioxidants is justified only in people, which are suffering from advanced form of AMD.

Unfortunately, people suffering from the early form of AMD will not reduce the chances of this disease by increasing intake of antioxidants. In turn, lipids (cholesterol, triglycerides) may accumulate in Bruch's membrane, which may result in the formation of drusen. Therefore, a diet rich in fats increase the risk of developing AMD [10,119]. Additionally, eating high-glycemic-index products may be a risk factor for the development of AMD [10].

Studies showed that a higher BMI correlates with the development of AMD [10,107, 114,119]. Individuals with BMI higher than 30 have an especially higher risk of developing late AMD than individuals with BMI in normal range (20–25). Obese individuals also tend to demonstrate elevated level of certain pro-inflammatory factors, as well as downgraded antioxidative possibilities that perturbs the functionality of RPE and escalate the development of AMD [10]. Therefore, overweight has a significant contribution in the increasing of risk development of AMD, because obesity is a pro-inflammatory state, promoting the development of oxidative stress [120]. Toxic metals tend to accumulate in the fatty tissue [121], which also promote the development of AMD [59,69,122].

5.4. Other Risk Factors

The strongest demographic risk factor for the development of AMD is age because structural and functional alternations of the retina, as well as additive effects of other pathological processes, are undeniably augmented as a consequence of aging. Numerous studies showed that risk of AMD increase with age [10,115,123–129]. In eye response to the damages relies on the elimination of senescent cells. The production of senescent cells or the mechanisms of the removal cells lead to a detrimental hoarding of senescent cells increases with age. In elder cells many harmful changes, such as telomere erosion, damages in DNA, increase ROS production, and metabolic disturbance are observed. The long-term accumulation of senescent cells in the effect of chronic stress is a result of aging and is characteristic for age-related diseases [130].

It has been confirmed that women are more likely to suffer from AMD [118,131–133]. However, in cases of female sex tendency of higher progression rate to early or late AMD may be confused by such factors as differences in life expectancy or influence of sex hormones (estrogen exposure in females may even exert some protective effects as well as may be connected with favorable changes in relation to serum lipids or activity of antioxidative agents) [10]. Furthermore, the researchers indicated that the incidence of women is higher only among patients over 75 years of age. In younger groups, the incidence is the same in men and women [134]. This tendency results from the influence of hormones, especially estrogen, which most likely protects against the development of AMD [118,135].

Studies suggested that also physical activity can influence on the development of AMD. Regular exercise has positive impact on the activity of antioxidative enzymes, which may protect against the development of AMD [118,136]. Thus, physical activity generally favors antioxidative possibilities and subsequently decreases the incidence of early or late AMD. Nevertheless, this factor remains highly unreliable in the prediction of AMD progression (difficulties in precise quantification of physical activity) [10]. Even educational status has been studied; this risk factor only in a few studies was associated with the development of AMD. Education level is inversely related with the development of early [137] and late AMD [138,139]. It may be connected with the fact that higher education level usually favors better nutrition, usage of eye care services and voluntary cessation of smoking [10].

Another very important risk factor is ethnicity: in multi-ethnic studies, it has been calculated that prevalence rates of any form of AMD for different ethnic groups. In effect the lowest risk of development of AMD was in African group (2.4%) and the highest was in Caucasians (5.4%). The remaining groups have medium rates: 4.2% Hispanics and 4.6% Chinese [140]. In the African population, AMD occur less frequently than in other ethnic groups. This is probably caused by increased melanin in RPE in Africans. This higher concentration may play role of UV filters and antioxidant free radical scavengers, which may protect by the development of AMD and subsequently act as alleviating factor in the matter of AMD development and progression. Conversely, AMD appears to be more common in the Chinese population when compared to Caucasians [10,141]. The differences in the chance of progression of this disease may also result from changes in DNA in genes susceptibility for AMD [142].

Associated diseases play important role in the development of AMD. There are several diseases, which may affect the progression of this illness. Several diseases confirmed the influence in AMD progression. In a few studies it has been observed that patients after cataract surgery have higher risk of late AMD development [143–145]. Research suggests that during and after cataract surgery, the process of inflammation is induced and after this, medical procedure the eye is more exposed on UV light [146]. Other hypothesis points to the same risk factors determining cataract and AMD [147]. The possible influence of hypertension has been well described [125,148,149], but this disease is not always treat as risk factor of AMD [150–152]. Hypertension may be a modulating factor because high blood pressure is connected with lower choroidal blood flow and troubled vascular homeostasis [10]. Some studies reported the possible influence of chronic kidney disease in the development of AMD, but this correlation is unclear [149,153]. In chronic kidney disease, a decrease in the estimated glomerular filtration rate may be connected with increased possibility of developing early AMD; however, only in patients older than 65 years. Additionally a certain serum biomarker for kidney function (cystatin C) appears to be associated with the development of nAMD and early AMD [10].

A different disease, hyperthyroidism, was presented in long-term population studies [154,155]. Research showed that an increase concentration of free thyroxine (FT4) in serum was associated with the development of AMD [155]. Conversely, an increased level of the thyroid-stimulating hormone (TSH) was not correlated with this disease [154]. Hyperthyroidism is probably brought about by increasing the level of FT4 and inducing oxidation stress [156]. Therefore, hyperthyroidism may be considered an accelerative factor for oxidative metabolism and basal metabolic rate (through induction of certain mitochondrial enzymes). Some studies showed that increase levels of thyroid hormones can damage RPE cells [10,157,158]. Other studies found that diabetes correlates with the development of AMD. Usually, there is no correlation between these two diseases [152,159]. It is possible that, in diabetic patients, hyperglycemia and dyslipidemia may induce oxidative stress in retinal tissue, which disbalance the homeostasis in this eye structure [160]. Thus, hyperglycemia and dyslipidemia accompanying diabetic individuals may act as a trigger factor for inflammatory responses in retinal tissue [10].

The correlation between Alzheimer's disease (AD) and AMD was studied. Due to the fact that in both diseases extracellular amyloid β-peptide deposits are produced (in AMD in drusen, in AD in extracellular amyloid β-peptide) [10,161] researchers studied the correlation between these diseases. Moreover, in both diseases, several isoforms of apolipoprotein E (APOE) are engaged. In AD, isoforms APOE ε 2 and ε 4 isoforms are associated with decreased and increased risk for AD and AMD, respectively.

The opposite directions are reported for AMD [162,163]. Despite these similarities, studies did not show associations between these diseases. Unclear associations also occur between AMD and Parkinson's disease (PD). In studies of PD patients treated with levodopa, the risk of AMD was lower than in patients not treated [164]. Levodopa activates the GRP143 receptor, which reduces the release of VEGF and inflammatory factors from RPE cells. This observation showed that levodopa may reduce inflammatory reactions in retina and delay the progression of AMD [10,165]. Ultimately, early AMD development may be accelerated in the event of heavy alcohol consumption and mentioned hypertension. In this context antioxidant disturbances and augmented oxidative stress persistently play a major role [113].

6. The Influence of Genetic Factors on the Development of AMD

Studies shown that many genetic factors affect in small or moderate morbidity of AMD. This is because of the interaction of both genetic and environmental factors. In specific genes heredity of this illness increases to 71% [114]. In 50–60% cases, progression of AMD is determined by polymorphisms in genes encoding *CFH*, *ARMS2*, and *Il-8* [166]. The oldest studies into the genetic background of AMD was conducted in 1988, on monozygotic twins. That was first research where severe AMD in both twins was observed [167]. In 1994, prospective studies (1986–1993) were examinated monozigotic and dizigotic twins. In effect, AMD occurred among 23 pairs of monozygotic twins [168].

Other studies (e.g., Familial Aggregation Studies) also confirmed genetical background. Studies showed that frequency of AMD occurrence was twice as high in people who had in first-degree relatives with that illness than in people who had no one in family tree with AMD [92]. In publications from 1999, 50 pairs of monozygotic twins and their 47 spouses were examined. All participants were from Iceland and exposure to environmental factors was similar for twins and spouses. Results showed that AMD occurred more frequently among twins than among spouses [169]. Other studies published in 2001 (Beaver Dam Eye) were conducted on patients with AMD who had siblings. After 5 years, healthy siblings which had no signs of disease before were again tested. The odds ratios (Ors) for patients were 10.32 for exudative AMD compared to controls [170].

Further important studies were published in 2005 in the USA. They examined pairs of twins (210 monozygotic, 181 dizygotic, and 58 singletons). Studies have divided the factors induce the development of AMD into environmental and genetic ones. The team confirmed the significance and role of genetic factors in the development of AMD. It is estimated that this influence is important for severity of AMD (46–71%) [95]. These studies confirm the involvement of genetic factors in AMD. Now, many studies have been carried out in search of specific genetic factors. These studies are presented in Table 5.

Table 5. Most significant polymorphisms engaged in the progression of AMD.

Gene	Polymorphism	Mechanism	Role in the AMD	Reference
CFH	*rs1061170 (Y402H)*	Deregulation of proinflammatory cytokines, changes in alternative complement pathway	Progression of AMD	[166,171–181]
	rs6677604	Regulation of complement system	Protection against AMD	[182–184]
CFHR1	*rs6677604*	Decrease level of CFHR1 in serum	Protection against AMD	[183]
CFHR3/CFHR1	*delCFHR1/CFHR3 (CNP147)*	Enhanced complement inhibitory activity	Protection against AMD	[185–187]
ARMS2	*rs10490924*	Activation of intraocular complement	Progression of AMD	[166,188]
HTRA1	*rs11200638*	Increased the expression of HtrA1	Progression of AMD	[189,190]
C3	*rs2230199 (R102G)*	Changes in the expression level and binding affinity of complement factors	Progression of AMD	[28,191–195]
VEGFA	*rs3025039*	Lower concentration of VEGFA	Progression of AMD	[196–198]
	rs3025020	Elevated level of VEGFA	Progression of AMD	[196]
CFB	*rs4151667, rs641153*	Suppressing the inflammation process	Protection against AMD	[166,190,199,200]
SKIV2L	*rs429608*	Unknown	Protection against AMD	[201,202]
LIPC	*rs493258, rs10468017*	Elevated HDL levels	Protection against AMD	[190,191,203–207]
CETP	*rs2230199, rs3764261*	Changes in the concentration of cholesteryl esters in HDL	Progression of AMD	[44,179,199,204,206, 208–212]
CFI	*rs10033900*	Changes in CFI production	Progression of AMD	[213,214]
Il-8	*IL8 −251AA, IL-8 +781C/T, rs2227306*	Increase in inflammation and angiogenesis	Progression of AMD	[166,215–217]
GSTM1	*rs1183423000*	Weaker antioxidant protection		[218–222]
GSTT1	*rs1601993659*	Weaker antioxidant protection		[218,222]
GSTP1	*rs1695, rs1138272*	Weaker antioxidant protection		[218,222,223]

6.1. CFH

This gene is located in locus 1q32-q32.1 [171], and encodes two proteins: CFH and factor H-like 1 protein (FHL-1), which are part of complement system [224,225]. Studies have shown that changes in noncoding parts of this gene are stronger correlated with the development of AMD than polymorphisms occurring in coding parts [183]. Many studies have identified a numerous of polymorphisms in this gene, which may promote the development of AMD. The most significant is *rs1061170* (named *Y402H*), which was detected in 2005 [171–174]. Later studies confirmed that the presence of C allele in this polymorphism significantly increases the risk of progression of AMD [175]. Furthermore, it is estimated that the proportion of this polymorphism in the development of this disease is 60%, and is the most important risk factor [176]. Other studies from 2008 indicated that *Y402H* polymorphism correlates with another (*LOC387715, C3*) and cigarette smoking increasing the risk of AMD [177]. Studies also indicated that *CC* variant in the *Y402H* in patients with dry AMD may have a significant effect on inflammatory regulation and affect the development of AMD [178]. Allele *CC* also affects the accumulation of macrophages after death in the tissues of eye. It is possible that polymorphism *rs1061170* affects deregulation of proinflammatory cytokines [179].

It was suggested that the occurrence of *rs1061170* allele does not affect neovascularization in healthy individuals, because studies have shown that the occurrence of this polymorphism does not affect changes in retinal blood vessels in healthy individuals [226]. Similar unexpected results were reported in 2015, which examined how the presence of extramacular drusen and genetic load of *rs1061170* correlated with the development of AMD. The results showed that drusen strongly correlates with AMD development, but the presence of polymorphism has no effect on the formation of drusen. The authors also stated that patients with extramacular drusen, but without AMD, could be in the control group in research on AMD risk factors [227].

Studies indicated correlation between two polymorphisms in genes *CFH rs1061170* and *HTRA1 rs3793917*, and the progression of exudative AMD [180,181]. Other polymorphisms that may increase the risk of AMD are *rs1410996*, which was found in a Caucasian population [191,228], *rs1329428*, and *rs1329421*, which were indicated by examining Spanish population [229], and also *rs551397, rs800292, rs1410996, rs2274700, rs1329424, rs10801555, rs10737680, rs12124794*, and *rs10733086*, which were obtained in the Chinese population [209,230]. Studies by an Iranian research group have found polymorphisms of *rs800292, rs2274700, rs1061170*, and the new *rs3753395*. All of them were correlated with AMD development [231]. Another research on the Spanish population confirmed role of haplotypes *rs800292, rs1061170*, and *rs800292* and show new ones—*rs529825* and *rs203674* —in the development of AMD [232]. Important polymorphisms were also obtained in *rs121913059* [224,225].

In the *CFH* gene, the protection of polymorphisms against the development of AMD are also known. In 2013, *rs6677604* was detected, which ensures correct regulation of complement system. This polymorphism is linked with lower concentration of CFH in serum [182–184].

6.2. CFHR1 and CFHR3/CFHR1

There are five CFH-related proteins, all involved in complement control. Genes encoding these proteins are located in the *1q32* locus within the tandem 355 kb of genomic region [233]. A study from 2013 demonstrated positive effect of deletion of CNP147 (*rs6677604*) on the protection against the development of AMD. This polymorphism correlates with a decrease level of CFHR1 in serum. This reduces competitive inhibition of CFH binding by competitive inhibition of CFH binding to tissue surfaces. As a result, CFH inhibits C3 convertase and protects against inflammation process [183].

Correctly, this polymorphism is *delCFHR1/CFHR3* deletion (*CNP147*), and this change affects the *CFHR3* and *CFHR1* genes and its other protective polymorphism [185]. The change was also detected in 2006 by studying European and African population [186].

Later, other studies also confirmed that this polymorphism protects against the development of AMD. An enhanced complement inhibitory effect was obtained in the presence of *delCFHR1/CFHR3*, induced by CFH, as the effect of alternative inhibited pathway activation [185–187].

6.3. ARMS2 and ARMS2/HTRA1

In this gene, the most common appears to be the polymorphism of LOC387715 (rs10490924 (A69S), located in exon 1 10q26 [234]. Numerous studies in many populations have shown that this change strongly correlates with the development of late AMD [173,174,191,192,209,229]. Other studies have shown that carriers of *TT* genotype for *rs10490924* have a 10 times higher risk for progression of AMD [172], because of the intraocular complement activation. However, precise mechanism is still unknown [188]. Studies have also confirmed the correlation of polymorphisms of *ARMS2*, *CHF*, and *C3* in increasing risk of development of AMD [177]. Important results were found in 2015, which confirmed that the presence of drusen correlates with AMD development. However, the presence of polymorphism of *ARMS2 rs10490924* is not related with the occurrence of drusen [227].

Studies confirmed that single nucleotide polymorphisms (SNPs) in the promoter region of *HtrA1* strongly correlates with the development of AMD [189,191,235]. This polymorphism increased the expression of HtrA1, but further influence on the development of AMD is unclear [189]. Studies conducted in Chinese population showed, that polymorphisms in these genes (*rs11200638* and *rs3793917*) play role in progression of AMD [209]. Other studies on the Mexican population have shown that *rs10490924* may affect the development of AMD [236]. Research conducted on the Spanish population in 2014 presented polymorphisms of *rs10490923* in *ARMS2* and *HTRA1 rs11200638*, which correlate with the development of AMD. Interestingly, both *ARMS2/HTRA1* (*rs10490923/rs11200638*) polymorphisms of both alleles increase the risk of AMD progression nine times [228].

6.4. Complement Factor C3

The third complement factor (*C3*) is located in the chromosome 19. Studies from 1990 have shown that gene *C3* can occur in two variants of *C3 F* and *C3 S*. *C3 S* is most common, *C3 F* is not so popular, but occur more frequently in Caucasians [28,237,238]. The presence of *rs2230199* (*Arg80Gly, R102G*) polymorphism within *C3* gene correlates with the development of AMD [28,192], because of the changes in the expression level and binding ability of complement factors to the ligands [195]. Other studies have also shown the effect of polymorphisms of *R102G* [191,193,194], *P314L* [194], *rs2241394* [239], *K155Q* [224], *rs6795735* [192], and *rs1047286* [236] in the development of AMD.

6.5. Vascular Endothelial Growth Factor VEGFA

Vascular endothelial growth factor VEGFA has locus in *6p12*. It plays an important role in angiogenesis, vasculogenesis and lymphangiogenesis. This factor is responsible for maintaining proper retinal structure by controlling the growth and maintenance of blood vessels. Polymorphisms in this gene lead to neovascularization and development of AMD [240]. Studies have shown that polymorphisms of *rs3025039* (*+936C/T*) [196,198], and *rs833061* (*−460C*>), *rs2010963* (*−634G*>C) [241], and *rs4711751* [242], correlate with the development of AMD because of the lower concentration of VEGF in serum. Conversely, polymorphism of *rs3025020* is associated with elevated VEGF level [196,197]. It is also possible that *rs699946* promotes the progression of AMD [192], while *rs699947* polymorphism is correlated with an increased concentration of VEGF, which is a risk factor for AMD [243]. Other studies suggest, that *rs943080* polymorphism protects and decreases the expression of this gene and protects against the development of this disease [242].

6.6. *CFB*

CFB gene is located in region 6p21.3, codes complement factor B. Studies showed that these polymorphisms: *rs641153* [244], *rs12614* (*32W*) may have protective effects against development of AMD [166,190,200]. Polymorphisms of *rs4151667* and *rs641153* (*32Q*) have different effects, identified as protective [192]. Other studies have suggested that the presence of these polymorphisms may promote the development of AMD [230,245].

6.7. *SKIV2L*

SKIV2L gene encodes helicase superkiller viralicidic activity 2-like enzyme. It is located in the 6p21. It participates in RNA degradation and play a role in the control of autophagy [201]. Studies have shown that *rs429608* polymorphism appeared more frequently in the control group, which suggest that it may have a protective effect against AMD development, but this mechanism is unknown [201,202].

6.8. *LIPC*

LIPC gene is located in locus *15q21.3*, encodes hepatic triglyceride lipase, an enzyme which participates in lipid metabolism converting HDL to LDL in the liver [203]. It was confirmed that the polymorphisms of *rs10468017* [190,203–206] and *rs493258* [190,191,207] correlate with high serum HDL levels, which protect against the progression of AMD [190].

6.9. *CETP*

CETP is the gene encoding cholesteryl ester transfer protein, which is located in the locus of 16q21 [192]. Studies have shown the contribution of two polymorphisms, which increase the risk of development of AMD, i.e., *rs2230199* [192] and *rs3764261* [179,199,204,206,208–211].

6.10. *CFI*

CFI gene encodes complement factor I (CFI) and is located in the locus of *4q25*. It plays a role in the inactivation of C3b factor, but its action is regulated by CFH. The study conducted in 2009 covered genotyping and sequencing, demonstrating the presence of *rs10033900* and its correlation with the development of AMD [213]. Next studies demonstrated the significant contribution of *rs10033900* of *TT* polymorphism in the development of AMD [191]. Additionally, the latest meta-analysis indicated that the polymorphisms of *rs10033900 T>C* and *rs2285714 C>T* could promote the progression of AMD [214].

6.11. *IL-8*

IL-8 gene is in the *4q12-q21* locus. Studies have shown that *IL8 -251AA* polymorphism is proinflammatory and may be a risk factor for the development of AMD [215]. Later, it was also indicated that the carriers of *IL-8 +781 C/T* of SNP have higher risk of the progression of wet AMD [216]. Other studies showed that polymorphism of *rs2227306* elevates inflammation and increase angiogenesis, which may lead to the development of AMD [166,217].

6.12. *GST Polymorphisms*

Glutathione S-transferase (GST) enzymes, as divided into four groups (GSTA-α, GSTM-μ, GSTT-θ, GSTP-π), are responsible for catalyzing the bonding reaction of glutathione with ROS [246]. This results in water-soluble products, that can be removed from the body. The first study of the polymorphism of this gene was from 2006. The research team analyzed DNA from patients' blood to determine changes in genes of *GSTP1*, *GSTM1* and *GSTT1*. The results from patients were significantly different from these from controls [218]. Also in 2006, another team tested a number of genes, including *GSTM1* microsomal glutathione-S-transferase 1. The results did not show *GSTM1* association with AMD development. However, as in the previous study the group was too small to explicitly deny the participation of *GSTM1* in the development of this disease [219].

Further studies on a larger number of individuals have shown that *GST* polymorphisms may correlate with the development of AMD. Studies confirmed that *GSTM1-null* polymorphism correlates with both dry and wet AMD [220]. In 2012, Chinese patients were conducted on the occurrence of variant *rs1695* of *GSTP1*. The results demonstrated that this allele is correlated with the development of wet AMD [223]. In the same year, *GST* polymorphism study was conducted in Iran. The results showed that *GSTM1* and *GSTT1* correlate with the higher risk of progression of AMD. *GSTM1* polymorphism is associated with the decline in the expression of protein and is a risk factor, if the activity of GST enzyme is lower, ROS cannot be effectively neutralized and affect the development of AMD [221,222].

6.13. Other Polymorphisms Promoting the Development of AMD

Studies indicated polymorphisms promoting the development of AMD, i.e., *rs8135665 (SLC16A8)*, *rs6795735 (ADAMTS9-AS2)* [192], *rs17810398* and *rs17810816 (DAPL1)* [247], *rs13095226 (COL8A1)* [206,242], *rs1999930 (FRK/COL10A1)* [242], *rs3094111 (DDR1)* [248], *rs13278062 (TNFRSF10A)* [249], *rs5888 (SCARB1)* [250], *rs2243250 (IL-4 -590)* [251], and *Pro197Leu* in *GPx1* [11]. Studies confirmed that polymorphisms of *rs547154 (C2)* [244], copy number variation of *C4A* and *C4B* [247], *rs2679798 (MYRIP)* [201], *rs174547 (FADS1)* [192], *rs429358 (APOE)*, and *rs9621532 (TIMP3)* [192] can play a protective role against the development of AMD.

Studies on the mitochondrial DNA show that the JTU haplogroup cluster (haplogroups in mtDNA; and changes in groups of J, T, and U) strongly correlates with the development of AMD. The SNP defining the JTU haplogroup cluster interferes with bioenergetics within the retina and may play a role in the development of AMD [173]. Studies conducted on eye tissues in 2015 showed that mtDNA damage in mice was significantly higher in mRNAs [252]. In addition, meta-analysis has shown that the polymorphisms of *mt16111*, *mt16362*, *mt16319*, *mt1736*, and *mt12007* were characteristic of the Mexican part of American population [253].

7. Summary and Conclusions

This review shows current knowledge about AMD. It focused on risk factors and current mechanisms of the development of AMD. This disease may develop as a result of various factors, focused on different polymorphisms which occur in the inhabitants of different regions and environments (also in those we examined directly) and they correlate with the place of residence and the development of AMD. These results will help with future diagnosing patients from different places of residence.

The action of chemical elements and oxidative stress are closely linked to the development of early and late AMD. Pb, Cd, and Hg had a clear impact on the late AMD. Mn and Zn have a positive impact on this AMD phase. The early phase is influenced by Pb. AMD is age-related and causes blindness in approximately 70% of patients. Pb and Cd accumulate in retinal tissues and may cause damages with participation of oxidative stress, which has consequences for AMD. Pb and Cd through DNA damage increase oxidative stress and produce reactive oxygen species in retinal pigment epithelial cells. These cells interact with basal metals, as well as toxic heavy metals. Cd in the eyes is mainly deposited in people exposed to prolonged exposure and cigarette smokers. Cd and Pb have the ability to appear at a higher concentration in the eye tissue than in the blood.

Currently, exposure to toxic metals is lower than in the past, but Pb and Cd deposited in the eye and can promote AMD. The occurrence of chronic oxidative stress has a big impact on the human eye, because the level of antioxidant enzymes decreases with age. The eye becomes more susceptible to stress and free radicals leading to the development of inter alia AMD.

Retina and choroid have the ability to assimilate heavy metals. Sometimes, elements such as Al, Se, and Tl are also deposited. Al concentration correlates with the development of AMD, causes damage, and the breakage of lysosomes, migration of immune cells, and inflammation, leading to tissue damage and tumor growth. Changes in the eye also cause Hg, which affect the late pathogenesis of AMD. Mn and Zn have a protective effect on the development of AMD.

A favorable activity of metallothioneins as low molecular weight intracellular proteins is able to bind metals and scavenging of free radicals encloses counteraction against destructive changes in the cell structures and preventing apoptosis of cells which happen as after-effects of oxidative stress. Metallothioneins stimulates the processes of cellular growth, favor sedate metabolism of cells and counteract inflammatory processes. They may also play protective role in the course of degenerative diseases that grow more intense in advanced age, like AMD. Retina itself is unceasingly subjected for various types of radiation that may contribute to uncontrolled generation of free radicals.

Three isoforms of metallothioneins, MT-1, MT-2, MT-3, were detected in retina in photoreceptors, pigment epithelium, and neurons. Especially the expression of MT-1 and MT-2 increases is due to the intensification of oxidative stress and escalation of inflammatory processes. Metallothioneins exhibit an ability of binding and detoxification of heavy metal ions and subsequently contribute to minimization of negative effects of their toxicity in cells. Conversely, the binding of Zn or Cu may act as a regulator of homeostasis and bioavailability of these microelements of certain antioxidative significance.

The facts about MT attest the profitable activity of metallothioneins in range of protection of structures of organism (nervous system, retina) against oxidative stress and their potential to counteraction negative after-effects of AMD, as well as to the elimination of many risk factors of this disease. The presence of different protein isoforms is an important aspect of antioxidative protection beside typical antioxidative enzymes.

AMD development is mainly correlated with changes in two types of genes: regulating inflammation process and coding proteins. In this way, the first group contains polymorphisms within *CFH, C3, CFI*. In the second group are *ARMS2* and *HTRA1*. Polymorphisms of *CETP* are correlated with AMD development. Many synergies have been identified between changes in genes of *CFH, HTRA1, CFH*, and *CRP, ARMS2* and *HTRA1*. Many other polymorphisms have been identified so far in both DNA and mtDNA that could promote the development of AMD. Today some polymorphisms that prevent the progression of AMD are known, e.g., in genes of *CFH, CFHR1, CFHR3/CFHR1, ARMS2, C3, VEGF, CFB, SKIV2L*, and *LIPC*.

We examined the polymorphisms of *rs3025039* (*VEGF*), *rs2243250* (*Il-4*), and *GST* to determine how they affect the development of this disease. The role of these mutation is still in question, so we should check how often they appear in patients. This information will allow us to get to know better the mechanism of AMD development. Our research aims to detect new markers of AMD and better understand the etiology of this disease. Although we expect a larger number of patients in the population of large cities, it is certain that we will encounter the effects of environmental pollution in this population. Here, such contaminants and the different types of defense mechanisms correlated with the immune response are described.

Author Contributions: Conceptualization, S.B., P.K., N.K., H.T., G.M. and A.W.; original draft preparation, S.B., K.R., K.H.-B., E.S., J.B. and P.K.; writing review and editing, S.B., J.B. and P.K. All authors have read and agreed to the published version of the manuscript.

Funding: This research received no external funding. The authors declare no competing financial interests. The authors declare that this research did not receive any specific funding.

Acknowledgments: We thank Brendan P. Kavanagh (RCSI, Dublin, Ireland) for his help with improving English.

Conflicts of Interest: The authors declare no conflicts of interest.

References

1. Congdon, N.; O'Colmain, B.; Klaver, C.C.; Klein, R.; Muñoz, B.; Friedman, D.S.; Kempen, J.; Taylor, H.R.; Mitchell, P. Eye Diseases Prevalence Research Group. Causes and prevalence of visual impairment among adults in the United States. *Arch. Ophthalmol.* **2004**, *122*, 477–485.
2. Tuo, J.; Bojanowski, C.M.; Chan, C.C. Genetic factors of age-related macular degeneration. *Prog. Retin. Eye Res.* **2004**, *23*, 229–249. [CrossRef] [PubMed]
3. Qin, S.; Rodrigues, G.A. Progress and perspectives on the role of RPE cell inflammatory responses in the development of age-related macular degeneration. *J. Inflam. Res.* **2008**, *1*, 49–65. [CrossRef] [PubMed]
4. Handa, J.T. How Does the Macula Protect Itself from Oxidative Stress? *Mol. Aspects Med.* **2012**, *33*, 418–435. [CrossRef]
5. Zeng, S.; Hernández, J.; Mullins, R.F. Effects of Antioxidant Components of AREDS Vitamins and Zinc Ions on Endothelial Cell Activation: Implications for Macular Degeneration. *Investig. Ophthalmol. Vis. Sci.* **2012**, *53*, 1041–1047. [CrossRef]
6. Tokarz, P.; Kaarniranta, K.; Błasiak, J. Role of antioxidant enzymes and small molecular weight antioxidants in the pathogenesis of age-related macular degeneration (AMD). *Biogerontology* **2013**, *14*, 461–482. [CrossRef]
7. Błasiak, J.; Petrovski, G.; Veréb, Z.; Facskó, A.; Kaarniranta, K. Oxidative Stress, Hypoxia, and Autophagy in the Neovascular Processes of Age-Related Macular Degeneration. *Biomed. Res. Int.* **2014**, *2014*, 768026. [CrossRef]
8. Plestina-Borjan, I.; Katusic, D.; Medvidovic-Grubisic, M.; Supe-Domic, D.; Bucan, K.; Tandara, L.; Rogosic, V. Association of Age-Related Macular Degeneration with Erythrocyte Antioxidant Enzymes Activity and Serum Total Antioxidant Status. *Oxid. Med. Cell. Longev.* **2015**, *2015*, 804054. [CrossRef] [PubMed]
9. Pugazhendhi, A.; Hubbell, M.; Jairam, P.; Ambati, B. Neovascular Macular Degeneration: A Review of Etiology, Risk Factors, and Recent Advances in Research and Therapy. *Int. J. Mol. Sci.* **2021**, *22*, 1170. [CrossRef]
10. Heesterbeek, T.J.; Lores-Motta, L.; Hoyng, C.B.; Lechanteur, Y.T.E.; den Hollander, A.I. Risk factors for progression of age-related macular degeneration. *Ophthalmic. Physiol. Opt.* **2020**, *40*, 140–170. [CrossRef]
11. Shughoury, A.; Sevgi, D.D.; Ciulla, T.A. Molecular Genetic Mechanisms in Age-Related Macular Degeneration. *Genes* **2022**, *13*, 1233. [CrossRef] [PubMed]
12. Kushwah, N.; Bora, K.; Maurya, M.; Pavlovich, M.C.; Chen, J. Oxidative Stress and Antioxidants in Age-Related Macular Degeneration. *Antioxidants* **2023**, *12*, 1379. [CrossRef] [PubMed]
13. Fernandes, A.R.; Zielińska, A.; Sanchez-Lopez, E.; dos Santos, T.; Garcia, M.L.; Silva, A.M.; Karczewski, J.; Souto, E.B. Exudative versus Nonexudative Age-Related Macular Degeneration: Physiopathology and Treatment Options. *Int. J. Mol. Sci.* **2022**, *23*, 2592. [CrossRef]
14. Deng, Y.; Qiao, L.; Du, M.; Qu Ch Wan, L.; Li, J.; Huang, L. Age-related macular degeneration: Epidemiology, genetics, pathophysiology, diagnosis, and targeted therapy. *Genes Dis.* **2022**, *9*, 6279. [CrossRef] [PubMed]
15. Hammadi, S.; Tzoumas, N.; Ferrara, M.; Porpino Meschede, I.; Lo, K.; Harris, C.; Lako, M.; Steel, D.H. Bruch's Membrane: A Key Consideration with Complement-Based Therapies for Age-Related Macular Degeneration. *J. Clin. Med.* **2023**, *12*, 2870. [CrossRef]
16. Huang, X.; Zhang, L.; Fu, Y.; Zhang, M.; Yang, Q.; Peng, J. Rethinking the potential and necessity of drug delivery systems in neovascular age-related macular degeneration therapy. *Front. Bioeng. Biotechnol.* **2023**, *11*, 1199922. [CrossRef] [PubMed]
17. Arslan, S.; Kadayifcilar, S.; Samur, G. The Potential Role of Dietary Antioxidant Capacity in Preventing Age-Related Macular Degeneration. *J. Am. Coll. Nutr.* **2018**, *38*, 424–432. [CrossRef] [PubMed]
18. Zarbin, M.A. Current concepts in the pathogenesis of age-related macular degeneration. *Arch. Ophthalmol.* **2004**, *122*, 598–614. [CrossRef] [PubMed]
19. Ambati, J.; Ambati, B.K.; Yoo, S.H.; Ianchulev, S.; Adamis, A.P. Age-related macular degeneration: Etiology, pathogenesis, and therapeutic strategies. *Surv. Ophthalmol.* **2003**, *48*, 257–293. [CrossRef]
20. Nachman-Clewner, M.; Giblin, F.J.; Dorey, C.K.; Blanks, R.H.I.; Dang, L.; Dougherty, C.J.; Blanks, J.C. Selective Degeneration of Central Photoreceptors after Hyperbaric Oxygen in Normal and Metallothionein-Knockout Mice. *Investig. Ophthalmol. Vis. Sci.* **2008**, *49*, 3207–3215. [CrossRef]
21. Bartosz, G. *Druga Twarz Tlenu. Wolne Rodniki w Przyrodzie*; PWN-Polish Scientific Publishers: Warszawa, Poland, 2019; 447p.
22. Alvarez, L.; Gonzalez-Iglesias, H.; Garcia, M.; Ghosh, S.; Sanz-Medel, A.; Coca-Prados, M. The Stoichiometric Transition from Zn_6Cu_1-Metallothionein to Zn_7-Metallothionein Underlies the Up-regulation of Metallothionein (MT) Expression—Quantitative analysis of MT-metal load in eye cells. *J. Biol. Chem.* **2012**, *287*, 28456–28469. [CrossRef]
23. Honda, S.; Farboud, B.; Hjelmeland, L.M.; Handa, J.T. Induction of an aging mRNA retinal pigment epithelial cell phenotype by matrix-containing advanced glycation end products in vitro. *Investig. Ophthalmol. Vis. Sci.* **2001**, *42*, 2419–2425.
24. Cano, M.; Thimmalappula, R.; Fujihara, M.; Nagai, N.; Sporn, M.; Wang, A.L.; Neufeld, A.H.; Biswal, S.; Handa, J.T. Cigarette smoking, oxidative stress, the anti-oxidant response through Nrf2 signaling, and Age-related Macular Degeneration. *Vis. Res.* **2010**, *50*, 652–664. [CrossRef]
25. Nguyen, T.; Sherratt, P.J.; Pickett, C.B. Regulatory mechanisms controlling gene expression mediated by the antioxidant response element. *Ann. Rev. Pharmacol. Toxicol.* **2003**, *43*, 233–260. [CrossRef]
26. Rangasamy, T.; Cho, C.Y.; Thimmulappa, R.K.; Zhen, L.; Srisuma, S.S.; Kensler, T.W.; Yamamoto, M.; Petrache, I.; Tuder, R.M.; Biswal, S. Genetic ablation of Nrf2 enhances susceptibility to cigarette smoke induced emphysema in mice. *J. Clin. Investig.* **2004**, *114*, 1248–1259. [CrossRef]

27. Thimmulappa, R.K.; Mai, K.H.; Srisuma, S.; Kensler, T.W.; Yamamoto, M.; Biswal, S. Identification of Nrf2-regulated genes induced by the chemopreventive agent sulforaphane by oligonucleotide microarray. *Cancer Res.* **2002**, *62*, 5196–5203.
28. Yates, J.R.; Sepp, T.; Matharu, B.K.; Khan, J.C.; Thurlby, D.A.; Shahid, H.; Clayton, D.G.; Hayward, C.; Morgan, J.; Wright, A.F.; et al. Complement C3 variant and the risk of age-related macular degeneration. *N. Eng. J. Med.* **2007**, *357*, 553–561. [CrossRef]
29. Biswas, S.K.; Rahman, I. Environmental toxicity, redox signaling and lung inflammation: The role of glutathione. *Mol. Asp. Med.* **2009**, *30*, 60–76. [CrossRef]
30. Walsh, A.C.; Michaud, S.G.; Malossi, J.A.; Lawrence, D.A. Glutathione depletion in human T lymphocytes: Analysis of activation-associated gene expression and the stress response. *Toxicol. Appl. Pharmacol.* **1995**, *133*, 249–261. [CrossRef]
31. Will, O.; Mahler, H.C.; Arrigo, A.P.; Epe, B. Influence of glutathione levels and heat-shock on the steady-state levels of oxidative DNA base modifications in mammalian cells. *Carcinogenesis* **1999**, *20*, 333–337. [CrossRef]
32. Handa, J.T.; Verzijl, N.; Matsunaga, H.; Aotaki-Keen, A.; Lutty, G.A.; Te Koppele, J.M.; Miyata, T.; Hjelmeland, L.M. Increase in the advanced glycation end product pentosidine in Bruch's membrane with age. *Investig. Ophthalmol. Vis. Sci.* **1999**, *40*, 775–779.
33. Nowak, J.Z. Age-related macular degeneration (AMD): Pathogenesis and therapy. *Pharmacol. Rep.* **2006**, *58*, 353–363.
34. Fisher, R.F. The influence of age on some ocular basement membranes. *Eye* **1987**, *1*, 184–189. [CrossRef]
35. Bok, D. Retinal photoreceptor-pigment epithelium interactions. Friedenwald lecture. *Investig. Ophthalmol. Vis. Sci.* **1985**, *26*, 1659–1694.
36. Kell, D.B.; Pretorius, E. Serum ferritin is an important inflammatory disease marker, as it is mainly a leakage product from damaged cells. *Metallomics* **2014**, *6*, 748–773. [CrossRef]
37. Klein, R.J.; Zeiss, C.; Chew, E.Y.; Tsai, J.Y.; Sackler, R.S.; Haynes, C.; Henning, A.K.; San Giovanni, J.P.; Mane, S.M.; Mayne, S.T.; et al. Complement Factor H Polymorphism in Age-Related Macular Degeneration. *Science* **2005**, *308*, 385–389. [CrossRef]
38. Wang, Y.; Kinzie, E.; Berger, F.G.; Lim, S.K.; Baumann, H. Haptoglobin, an inflammation-inducible plasma protein. *Redox Rep.* **2001**, *6*, 379–385. [CrossRef]
39. Fenkci, V.; Fenkci, S.; Yilmazer, M.; Serteser, M. Decreased total antioxidant status and increased oxidative stress in women with polycystic ovary syndrome may contribute to the risk of cardiovascular disease. *Fertil. Steril.* **2003**, *80*, 123–127. [CrossRef]
40. Piotrowski, J. *Podstawy Toksykologii*; WNT: Warszawa, Poland, 2008; 467p.
41. Haddad, H.H. The Effect of Heavy Metals Cadmium, Chromium and Iron Accumulation in Human Eyes. *Am. J. Analyt. Chem.* **2012**, *3*, 710–713. [CrossRef]
42. Lutz, W.; Pałczyński, C. *Immunotoksykologia*; Publishing House, Institute of Occupational Medicine named after Professor George Nofer: Łódź, Poland, 2005; 395p.
43. Kędryna, T. *Chemia Ogólna z Elementami Biochemii*; ZamKor: Kraków, Poland, 2006; 530p.
44. Tanito, M.; Masutani, H.; Kim, Y.C.; Nishikawa, M.; Ohira, A.; Yodoi, J. Sulforaphane induces thioredoxin through the antioxidant-responsive element and attenuates retinal light damage in mice. *Investig. Ophthalmol. Vis. Sci.* **2005**, *46*, 979–987. [CrossRef]
45. Gao, X.; Talalay, P. Induction of phase 2 genes by sulforaphane protects retinal pigment epithelial cells against photooxidative damage. *Proc. Nat. Acad. Sci. USA* **2004**, *101*, 10446–10451. [CrossRef] [PubMed]
46. Nelson, K.C.; Armstrong, J.S.; Moriarty, S.; Cai, J.; Wu, M.W.; Sternberg, P. Jr, Jones D.P. Protection of retinal pigment epithelial cells from oxidative damage by oltipraz, a cancer chemo-preventive agent. *Investig. Ophthalm. Vis. Sci.* **2002**, *43*, 3550–3554.
47. Ha, K.N.; Chen, Y.; Cai, J.; Sternberg, P., Jr. Increased glutathione synthesis through an ARE-Nrf2-dependent pathway by zinc in the RPE: Implication for protection against oxidative stress. *Investig. Ophthalmol. Vis. Sci.* **2006**, *47*, 2709–2715. [CrossRef]
48. Chlebda, E.; Antonowicz-Juchniewicz, E.; Andrzejak, R. Wpływ ekspozycji zawodowej na ołów i arsen na stężenie karotenoidów w surowicy u pracowników huty miedzi. *Med. Pr.* **2004**, *55*, 389–401.
49. Ociepa-Kubicka, A.; Ociepa, E. Toksyczne oddziaływanie metali ciężkich na rośliny, zwierzęta i ludzi. *Inż. Ochr. Środ.* **2012**, *15*, 169–180.
50. Wojciechowska, M.; Kołodziejczyk, J.; Gocki, J.; Bartuzi, Z. Nadwrażliwość na nikiel. *Alerg. Astma Immunol.* **2008**, *13*, 136–140.
51. Brzóska, M.; Jurczuk, M.; Moniuszko-Jakoniuk, J. Interakcje kadmu z wybranymi biopierwiastkami. *Terapia* **1997**, *7*, 28–30.
52. Bridges, C.C.; Zalups, R.K. Molecular and ionic mimicry and the transport of toxic metals. *Toxicol. Appl. Pharmacol.* **2005**, *204*, 274–308. [CrossRef]
53. Klassen, C.D.; Liu, J.; Choudhuri, S. Metallothionein: An intracelluar protein to protect against cadmium toxicity. *Ann. Rev. Pahrmacol. Toxicol.* **1999**, *39*, 267–294. [CrossRef]
54. Swindell, W.R. Metallothionein and the Biology of Aging. *Ageing Res. Rev.* **2011**, *10*, 132–145. [CrossRef]
55. Krzywy, I.; Krzywy, E.; Pastuszak-Gabinowska, M.; Brodkiewicz, A. Ołów—Czy jest się czego obawiać? *Ann. Acad. Med. Stet.* **2010**, *56*, 118–128.
56. Erie, J.C.; Butz, J.A.; Good, J.A.; Erie, E.A.; Burritt, M.F.; Cameron, J.D. Heavy Metal Concentrations in Human Eyes. *Am. J. Ophthalmol.* **2005**, *139*, 888–893. [CrossRef] [PubMed]
57. Wills, N.K.; Ramanujam, V.M.; Kalariya, N.; Lewis, J.R.; van Kuijk, F.J. Copper and zinc distribution in the human retina: Relationship to cadmium accumulation, age, and gender. *Exp. Eye Res.* **2008**, *87*, 80–88. [CrossRef]
58. Brodzka, S. Środowiskowe i Immunogenetyczne Uwarunkowania Zmian Patofizjologicznych w Zwyrodnieniu Plamki Żółtej. Environmental and Immunogenetic Determinants of Pathophysiological Changes in Macular Degeneration. Ph.D. Thesis, University of Zielona Góra, Zielona Góra, Poland, 2021; 388p.

59. Park, S.J.; Lee, J.H.; Woo, S.J.; Kang, S.W.; Park, K.H. Five Heavy Metallic Elements and Age-Related Macular Degeneration: Korean National Health and Nutrition Examination Survey, 2008–2011. *Ophthalmology* **2015**, *122*, 129–137. [CrossRef]
60. Bhattacharyya, M.H.; Wilson, A.K.; Rajan, S.S.; Jonah, M. Biochemical pathways in cadmium toxicity. In *Molecular Biology and Toxicology of Metals*; Zalups, R.K., Koropatnick, J., Eds.; Taylor and Francis: London, UK, 2000; pp. 34–74.
61. Potts, A.M.; Au, P.C. The affinity of melanin for inorganic ions. *Exp. Eye Res.* **1976**, *22*, 487–491. [CrossRef] [PubMed]
62. Ulshafer, R.J.; Allen, C.B.; Rubin, M.L. Distributions of elements in the human retinal pigment epithelium. *Arch. Ophthalmol.* **1990**, *108*, 113–117. [CrossRef] [PubMed]
63. Pamphlett, R.; Cherepanoff, S.; Khoon Too, L.; Kum Jew, S.; Doble, P.A.; Bishop, D.P. The distribution of toxic metals in the human retina and optic nerve head: Implications for age-related macular degeneration. *PLoS ONE* **2020**, *15*, e0241054. [CrossRef] [PubMed]
64. Filipoiu, D.C.; Bungau, S.G.; Endres, L.; Negru, P.A.; Bungau, A.F.; Pasca, B.; Radu, A.-F.; Tarce, A.G.; Bogdan, M.A.; Behl, T.; et al. Characterization of the Toxicological Impact of Heavy Metals on Human Health in Conjunction with Modern Analytical Methods. *Toxics* **2022**, *10*, 716. [CrossRef]
65. Eichenbaum, J.W.; Zheng, W. Distribution of lead and transthyretin in human eyes. *Clin. Toxicol.* **2000**, *38*, 377–381. [CrossRef]
66. Valko, M.; Morris, H.; Cronin, M.T. Metals, toxicity and oxidative stress. *Curr. Med. Chem.* **2005**, *12*, 1161–1208. [CrossRef] [PubMed]
67. Dantzig, P.I. Age related macular degeneration and cutaneous signs of mercury. *Toxic Cutan. Ocular Toxicol.* **2005**, *24*, 3–9. [CrossRef]
68. Hornung, V.; Bauernfeind, F.; Halle, A.; Samstad, E.O.; Kono, H.; Rock, K.L.; Fitzgerald, K.A.; Latz, E. Silica crystals and aluminum salts mediate NALP-3 inflammasome activation via phagosomal destabilization. *Nat. Immunol.* **2008**, *9*, 847–856. [CrossRef] [PubMed]
69. Erie, J.C.; Good, J.A.; Butz, J.A.; Pulido, J.S. Reduced zinc and copper in the retinal pigment epithelium and choroid in age-related macular degeneration. *Am. J. Ophthalmol.* **2009**, *147*, 276–282. [CrossRef] [PubMed]
70. Awh, C.C.; Lane, A.M.; Hawken, S.; Zanke, B.; Kim, I.K. CFH and ARMS2 genetic polymorphisms predict response to antioxidants and zinc in patients with age-related macular degeneration. *Ophthalmology* **2013**, *120*, 2317–2323. [CrossRef]
71. Finkel, T. Signal transduction by reactive oxygen species. *J. Cell Biol.* **2011**, *194*, 7–15. [CrossRef] [PubMed]
72. Evans, M.D.; Dizdaroglu, M.; Cooke, M.S. Oxidative DNA damage and disease: Induction, repair and significance. *Mutation Res.* **2004**, *567*, 1–61. [CrossRef] [PubMed]
73. Shen, J.K.; Dong, A.; Hackett, S.F.; Bell, W.R.; Green, W.R.; Campochiaro, P.A. Oxidative damage in age-related macular degeneration. *Histol. Histopathol.* **2007**, *22*, 1301–1308.
74. Beatty, S.; Koh, H.; Phil, M.; Henson, D.; Boulton, M. The role of oxidative stress in the pathogenesis of age-related macular degeneration. *Surv. Ophthalmol.* **2000**, *45*, 115–134. [CrossRef] [PubMed]
75. Hollyfield, J.G.; Bonilha, V.L.; Rayborn, M.E.; Yang, X.; Shadrach, K.G.; Lu, L.; Ufret, R.L.; Salomon, R.G.; Perez, V.L. Oxidative damage-induced inflammation initiates age-related macular degeneration. *Nat. Med.* **2008**, *14*, 194–198. [CrossRef]
76. Thurman, J.M.; Renner, B.; Kunchithapautham, K.; Ferreira, V.P.; Pangburn, M.K.; Ablonczy, Z.; Tomlinson, S.; Holers, V.M.; Rohrer, B. Oxidative stress renders retinal pigment epithelial cells susceptible to complement-mediated injury. *J. Biol. Chem.* **2009**, *284*, 16939–16947. [CrossRef]
77. Wu, Z.; Lauer, T.W.; Sick, A.; Hackett, S.F.; Campochiaro, P.A. Oxidative stress modulates complement factor H expression in retinal pigmented epithelial cells by acetylation of FOXO3. *J. Biol. Chem.* **2007**, *282*, 22414–22425. [CrossRef] [PubMed]
78. Mitter, S.K.; Song, C.; Qi, X.; Mao, H.; Rao, H.; Akin, D.; Lewin, A.; Grant, M.; Dunn, W.; Ding, J.; et al. Dysregulated autophagy in the RPE is associated with increased susceptibility to oxidative stress and AMD. *Autophagy* **2014**, *10*, 1989–2005. [CrossRef] [PubMed]
79. Rose, R.C.; Richer, S.P.; Bode, A.M. Ocular oxidants and antioxidant protection. *Proc. Soc. Exp. Biol. Med.* **1998**, *217*, 397–407. [CrossRef] [PubMed]
80. Winkler, B.S.; Boulton, M.E.; Gottsch, J.D.; Sternberg, P. Oxidative damage and age-related macular degeneration. *Mol. Vis.* **1999**, *5*, 32.
81. Schneeweis, D.M.; Schnapf, J.L. Photovoltage of rods and cones in the macaque retina. *Science* **1995**, *268*, 1053–1056. [CrossRef] [PubMed]
82. Rogers, B.S.; Symons, R.C.; Komeima, K.; Shen, J.; Xiao, W.; Swaim, M.E.; Gong, Y.Y.; Kachi, S.; Campochiaro, P.A. Differential sensitivity of cones to iron-mediated oxidative damage. *Investig. Ophthalmol. Vis. Sci.* **2007**, *48*, 438–445. [CrossRef] [PubMed]
83. Roth, F.; Bindewald, A.; Holz, F.G. Key pathophysiologic pathways in age-related macular disease. *Graefe's Arch. Clin. Exp. Ophthalmol.* **2004**, *242*, 710–716. [CrossRef] [PubMed]
84. Tate Jr, D.J.; Miceli, M.V.; Newsome, D.A. Phagocytosis and H_2O_2 induce catalase and metallothionein gene expression in human retinal pigment epithelial cells. *Investig. Ophthalmol. Vis. Sci.* **1995**, *36*, 1271–1279.
85. Kennedy, C.J.; Rakoczy, P.E.; Constable, I.J. Lipofuscin of the retinal pigment epithelium: A review. *Eye* **1995**, *9*, 763–771. [CrossRef]
86. Murdaugh, L.S.; Avalle, L.B.; Mandal, S.; Dill, A.E.; Dillon, J.; Simon, J.D.; Gaillard, E.R. Compositional studies of human RPE lipofuscin. *J. Mass Spectrom.* **2010**, *45*, 1139–1147. [CrossRef]

87. Iliescu, D.A.; Ghita, A.C.; Ilie, L.A.; Voiculescu, S.E.; Geamanu, A.; Ghita, A.M. Non-Neovascular Age-Related Macular Degeneration Assessment: Focus on Optical Coherence Tomography Biomarkers. *Diagnostics* **2024**, *14*, 764. [CrossRef] [PubMed]
88. Jamrozik, D.; Dutczak, R.; Machowicz, J.; Wojtyniak, A.; Smędowski, A.; Pietrucha-Dutczak, M. Metallothioneins, a Part of the Retinal Endogenous Protective System in Various Ocular Diseases. *Antioxidants* **2023**, *12*, 1251. [CrossRef] [PubMed]
89. Nakamura, S.; Shimazawa, M.; Hara, H. Physiological Roles of Metallothioneins in Central Nervous System Diseases. *Biol. Pharm. Bull.* **2018**, *41*, 1006–1013. [CrossRef] [PubMed]
90. Velilla, S.; García-Medina, J.J.; García-Layana, A.; Dolz-Marco, R.; Pons-Vázquez, S.; Pinazo-Durán, M.D.; Gómez-Ulla, F.; Arévalo, J.F.; Díaz-Llopis, M.; Gallego-Pinazo, R. Smoking and Age-Related Macular Degeneration: Review and Update. *J. Ophthalmol. Hindawi.* **2013**, *2013*, 895147. [CrossRef] [PubMed]
91. Thornton, J.; Edwards, R.; Mitchell, P.; Harrison, R.A.; Buchan, I.; Kelly, S.P. Smoking and age-related macular degeneration: A review of association. *Eye* **2005**, *19*, 935–944. [CrossRef] [PubMed]
92. Seddon, J.M.; Ajani, U.A.; Mitchell, B.D. Familial aggregation of age-related maculopathy. *Am. J. Ophthalmol.* **1997**, *123*, 199–206. [CrossRef] [PubMed]
93. Vingerling, J.R.; Hofman, A.; Grobbee, D.E.; de Jong, P.T. Age-related macular degeneration and smoking. The Rotterdam Study. *Arch. Ophthalmol.* **1996**, *114*, 1193–1196. [CrossRef] [PubMed]
94. Chaudhuri, M.; Hassan, Y.; Bakka Vemana, P.P.S.; Bellary Pattanashetty, M.S.; Abdin, Z.U.; Siddiqui, H.F. Age-Related Macular Degeneration: An Exponentially Emerging Imminent Threat of Visual Impairment and Irreversible Blindness. *Cureus* **2023**, *15*, e39624. [CrossRef]
95. Seddon, J.M.; Gensler, G.; Milton, R.C.; Klein, M.L.; Rifai, N. Association between C-reactive protein and age-related macular degeneration. *J. Am. Med. Assoc.* **2004**, *291*, 704–710. [CrossRef]
96. Molins, B.; Romero-Vázquez, S.; Fuentes-Prior, P.; Adan, A.; Dick, A.D. C-Reactive Protein as a Therapeutic Target in Age-Related Macular Degeneration. *Front Immunol.* **2018**, *9*, 808. [CrossRef]
97. Khan, J.C.; Thurlby, D.A.; Shahid, H.; Clayton, D.G.; Yates, J.R.; Bradley, M.; Moore, A.T.; Bird, A.C.; Genetic Factors in AMD Study. Smoking and age related macular degeneration: The number of pack years of cigarette smoking is a major determinant of risk for both geographic atrophy and choroidal neovascularisation. *Br. J. Ophthalmol.* **2006**, *90*, 75–80. [CrossRef] [PubMed]
98. Sowa, P.; Rutkowska-Talipska, J.; Rutkowski, K.; Kosztyła-Hojna, B.; Rutkowski, R. Optical radiation in modern medicine. *Post. Dermatol. Alergol.* **2013**, *30*, 246–251. [CrossRef] [PubMed]
99. Youssef, P.N.; Sheibani, N.; Albert, D.M. Retinal light toxicity. *Eye* **2011**, *25*, 1–14. [CrossRef]
100. Neelam, K.; Zhou, S.W.; Au Eong, K.G. The role of blue light in the pathogenesis of Age-related Macular Degeneration. *Points De Vue* **2014**, *71*, 77.
101. Wu, M.; Guo, Y.; Ma, Y.; Zheng, Z.; Wang, Q.; Zhou, X. Association of Two Polymorphisms, rs1061170 and rs1410996, in Complement Factor H with Age-Related Macular Degeneration in an Asian Population: A Meta-Analysis. *Ophthalm. Res.* **2016**, *55*, 135–144. [CrossRef] [PubMed]
102. Fletcher, A.E.; Bentham, G.C.; Agnew, M.; Young, I.S.; Augood, C.; Chakravarthy, U.; de Jong, P.T.; Rahu, M.; Seland, J.; Soubrane, G.; et al. Sunlight exposure, antioxidants, and age-related macular degeneration. *Arch. Ophthalmol.* **2008**, *126*, 1396–1403. [CrossRef]
103. Brodzka, S.; Baszyński, J.; Rektor, K.; Hołderna-Bona, K.; Stanek, E.; Kurhaluk, N.; Tkaczenko, H.; Malukiewicz, G.; Woźniak, A.; Kamiński, P. The Role of Glutathione in Age-Related Macular Degeneration (AMD). *Int. J. Mol. Sci.* **2024**, *25*, 4158. [CrossRef] [PubMed]
104. Rapp, L.M.; Smith, S.C. Morphologic comparisons between rhodopsin-mediated and short wavelength classes of retinal light damage. *Investig. Ophthalmol. Vis. Sci.* **1992**, *33*, 3367–3377.
105. Braunstein, R.E.; Sparrow, J.R. A blue-blocking intraocular lens should be used in cataract surgery. *Arch. Ophthalmol.* **2005**, *123*, 547–549. [CrossRef]
106. Remé, C.E.; Grimm, C.; Hafezi, F.; Wenzel, A.; Williams, T.P. Apoptosis in the Retina: The Silent Death of Vision. *News Physiol. Sci.* **2000**, *15*, 120–124. [CrossRef]
107. Connell, P.P.; Keane AO'Neill, C.; Altaie, W.R.; Loane, E.; Neelam, K.; Nolan, M.J.; Beatty, S. Risk Factors for Age-Related Maculopathy. *J. Ophthalmol.* **2009**, *2009*, 360764. [CrossRef] [PubMed]
108. Sui, G.Y.; Liu, G.C.; Liu, G.Y.; Gao, Y.Y.; Deng, Y.; Wang, W.Y.; Tong, S.H.; Wang, L. Is sunlight exposure a risk factor for age-related macular degeneration? A systematic review and meta-analysis. *Br. J. Ophthalmol.* **2013**, *97*, 389–394. [CrossRef] [PubMed]
109. Mainster, M.A. Violet and blue light blocking intraocular lenses: Photoprotection versus photoreception. *Br. J. Ophthalmol.* **2006**, *90*, 784–792. [CrossRef] [PubMed]
110. Taylor, H.R.; Muñoz, B.; West, S.; Bressler, N.M.; Bressler, S.B.; Rosenthal, F.S. Visible light and risk of age-related macular degeneration. *Trans. Am. Ophthalmol. Soc.* **1990**, *88*, 163–178. [PubMed]
111. Tomany, S.C.; Cruickshanks, K.J.; Klein, R.; Klein, B.E.; Knudtson, M.D. Sunlight and the 10-year incidence of age-related maculopathy: The Beaver Dam Eye Study. *Arch. Ophthalmol.* **2004**, *122*, 750–757. [CrossRef]
112. Delcourt, C.; Carrière, I.; Ponton-Sanchez, A.; Fourrey, S.; Lacroux, A.; Papoz, L. POLA Study Group. Light exposure and the risk of age-related macular degeneration: The Pathologies Oculaires Liées à l'Age (POLA) study. *Arch. Ophthalmol.* **2001**, *119*, 1463–1468. [CrossRef] [PubMed]

113. Maurya, M.; Bora, K.; Blomfield, A.K.; Pavlovich, M.C.; Huang, S.; Liu, C.h.-H.; Chen, J. Oxidative stress in retinal pigment epithelium degeneration: From pathogenesis to therapeutic targets in dry age-related macular degeneration. *Neural Regen. Res.* **2023**, *18*, 2173–2181. [PubMed]
114. Seddon, J.M.; Sobrin, L. Epidemiology of age-related macular degeneration. In *Retina*, 5th ed.; Ryan, S.J., Ed.; Saunders: London, UK; Elsevier: Amsterdam, The Netherlands, 2013; pp. 1134–1144.
115. WHO. *WHO World Report on Vision*; WHO: Geneva, Switzerland, 2019; 180p.
116. Chew, E.Y. Age-related Macular Degeneration: Nutrition, Genes and Deep Learning—The LXXVI Edward Jackson Memorial Lecture. *Am. J. Ophthalmol.* **2020**, *217*, 335–347. [CrossRef] [PubMed]
117. Merle, B.M.J.; Colijn, J.M.; Cougnard-Gregoire, A.; de Koning-Backus, A.P.; Delyfer, M.N.; Kiefte-de Jong, J.C.; Meester-Smoor, M.; Féart, C.; Verzijden, T.; Samieri, C.; et al. Mediterranean diet and incidence of advanced age-related macular degeneration: The EYE-RISK Consortium. *Ophthalmology* **2019**, *126*, 381–390. [CrossRef]
118. McGuinness, M.B.; Karahalios, A.; Simpson, J.A.; Guymer, R.H.; Robman, L.D.; Hodge, A.M.; Cerin, E.; Giles, G.G.; Finger, R.P. Past physical activity and age-related macular degeneration: The Melbourne Collaborative Cohort Study. *Br. J. Ophthalmol.* **2016**, *100*, 1353–1358. [CrossRef]
119. Bringmann, A.; Hollborn, M.; Kohen, L.; Wiedemann, P. Intake of dietary salt and drinking water: Implications for the development of age-related macular degeneration. *Mol. Vis.* **2016**, *22*, 1437–1454.
120. Jaisankar, D.; Swaminathan, G.; Roy, R.; Kulothungan, V.; Sharma, T.; Raman, R. Association of obesity and age-related macular degeneration in Indian population. *Indian J. Ophthalmol.* **2018**, *66*, 976–983. [PubMed]
121. Nielsen, F.H. The emergence of boron as nutritionally important throughout the life cycle. *Nutrition* **2000**, *16*, 512–514. [CrossRef]
122. Kim, M.H.; Zhao, D.; Cho, J.; Guallar, E. Cadmium exposure and age-related macular degeneration. *J. Exp. Sci. Environ. Epidemiol.* **2016**, *26*, 214–218. [CrossRef]
123. Zhang, M.; Jiang, N.; Chu, Y.; Postnikova, O.; Varghese, R.; Horvath, A.; Cheema, A.K.; Golestaneh, N. Dysregulated metabolic pathways in age-related macular degeneration. *Sci. Rep.* **2020**, *10*, 2464. [CrossRef] [PubMed]
124. Buch, H.; Vinding, T.; la Cour, M.; Jensen, G.B.; Prause, J.U.; Nielsen, N.V. Risk factors for age-related maculopathy in a 14-year follow-up study: The Copenhagen City Eye Study. *Acta Ophthalmol. Scand.* **2005**, *83*, 409–418. [CrossRef]
125. Choudhury, F.; Varma, R.; McKean-Cowdin, R.; Klein, R.; Azen, S.P. Los Angeles Latino Eye Study G. Risk factors for four-year incidence and progression of age-related macular degeneration: The los angeles latino eye study. *Am. J. Ophthalmol.* **2011**, *152*, 385–395. [CrossRef] [PubMed]
126. Erke, M.G.; Bertelsen, G.; Peto, T.; Sjolie, A.K.; Lindekleiv, H.; Njolstad, I. Prevalence of age-related macular degeneration in elderly Caucasians: The Tromso Eye Study. *Ophthalmology* **2012**, *119*, 1737–1743. [CrossRef]
127. Klein, R.; Klein, B.E.; Knudtson, M.D.; Meuer, S.M.; Swift, M.; Gangnon, R.E. Fifteen-year cumulative incidence of age-related macular degeneration: The Beaver Dam Eye Study. *Ophthalmology* **2007**, *114*, 253–262. [CrossRef] [PubMed]
128. Woo, S.J.; Ahn, J.; Morrison, M.A.; Ahn, S.Y.; Lee, J.; Kim, K.W.; DeAngelis, M.M.; Park, K.H. Analysis of Genetic and Environmental Risk Factors and Their Interactions in Korean Patients with Age-Related Macular Degeneration. *PLoS ONE* **2015**, *10*, e0132771. [CrossRef]
129. Labarrere, C.A.; Kassab, G.S. Glutathione: A Samsonian life-sustaining small molecule that protects against oxidative stress, aging and damaging inflammation. *Front. Nutr.* **2022**, *9*, 1007816. [CrossRef]
130. Lee, K.S.; Lin, S.; Copland, D.A.; Dick, A.D.; Liu, J. Cellular senescence in the aging retina and developments of genotherapies for age-related macular degeneration. *J. Neuroinflam.* **2021**, *18*, 32. [CrossRef] [PubMed]
131. Lim, J.I. *Age-Related Macular Degeneration*, 2nd ed.; Informa Healthcare: New York, NY, USA, 2008; 386p.
132. Zetterberg, M. Age-related eye disease and gender. *Maturitas* **2016**, *83*, 19–26. [CrossRef] [PubMed]
133. Lin, X.; Lou, L.; Miao, Q.; Wang, Y.; Jin, K.; Shan, P.; Xu, Y. The pattern and gender disparity in global burden of age-related macular degeneration. *Eur. J. Ophthalmol.* **2020**, *31*, 1120672120927256. [CrossRef] [PubMed]
134. Jankowska-Lech, I.; Grabska-Liberek, I.; Krzyżewska-Niedziałek, A.; Pietruszyńska, M. Zwyrodnienie plamki związane z wiekiem (AMD)—choroba starzejących się społeczeństw. *Post. Nauk Med.* **2013**, *26*, 868–873.
135. Pennington, K.L.; DeAngelis, M.M. Epidemiology of age-related macular degeneration (AMD): Associations with cardiovascular disease phenotypes and lipid factors. *Eye Vis.* **2016**, *3*, 34. [CrossRef] [PubMed]
136. Rim, T.H.; Kim, H.K.; Kim, J.W.; Lee, J.S.; Kim, D.W.; Kim, S.S. A Nationwide Cohort Study on the Association Between Past Physical Activity and Neovascular Age-Related Macular Degeneration in an East Asian Population. *JAMA Ophthalmol.* **2018**, *136*, 132–139. [CrossRef] [PubMed]
137. Klein, R.; Knudtson, M.D.; Lee, K.E.; Gangnon, R.E.; Klein, B.E. Age-period-cohort effect on the incidence of age-related macular degeneration: The Beaver Dam Eye Study. *Ophthalmology* **2008**, *115*, 1460–1467. [CrossRef] [PubMed]
138. Merle, B.M.J.; Silver, R.E.; Rosner, B.; Seddon, J.M. Associations between vitamin D intake and progression to incident advanced age-related macular degeneration. *Investig. Ophthalmol. Vis. Sci.* **2017**, *58*, 4569–4578. [CrossRef] [PubMed]
139. Klein, R.; Knudtson, M.D.; Cruickshanks, K.J.; Klein, B.E. Further observations on the association between smoking and the long-term incidence and progression of age-related macular degeneration: The Beaver Dam Eye Study. *Arch. Ophthalmol.* **2008**, *126*, 115–121. [CrossRef]

140. Klein, R.; Klein, B.E.; Knudtson, M.D.; Wong, T.Y.; Cotch, M.F.; Liu, K.; Burke, G.; Saad, M.F.; Jacobs, D.R., Jr. Prevalence of age-related macular degeneration in 4 racial/ethnic groups in the multi-ethnic study of atherosclerosis. *Ophthalmology* **2006**, *113*, 373–380. [CrossRef]
141. Age-Related Eye Disease Study Research Group. Risk factors associated with age-related macular degeneration. A case-control study in the age-related eye disease study: Age-Related Eye Disease Study Report Number 3. *Ophthalmology* **2000**, *107*, 2224–2232.
142. Pang, C.P.; Baum, L.; Chan, W.M.; Lau, T.C.; Poon, P.M.; Lam, D.S. The apolipoprotein E epsilon4 allele is unlikely to be a major risk factor of age-related macular degeneration in Chinese. *Ophthalmologica* **2000**, *214*, 289–291. [CrossRef]
143. Krishnaiah, S.; Das, T.; Nirmalan, P.K.; Nutheti, R.; Shamanna, B.R.; Rao, G.N.; Thomas, R. Risk factors for age-related macular degeneration: Findings from the Andhra Pradesh eye disease study in South India. *Investig. Ophthalmol Vis Sci.* **2005**, *46*, 4442–4449. [CrossRef]
144. Cugati SMitchell, P.; Rochtchina, E.; Tan, A.G.; Smith, W.; Wang, J.J. Cataract surgery and the 10-year incidence of age-related maculopathy: The Blue Mountains Eye Study. *Ophthalmology* **2006**, *113*, 2020–2025. [CrossRef]
145. Ho, L.; Boekhoorn, S.S.; Liana van Duijn, C.M.; Uitterlinden, A.G.; Hofman, A.; de Jong, P.T.; Stijnen, T.; Vingerling, J.R. Cataract surgery and the risk of aging macula disorder: The Rotterdam study. *Investig. Ophthalmol. Vis. Sci.* **2008**, *49*, 4795–4800. [CrossRef]
146. Rosen, E.S. Age-related macular degeneration and cataract surgery. *J. Cataract Refract. Surg.* **2014**, *40*, 173–174. [CrossRef]
147. Casparis, H.; Lindsley, K.; Kuo, I.C.; Sikder, S.; Bressler, N.M. Surgery for cataracts in people with age-related macular degeneration. *Cochrane Database Syst. Rev.* **2017**, *2*, CD006757. [CrossRef]
148. Shim, S.H.; Kim, S.G.; Bae, J.H.; Yu, H.G.; Song, S.J. Risk factors for progression of early age-related macular degeneration in Koreans. *Ophthalmic Epidemiol.* **2016**, *23*, 80–87. [CrossRef]
149. Wang, I.K.; Lin, H.J.; Wan, L.; Lin, C.L.; Yen, T.H.; Sung, F.C. Risk of age-related macular degeneration in end-stage renal disease patients receiving long-term dialysis. *Retina* **2016**, *36*, 1866–1873. [CrossRef]
150. Klein, R.; Klein, B.E.; Knudtson, M.D.; Cotch, M.F.; Wong, T.Y.; Liu, K.; Burke, G.L.; Saad, M.F.; Jacobs, D.R., Jr.; Sharrett, A.R. Subclinical atherosclerotic cardiovascular disease and early age-related macular degeneration in a multiracial cohort: The Multiethnic Study of Atherosclerosis. *Arch. Ophthalmol.* **2007**, *125*, 534–543. [CrossRef]
151. Yip, J.L.; Khawaja, A.P.; Chan, M.P.; Broadway, D.C.; Peto, T.; Tufail, A.; Luben, R.; Hayat, S.; Bhaniani, A.; Wareham, N.J.; et al. Cross sectional and longitudinal associations between cardiovascular risk factors and age related macular degeneration in the EPIC-Norfolk Eye Study. *PLoS ONE* **2015**, *10*, e0132565. [CrossRef]
152. Saunier, V.; Merle, B.M.J.; Delyfer, M.N.; Cougnard-Grégoire, A.; Rougier, M.B.; Amouyel, P.; Lambert, J.C.; Dartigues, J.F.; Korobelnik, J.F.; Delcourt, C. Incidence of and Risk Factors Associated with Age-Related Macular Degeneration: Four-Year Follow-up From the ALIENOR Study. *JAMA Ophthalmol.* **2018**, *136*, 473–481. [CrossRef]
153. Chen, Y.J.; Yeung, L.; Sun, C.C.; Huang, C.C.; Chen, K.S.; Lu, Y.H. Age-related macular degeneration in chronic kidney disease: A meta-analysis of observational studies. *Am. J. Nephrol.* **2018**, *48*, 278–291. [CrossRef]
154. Gopinath, B.; Liew, G.; Kifley, A.; Mitchell, P. Thyroid dysfunction and ten-year incidence of age-related macular degeneration. *Investig. Ophthalmol. Vis. Sci.* **2016**, *57*, 5273–5277. [CrossRef]
155. Chaker, L.; Buitendijk, G.H.; Dehghan, A.; Medici, M.; Hofman, A.; Vingerling, J.R.; Franco, O.H.; Klaver, C.C.; Peeters, R.P. Thyroid function and age-related macular degeneration: A prospective population-based cohort study–The Rotterdam Study. *BMC Med.* **2015**, *13*, 94. [CrossRef]
156. Tsai, C.C.; Kao, S.C.; Cheng, C.Y.; Kau, H.C.; Hsu, W.M.; Lee, C.F.; Wei, Y.H. Oxidative stress change by systemic corticosteroid treatment among patients having active graves ophthalmopathy. *Arch. Ophthalmol.* **2007**, *125*, 1652–1656. [CrossRef]
157. Duncan, K.G.; Bailey, K.R.; Baxter, J.D.; Schwartz, D.M. The human fetal retinal pigment epithelium: A target tissue for thyroid hormones. *Ophthalmic Res.* **1999**, *31*, 399–406. [CrossRef]
158. Ma, H.W.; Thapa, A.; Morris, L.; Redmond, T.M.; Baehr, W.; Ding, X.Q. Suppressing thyroid hormone signaling preserves cone photoreceptors in mouse models of retinal degeneration. *Proc. Nat. Acad. Sci. USA* **2014**, *111*, 3602–3607. [CrossRef]
159. Jonasson, F.; Fisher, D.E.; Eiriksdottir, G.; Sigurdsson, S.; Klein, R.; Launer, L.J.; Harris, T.; Gudnason, V.; Cotch, M.F. Five-year incidence, progression, and risk factors for age-related macular degeneration: The age, gene/environment susceptibility study. *Ophthalmology* **2014**, *121*, 1766–1772. [CrossRef]
160. Zhang, W.; Liu, H.; Al-Shabrawey, M.; Caldwell, R.W.; Caldwell, R.B. Inflammation and diabetic retinal microvascular complications. *J. Cardiovasc. Dis. Res.* **2011**, *2*, 96–103. [CrossRef] [PubMed]
161. Ratnayaka, J.A.; Serpell, L.C.; Lotery, A.J. Dementia of the eye: The role of amyloid beta in retinal degeneration. *Eye* **2015**, *29*, 1013–1026. [CrossRef] [PubMed]
162. Bojanowski, C.M.; Shen, D.; Chew, E.Y.; Ning, B.; Csaky, K.G.; Green, W.R.; Chan, C.C.; Tuo, J. An apolipoprotein E variant may protect against age-related macular degeneration through cytokine regulation. *Environ. Mol. Mutagen.* **2006**, *47*, 594–602. [CrossRef] [PubMed]
163. Keenan, T.D.; Goldacre, R.; Goldacre, M.J. Associations between age-related macular degeneration, Alzheimer disease, and dementia: Record linkage study of hospital admissions. *JAMA Ophthalmol.* **2014**, *132*, 63–68. [CrossRef] [PubMed]
164. Brilliant, M.H.; Vaziri, K.; Connor, T.B., Jr.; Schwartz, S.G.; Carroll, J.J.; McCarty, C.A.; Schrodi, S.J.; Hebbring, S.J.; Kishor, K.S.; Flynn, H.W., Jr.; et al. Mining Retrospective Data for Virtual Prospective Drug Repurposing: L-DOPA and Age-related Macular Degeneration. *Am. J. Med.* **2016**, *129*, 292–298. [CrossRef]

165. Locke, C.J.; Congrove, N.R.; Stamer, W.D.; Bowen, T.J.; Stamer, W.D.; McKay, B.S. Controlled exosome release from the retinal pigment epithelium in situ. *Exp. Eye Res.* **2014**, *129*, 1–4. [CrossRef] [PubMed]
166. Cascella, R.; Ragazzo, M.; Strafella, C.; Missiroli, F.; Borgiani, P.; Angelucci, F.; Marsella, L.T.; Cusumano, A.; Novelli, G.; Ricci, F.; et al. Age-related macular degeneration: Insights into inflammatory genes. *J. Ophthalmol.* **2014**, *2014*, 582842. [CrossRef] [PubMed]
167. Meyers, S.M.; Zachary, A.A. Monozygotic twins with age-related macular degeneration. *Arch. Ophthalmol.* **1988**, *106*, 651–653. [CrossRef] [PubMed]
168. Meyers, S.M. A twin study on age-related macular degeneration. *Trans Am. Ophthalmol. Soc.* **1994**, *92*, 775–843. [CrossRef]
169. Gottfredsdottir, M.S.; Sverrisson, T.; Musch, E.; Stefansson, D.C. Age related macular degeneration in monozygotic twins and theirspouses in Iceland. *Acta Ophthalmol. Scand.* **1999**, *77*, 422–425. [CrossRef]
170. Klein, B.E.K.; Klein, R.; Lee, K.E.; Moore, E.L.; Danforth, L. Risk of incident age-related eye diseases in people with an affected sibling: The Beaver Dam Eye Study. *Am. J. Epidemiol.* **2001**, *154*, 207–211. [CrossRef] [PubMed]
171. Haines, J.H.; Hauser, M.A.; Schmidt, S. Complement factor H variant increases the risk of age-related macular degeneration. *Science* **2005**, *308*, 419–421. [CrossRef] [PubMed]
172. Chakravarthy, U.; McKay, G.J.; de Jong, P.T.; Rahu, M.; Seland, J.; Soubrane, G.; Tomazzoli, L.; Topouzis, F.; Vingerling, J.R.; Vioque, J.; et al. ARMS2 increases the risk of early and late age-related macular degeneration in the European Eye Study. *Ophthalmology* **2013**, *120*, 342–348. [CrossRef] [PubMed]
173. Kenney, M.C.; Hertzog, D.; Chak, G.; Atilano, S.R.; Khatibi, N.; Soe, K.; Nobe, A.; Yang, E.; Chwa, M.; Zhu, F.; et al. Mitochondrial DNA haplogroups confer differences in risk for age-related macular degeneration: A case control study. *BMC Med. Genet.* **2013**, *14*, 4. [CrossRef] [PubMed]
174. Nazari Khanamiri, H.; Ghasemi Falavarjani, K.; Sanati, M.H.; Aryan, H.; Irani, A.; Hashemi, M.; Modarres, M.; Parvaresh, M.M.; Nikeghbali, A. Complement Factor H Y402H and LOC387715 A69S Polymorphisms in Association with Age-Related Macular Degeneration in Iran. *J. Ophthalm. Vis. Res.* **2014**, *9*, 181–187.
175. Francis, P.J.; Schultz, D.W.; Hamon, S.; Ott, J.; Weleber, R.G.; Klein, M.L. Haplotypes in the complement factor H (CFH) gene: Associations with drusen and advanced age-related macular degeneration. *PLoS ONE* **2007**, *2*, e1197. [CrossRef] [PubMed]
176. Thakkinstian, A.; Han, P.; McEvoy, M.; Smith, W.; Hoh, J.; Magnusson, K.; Zhang, K.; Attia, J. Systematic review and meta-analysis of the association between complement factor H Y402H polymorphisms and age-related macular degeneration. *Human Mol. Genet.* **2006**, *15*, 2784–2790. [CrossRef] [PubMed]
177. Seitsonen, S.P.; Onkamo, P.; Peng, G.; Xiong, M.; Tommila, P.V.; Ranta, P.H.; Holopainen, J.M.; Moilanen, J.A.; Palosaari, T.; Kaarniranta, K.; et al. Multifactor Effects and Evidence of Potential Interaction between Complement Factor H Y402H and LOC387715 A69S in Age-Related Macular Degeneration. *PLoS ONE* **2008**, *3*, e3833. [CrossRef]
178. Cao, S.; Ko, A.; Partanen, M.; Pakzad-Vaezi, K.; Merkur, A.B.; Albiani, D.A.; Kirker, A.W.; Wang, A.; Cui, J.Z.; Forooghian, F.; et al. Relationship between systemic cytokines and complement factor H Y402H polymorphism in patients with dry age-related macular degeneration. *Am. J. Ophthalmol.* **2013**, *156*, 1176–1183. [CrossRef]
179. Wang, J.C.; Cao, S.; Wang, A.; To, E.; Law, G.; Gao, J.; Zhang, D.; Cui, J.Z.; Matsubara, J.A. CFH Y402H polymorphism is associated with elevated vitreal GM-CSF and choroidal macrophages in the postmortem human eye. *Mol. Vis.* **2015**, *21*, 264–272.
180. Fang, K.; Gao, P.; Tian, J.; Qin, X.; Yu, W.; Li, J.; Chen, Q.; Huang, L.; Chen, D.; Hu, Y.; et al. Joint Effect of CFH and ARMS2/HTRA1 Polymorphisms on Neovascular Age-Related Macular Degeneration in Chinese Population. *J. Ophthalmol.* **2015**, *2015*, 821918. [CrossRef]
181. Meyers, K.J.; Liu, Z.; Millen, A.E.; Iyengar, S.K.; Blodi, B.A.; Johnson, E.; Snodderly, D.M.; Klein, M.L.; Gehrs, K.M.; Tinker, L.; et al. Joint Associations of Diet, Lifestyle, and Genes with Age-Related Macular Degeneration. *Ophthalmology* **2015**, *122*, 2286–2294. [CrossRef]
182. Hughes, A.E.; Orr, N.; Esfandiary, H.; Diaz-Torres, M.; Goodship, T.; Chakravarthy, U. A common CFH haplotype, with deletion of CFHR1 and CFHR3, is associated with lower risk of age-related macular degeneration. *Nat. Genet.* **2006**, *38*, 1173–1177. [CrossRef] [PubMed]
183. Ansari, M.; McKeigue, P.M.; Skerka, C.; Hayward, C.; Rudan, I.; Vitart, V.; Polasek, O.; Armbrecht, A.M.; Yates, J.R.; Vatavuk, Z.; et al. Genetic influences on plasma CFH and CFHR1 concentrations and their role in susceptibility to age-related macular degeneration. *Human Mol. Genet.* **2013**, *22*, 4857–4869. [CrossRef]
184. Emilsson, V.; Gudmundsson, E.F.; Jonmundsson, T.; Jonsson, B.G.; Twarog, M.; Gudmundsdottir, V.; Li, Z.; Finkel, N.; Poor, S.; Liu, X.; et al. A proteogenomic signature of age-related macular degeneration in blood. *Nat. Commun.* **2022**, *13*, 3401. [CrossRef] [PubMed]
185. Fritsche, L.G.; Lauer, N.; Hartmann, A.; Stippa, S.; Keilhauer, C.N.; Oppermann, M.; Pandey, M.K.; Köhl, J.; Zipfel, P.F.; Weber, B.H.; et al. An imbalance of human complement regulatory proteins CFHR1, CFHR3 and factor H influences risk for age-related macular degeneration (AMD). *Human Mol. Genet.* **2010**, *19*, 4694–4704. [CrossRef] [PubMed]
186. Hageman, G.S.; Hancox, L.S.; Taiber, A.J.; Gehrs, K.M.; Anderson, D.H.; Johnson, L.V.; Radeke, M.J.; Kavanagh, D.; Richards, A.; Atkinson, J.; et al. AMD Clinical Study Group. Extended haplotypes in the complement factor H (CFH) and CFH-related (CFHR) family of genes protect against age-related macular degeneration: Characterization, ethnic distribution and evolutionary implications. *Ann. Med.* **2006**, *38*, 592–604. [CrossRef]

187. Sawitzke, J.; Im, K.M.; Kostiha, B.; Dean, M.; Gold, B. Association Assessment of Copy Number Polymorphism and Risk of Age-Related Macular Degeneration. *Ophthalmology* **2011**, *118*, 2442–2446. [CrossRef]
188. Kato, Y.; Oguchi, Y.; Omori, T.; Kasai, A.; Ogasawara, M.; Sugano, Y.; Itagaki, K.; Ojima, A.; Ishida, Y.; Machida, T.; et al. Age-Related Maculopathy Susceptibility 2 and Complement Factor H Polymorphism and Intraocular Complement Activation in Neovascular Age-Related Macular Degeneration. *Ophthalmol. Sci.* **2022**, *2*, 100167. [CrossRef]
189. He, F.; Li, X.; Cai, S.; Lu, L.; Zhang, T.; Yang, M.; Fan, N.; Wang, X.; Liu, X. Polymorphism rs11200638 enhanced HtrA1 responsiveness and expression are associated with age-related macular degeneration. *Eye* **2022**, *36*, 1631–1638. [CrossRef]
190. Anand, A.; Sharma, K.; Sharma, S.K.; Singh, R.; Sharma, N.K.; Prasad, K. AMD genetics in India: The missing links. *Front. Aging Neurosci.* **2016**, *8*, 115. [CrossRef] [PubMed]
191. Peter, I.; Huggins, G.S.; Ordovas, J.M.; Haan, M.; Seddon, J.M. Evaluation of new and established age-related macular degeneration susceptibility genes in the Women's Health Initiative Sight Exam (WHI-SE) Study. *Am. J. Ophthalmol.* **2011**, *152*, 1005–1013. [CrossRef] [PubMed]
192. Ristau, T.; Paun, C.; Ersoy, L.; Hahn, M.; Lechanteur, Y.; Hoyng, C.; de Jong, E.K.; Daha, M.R.; Kirchhof, B.; den Hollander, A.I.; et al. Impact of the common genetic associations of age-related macular degeneration upon systemic complement component C3d levels. *PLoS ONE* **2014**, *9*, e93459. [CrossRef] [PubMed]
193. Zerbib, J.; Richard, F.; Puche, N.; Leveziel, N.; Cohen, S.Y.; Korobelnik, J.F.; Sahel, J.; Munnich, A.; Kaplan, J.; Rozet, J.M.; et al. R102G polymorphism of the C3 gene associated with exudative age-related macular degeneration in a French population. *Mol. Vis.* **2010**, *16*, 1324–1330. [PubMed]
194. Despriet, D.D.; van Duijn, C.M.; Oostra, B.A. Complement component C3 and risk of age-related macular degeneration. *Ophthalmology* **2009**, *116*, 474–480.e2. [CrossRef] [PubMed]
195. Neto, J.M.; Viturino, M.G.; Ananina, G.; Bajano, F.F.; Costa, S.M.D.S.; Roque, A.B.; Borges, G.F.; Franchi, R.; Rim, P.H.; Medina, F.M.; et al. Association of genetic variants rs641153 (CFB), rs2230199 (C3), and rs1410996 (CFH) with age-related macular degeneration in a Brazilian population. *Exp. Biol. Med.* **2021**, *246*, 2290–2296. [CrossRef] [PubMed]
196. Shimizu, Y.; Arima, K.; Noguchi, Y.; Yamanashi, H.; Kawashiri, S.Y.; Nobusue, K.; Nonaka, F.; Aoyagi, K.; Nagata, Y.; Maeda, T. Vascular endothelial growth factor (VEGF) polymorphism rs3025039 and atherosclerosis among older with hypertension. *Sci. Rep.* **2022**, *12*, 5564. [CrossRef] [PubMed]
197. Villani, L.; Carolei, A.; Rosti, V.; Massa, M.; Campanelli, R.; Catarsi, P.; Abbà, C.; Gale, R.P.; Barosi, G. Clinical Relevance of VEGFA (rs3025039) +936 C>T Polymorphism in Primary Myelofibrosis: Susceptibility, Clinical Co-Variates, and Outcomes. *Genes* **2021**, *12*, 1271. [CrossRef]
198. Lin, J.M.; Wan, L.; Tsai, Y.Y.; Lin, H.J.; Tsai, Y.; Lee, C.C.; Tsai, C.H.; Tseng, S.H.; Tsai, F.J. Vascular endothelial growth factor gene polymorphisms in age-related macular degeneration. *Am. J. Ophthalmol.* **2008**, *145*, 1045–1051. [CrossRef]
199. Wang, D.; Zhou, J.; Hou, X.; Nguyen, D.H.; Cao, G.; Li, G.; Qiu, G.; Zhang, K.; Zhang, M.; Su, Z. CETP Gene may be Associated with Advanced Age-Related Macular Degeneration in the Chinese Population. *Ophthalmic Genet.* **2015**, *36*, 303–308. [CrossRef]
200. Hughes, A.E.; Mullan, G.M.; Bradley, D.T. Complement factor B polymorphism 32W protects against age-related macular degeneration. *Mol. Vis.* **2011**, *20*, 983–988.
201. Kopplin, L.J.; Igo, R.P., Jr.; Wang, Y.; Sivakumaran, T.A.; Hagstrom, S.A.; Peachey, N.S.; Francis, P.J.; Klein, M.L.; SanGiovanni, J.P.; Chew, E.Y.; et al. Genome-wide association identifies SKIV2L and MYRIP as protective factors for age-related macular degeneration. *Genes Immunol.* **2010**, *11*, 609–621. [CrossRef] [PubMed]
202. Lu, F.; Shi, Y.; Qu, C.; Zhao, P.; Liu, X.; Gong, B.; Ma, S.; Zhou, Y.; Zhang, Q.; Fei, P.; et al. A genetic variant in the SKIV2L gene is significantly associated with age-related macular degeneration in a Han Chinese population. *Investig. Ophthalmol. Vis. Sci.* **2013**, *54*, 2911–2917. [CrossRef] [PubMed]
203. Reynolds, R.; Rosner, B.; Seddon, J.M. Serum Lipid Biomarkers and Hepatic Lipase Gene Associations with Age-Related Macular Degeneration. *Ophthalmology* **2010**, *117*, 1989–1995. [CrossRef] [PubMed]
204. Yu, Y.; Reynolds, R.; Fagerness, J.; Rosner, B.; Daly, M.J.; Seddon, J.M. Association of variants in the LIPC and ABCA1 genes with intermediate and large drusen and advanced age-related macular degeneration. *Investig. Ophthalmol. Vis. Sci.* **2011**, *52*, 4663–4670. [CrossRef] [PubMed]
205. Seddon, J.M.; Reynolds, R.; Rosne, B. Associations of smoking, body mass index, dietary lutein, and the LIPC gene variant rs10468017 with advanced age-related macular degeneration. *Mol. Vis.* **2010**, *16*, 2412–2424. [PubMed]
206. Neale, B.M.; Fagerness, J.; Reynolds, R.; Sobrin, L.; Parker, M.; Raychaudhuri, S.; Tan, P.L.; Oh, E.C.; Merriam, J.E.; Souied, E.; et al. Genome-wide association study of advanced age-related macular degeneration identifies a role of the hepatic lipase gene (LIPC). *Proc. Nat. Acad. Sci. USA* **2010**, *107*, 7395–7400. [CrossRef] [PubMed]
207. Merle, B.M.J.; Maubaret, C.; Korobelnik, J.F.; Delyfer, M.N.; Rougier, M.B. Association of HDL-Related Loci with Age-Related Macular Degeneration and Plasma Lutein and Zeaxanthin: The Alienor Study. *PLoS ONE* **2013**, *8*, e79848. [CrossRef] [PubMed]
208. Restrepo, N.A.; Spencer, K.L.; Goodloe, R.; Garrett, T.A.; Heiss, G.; Bůžková, P.; Jorgensen, N.; Jensen, R.A.; Matise, T.C.; Hindorff, L.A.; et al. Genetic Determinants of Age-Related Macular Degeneration in Diverse Populations from the PAGE Study. *Investig. Ophthalmol. Vis. Sci.* **2014**, *55*, 6839–6850. [CrossRef]
209. Tian, J.; Yu, W.; Qin, X.; Fang, K.; Chen, Q.; Hou, J.; Li, J.; Chen, D.; Hu, Y.; Li, X. Association of genetic polymorphisms and age-related macular degeneration in Chinese population. *Investig. Ophthalmol. Vis. Sci.* **2012**, *53*, 4262–4269. [CrossRef]

210. Chen, W.; Stambolian, D.; Edwards, A.O.; Branham, K.E.; Othman, M.; Jakobsdottir, J.; Tosakulwong, N.; Pericak-Vance, M.A.; Campochiaro, P.A.; Klein, M.L.; et al. Genetic variants near TIMP3 and high-density lipoprotein-associated loci influence susceptibility to age-related macular degeneration. *Proc. Natl. Acad. Sci. USA* **2010**, *107*, 7401–7406. [CrossRef] [PubMed]
211. Cipriani, V.; Leung, H.T.; Plagnol, V.; Bunce, C.; Khan, J.C.; Shahid, H.; Moore, A.T.; Harding, S.P.; Bishop, P.N.; Hayward, C.; et al. French AMD Investigators; Cree A.J., Gibson J., Ennis S., Lotery A.J., Wright A.F., Clayton D.G., Yates J.R. Genome-wide association study of age-related macular degeneration identifies associated variants in the TNXB-FKBPL-NOTCH4 region of chromosome 6p21.3. *Human Mol. Genet.* **2012**, *21*, 4138–4150. [CrossRef] [PubMed]
212. Ghorbani, R.; Rasouli, M.; Sefat, F.; Heidari Keshel, S. Pathogenesis of Common Ocular Diseases: Emerging Trends in Extracellular Matrix Remodeling. *Semin. Ophthalmol.* **2023**, *39*, 27–39. [CrossRef] [PubMed]
213. Fagerness, J.A.; Maller, J.B.; Neale, B.M.; Reynolds, R.C.; Daly, M.J.; Seddon, J.M. Variation near complement factor I is associated with risk of advanced AMD. *Eur. J. Human Genet.* **2009**, *17*, 100–104. [CrossRef] [PubMed]
214. Wang, Q.; Zhao, H.S.; Li, L. Association between complement factor I gene polymorphisms and the risk of age-related macular degeneration: A Meta-analysis of literature. *Int. J. Ophthalmol.* **2016**, *9*, 298–305. [PubMed]
215. Goverdhan, S.V.; Ennis, S.; Hannan, S.R.; Madhusudhana, K.C.; Cree, A.J.; Luff, A.J.; Lotery, A.J. Interleukin-8 promoter polymorphism -251A/T is a risk factor for age-related macular degeneration. *Br. J. Ophthalmol.* **2008**, *92*, 537–540. [CrossRef] [PubMed]
216. Tsai, Y.Y.; Lin, J.M.; Wan, L.; Lin, H.J.; Tsai, Y.; Lee, C.C.; Tsai, C.H.; Tsai, F.J.; Tseng, S.H. Interleukin gene polymorphisms in age-related macular degeneration. *Investig. Ophthalmol. Vis. Sci.* **2008**, *49*, 693–698. [CrossRef] [PubMed]
217. Ricci, F.; Staurenghi, G.; Lepre, T.; Missiroli, F.; Zampatti, S.; Cascella, R.; Borgiani, P.; Marsella, L.T.; Eandi, C.M.; Cusumano, A.; et al. Haplotypes in IL-8 gene are associated to age-related macular degeneration: A case-control study. *PLoS ONE* **2013**, *8*, e66978. [CrossRef]
218. Oz, O.; Aras Ates, N.; Tamer, L.; Yildirim, O.; Adigüzel, U. Glutathione S-transferase M1, T1, and P1 gene polymorphism in exudative age-related macular degeneration: A preliminary report. *Eur. J. Ophthalmol.* **2006**, *16*, 105–110. [CrossRef]
219. Haines, J.L.; Schnetz-Boutaud, N.; Schmidt, S.; Scott, W.K.; Agarwal, A.; Postel, E.A.; Olson, L.; Kenealy, S.J.; Hauser, M.; Gilbert, J.R.; et al. Functional Candidate Genes in Age-Related Macular Degeneration: Significant Association with VEGF, VLDLR, and LRP6, IOVS. *Investig. Ophthalmol. Vis. Sci.* **2006**, *47*, 329–335. [CrossRef]
220. Güven, M.; Görgün, E.; Ünal, M.; Yenerel, M.; Batar, B.; Küçümen, B.; Dinç, U.A.; Güven, G.S.; Ulus, T.; Yüksel, A. Glutathione S-transferase M1, GSTT1 and GSTP1 genetic polymorphisms and the risk of age-related macular degeneration. *Ophthalm. Res.* **2011**, *46*, 31–37. [CrossRef]
221. Othman, H.; Gholampour, A.R.; Saadat, I.; Farvardin-Jahromoi, M.; Saadat, M. Age-related macular degeneration and genetic polymorphisms of glutathione S-transferases M1 (GSTM1) and T1 (GSTT1). *Mol. Biol. Rep.* **2012**, *39*, 3299–3303. [CrossRef]
222. Chen, X.; Luo, Y. Association of GSTM1, GSTT1, and GSTP1 Ile105Val polymorphisms with risk of age-related macular degeneration: A meta-analysis. *Ophthalmic Genet.* **2022**, *43*, 615–621. [CrossRef]
223. Gu, H.; Sun, E.; Cui, L.; Yang, X.; Lim, A.; Xu, J.; Snellingen, T.; Liu, X.; Wang, N.; Liu, N. Association of glutathione S-transferase pi isoform single-nucleotide polymorphisms with exudative age-related macular degeneration in a Chinese population. *Retina* **2012**, *32*, 1967–1972. [CrossRef]
224. Zhan, X.; Larson, D.E.; Wang, C.; Koboldt, D.C.; Sergeev, Y.V.; Fulton, R.S.; Fulton, L.L.; Fronick, C.C.; Branham, K.E.; Bragg-Gresham, J.; et al. Identification of a rare coding variant in complement 3 associated with age-related macular degeneration. *Nat. Genet.* **2013**, *45*, 1375–1379. [CrossRef]
225. Raychaudhuri, S.; Iartchouk, O.; Chin, K.; Tan, P.L.; Tai, A.K.; Ripke, S.; Gowrisankar, S.; Vemuri, S.; Montgomery, K.; Yu, Y.; et al. A rare penetrant mutation in CFH confers high risk of age-related macular degeneration. *Nat. Genet.* **2011**, *43*, 1232–1236. [CrossRef]
226. Told, R.; Palkovits, S.; Boltz, A.; Schmidl, D.; Napora, K.J.; Werkmeister, R.M.; Haslacher, H.; Frantal, S.; Popa-Cherecheanu, A.; Schmetterer, L.; et al. Flicker-induced retinal vasodilatation is not dependent on complement factor H polymorphism in healthy young subjects. *Acta Ophthalmol.* **2014**, *92*, e540–e545. [CrossRef]
227. Ersoy, L.; Schick, T.; de Graft, D.; Felsch, M.; Hoyng, C.B.; den Hollander, A.I.; Kirchhof, B.; Fauser, S.; Liakopoulos, S. Extramacular drusen are highly associated with age-related macular degeneration, but not with CFH and ARMS2 genotypes. *Br. J. Ophthalmol.* **2015**, *100*, 1047–1051. [CrossRef]
228. Cruz-González, F.; Cieza-Borrella, C.; López Valverde, G.; Lorenzo-Pérez, R.; Hernández-Galilea, E.; González-Sarmiento, R. CFH (rs1410996), HTRA1 (rs112000638) and ARMS2 (rs10490923) gene polymorphisms are associated with AMD risk in Spanish patients. *Ophthalm. Genet.* **2014**, *35*, 68–73. [CrossRef] [PubMed]
229. Brión, M.; Sanchez-Salorio, M.; Cortón, M.; de la Fuente, M.; Pazos, B.; Othman, M.; Swaroop, A.; Abecasis, G.; Sobrino, B.; Carracedo, A. Genetic association study of age-related macular degeneration in the Spanish population. *Acta Ophthalmol.* **2011**, *89*, e12–e22. [CrossRef] [PubMed]
230. Wu, L.; Tao, Q.; Chen, W.; Wang, Z.; Song, Y.; Sheng, S.; Li, P.; Zhou, J. Association between Polymorphisms of Complement Pathway Genes and Age-Related Macular Degeneration in a Chinese Population. *Investig. Ophthalmol Vis. Sci.* **2013**, *54*, 170–174. [CrossRef]
231. Babanejad, M.; Moein, H.; Akbari, M.R. Investigating the CFH Gene Polymorphisms as a Risk Factor for Age-related Macular Degeneration in an Iranian Population. *Ophthalmol. Genet.* **2015**, *21*, 1–6. [CrossRef]

232. García, M.; Álvarez, L.; Nogacka, A.M.; González-Iglesias, H.; Escribano, J.; Fernández-Vega, B.; Fernández-Vega, Á.; Fernández-Vega, L.; Coca-Prados, M. CFH polymorphisms in a Northern Spanish population with neovascular and dry forms of age-related macular degeneration. *Acta Ophthalmol.* **2015**, *93*, e658–e666. [CrossRef]
233. Hyttinen, J.M.T.; Błasiak, J.; Kaarniranta, K. Non-Coding RNAs Regulating Mitochondrial Functions and the Oxidative Stress Response as Putative Targets against Age-Related Macular Degeneration (AMD). *Int. J. Mol. Sci.* **2023**, *24*, 2636. [CrossRef]
234. Seddon, J.M.; Francis, P.J.; George, S.; Schultz, D.W.; Rosner, B.; Klein, M.L. Association of CFH Y402H and LOC387715 A69S with progression of age-related macular degeneration. *JAMA* **2007**, *297*, 1793–1800. [CrossRef]
235. Tuo, J.; Ross, R.J.; Reed, G.F.; Yan, Q.; Wang, J.J.; Bojanowski, C.M.; Chew, E.Y.; Feng, X.; Olsen, T.W.; Ferris, F.L., 3rd; et al. The HtrA1 promotor polymorphism, smoking, and age-related macular degeneration in multiple case-control samples. *Ophthalmology* **2008**, *115*, 1891–1898. [CrossRef]
236. Contreras, A.V.; Zenteno, J.C.; Fernández-López, J.C.; Rodríguez-Corona, U.; Falfán-Valencia, R.; Sebastian, L.; Morales, F.; Ochoa-Contreras, D.; Carnevale, A.; Silva-Zolezzi, I. CFH haplotypes and ARMS2, C2, C3, and CFB alleles show association with susceptibility to age-related macular degeneration in Mexicans. *Mol. Vis.* **2014**, *20*, 105–116.
237. Botto, M.; Fong, K.Y.; So, A.K.; Koch, C.; Walport, M.J. Molecular basis of polymorphisms of human complement component C3. *J. Exp. Med.* **1990**, *172*, 1011–1017. [CrossRef]
238. Thakkinstian, A.; McKay, G.J.; McEvoy, M.; Chakravarthy, U.; Chakrabarti, S.; Silvestri, G.; Kaur, I.; Li, X.; Attia, J. Systematic review and meta-analysis of the association between complement component 3 and age-related macular degeneration: A HuGE review and meta-analysis. *Am. J. Epidemiol.* **2011**, *173*, 1365–1379. [CrossRef] [PubMed]
239. Yanagisawa, S.; Kondo, N.; Miki, A.; Matsumiya, W.; Kusuhara, S.; Tsukahara, Y.; Honda, S.; Negi, A. A Common Complement C3 Variant Is Associated with Protection against Wet Age-Related Macular Degeneration in a Japanese Population. *PLoS ONE* **2011**, *6*, e28847. [CrossRef] [PubMed]
240. Penn, J.S.; Madanet, A.; Caldwell, R.B.; Bartoli, M.; Caldwell, R.W.; Hartnett, M.E. Vascular endothelial growth factor in eye disease. *Progr. Retin. Eye Res.* **2008**, *27*, 331–371. [CrossRef] [PubMed]
241. Janik-Papis, K.; Zaras, M.; Krzyżanowska, A.; Woźniak, K.; Błasiak, J.; Szaflik, J.; Szaflik, J.P. Association between vascular endothelial growth factor gene polymorphisms and age-related macular degeneration in a Polish population. *Exp. Mol. Pathol.* **2009**, *87*, 234–238. [CrossRef] [PubMed]
242. Yu, Y.; Bhangale, T.R.; Fagerness, J.; Ripke, S.; Thorleifsson, G.; Tan, P.L.; Souied, E.H.; Richardson, A.J.; Merriam, J.E.; Buitendijk, G.H.; et al. Common variants near FRK/COL10A1 and VEGFA are associated with advanced age-related macular degeneration. *Human Mol. Genet.* **2011**, *20*, 3699–3709. [CrossRef]
243. Ambreen, F.; Ismail, M.; Qureshi, I.Z. Association of gene polymorphism with serum levels of inflammatory and angiogenic factors in Pakistani patients with age-related macular degeneration. *Mol. Vis.* **2015**, *21*, 985–999.
244. Kaur, I.; Katta, S.; Reddy, R.K.; Narayanan, R.; Mathai, A.; Majji, A.B.; Chakrabarti, S. The involvement of complement factor B and complement component C2 in an Indian cohort with age-related macular degeneration. *Investig. Ophthalmol. Vis. Sci.* **2010**, *51*, 59–63. [CrossRef]
245. Mantel, I.; Ambresin, A.; Moetteli, L.; Droz, I.; Roduit, R.; Munier, F.L.; Schorderet, D.F. Complement factor B polymorphism and the phenotype of early age-related macular degeneration. *Ophthalm. Genet.* **2014**, *35*, 12–17. [CrossRef]
246. Nasr, A.S.; Sami, R.M.; Ibrahim, N.Y.; Darwish, D.O. Glutathione S transferase (GSTP 1, GSTM 1, and GSTT 1) gene polymorphisms in Egyptian patients with acute myeloid leukemia. *Indian J. Cancer.* **2015**, *52*, 490–495. [CrossRef]
247. Grassmann, F.; Friedrich, U.; Fauser, S.; Schick, T.; Milenkovic, A.; Schulz, H.L.; von Strachwitz, C.N.; Bettecken, T.; Lichtner, P.; Meitinger, T.; et al. A Candidate Gene Association Study Identifies DAPL1 as a Female-Specific Susceptibility Locus for Age-Related Macular Degeneration (AMD). *Neuromol. Med.* **2015**, *17*, 111–120. [CrossRef]
248. Grizzard, S.W.; Arnett, D.; Haag, S.L. Twin study of age-related macular degeneration. *Ophthalm. Epidemiol.* **2003**, *10*, 315–322. [CrossRef]
249. Arakawa, S.; Takahashi, A.; Ashikawa, K.; Hosono, N.; Aoi, T.; Yasuda, M.; Oshima, Y.; Yoshida, S.; Enaida, H.; Tsuchihashi, T.; et al. Genome-wide association study identifies two susceptibility loci for exudative age-related macular degeneration in the Japanese population. *Nat. Genet.* **2011**, *43*, 1001–1004. [CrossRef] [PubMed]
250. Zerbib, J.; Seddon, J.M.; Richard, F.; Reynolds, R.; Leveziel, N.; Benlian, P.; Borel, P.; Feingold, J.; Munnich, A.; Soubrane, G.; et al. rs5888 variant of SCARB1 gene is a possible susceptibility factor for age-related macular degeneration. *PLoS ONE* **2009**, *4*, e7341. [CrossRef]
251. Sheu, S.J.; Ger, L.P.; Kuo, N.W.; Liu, N.C.; Wu, T.T.; Lin, M.C. Association of IL-4 gene polymorphism and age-related macular degeneration in Taiwanese adults. *Taiwan J. Ophthalmol.* **2012**, *2*, 51–55. [CrossRef]
252. Terluk, M.R.; Kapphahn, R.J.; Soukup, L.M.; Gong, H.; Gallardo, C.; Montezuma, S.R.; Ferrington, D.A. Investigating mitochondria as a target for treating age-related macular degeneration. *J. Neurosci.* **2015**, *35*, 7304–7311. [CrossRef] [PubMed]
253. Restrepo, N.A.; Mitchell, S.L.; Goodloe, R.J.; Murdock, D.G.; Haines, J.L.; Crawford, D.C. Mitochondrial variation and the risk of age-related macular degeneration across diverse populations. *Pac. Symp. Biocomput.* **2015**, *2015*, 243–254.

International Journal of
Molecular Sciences

MDPI

Review

Unraveling the Role of Reactive Oxygen Species in T Lymphocyte Signaling

Karsten Gülow *, Deniz Tümen, Philipp Heumann, Stephan Schmid, Arne Kandulski, Martina Müller and Claudia Kunst

Department of Internal Medicine I, Gastroenterology, Hepatology, Endocrinology, Rheumatology, Immunology, and Infectious Diseases, University Hospital Regensburg, 93053 Regensburg, Germany; deniz.tuemen@ukr.de (D.T.); philipp.heumann@klinik.uni-regensburg.de (P.H.); stephan.schmid@ukr.de (S.S.); arne.kandulski@ukr.de (A.K.); martina.mueller-schilling@ukr.de (M.M.); claudia.kunst@ukr.de (C.K.)

* Correspondence: karsten.guelow@ukr.de

Abstract: Reactive oxygen species (ROS) are central to inter- and intracellular signaling. Their localized and transient effects are due to their short half-life, especially when generated in controlled amounts. Upon T cell receptor (TCR) activation, regulated ROS signaling is primarily initiated by complexes I and III of the electron transport chain (ETC). Subsequent ROS production triggers the activation of nicotinamide adenine dinucleotide phosphate oxidase 2 (NADPH oxidase 2), prolonging the oxidative signal. This signal then engages kinase signaling cascades such as the mitogen-activated protein kinase (MAPK) pathway and increases the activity of REDOX-sensitive transcription factors such as nuclear factor-kappa B (NF-κB) and activator protein-1 (AP-1). To limit ROS overproduction and prevent oxidative stress, nuclear factor erythroid 2-related factor 2 (Nrf2) and antioxidant proteins such as superoxide dismutases (SODs) finely regulate signal intensity and are capable of terminating the oxidative signal when needed. Thus, oxidative signals, such as T cell activation, are well-controlled and critical for cellular communication.

Keywords: T lymphocytes; reactive oxygen species (ROS); oxidative signal; T cell activation; T cell receptor (TCR); electron transport chain (ETC); glycolysis; metabolic shift

1. Introduction

Prior to the emergence of photosynthetic archaebacteria approximately 2.7 billion years ago, the Earth's atmosphere was predominantly anoxic, lacking significant levels of oxygen. At that time, life on Earth was completely anaerobic and, thus, the advent of oxygen (O_2) generated by oxygenic photosynthesis posed a significant evolutionary challenge to terrestrial life forms on Earth. The absence of defense mechanisms against oxidative damage rendered O_2 toxic to many organisms, precipitating widespread extinctions across numerous species. Nevertheless, life found adaptive pathways for survival and evolved protective mechanisms, eventually harnessing oxygen for energy generation via oxidative phosphorylation [1–4]. However, managing the utilization of oxygen for energy production posed additional hazards, notably the generation of reactive oxygen species (ROS). Once again, life adapted and evolved complex systems to build a defense against these highly reactive toxic agents [5–7].

Over time, reactive oxygen species (ROS) have transitioned from being undesired byproducts of enzymatic reactions to molecules intentionally produced and by regulation, serving as signaling molecules or weapons to eliminate pathogens [8–10]. To comprehend the utilization of ROS as signaling molecules, one should (i) be familiar with the chemical properties of the major ROS, (ii) understand the key mechanisms governing their control and removal, and (iii) be aware of the primary sources of ROS production in eukaryotic cells.

1.1. Key ROS Vital for T Cell Signaling

The primary ROS crucial for cellular signaling include superoxide anions ($O_2^{\bullet-}$), hydroxyl radicals ($^{\bullet}OH$), and hydrogen peroxide (H_2O_2) [8–12]. Cellular ROS production typically begins with electron transfer to O_2, forming $O_2^{\bullet-}$. Due to their energetically unstable state, these molecules exhibit high reactivity and possess a short half-life of approximately 1 µs. Furthermore, their charged nature impedes free diffusion through cellular membranes. Consequently, $O_2^{\bullet-}$ exerts a locally constrained impact, primarily contributing to oxidative damage rather than serving as a signaling molecule [9,10].

In an aqueous environment, $O_2^{\bullet-}$ rapidly transforms into H_2O_2. Intracellularly, this conversion is expedited by superoxide dismutases (SODs) [10,13–15]. Despite H_2O_2 not being a radical, it is classified as an ROS. H_2O_2 exhibits a longer half-life of approximately 1 ms compared to $O_2^{\bullet-}$. Moreover, H_2O_2 can freely diffuse through membranes, and it primarily targets free thiols. This oxidation is generally reversible. Thus, H_2O_2 meets the criteria to serve as a second messenger [9,10,16,17].

However, an accumulation of H_2O_2 also poses risks of cellular damage. To ensure proper signaling, concentrations of H_2O_2 must be maintained within a specific range. Additionally, an accumulation of H_2O_2 increases the likelihood of its interaction with free iron^{2+} (Fe^{2+}). While Fe^{2+} generally exists in a bound form in cells, a small free fraction exists called the labile iron pool. When H_2O_2 encounters free Fe^{2+}, highly reactive hydroxyl radicals ($^{\bullet}OH$) are generated. These radicals have a half-life of less than 1 µs and can lead to massive cell damage, including lipid peroxidation [9,10,18,19].

1.2. The Essential Components of the Oxidative Defense

To prevent the accumulation of ROS, a complex network of antioxidative enzymes and ROS scavengers exists [8,10]. Superoxide dismutases (SODs) are essential enzymes that catalyze the conversion of $O_2^{\bullet-}$ into H_2O_2. SODs play a vital role in the disproportionation of $O_2^{\bullet-}$ both intracellularly and extracellularly, thus safeguarding cell membranes and DNA from ROS-mediated damage. In the cytosol and mitochondrial intermembrane space, SOD1 catalyzes the conversion of $O_2^{\bullet-}$ into H_2O_2; in the mitochondrial matrix, SOD2 performs this function; and extracellularly, SOD3 fulfills this task [13,20,21].

Catalase eliminates H_2O_2, transforming it into O_2 and water (H_2O). The chemical equation is $2H_2O_2 \rightarrow O_2 + 2H_2O$. Catalase is primarily active in peroxisomes, where it neutralizes excessive H_2O_2. Catalase can also be extracellularly secreted, where it may associate with the plasma membrane or disperse in the extracellular milieu [22–24].

Another important factor in H_2O_2 removal is the tripeptide glutathione, which acts as an effective oxidative scavenger in its reduced form (GSH). With concentrations typically ranging from 1 to 10 mM, GSH emerges as the primary antioxidant within cells, playing a pivotal role in the efficient removal of H_2O_2 [25]. Abundantly distributed across various cellular compartments, glutathione is present in the cytosol, endoplasmic reticulum (ER), mitochondria, vacuoles, and peroxisomes. The scavenging ability of GSH is attributed to its high reductive potential. A central cysteine residue with nucleophilic characteristics is the source of its reducing power. GSH can directly reduce H_2O_2 or reverse thiol oxidations on proteins [8]. In addition to GSH, other thiol scavengers are thioredoxin 1 and 2. Thioredoxins serve as key disulfide (-S-S-) reductases within the cell, primarily reversing thiol oxidations [9,26].

1.3. The Sources of ROS Production in T cells

ROS are generated at multiple cellular sites, including the mitochondria, peroxisomes, endoplasmic reticulum (ER), peroxisomes, and plasma membrane. Among these, nicotinamide adenine dinucleotide phosphate oxidases (NADPH oxidases) at the plasma membrane and mitochondria stand out as prominent sites for ROS generation in immune cells [27].

The NOX family comprises inducible NOX complexes 1–5 associated with the plasma membrane and two dual oxidases (DUOX)1/2. NOX2 (respiratory burst NOX), the most extensively studied isoform, is primarily expressed in phagocytes (e.g., at high levels—neutrophils, macrophages, dendritic cells; at low levels—B cells, mast cells, eosinophils, natural killer cells), where it facilitates ROS-dependent pathogen killing [28,29]. It is also found in certain non-phagocytic cell types, including T cells [30–32]. NOX2 consists of various components: the transmembrane flavocytochrome b558 (a heterodimer containing gp91phox and p22phox) as well as the four cytoplasmic proteins p47phox, p67phox, p40phox, and small G protein Rac1/2. Upon activation, the NOX2 complex generates membrane-impermeant $O_2^{\bullet -}$ towards the inside of the phagosomes or the extracellular side of the plasma membrane. There, $O_2^{\bullet -}$ can then be converted into H_2O_2 and re-enter the cell to work as a second messenger [9,29].

Aerobic respiration leads to substantial ROS production by mitochondria. Mitochondrial ROS are primarily released at two sites within the electron transport chain (ETC): Complex I (NADH dehydrogenase) and Complex III (ubiquinone-cytochrome c reductase) [10,29,33,34]. Mitochondrial ROS serve as crucial signals in various cellular processes such as adaptation to hypoxia, regulation of innate or adaptive immunity, differentiation, or autophagy [10,27,29,35,36]. Mitochondrial ROS act as important signaling molecules, particularly in T cells, determining the activation and deactivation of T cell-mediated immune responses [29].

In summary, the nature of produced ROS, the cellular oxidative defense, and ROS localization are critical factors determining the induction of the generation of an oxidative signal [10,11]. To enable oxidative signaling, a constant oxidative environment must be established [8].

Oxidative processes, once hazardous in early life's development, are utilized not only by the majority of organisms for energy production but also for signal transduction. This review focuses on oxidative signaling cascades in T lymphocytes and explores the mechanisms driving their formation.

2. T Cell Receptor (TCR) Stimulation-Induced Oxidative Signaling

T cells are an integral part of the adaptive immune response. Like all blood cells, T cells originate in the bone marrow. From there, they migrate to the thymus. In the thymus, T cells undergo differentiation and selection to specifically recognize a particular antigen. Upon exiting the thymus, these cells circulate throughout the body and scan major histocompatibility (MHC) complexes displayed on the surface of antigen-presenting cells (APCs) to locate their target antigen. A target antigen binding to the T cell receptor (TCR) subsequently triggers a complex signaling cascade that initiates T cell activation.

Antioxidants were observed to inhibit T cell activation, suggesting a regulatory role of ROS [37–39]. Subsequently, direct evidence of TCR-triggered ROS production is demonstrated in various studies [40–44]. This supports the specific involvement of ROS in T cell activation signaling pathways, emphasizing the intricate role of oxidative processes in cellular responses [29,45,46].

2.1. *The Proximal TCR Machinery*

Upon TCR stimulation, receptor-bound tyrosine kinases are activated, including the Zeta-chain-associated protein kinase 70 (ZAP70). ZAP70 subsequently phosphorylates the adapter protein Linker for activation of T cells (LAT), to which Phosphoinositid-phospholipase (PLC)γ1 is recruited and activated [47–49]. Activation of PLCγ1 results in the generation of inositol 3,4,5-triphosphate (IP_3) and diacylglycerol (DAG). IP_3 increases cytosolic calcium (Ca^{2+}) levels, while DAG activates protein kinase C θ (PKCθ). A subpopulation of activated PKCθ is translocated to the mitochondria and induces controlled ROS generation by the ETC [29,32,46,50,51]. Concurrently, activation of NADPH oxidase 2 is induced, ensuring a sustained ROS signal [30]. Thus, mitochondria are crucial for initiating the oxidative signal, while NADPH oxidase 2 is responsible for its maintenance [32] (Figure 1 and Table 1).

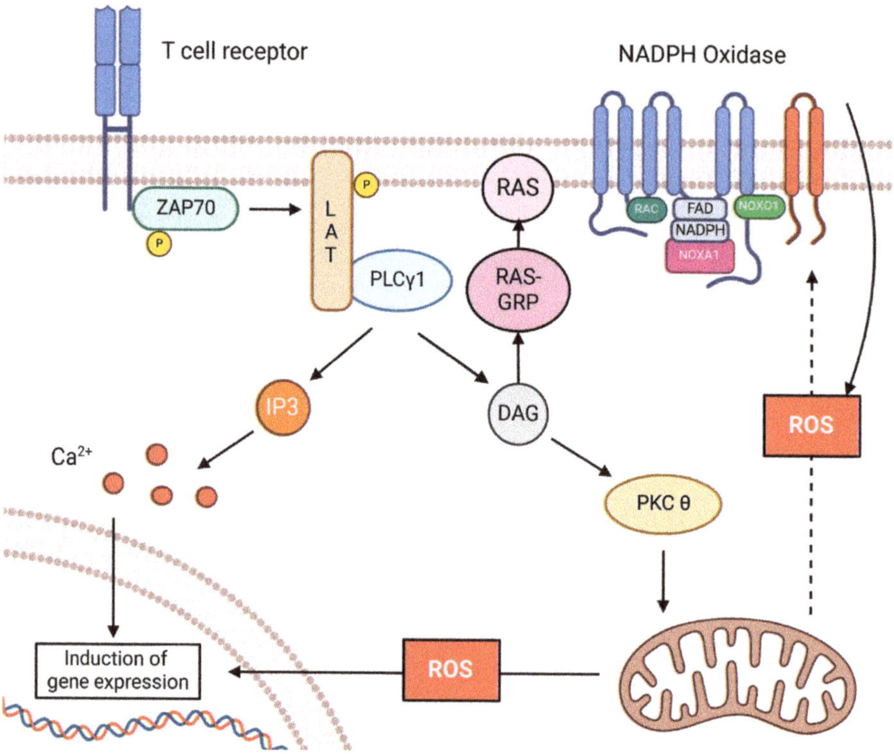

Figure 1. Schematic representation of oxidative signaling in T cells. Stimulation of the T cell receptor (TCR) leads to phosphorylation and activation of tyrosine kinase ZAP70, which phosphorylates LAT. Consequently, LAT recruits PLCγ1, which generates inositol 3,4,5-triphosphate (IP_3) and diacylglycerol (DAG). At this point, the activation-induced signal diverges. IP_3 is responsible for releasing Ca^{2+} into the cytosol. DAG activates the RAS guanyl-releasing protein 1 (RAS-GRP) and PKCθ. RAS-GRP activates Rat sarcoma (RAS) and subsequent kinase signaling. PKCθ induces ROS release via the mitochondria. Both signals are essential for the induction of activation-induced gene expression in T cells. Mitochondrial ROS release is a prerequisite for the induction of NADPH oxidase 2. ROS generation via NADPH oxidase 2 leads to a sustained oxidative signal. The figure was created with BioRender.com (accessed on 31 May 2024).

Importantly, both signals, Ca^{2+} release and ROS production, constitute the minimal requirement for inducing the expression of essential interleukins crucial for initiating an immune response [29,41,50].

2.2. Mitochondria: Oxidative Signaling Hub in TCR Activation

Following TCR stimulation, mitochondria serve as the central oxidative signaling platform. Upon translocation of a subpopulation of PKCθ to the mitochondria, there is a controlled release of $O_2^{\bullet-}$ into the mitochondrial matrix, mediated by Complex I [29,32,36], and of Complex III of the ETC to the mitochondrial intermembrane space [33,35,51–53] (Table 1).

During later phases of TCR signaling, upregulation of MnSOD (SOD2) expression occurs to mitigate potential oxidative damage to the mitochondria. Elevated MnSOD levels lead to reduced oxidative signaling, contributing to the cessation of TCR-triggered T cell activation. The precise mechanism by which MnSOD blocks the oxidative signal remains incompletely understood. It is hypothesized that the high efficiency of MnSOD in scavenging oxygen reduces the probability of oxygen participating in other reactions that could generate H_2O_2 (e.g., reduction in the Aconitase [4Fe–4S] cluster) and therefore produces less H_2O_2 per oxygen molecule than the non-catalyzed conversion of $O_2^{\bullet-}$ to H_2O_2 [20] (Figure 2). This is in line with two studies emphasizing the critical role of MnSOD in T cell homeostasis. The first study demonstrated that T cell-specific MnSOD knockout increases mitochondrial superoxide anion levels, leading to the hyperactivation of thymocytes and peripheral T cells [54]. The second study shows that reduced MnSOD activity and content in mouse T cells lacking the interferon (IFN)-γ-inducible lysosomal thiol reductase ($Gilt^{-/-}$) result in increased TCR-triggered ERK activation and proliferation by elevated intracellular ROS levels [55].

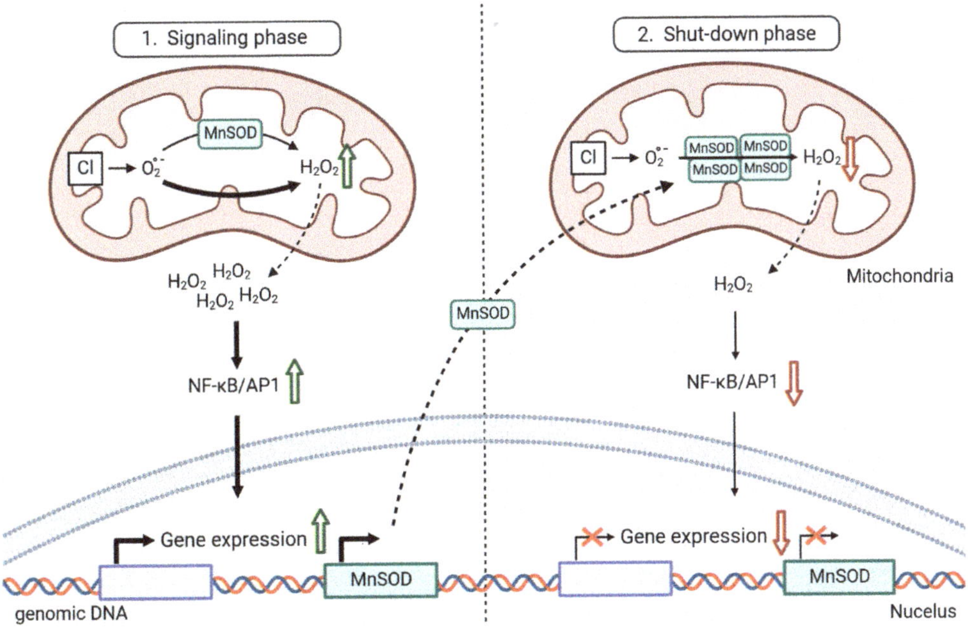

Figure 2. Complex I (CI) releases $O_2^{\bullet-}$ into the mitochondrial matrix, where most of it is converted to H_2O_2. However, certain conditions can lead to increased H_2O_2 production through other reactions. This high H_2O_2 production induces MnSOD, which further accelerates the conversion of $O_2^{\bullet-}$ to H_2O_2. As a result, fewer alternative reactions leading to H_2O_2 formation occur. Overall, this leads to reduced H_2O_2 production and a downregulation of the oxidative signal. The figure was created with BioRender.com.

A study by Sena et al. highlights the role of Complex III as an additional ROS source following TCR stimulation. The authors report the necessity of intact Complex III for T cell activation and mitochondrial ROS production through this complex. Using a T cell-specific knockout mouse with a deletion of the Rieske iron-sulfur protein (RISP) subunit of Complex III, they demonstrated impaired CD4+ and CD8+ T cell responses in murine models of asthma and listeriosis, respectively [51].

Complex III releases ROS into both the mitochondrial matrix and the mitochondrial intermembrane space. Here, $O_2^{\bullet-}$ converts into H_2O_2 and can act as second messenger in the cytosol. Cu/ZnSOD (SOD1) could play a role similar to MnSOD in the matrix by protecting the cell against oxidative damage and terminating the oxidative signal.

Table 1. Overview of key signaling molecules responsible for ROS induction and complexes generating the oxidative signal.

<table>
<tr><th colspan="4">TCR Activation-Induced Oxidative Signaling</th></tr>
<tr><th colspan="4">Proximal Proteins of the TCR Machinery Essential for Oxidative Signal Induction</th></tr>
<tr><th>Protein Name</th><th colspan="2">Function</th><th>References Demonstrating the Essential Role of These Proteins in Oxidative Signaling.</th></tr>
<tr><td>Zeta-chain-associated protein kinase 70 (ZAP70)</td><td>phosphorylation of LAT</td><td rowspan="4">signal cascade from the T cell receptor (TCR) to the mitochondria (essential proteins for the induction of activation-induced oxidative signals)</td><td rowspan="4">Kaminski et al., 2007 [32],
Kaminski et al., 2010 [50],
Kaminski et al., 2012 [36]</td></tr>
<tr><td>Linker for activation of T cells (LAT)</td><td>scaffold protein recruiting PLCγ1 upon phosphorylation</td></tr>
<tr><td>Phosphoinositide-phospholipase 1 (PLCγ1)</td><td>PLCγ1 generates inositol 3,4,5-triphosphate (IP_3) and diacylglycerol (DAG)</td></tr>
<tr><td>Protein kinase C θ (PKCθ)</td><td>a subpopulation DAG-activated PKCθ translocates to mitochondria and induces ROS release</td></tr>
<tr><th colspan="4">Initial Oxidative Signaling: ETC Complexes Releasing ROS upon TCR Triggering</th></tr>
<tr><th>Name of the ETC Complex</th><th colspan="2">ROS Is Released Into:</th><th>References Demonstrating ROS Production via the ETC</th></tr>
<tr><td>Complex I</td><td colspan="2">the mitochondrial matrix</td><td>Kaminski et al., 2007 [32],
Kaminski et al., 2010 [50],
Kaminski et al., 2012 [36]</td></tr>
<tr><td>Complex II</td><td colspan="2">the mitochondrial intermembrane space</td><td>Sena et al., 2013 [51]</td></tr>
<tr><th colspan="4">Protein Complexes Involved in Sustained Oxidative Signaling</th></tr>
<tr><th>Name of the Complex</th><th colspan="2">ROS Is Released Into:</th><th>References Demonstrating ROS Production via NADPH Oxidases</th></tr>
<tr><td>NADPH oxidase 2</td><td colspan="2">the extracellular space</td><td>Jackson et al., 2004 [30],
Krammer et al., 2007 [56]
Kaminski et al. 2007 [32]</td></tr>
</table>

3. Interplay between Glucose Metabolism and Oxidative Signaling

Mitochondria serve as aerobic energy producers, with the respiratory chain requiring a precise balance between pro-oxidative and anti-oxidative systems for optimal operation. ROS are natural byproducts of this process. However, during the generation of oxidative signals, specific and finely regulated quantities of ROS are deliberately released [29,57]. As described in Section 2, mitochondrial ROS are generated in a regulated manner by Complexes I and III following TCR stimulation. Complex I is the site where electrons from NADH are fed into the respiratory chain. Following TCR stimulation, mitochondrial ROS are primarily generated by Complex I most likely through reverse electron transport (RET). RET occurs in a two-step process involving reduced coenzyme Q (CoQ) and a change in proton motive force that drives electrons back into Complex I, where ROS are released into the mitochondrial matrix [29,32,57–59]. Complex III generates ROS as well. Under normal conditions, electrons flow from the CoQ pool to cytochrome C. However, alterations within the ubiquinone binding site of Complex III can prompt O_2 to react with ubisemiquinone, resulting in $O_2^{\bullet-}$ formation. $O_2^{\bullet-}$ generated by Complex III is primarily released into the intermembrane space. After disproportionation, H_2O_2 can diffuse freely into the cytosol either from the matrix or from the intermembrane space [60,61] (Figure 3).

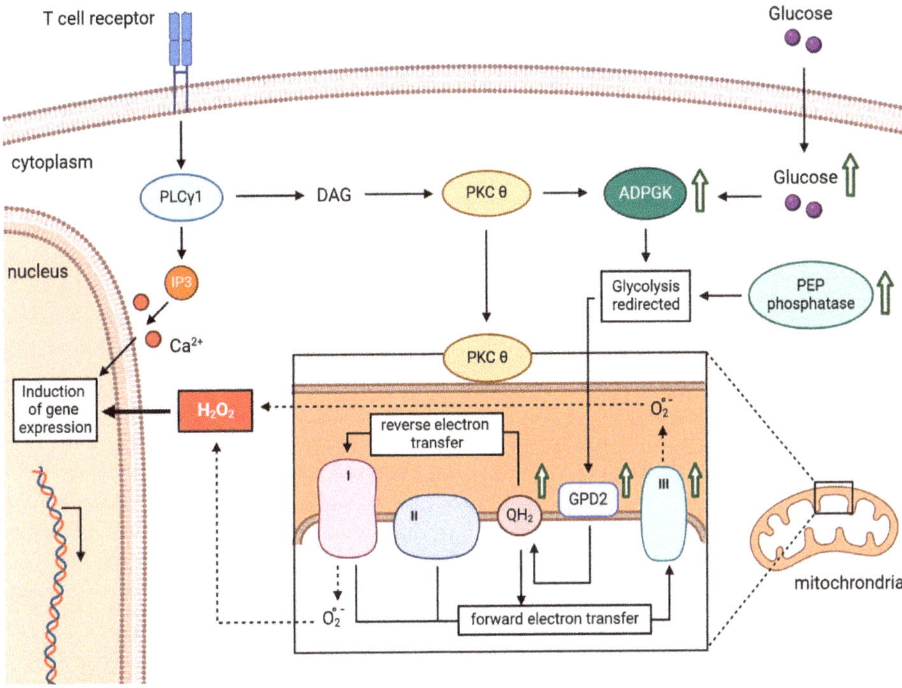

Figure 3. After TCR stimulation, PLCγ1 generates 3,4,5-triphosphate (IP_3) and diacylglycerol (DAG). IP_3 induces Ca^{2+} influx into the cytosol and activates NF-AT via calcineurin. Simultaneously, DAG activates PKCθ. PKCθ most likely phosphorylates ADPGK. ADPGK induces a redirection of glycolysis, allowing electrons to be directly transferred to ubiquinone via GPD2. From there, electrons can be retrogradely directed towards Complex I of the ETC and forward directed to Complex III. At these complexes, ROS is released, activating REDOX-sensitive transcription factors and inducing the expression of specific genes essential for T cell activation. The figure was created with BioRender.com.

A shift in cellular metabolism is essential to modify electron flow, which is accompanied by increased glucose uptake. Only this metabolic adjustment ensures the targeted release of oxidative signals through mitochondria.

3.1. Cellular Metabolic Alterations Are Essential for Oxidative Signal Generation

The metabolic program of T cell activation involves a transition from quiescence to increased energy demand upon TCR engagement. Naïve T cells primarily rely on oxidative phosphorylation for ATP production, fueled by pyruvate oxidation and fatty acid oxidation, with low glycolytic activity. TCR engagement leads to activation, proliferation, and the differentiation of naïve T cells into effector, memory, and central memory T cells, accompanied by the upregulation of key metabolic enzymes such as glucose transporter GLUT1 and acetyl-CoA carboxylase 1 (ACC1) [62].

TCR signaling relies on glucose uptake [63,64] and is accompanied by a metabolic shift from mitochondrial ATP production to aerobic glycolysis, comparable to the Warburg effect [65–67], a phenomenon characteristic of rapidly proliferating cells [66,68]. Therefore, upon T cell activation, the mitochondrial respiratory chain switches from an ATP-producing to an oxidative signaling function, while glycolysis fulfills cellular energy demands.

In this context, TCR triggering has been shown to activate ADP-dependent glucokinase (ADPGK), an ER-localized glycolytic enzyme [36,69]. ADPGK activation is associated with a rapid glucose uptake, downregulation of mitochondrial oxygen consumption, and redirection of glycolysis via the Glycerin-3-phosphat-Dehydrogenase (GPD) shuttle. Glycerol-3-phosphate dehydrogenase 2 (GPD2), located proximal to the ETC at the inner mitochondrial membrane, oxidizes glycerol-3-phosphate to dihydroxyacetone phosphate. This process directs electrons into the CoQ pool, leading to the generation of ubiquinol and the onset of an electron backlog. This backlog results in RET, with the subsequent release of $O_2^{\bullet-}$ from Complex I [36,70]. Thus, this process represents the initial step in the generation of the activation-induced oxidative signal (Figure 3).

In addition to Complex I, Complex III also contributes to activation-induced oxidative signaling in T cells. The release of ROS via Complex III has been demonstrated through knockout of the Rieske iron-sulfur protein (RISP), a component of Complex III. RISP knockout can impair both the oxidative signal and the activation of T cells [51,70]. The release of ROS via Complex III may be induced by enhanced electron transport from ubiquinol toward Complex III, leading to electron flow from ubiquinol both backward to Complex I [36] and forward to Complex III (Figure 3). An alternative explanation could involve an increased electron flow through the ETC due to an enhanced production of reducing equivalents NADH and $FADH_2$. These could be generated by heightened activity of the tricarboxylic acid cycle (TCA cycle) induced by the activation-induced Ca^{2+} signal [35]. However, the exact mechanism requires further detailed investigation.

The individual complexes of the mitochondrial ETC play distinct roles in generating ROS and regulating cellular functions, such as inflammation, epigenetic changes, and T cell differentiation [71]. A fundamental prerequisite for mitochondria to become oxidative signaling platforms is a metabolic alteration associated with increased glucose uptake, irrespective of whether Complex I, Complex III, or both complexes generate the oxidative signal.

3.2. *Elevated Glucose Uptake in Different T Cell Subsets*

An increase in glucose uptake and a shift from respiration and ATP production via oxidative phosphorylation to aerobic glycolysis is observed in many T cell subsets following TCR stimulation [72]. Elevated glucose uptake induces heightened glycolytic flux, leading to elevated ROS levels. This heightened glycolytic flux typically correlates with increased activity in the TCA cycle, resulting in the accumulation of ETC substrates and elevated $NADH/NAD^+$ and FADH2/FAD ratios [73]. Excessive mitochondrial NADH levels promote an elevated electron flow through the ETC, triggering the formation of $O_2^{\bullet-}$ by Complex III [51]. Additionally, the combination of an increased glycolytic flux, GPD2-dependent hyper-reduction in ubiquinone, and a highly negative membrane potential ($\Delta\psi$) lead to $O_2^{\bullet-}$ production by Complex I via RET [29]. The metabolic shift is thus a prerequisite for the generation of the oxidative signal and for T cell activation. To enable the shift toward glycolysis, increased glucose import is required. Only through this metabolic shift can controlled ROS release from Complex I and Complex III occur. This oxidative signal then regulates the activation and differentiation of T cells.

3.2.1. Increased Glucose Uptake in $CD4^+$ T Cells

Upon activation of and exposure to lineage-specific cytokines, naive $CD4^+$ T cells undergo differentiation into different lineages (Figure 4). This differentiation process is tightly linked with a metabolic shift towards aerobic glycolysis [72,74,75], with the regulation of glycolysis rates playing a pivotal role. An activation of $CD4^+$ T cells promotes glucose uptake and aerobic glycolysis, influencing the differentiation of T cell subsets. Several glucose transporters (Glut), including Glut1, Glut3, Glut6, and Glut8, are upregulated and involved in this process [76]. The specific significance of inducing glucose import by increased Glut transporter expression in the differentiation of $CD4^+$ T cells becomes particularly evident in cells with mutations or deletions in the genes for Glut1 and Glut3.

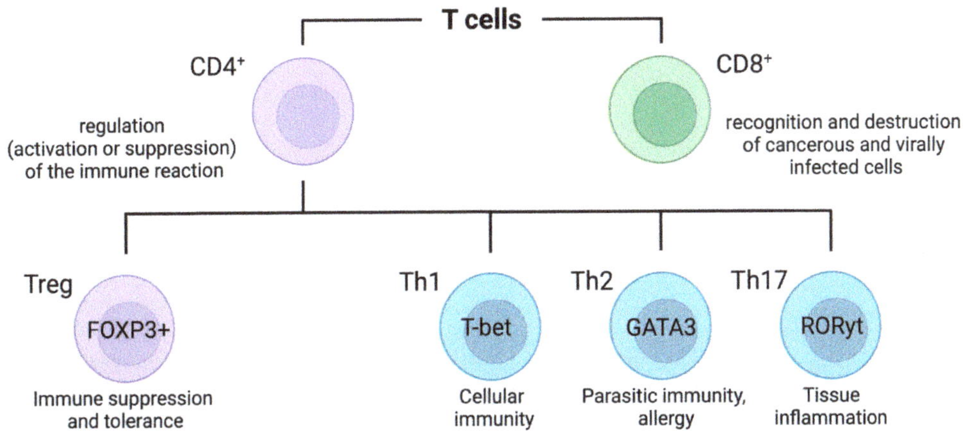

Figure 4. Overview of T cell subsets. CD8$^+$ T cells, also known as cytotoxic T lymphocytes (CTLs), are a vital component of the adaptive immune system. Their primary function is to identify and destroy infected or malignant cells. CD4$^+$ T cells, also known as helper T cells, are crucial for orchestrating the immune response. They assist other immune cells by releasing cytokines that regulate the activity, growth, and differentiation of various immune cells. CD4$^+$ can further be divided into Th1 cells (expressing T-box expressed in T cells [T-bet]), Th2 cells (expressing the GATA binding protein 3 [GATA3]), Th17 cells (expressing the RAR-related orphan receptor gamma t [RORγt]), and T regulatory cells (Treg; expressing elevated levels of (Forkhead-Box-Protein P3 [FOXP3]). Th1 cells promote cell-mediated immunity by activating macrophages and cytotoxic T cells. They are crucial for defending against intracellular pathogens like viruses and certain bacteria. Th2 cells support humoral immunity by stimulating B cells to produce antibodies. They are important for combating extracellular parasites and allergens. Th17 cells are involved in defending against extracellular bacteria and fungi. They also play a role in autoimmune diseases by promoting inflammation. Tregs maintain immune tolerance by suppressing immune responses, preventing autoimmune diseases, and controlling inflammation. The figure was created with BioRender.com.

A deficiency in Slc2a1 (encoding the Glut1 protein) reduces thymocyte numbers, inhibits effector CD4$^+$ T cell expansion, and diminishes cytokine production across various CD4$^+$ T cell subsets. Notably, while Glut1 is dispensable for the lineage commitment of CD4$^+$ Treg cells, the cell surface expression of Glut1 is high in Th1, Th2, and Th17 cells. Consistent with Glut1 expression patterns, Th1, Th2, and Th17 cells exhibit increased glycolysis rates, while Treg cells display increased rates of fatty acid oxidation [76,77].

Glut3 plays a crucial role in Th17 cell differentiation, as deficiency in Slc2a3 (encoding the Glut3 protein) inhibits this process [78]. Glut3-mediated glucose uptake is critical for mitochondrial glucose oxidation and subsequent acetyl-CoA production. In addition to its central function during energy production, acetyl-CoA plays a pivotal role in regulating the chromatin accessibility of loci associated with inflammatory genes, including IL17a [78].

Treg cells exhibit distinct glucose metabolic patterns compared to Th1, Th2, and Th17 cells, with significantly lower levels of Glut1 and reduced glycolytic rates [77].

In summary, it can be concluded that Th1, Th2, and Th17 cells exhibit elevated Glut levels and consequently an increased glycolytic flux. Conversely, Treg cells do not show an elevated glycolytic flux and tend to shift toward lipid metabolism.

3.2.2. Increased Glucose Uptake in CD8$^+$ T Cells

Similar to CD4$^+$ T cells, the activation and differentiation of CD8$^+$ T cells (Figure 4) are associated with an upregulation of Glut transporters [72]. However, there are characteristic differences. Unlike CD4$^+$ T cells, CD8$^+$ T cells do not rely on Glut1 for proliferation or granzyme B production. Instead, CD8$^+$ T cells depend on Glut2 for optimal proliferation and effector cytokine production [72,76]. A deficiency in Slc2a2 (which encodes the Glut2 protein) affects $CD8^+$ T cell activation [79]. Thus, Glut2 emerges as the primary factor involved in $CD8^+$ T cell activation.

Increased glucose import via the induction of Glut transporters is a prerequisite for heightened glycolytic flux. Through enhanced glycolysis and the consequent increased activity of the TCA cycle, more reducing equivalents (NADH and $FADH_2$) can be produced. By promoting glycolysis and channeling electrons through GPD2 to ubiquinone, RET is induced, enabling the generation of an oxidative signal from Complex I. The augmented electron influx into the ETC via the reducing equivalents NADH (Complex I) and $FADH_2$ (Complex II) leads to the release of ROS at Complex III. Consequently, glucose metabolism is intricately associated with the generation of the oxidative signal.

3.3. *Modulation of TCR Activation-Induced Signaling by Co-Stimulation*

T cell activation is a complex process that requires multiple signals to ensure a robust and controlled immune response. TCR activation alone is not sufficient for complete T cell activation. Co-stimulatory signals are also required. Co-stimulatory receptors on T cells, such as CD28, the cytotoxic T lymphocyte-associated protein 4 (CTLA-4), and the programed cell death protein 1 (PD-1), interact with their respective ligands on APCs (e.g., CD80/CD86 for CD28 and CTLA-4, and PD-1L for PD-1). The balance between activating (e.g., CD28) and inhibitory (e.g., CTLA-4, PD-1) signals determines T cell activation, anergy, or exhaustion.

3.3.1. CD28

CD28 is a critical co-stimulatory receptor on T cells that plays a vital role in their activation and survival. It works in tandem with the TCR to ensure a full and effective immune response. The interaction between CD28 on T cells and its ligands, CD80 and CD86, on antigen-presenting cells (APCs) is essential for the proper functioning of the immune system. Simultaneously to TCR stimulation, CD28 binds to its ligands CD80 or CD86 on the APC. This interaction provides a necessary secondary signal. The co-stimulatory signal from CD28 amplifies the TCR signal, leading to full T cell activation. Activation of CD28 initiates several intracellular signaling pathways, including the activation of PI3K/AKT, MAPK, and NF-κB pathways.

CD28 stimulation leads to the induction of ROS [8,80], which amplifies the oxidative signal induced by TCR activation. Unlike the mitochondrial origin of the TCR-derived oxidative signal, the CD28-depenent signal is generated by Arachidonate 5-Lipoxygenase (5-LO) [8,80,81].

3.3.2. Cytotoxic T Lymphocyte-Associated Protein 4 (CTLA-4)

Cytotoxic T lymphocyte-associated protein 4 (CTLA-4) is an inhibitory receptor primarily expressed on activated T cells. Its main function is to downregulate immune responses, thereby maintaining immune homeostasis and preventing autoimmunity. CTLA-4 achieves this by binding to its ligands, CD80 and CD86, which are expressed on antigen-presenting cells. This binding competes with the costimulatory receptor CD28, which also binds to CD80 and CD86, but promotes T cell activation. By outcompeting CD28 for these ligands, CTLA-4 effectively inhibits T cell proliferation and cytokine production, leading to immune suppression [82].

The fact that CTLA-4 competes with CD80 and CD86 for binding explains its ability to block CD28-dependent ROS production via 5-LO. This was confirmed using a soluble CTLA-4-Ig protein, a chimeric protein consisting of the extracellular domain of human CTLA-4 and a fragment (hinge and constant region) of the Fc portion of human IgG1 [83]. CTLA-4-Ig successfully inhibited CD28-induced ROS production, thereby preventing T cell activation [84].

3.3.3. Programmed Cell Death Protein 1 (PD-1)

Programmed cell death protein 1 (PD-1) is an inhibitory receptor expressed primarily on the surface of T cells, B cells, and myeloid cells. Its primary function is to regulate immune responses by maintaining self-tolerance and preventing autoimmunity. PD-1 achieves this by binding to its ligands, PD-L1 and PD-L2, which are often upregulated in various tissues, including tumor cells. This interaction inhibits T cell activation and proliferation, leading to reduced cytokine production and immune suppression. The PD-1/PD-L1 pathway is a critical mechanism for cancer cells to evade the immune system, making it a significant target for immunotherapy treatments.

Surprisingly, PD-1 stimulation or upregulation can generate ROS, probably through mitochondrial or mitochondria-related processes [85,86]. This ROS production is not associated with a proliferation or differentiation of T cells, but it is associated with the induction of apoptosis [85]. The CD95 death ligand (CD95L) is regulated by NFκB [56]. Naive or resting T cells are resistant to apoptosis, with CD95L acting as a weapon to induce apoptosis in neighboring (malignant or damaged) cells. Activated T cells become sensitive to apoptosis, where CD95L can induce fratricide in neighboring T cells or suicide within the same cell through being cleaved by the metalloprotease a disintegrin and metalloproteinase domain-containing protein 10 (ADAM10). This process, known as activation-induced cell death (AICD), is thought to eliminate autoreactive (persistently activated) T cells, thereby preventing autoimmunity [41,56,87–89].

4. Oxidative Signaling Modulates Signal Cascades and the Expression Profile of Activated T Cells

The oxidative signal can oxidize various proteins, including phosphatases, kinases, and transcription factors. This oxidation can either inhibit or induce protein activity and thereby modulate signal pathways.

4.1. Regulation of Phosphatases

The protein tyrosine phosphatases (PTPs) comprise a crucial group of redox-regulated proteins. Within their catalytic center, all PTPs have a redox-regulated cysteine. Oxidation of this thiol group by H_2O_2 leads to PTP inactivation. Given the reversibility of PTP oxidation, these proteins exist in two alternative states: an active state with a reduced cysteine or an inactive state with an oxidized cysteine [9,90]. ROS generated upon TCR stimulation rapidly convert to H_2O_2, which constitutes the primary oxidative signal. Thus, activation-induced oxidative signaling can inhibit PTPs, thereby regulating the signaling pathways they govern.

A particularly important signaling pathway in T cells is the mitogen-activated protein kinase (MAPK) pathway. These kinases are activated via phosphorylation and deactivated by dephosphorylation through MAPK phosphatases (MKPs). The activation-induced oxidative signal inactivates MKPs, thereby activating MAPKs (Figure 5) such as extracellular signal-regulated kinase (ERK), c-Jun N-terminal kinases (JNK), and p38, inducing cell proliferation and differentiation [90–92]. Another critical regulator of T cell activation and function is the Phosphatase and Tensin Homolog (PTEN). It acts as a tumor suppressor by dampening the Phosphoinositid-3-Kinase (PI3K)–Protein kinase B (PKB/AKT) signaling axis crucial for T cell activation, growth, and viability. Through its action of dephosphorylating phosphatidylinositol (3,4,5)-trisphosphate (PIP_3), PTEN counteracts the PI3K pathway, thus restraining downstream AKT activation. This control aids in preserving T cell balance and preventing hyperactive immune reactions. ROS can oxidize PTEN at its active site cysteine residue, leading to its inhibition and therefore opening the AKT pathway, enabling T cell activation [93].

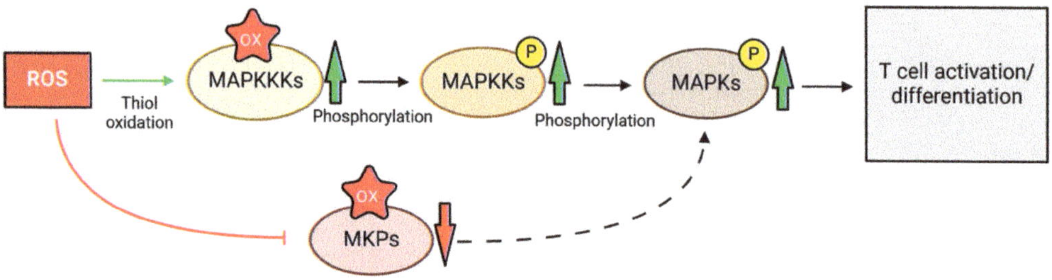

Figure 5. ROS-induced oxidation of free thiol groups activates MAP kinase kinase kinases (MAPKKKs), which phosphorylate MAPKK and, subsequently, MAPK. This signaling pathway is further amplified by the oxidation of MAPK phosphatases, which are inhibited by oxidation of a thiol residue in their active site, resulting in full activation of the MAPK cascade. OX = oxidation, P = phosphorylation. The figure was created with BioRender.com.

4.2. *Regulation of Kinases*

Kinases can be activated by oxidation. Again, the MAPK signaling pathway is the most prominent example in T cells. Activation of MAPKs typically involves a cascade of kinase reactions, progressing from MAP kinase kinase kinases (MAPKKKs) to MAP kinase kinases (MAPKKs), culminating in the final activation of MAPK [90,92,94]. ROS-induced oxidation of free thiol groups activates MAPKKKs, promoting kinase activity through conformational changes [95,96]. This, along with PTP inhibition, induces the MAPK signaling pathway, fostering T cell proliferation and differentiation (Figure 5).

4.3. *Essential Transcription Factors for T Cell Activation*

Oxidative signals not only activate or deactivate signaling pathways but also directly modulate the activity of transcription factors [97].

4.3.1. Activator Protein-1 (AP-1)

The AP-1 transcription factors comprise homodimers and heterodimers formed by members of the Jun (c-Jun, JunB, JunD), Fos (c-Fos, FosB, Fra-1, Fra-2), Maf (c-Maf, MafA, MafB, MafG/F/K, Nrl), and ATF (ATF2, ATF3, B-ATF, JDP1, JDP2) protein subfamilies [90]. AP-1 dimers vary in composition, offering functional diversity. Different family members have unique affinities for DNA sequences and interaction partners crucial for the induction of processes like proliferation, differentiation, and apoptosis [98]. Numerous studies have demonstrated a highly conserved characteristic of AP-1 from yeast to mammals, namely its redox regulation and activation by oxidative signals [90]. For instance, exposure to ROS enhances both gene expression and protein levels of c-Fos and c-Jun, leading to increased DNA-binding activity in AP-1 [99,100]. In addition, ROS scavengers or antioxidants can effectively inhibit AP-1 activity [99–102]. The DNA-binding activity of c-Fos and c-Jun are determined by the redox state of several conserved free thiol groups. Many studies have shown that AP-1 is also indirectly regulated by ROS signals. The AP-1 family of transcription factors is a target for phosphorylation by the MAPK cascades, which, as described above, respond to ROS, leading to an enhanced transcriptional activation of AP-1 [29,90,91,98,101,103].

4.3.2. Nuclear Factor Kappa B (NF-κB)

NF-κB represents a collective term for dimeric transcription factors within the Rel family [90]. In the absence of activating signals, NF-κB is sequestered in the cytoplasm through binding to the inhibitor of NF-κB (IκB). Upon stimulation, the IκB kinase (IKK) complex is activated, phosphorylating IκB and marking it for ubiquitination and subsequent degradation (Figure 6). This degradation exposes the nuclear localization sequence of NF-κB, facilitating its translocation into the nucleus [104,105]. In the early 1990s, Schreck et al. demonstrated that treatment with H_2O_2 induces nuclear translocation [106]. Subsequent studies have revealed the inhibition of NF-κB activation by antioxidants [8]. Although H_2O_2 does not universally induce NF-κB activation across all cell types, it is generally recognized to be reliant on ROS in activated T cells [29,107]. Mechanisms of H_2O_2-induced NF-κB induction are diverse and dose-dependent. In the cytosol, a pro-oxidative environment prompts IκB oxidation and degradation, releasing NF-κB for nuclear translocation. Free thiol groups in NF-κB subunits are also oxidized in the cytosolic milieu, with thioredoxin (TRX)-1 primarily reversing this process in the nucleus (Figure 6). Notably, NF-κB must be in its reduced form in the nucleus to effectively bind to DNA binding sites [29,90,108,109].

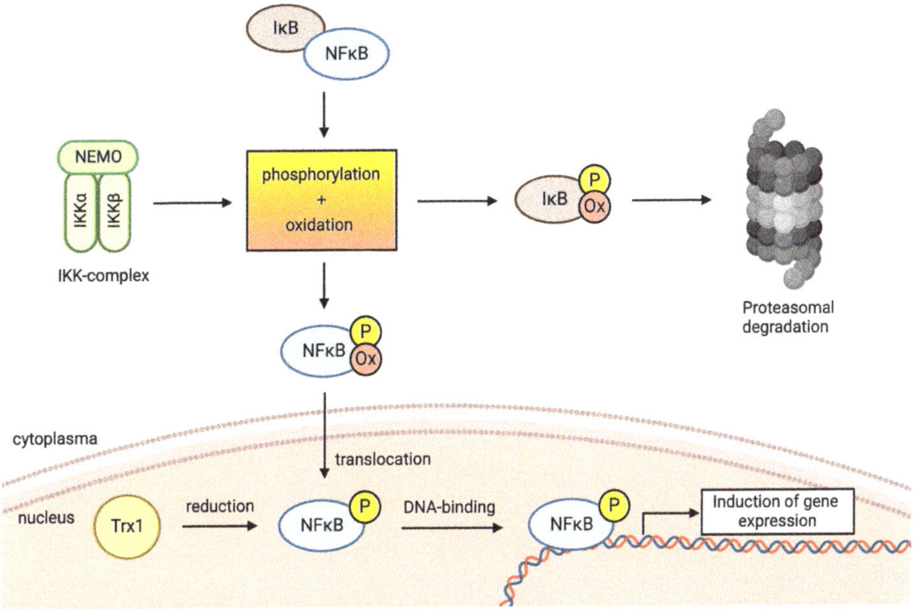

Figure 6. The role of oxidative signals in the activation of NF-κB. The inhibitor of NF-κB (IκB) is phosphorylated by the IκB kinase (IKK) complex. Additionally, a pro-oxidative environment in the cytosol causes IκB to become oxidized. Both phosphorylation and oxidation result in the accelerated degradation of IκB, so NF-κB is released. The free NF-κB can now translocate to the nucleus. Oxidation further accelerates this translocation. The oxidation of NF-κB is enabled by a shift in the cytosolic REDOX homeostasis towards a pro-oxidative state, induced by the oxidative signal. In the nucleus, NF-κB must be reduced again to enable optimal DNA binding. This is primarily achieved by Thioredoxin 1 (TRX1). The figure was created with BioRender.com 4.3.3. Nuclear Factor of Activated t cells (NF-AT).

NF-AT is primarily regulated by Ca^{2+}. The increase in intracellular Ca^{2+} levels stimulated by TCR triggering activates the Ca^{2+}-calmodulin-regulated phosphatase, calcineurin. Calcineurin, in turn, dephosphorylates the transcription factor NF-AT. Dephosphorylation exposes a nuclear localization signal, leading to the translocation of transcriptionally active NF-AT into the nucleus [56,110]. Additionally, it has been shown that ROS can also promote the nuclear translocation of NF-AT and enhance the transcriptional activity of NF-AT. However, the precise mechanisms involved in this process remain unclear [51].

4.3.3. Nuclear Factor Erythroid 2-Related Factor 2 (NRF2)

The Keap1-Nrf2 signaling pathway is crucial for maintaining cellular redox equilibrium and protecting cells from oxidative stress. ROS plays a central role in this mechanism, dynamically balancing Nrf2 activation and inhibition by Kelch-like ECH-associated protein 1 (Keap1).

Under the conditions of a balanced redox equilibrium, Nrf2 is sequestered in the cytoplasm through its interaction with Keap1, which facilitates the ubiquitination and subsequent degradation of Nrf2. Upon the induction of oxidative stress, Keap1 releases Nrf2, allowing Nrf2 to stabilize and translocate into the nucleus. Once in the nucleus, Nrf2 initiates an antioxidative response, protecting cells from oxidative stress and the accumulation of cellular damage [90,111–114]. In particular, Nrf2 is involved in modulating the anti-inflammatory response by controlling redox balance and activating Antioxidant Response Enzyme (ARE)-mediated anti-inflammatory genes. This includes upregulating the expression of antioxidant genes like NQO1, HO-1, and PRX1, all of which possess anti-inflammatory properties [114].

On the one hand, a misregulated oxidative defense and the accumulation of ROS lead to extensive oxidative damage to macromolecules, including proteins, lipids, and DNA. Damage to proteins and lipids not only disrupts intracellular signaling but also leads to cell death. DNA oxidation results in the accumulation of mutations, which can cause the development of malignancies [111,115]. If regulation of the redox balance is disrupted due to alterations in the Nrf2/Keap1 signaling pathway, it can lead to increased oxidative damage and mutations, thereby raising the risk of tumor development [115–118]. On the other hand, Nrf2 protects cells by preventing the accumulation of ROS and thus oxidative stress. Tumors often exhibit elevated ROS levels due to their higher energy demands. In this context, the deregulation of Nrf2 and/or Keap1 is detrimental [118–121] as the induction of antioxidant genes by Nrf2 enables the survival of these malignant cells, which would otherwise die from oxidative damage. Additionally, resistance to chemotherapy increases because many chemotherapeutic agents work by shifting the redox balance and generating ROS [116,118,119]. Thus, Nrf2 is indeed a double-edged sword. It fine-tunes the redox balance, allowing oxidative signaling to occur. Additionally, it prevents the accumulation of mutations and the development of tumors by curbing excessive ROS production. In contrast, Nrf2 enables malignant cells to adapt to elevated ROS levels, thereby protecting them from cell death. Therefore, Nrf2 is a crucial factor in the regulation of the intracellular redox balance.

Taken together, ROS can oxidize essential signal proteins such as phosphatases, kinases, and transcription factors, altering their activity. Oxidation of specific functional amino acid residues within these proteins can either inhibit or enhance their function. Thus, oxidative signaling directly controls T cell proliferation and differentiation by impairing the activity of protein phosphatases, kinases, and transcription factors.

These ROS-dependent changes in signaling pathways and transcription factor activity significantly alter the expression profile of T cells. The minimal requirements for inducing IL2 expression, an essential factor in T cell activation, involve the activity of the transcription factors AP-1, NF-κB, and NF-AT. The MAP kinase cascade, primarily via ERK activation, significantly contributes to the induction of T cell proliferation and differentiation [29,35,122].

5. Conclusions

Redox signaling has evolved from being seen as a mere byproduct of enzymatic reactions to intentionally produced molecules crucial for cellular signaling. In T cells, ROS play a pivotal role in signaling pathways, with H_2O_2 acting as a vital second messenger. The precise localization of ROS generation and the balance between ROS and antioxidant defense mechanisms are critical for enabling oxidative signaling. Furthermore, cellular metabolic alterations are essential for generating oxidative signals, emphasizing the intricate interplay between metabolism and signaling in T cell activation, proliferation, and differentiation. Overall, redox signaling is indispensable for fine-tuning cellular responses, particularly in the context of T cell function and immune responses [29,35].

Stimulation of the TCR by an antigen triggers activation of the proximal T cell machinery, leading to activation of the tyrosine kinase ZAP70. ZAP70 phosphorylates the scaffold protein LAT, which in turn recruits proteins like PLCγ1. Recruitment to LAT activates PLCγ1, leading to the production of IP_3 and DAG. IP_3 then induces the opening of calcium channels in the endoplasmic reticulum (ER). This initial influx of Ca^{2+} is followed by a stronger influx through calcium channels in the plasma membrane. This increase in cytosolic Ca^{2+} levels represents one arm of the TCR signaling pathway. The other arm of TCR signaling involves DAG. DAG activates PKC, with a subpopulation of PKCθ translocating to the outer mitochondrial membrane. Additionally, PKCθ likely phosphorylates ADPGK, inducing increased glucose influx and a metabolic switch. This results in an electron transfer to the CytQ pool, directly mediated by GPD2. At this point, hyper-reduced ubiquinol likely leads to electron flow in the ETC retrograde to Complex I. During this process, ROS are released into the mitochondrial matrix. Simultaneously, electrons may also flow forward from ubiquinol to Complex III, which then releases ROS into the intermembrane space. The release of ROS via Complexes I and III of the ETC constitutes the oxidative signal [36,51]. The Ca^{2+} signal activates the transcription factor NF-AT via calcineurin, an integral component of T cell activation. The oxidative signal enhances/enables the activation of kinase signaling cascades through the modification/oxidation of kinases and regulatory phosphatases (e.g., MAPK signaling cascade). Additionally, REDOX-sensitive transcription factors such as AP-1 and NF-κB are also modified/oxidized, thereby increasing their translocation into the nucleus. The oxidative signal also induces the transcription factor Nrf2. Nrf2 regulates a variety of antioxidant proteins, which then prevent an excess of ROS production, thus regulating/terminating the oxidative signal. The transcription factors NF-AT, AP-1, and NF-κB form the minimal requirement for T cell activation. Thus, the Ca^{2+} signal in combination with the oxidative signal is indispensable for the activation of T lymphocytes [8,29,35,36,51,90] (Figure 7).

Changes in the redox balance and oxidative signals are ubiquitous and play a fundamental role in both inter- and intracellular communication. In the case of T cell activation, these signals are indispensable. A more in-depth exploration of these signals, their formation, and the impact of metabolic changes on these signals or, conversely, how changes in the redox equilibrium affect metabolism, is imperative and requires further investigation.

The role of the activating co-stimulatory surface molecule CD28 was last studied in detail in the late 1990s [8]. New investigative methods could provide more precise insights into the production of oxidative signals through CD28 stimulation and elucidate how these signals interact.

CTLA-4, which competes with CD28 for its ligands, attenuates T cell activation. This competition naturally reduces the oxidative signal that can be induced by CD28. This has potential clinical applications; soluble CTLA-4 can be used to dampen T cell immune responses, thereby preventing excessive immune responses and autoimmunity [83,84]. However, further research is needed to investigate in detail how CTLA-4 stimulation affects TCR-induced oxidative signaling.

For PD-1, the situation is more complex. There are limited data on the generation of oxidative signals or changes in redox balance. The available data suggest that PD-1 stimulation initiates an oxidative signal. This is puzzling because oxidative signals contribute to T cell activation, whereas PD-1 limits or inhibits T cell activation [85,86]. However, oxidative signals are also crucial for terminating an immune response by inducing CD95L, which can trigger apoptosis (AICD). Naive and resting T cells are resistant to apoptosis. If CD95L is induced, these cells survive. In contrast, activated T cells are sensitive to apoptosis, and CD95L induction leads to apoptosis (AICD) in these cells [56,88]. The ability of oxidative signals in T cells to terminate an immune response is crucial for preventing autoimmune reactions.

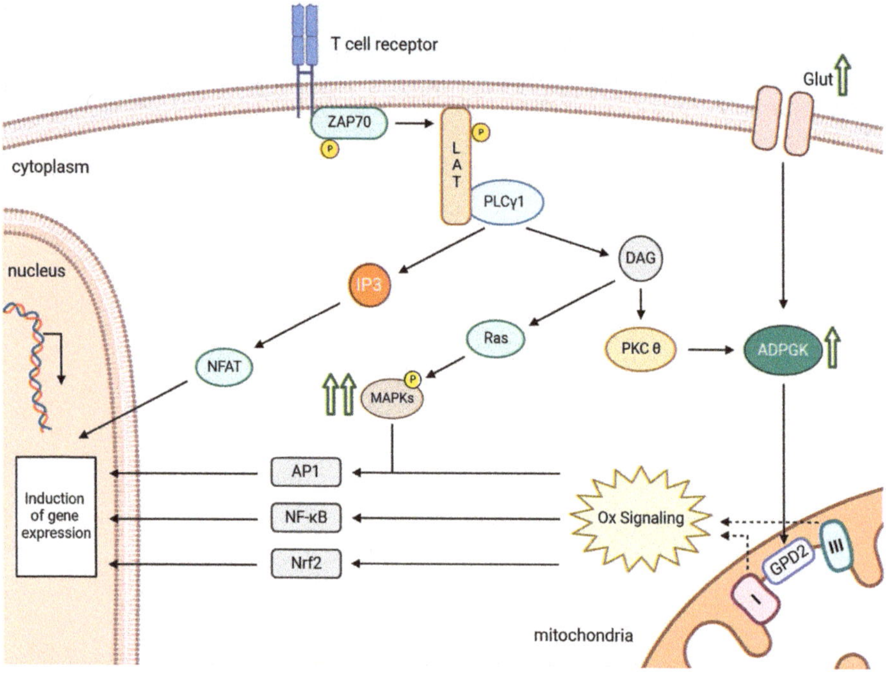

Figure 7. After TCR stimulation, two signaling pathways are induced via 3,4,5-triphosphate (IP_3) and diacylglycerol (DAG). The IP_3-dependent pathway leads to Ca^{2+} release into the cytosol, activating NF-AT. The DAG-dependent pathway activates PKCθ and induces an oxidative signal via Complex I and III of the mitochondrial electron transport chain (ETC). Additionally, the MAPK signaling pathway is induced via PKCθ. The oxidative signal originating from mitochondrial ETC amplifies the MAPK signaling pathway and activates the redox-sensitive transcription factors AP1, NF-κB, and Nrf2. AP1 and NF-κB, together with NF-AT, create the minimal requirement for T cell activation. Conversely, Nrf2, by inducing antioxidant proteins, contributes to the control and potential termination of the oxidative signal. The figure was created with BioRender.com Oxidative signals in T cells are essential for the initiation of a T cell immune response. Several publications and data have addressed the induction of oxidative signals by TCR stimulation and the mechanisms by which these signals are generated. However, the role of co-stimulation in oxidative signaling is not well understood. This represents a knowledge gap that urgently needs to be investigated in detail.

However, induction of AICD can be dysregulated in certain diseases. In Human Immunodeficiency Virus (HIV) infection, the trans-activating regulatory protein (Tat) induces an oxidative signal, sensitizing T cells to AICD [41,123,124]. This may be a mechanism that leads to T cell depletion following HIV infection and the development of the acquired immune deficiency syndrome (AIDS). This demonstrates how little we know about oxidative signals and highlights the importance of pursuing further research in this area.

However, information on how oxidative signals affect transcription factors in T cells has already been exploited clinically. In cutaneous T cell lymphoma (CTCL), especially in its incurable and aggressive leukemic variant, the Sézary syndrome, there is constitutive NF-κB activation. This leads to increased proliferation and apoptosis resistance in malignant T lymphocytes. In the cytosol, a pro-oxidative environment promotes NF-κB translocation to the nucleus. Within the nucleus, however, an anti-oxidative environment is required for NF-κB to be reduced and able to bind to DNA. The reduction in NF-κB is primarily mediated by Trx-1. When Trx-1 is inhibited in the nucleus, NF-κB can no longer be reduced, leading to the downregulation of important protective NF-κB target genes, such as inhibitors of apoptosis (IAPs). As a result, a cytosolic death platform known as the ripoptosome can form, inducing apoptosis and necroptosis in malignant cells [108]. Trx-1 can be specifically inhibited by the small molecule dimethyl fumarate (DMF). Treatment with DMF has been shown to significantly induce cell death in CTCL cells, both in vitro and in vivo. Healthy T cells were spared by DMF as they do not require constitutive NF-κB activation for survival [108,125,126]. These findings were also corroborated in a phase II clinical study [127].

These studies underscore the importance of understanding oxidative signals—their origins, mechanisms of action, and how they can be modulated. A precise understanding of the redox balance will allow us to modulate immune responses, understand disease, and develop new treatments.

Author Contributions: Conceptualization, K.G.; writing—original draft preparation, K.G., D.T. and C.K.; writing—review and editing, K.G., C.K., D.T., P.H., A.K., S.S. and M.M.; visualization, K.G. and D.T.; supervision, K.G.; funding acquisition, K.G. All authors have read and agreed to the published version of the manuscript.

Funding: This research was funded by the Wilhelm Sander-Stiftung (2022.096.1).

Acknowledgments: We would like to thank all members of the Department of Internal Medicine I for their support.

Conflicts of Interest: The authors declare no conflict of interest.

References

1. Bendall, D.S.; Howe, C.J.; Nisbet, E.G.; Nisbet, R.E.R. Photosynthetic and atmospheric evolution. Introduction. *Philos. Trans. R. Soc. Lond. B Biol. Sci.* **2008**, *363*, 2625–2628. [CrossRef]
2. Gumsley, A.P.; Chamberlain, K.R.; Bleeker, W.; Söderlund, U.; de Kock, M.O.; Larsson, E.R.; Bekker, A. Timing and tempo of the Great Oxidation Event. *Proc. Natl. Acad. Sci. USA* **2017**, *114*, 1811–1816. [CrossRef] [PubMed]
3. Sessions, A.L.; Doughty, D.M.; Welander, P.V.; Summons, R.E.; Newman, D.K. The continuing puzzle of the great oxidation event. *Curr. Biol.* **2009**, *19*, R567–R574. [CrossRef]
4. West, J.B. The strange history of atmospheric oxygen. *Physiol. Rep.* **2022**, *10*, e15214. [CrossRef] [PubMed]
5. Case, A.J. On the Origin of Superoxide Dismutase: An Evolutionary Perspective of Superoxide-Mediated Redox Signaling. *Antioxidants* **2017**, *6*, 82. [CrossRef]
6. Choudhury, F.K.; Rivero, R.M.; Blumwald, E.; Mittler, R. Reactive oxygen species, abiotic stress and stress combination. *Plant J.* **2017**, *90*, 856–867. [CrossRef] [PubMed]
7. Dröge, W. Redox regulation in anabolic and catabolic processes. *Curr. Opin. Clin. Nutr. Metab. Care* **2006**, *9*, 190–195. [CrossRef]
8. Dröge, W. Free radicals in the physiological control of cell function. *Physiol. Rev.* **2002**, *82*, 47–95. [CrossRef]
9. Reth, M. Hydrogen peroxide as second messenger in lymphocyte activation. *Nat. Immunol.* **2002**, *3*, 1129–1134. [CrossRef]
10. Kunst, C.; Schmid, S.; Michalski, M.; Tümen, D.; Buttenschön, J.; Müller, M.; Gülow, K. The Influence of Gut Microbiota on Oxidative Stress and the Immune System. *Biomedicines* **2023**, *11*, 1388. [CrossRef]
11. Flohé, L. Looking Back at the Early Stages of Redox Biology. *Antioxidants* **2020**, *9*, 1254. [CrossRef] [PubMed]
12. Zuo, J.; Zhang, Z.; Luo, M.; Zhou, L.; Nice, E.C.; Zhang, W.; Wang, C.; Huang, C. Redox signaling at the crossroads of human health and disease. *MedComm (2020)* **2022**, *3*, e127. [CrossRef] [PubMed]
13. McCord, J.M.; Fridovich, I. Superoxide dismutase. An enzymic function for erythrocuprein (hemocuprein). *J. Biol. Chem.* **1969**, *244*, 6049–6055. [CrossRef] [PubMed]
14. Brand, M.D. Riding the tiger-physiological and pathological effects of superoxide and hydrogen peroxide generated in the mitochondrial matrix. *Crit. Rev. Biochem. Mol. Biol.* **2020**, *55*, 592–661. [CrossRef] [PubMed]
15. Birben, E.; Sahiner, U.M.; Sackesen, C.; Erzurum, S.; Kalayci, O. Oxidative stress and antioxidant defense. *World Allergy Organ. J.* **2012**, *5*, 9–19. [CrossRef] [PubMed]
16. Sies, H.; Jones, D.P. Reactive oxygen species (ROS) as pleiotropic physiological signalling agents. *Nat. Rev. Mol. Cell Biol.* **2020**, *21*, 363–383. [CrossRef]
17. Andrés, C.M.C.; La Pérez de Lastra, J.M.; Andrés Juan, C.; Plou, F.J.; Pérez-Lebeña, E. Superoxide Anion Chemistry-Its Role at the Core of the Innate Immunity. *Int. J. Mol. Sci.* **2023**, *24*, 1841. [CrossRef]
18. Kruszewski, M. Labile iron pool: The main determinant of cellular response to oxidative stress. *Mutat. Res.* **2003**, *531*, 81–92. [CrossRef]
19. Kiessling, M.K.; Klemke, C.D.; Kaminski, M.M.; Galani, I.E.; Krammer, P.H.; Gülow, K. Inhibition of constitutively activated nuclear factor-kappaB induces reactive oxygen species- and iron-dependent cell death in cutaneous T-cell lymphoma. *Cancer Res.* **2009**, *69*, 2365–2374. [CrossRef]

20. Kamiński, M.M.; Röth, D.; Sass, S.; Sauer, S.W.; Krammer, P.H.; Gülow, K. Manganese superoxide dismutase: A regulator of T cell activation-induced oxidative signaling and cell death. *Biochim. Biophys. Acta* **2012**, *1823*, 1041–1052. [CrossRef]
21. Zheng, M.; Liu, Y.; Zhang, G.; Yang, Z.; Xu, W.; Chen, Q. The Applications and Mechanisms of Superoxide Dismutase in Medicine, Food, and Cosmetics. *Antioxidants* **2023**, *12*, 1675. [CrossRef] [PubMed]
22. Böhm, B.; Heinzelmann, S.; Motz, M.; Bauer, G. Extracellular localization of catalase is associated with the transformed state of malignant cells. *Biol. Chem.* **2015**, *396*, 1339–1356. [CrossRef] [PubMed]
23. Heinzelmann, S.; Bauer, G. Multiple protective functions of catalase against intercellular apoptosis-inducing ROS signaling of human tumor cells. *Biol. Chem.* **2010**, *391*, 675–693. [CrossRef]
24. Rampon, C.; Volovitch, M.; Joliot, A.; Vriz, S. Hydrogen Peroxide and Redox Regulation of Developments. *Antioxidants (Basel)* **2018**, *7*, 159. [CrossRef] [PubMed]
25. Meister, A.; Anderson, M.E. Glutathione. *Annu. Rev. Biochem.* **1983**, *52*, 711–760. [CrossRef] [PubMed]
26. AlOkda, A.; van Raamsdonk, J.M. Evolutionarily Conserved Role of Thioredoxin Systems in Determining Longevity. *Antioxidants* **2023**, *12*, 944. [CrossRef]
27. Finkel, T. Signal transduction by reactive oxygen species. *J. Cell Biol.* **2011**, *194*, 7–15. [CrossRef]
28. Sareila, O.; Kelkka, T.; Pizzolla, A.; Hultqvist, M.; Holmdahl, R. NOX2 complex-derived ROS as immune regulators. *Antioxid. Redox Signal.* **2011**, *15*, 2197–2208. [CrossRef]
29. Kamiński, M.M.; Röth, D.; Krammer, P.H.; Gülow, K. Mitochondria as oxidative signaling organelles in T-cell activation: Physiological role and pathological implications. *Arch. Immunol. Ther. Exp. (Warsz)* **2013**, *61*, 367–384. [CrossRef]
30. Jackson, S.H.; Devadas, S.; Kwon, J.; Pinto, L.A.; Williams, M.S. T cells express a phagocyte-type NADPH oxidase that is activated after T cell receptor stimulation. *Nat. Immunol.* **2004**, *5*, 818–827. [CrossRef]
31. Gelderman, K.A.; Hultqvist, M.; Pizzolla, A.; Zhao, M.; Nandakumar, K.S.; Mattsson, R.; Holmdahl, R. Macrophages suppress T cell responses and arthritis development in mice by producing reactive oxygen species. *J. Clin. Investig.* **2007**, *117*, 3020–3028. [CrossRef]
32. Kaminski, M.; Kiessling, M.; Süss, D.; Krammer, P.H.; Gülow, K. Novel role for mitochondria: Protein kinase Ctheta-dependent oxidative signaling organelles in activation-induced T-cell death. *Mol. Cell. Biol.* **2007**, *27*, 3625–3639. [CrossRef] [PubMed]
33. Dröse, S.; Brandt, U. Molecular mechanisms of superoxide production by the mitochondrial respiratory chain. *Adv. Exp. Med. Biol.* **2012**, *748*, 145–169. [CrossRef] [PubMed]
34. Lambert, A.J.; Brand, M.D. Reactive oxygen species production by mitochondria. *Methods Mol. Biol.* **2009**, *554*, 165–181. [CrossRef]
35. Sena, L.A.; Chandel, N.S. Physiological roles of mitochondrial reactive oxygen species. *Mol. Cell* **2012**, *48*, 158–167. [CrossRef] [PubMed]
36. Kamiński, M.M.; Sauer, S.W.; Kamiński, M.; Opp, S.; Ruppert, T.; Grigaravičius, P.; Grudnik, P.; Gröne, H.-J.; Krammer, P.H.; Gülow, K. T cell activation is driven by an ADP-dependent glucokinase linking enhanced glycolysis with mitochondrial reactive oxygen species generation. *Cell Rep.* **2012**, *2*, 1300–1315. [CrossRef]
37. Chaudhri, G.; Hunt, N.H.; Clark, I.A.; Ceredig, R. Antioxidants inhibit proliferation and cell surface expression of receptors for interleukin-2 and transferrin in T lymphocytes stimulated with phorbol myristate acetate and ionomycin. *Cell. Immunol.* **1988**, *115*, 204–213. [CrossRef]
38. Novogrodsky, A.; Ravid, A.; Rubin, A.L.; Stenzel, K.H. Hydroxyl radical scavengers inhibit lymphocyte mitogenesis. *Proc. Natl. Acad. Sci. USA* **1982**, *79*, 1171–1174. [CrossRef]
39. Bauer, M.K.; Vogt, M.; Los, M.; Siegel, J.; Wesselborg, S.; Schulze-Osthoff, K. Role of reactive oxygen intermediates in activation-induced CD95 (APO-1/Fas) ligand expression. *J. Biol. Chem.* **1998**, *273*, 8048–8055. [CrossRef]
40. Devadas, S.; Zaritskaya, L.; Rhee, S.G.; Oberley, L.; Williams, M.S. Discrete generation of superoxide and hydrogen peroxide by T cell receptor stimulation: Selective regulation of mitogen-activated protein kinase activation and fas ligand expression. *J. Exp. Med.* **2002**, *195*, 59–70. [CrossRef]
41. Gülow, K.; Kaminski, M.; Darvas, K.; Süss, D.; Li-Weber, M.; Krammer, P.H. HIV-1 trans-activator of transcription substitutes for oxidative signaling in activation-induced T cell death. *J. Immunol.* **2005**, *174*, 5249–5260. [CrossRef]
42. Gutscher, M.; Pauleau, A.-L.; Marty, L.; Brach, T.; Wabnitz, G.H.; Samstag, Y.; Meyer, A.J.; Dick, T.P. Real-time imaging of the intracellular glutathione redox potential. *Nat. Methods* **2008**, *5*, 553–559. [CrossRef] [PubMed]
43. Hildeman, D.A.; Mitchell, T.; Teague, T.K.; Henson, P.; Day, B.J.; Kappler, J.; Marrack, P.C. Reactive oxygen species regulate activation-induced T cell apoptosis. *Immunity* **1999**, *10*, 735–744. [CrossRef]
44. Tatla, S.; Woodhead, V.; Foreman, J.C.; Chain, B.M. The role of reactive oxygen species in triggering proliferation and IL-2 secretion in T cells. *Free Radic. Biol. Med.* **1999**, *26*, 14–24. [CrossRef]
45. Milkovic, L.; Cipak Gasparovic, A.; Cindric, M.; Mouthuy, P.-A.; Zarkovic, N. Short Overview of ROS as Cell Function Regulators and Their Implications in Therapy Concepts. *Cells* **2019**, *8*, 793. [CrossRef] [PubMed]
46. Yarosz, E.L.; Chang, C.-H. The Role of Reactive Oxygen Species in Regulating T Cell-mediated Immunity and Disease. *Immune Netw.* **2018**, *18*, e14. [CrossRef] [PubMed]
47. Courtney, A.H.; Lo, W.-L.; Weiss, A. TCR Signaling: Mechanisms of Initiation and Propagation. *Trends Biochem. Sci.* **2018**, *43*, 108–123. [CrossRef]
48. Chan, A.C.; Irving, B.A.; Fraser, J.D.; Weiss, A. The zeta chain is associated with a tyrosine kinase and upon T-cell antigen receptor stimulation associates with ZAP-70, a 70-kDa tyrosine phosphoprotein. *Proc. Natl. Acad. Sci. USA* **1991**, *88*, 9166–9170. [CrossRef]

49. Finco, T.S.; Kadlecek, T.; Zhang, W.; Samelson, L.E.; Weiss, A. LAT is required for TCR-mediated activation of PLCgamma1 and the Ras pathway. *Immunity* **1998**, *9*, 617–626. [CrossRef]
50. Kaminski, M.M.; Sauer, S.W.; Klemke, C.-D.; Süss, D.; Okun, J.G.; Krammer, P.H.; Gülow, K. Mitochondrial reactive oxygen species control T cell activation by regulating IL-2 and IL-4 expression: Mechanism of ciprofloxacin-mediated immunosuppression. *J. Immunol.* **2010**, *184*, 4827–4841. [CrossRef] [PubMed]
51. Sena, L.A.; Li, S.; Jairaman, A.; Prakriya, M.; Ezponda, T.; Hildeman, D.A.; Wang, C.-R.; Schumacker, P.T.; Licht, J.D.; Perlman, H.; et al. Mitochondria are required for antigen-specific T cell activation through reactive oxygen species signaling. *Immunity* **2013**, *38*, 225–236. [CrossRef] [PubMed]
52. Chowdhury, S.K.R.; Gemin, A.; Singh, G. High activity of mitochondrial glycerophosphate dehydrogenase and glycerophosphate-dependent ROS production in prostate cancer cell lines. *Biochem. Biophys. Res. Commun.* **2005**, *333*, 1139–1145. [CrossRef] [PubMed]
53. Tretter, L.; Takacs, K.; Hegedus, V.; Adam-Vizi, V. Characteristics of alpha-glycerophosphate-evoked H2O2 generation in brain mitochondria. *J. Neurochem.* **2007**, *100*, 650–663. [CrossRef] [PubMed]
54. Case, A.J.; McGill, J.L.; Tygrett, L.T.; Shirasawa, T.; Spitz, D.R.; Waldschmidt, T.J.; Legge, K.L.; Domann, F.E. Elevated mitochondrial superoxide disrupts normal T cell development, impairing adaptive immune responses to an influenza challenge. *Free Radic. Biol. Med.* **2011**, *50*, 448–458. [CrossRef]
55. Maric, M.; Barjaktarevic, I.; Bogunovic, B.; Stojakovic, M.; Maric, C.; Vukmanovic, S. Cutting edge: Developmental up-regulation of IFN-gamma-inducible lysosomal thiol reductase expression leads to reduced T cell sensitivity and less severe autoimmunity. *J. Immunol.* **2009**, *182*, 746–750. [CrossRef] [PubMed]
56. Krammer, P.H.; Kamiński, M.; Kiessling, M.; Gülow, K. No life without death. *Adv. Cancer Res.* **2007**, *97*, 111–138. [CrossRef]
57. Forrester, S.J.; Kikuchi, D.S.; Hernandes, M.S.; Xu, Q.; Griendling, K.K. Reactive Oxygen Species in Metabolic and Inflammatory Signaling. *Circ. Res.* **2018**, *122*, 877–902. [CrossRef] [PubMed]
58. Kaneto, H.; Katakami, N.; Matsuhisa, M.; Matsuoka, T. Role of reactive oxygen species in the progression of type 2 diabetes and atherosclerosis. *Mediators Inflamm.* **2010**, *2010*, 453892. [CrossRef] [PubMed]
59. Goncharov, N.V.; Avdonin, P.V.; Nadeev, A.D.; Zharkikh, I.L.; Jenkins, R.O. Reactive oxygen species in pathogenesis of atherosclerosis. *Curr. Pharm. Des.* **2015**, *21*, 1134–1146. [CrossRef]
60. Kussmaul, L.; Hirst, J. The mechanism of superoxide production by NADH:ubiquinone oxidoreductase (complex I) from bovine heart mitochondria. *Proc. Natl. Acad. Sci. USA* **2006**, *103*, 7607–7612. [CrossRef]
61. Murphy, M.P. How mitochondria produce reactive oxygen species. *Biochem. J.* **2009**, *417*, 1–13. [CrossRef] [PubMed]
62. Noble, J.; Macek Jilkova, Z.; Aspord, C.; Malvezzi, P.; Fribourg, M.; Riella, L.V.; Cravedi, P. Harnessing Immune Cell Metabolism to Modulate Alloresponse in Transplantation. *Transpl. Int.* **2024**, *37*, 12330. [CrossRef] [PubMed]
63. Stentz, F.B.; Kitabchi, A.E. Hyperglycemia-induced activation of human T-lymphocytes with de novo emergence of insulin receptors and generation of reactive oxygen species. *Biochem. Biophys. Res. Commun.* **2005**, *335*, 491–495. [CrossRef] [PubMed]
64. Jacobs, S.R.; Herman, C.E.; Maciver, N.J.; Wofford, J.A.; Wieman, H.L.; Hammen, J.J.; Rathmell, J.C. Glucose uptake is limiting in T cell activation and requires CD28-mediated Akt-dependent and independent pathways. *J. Immunol.* **2008**, *180*, 4476–4486. [CrossRef] [PubMed]
65. Wang, T.; Marquardt, C.; Foker, J. Aerobic glycolysis during lymphocyte proliferation. *Nature* **1976**, *261*, 702–705. [CrossRef] [PubMed]
66. Chelakkot, C.; Chelakkot, V.S.; Shin, Y.; Song, K. Modulating Glycolysis to Improve Cancer Therapy. *Int. J. Mol. Sci.* **2023**, *24*, 2606. [CrossRef] [PubMed]
67. WARBURG, O. On respiratory impairment in cancer cells. *Science* **1956**, *124*, 269–270. [CrossRef] [PubMed]
68. Barba, I.; Carrillo-Bosch, L.; Seoane, J. Targeting the Warburg Effect in Cancer: Where Do We Stand? *Int. J. Mol. Sci.* **2024**, *25*, 3142. [CrossRef]
69. Guixé, V.; Merino, F. The ADP-dependent sugar kinase family: Kinetic and evolutionary aspects. *IUBMB Life* **2009**, *61*, 753–761. [CrossRef]
70. Murphy, M.P.; Siegel, R.M. Mitochondrial ROS fire up T cell activation. *Immunity* **2013**, *38*, 201–202. [CrossRef]
71. Yin, M.; O'Neill, L.A.J. The role of the electron transport chain in immunity. *FASEB J.* **2021**, *35*, e21974. [CrossRef] [PubMed]
72. Ma, S.; Ming, Y.; Wu, J.; Cui, G. Cellular metabolism regulates the differentiation and function of T-cell subsets. *Cell. Mol. Immunol.* **2024**, *21*, 419–435. [CrossRef]
73. Liemburg-Apers, D.C.; Willems, P.H.G.M.; Koopman, W.J.H.; Grefte, S. Interactions between mitochondrial reactive oxygen species and cellular glucose metabolism. *Arch. Toxicol.* **2015**, *89*, 1209–1226. [CrossRef] [PubMed]
74. Jameson, S.C. Maintaining the norm: T-cell homeostasis. *Nat. Rev. Immunol.* **2002**, *2*, 547–556. [CrossRef]
75. Saravia, J.; Chapman, N.M.; Chi, H. Helper T cell differentiation. *Cell. Mol. Immunol.* **2019**, *16*, 634–643. [CrossRef] [PubMed]
76. Macintyre, A.N.; Gerriets, V.A.; Nichols, A.G.; Michalek, R.D.; Rudolph, M.C.; Deoliveira, D.; Anderson, S.M.; Abel, E.D.; Chen, B.J.; Hale, L.P.; et al. The glucose transporter Glut1 is selectively essential for CD4 T cell activation and effector function. *Cell Metab.* **2014**, *20*, 61–72. [CrossRef] [PubMed]
77. Michalek, R.D.; Gerriets, V.A.; Jacobs, S.R.; Macintyre, A.N.; Maciver, N.J.; Mason, E.F.; Sullivan, S.A.; Nichols, A.G.; Rathmell, J.C. Cutting edge: Distinct glycolytic and lipid oxidative metabolic programs are essential for effector and regulatory CD4+ T cell subsets. *J. Immunol.* **2011**, *186*, 3299–3303. [CrossRef] [PubMed]

78. Peng, M.; Yin, N.; Chhangawala, S.; Xu, K.; Leslie, C.S.; Li, M.O. Aerobic glycolysis promotes T helper 1 cell differentiation through an epigenetic mechanism. *Science* **2016**, *354*, 481–484. [CrossRef] [PubMed]
79. Fu, H.; Vuononvirta, J.; Fanti, S.; Bonacina, F.; D'Amati, A.; Wang, G.; Poobalasingam, T.; Fankhaenel, M.; Lucchesi, D.; Coleby, R.; et al. The glucose transporter 2 regulates CD8+ T cell function via environment sensing. *Nat. Metab.* **2023**, *5*, 1969–1985. [CrossRef]
80. Belikov, A.V.; Schraven, B.; Simeoni, L. T cells and reactive oxygen species. *J. Biomed. Sci.* **2015**, *22*, 85. [CrossRef]
81. Los, M.; Schenk, H.; Hexel, K.; Baeuerle, P.A.; Dröge, W.; Schulze-Osthoff, K. IL-2 gene expression and NF-kappa B activation through CD28 requires reactive oxygen production by 5-lipoxygenase. *EMBO J.* **1995**, *14*, 3731–3740. [CrossRef]
82. Kennel, K.B.; Greten, F.R. Immune cell-produced ROS and their impact on tumor growth and metastasis. *Redox Biol.* **2021**, *42*, 101891. [CrossRef] [PubMed]
83. Najafian, N.; Sayegh, M.H. CTLA4-Ig: A novel immunosuppressive agent. *Expert Opin. Investig. Drugs* **2000**, *9*, 2147–2157. [CrossRef] [PubMed]
84. Remans, P.H.J.; Wijbrandts, C.A.; Sanders, M.E.; Toes, R.E.; Breedveld, F.C.; Tak, P.P.; van Laar, J.M.; Reedquist, K.A. CTLA-4IG suppresses reactive oxygen species by preventing synovial adherent cell-induced inactivation of Rap1, a Ras family GTPASE mediator of oxidative stress in rheumatoid arthritis T cells. *Arthritis Rheum.* **2006**, *54*, 3135–3143. [CrossRef] [PubMed]
85. Tkachev, V.; Goodell, S.; Opipari, A.W.; Hao, L.-Y.; Franchi, L.; Glick, G.D.; Ferrara, J.L.M.; Byersdorfer, C.A. Programmed death-1 controls T cell survival by regulating oxidative metabolism. *J. Immunol.* **2015**, *194*, 5789–5800. [CrossRef] [PubMed]
86. Peng, H.-Y.; Lucavs, J.; Ballard, D.; Das, J.K.; Kumar, A.; Wang, L.; Ren, Y.; Xiong, X.; Song, J. Metabolic Reprogramming and Reactive Oxygen Species in T Cell Immunity. *Front. Immunol.* **2021**, *12*, 652687. [CrossRef] [PubMed]
87. Brenner, D.; Krammer, P.H.; Arnold, R. Concepts of activated T cell death. *Crit. Rev. Oncol. Hematol.* **2008**, *66*, 52–64. [CrossRef] [PubMed]
88. Krammer, P.H.; Arnold, R.; Lavrik, I.N. Life and death in peripheral T cells. *Nat. Rev. Immunol.* **2007**, *7*, 532–542. [CrossRef]
89. Baumann, S.; Krueger, A.; Kirchhoff, S.; Krammer, P.H. Regulation of T cell apoptosis during the immune response. *Curr. Mol. Med.* **2002**, *2*, 257–272. [CrossRef] [PubMed]
90. Hong, Y.; Boiti, A.; Vallone, D.; Foulkes, N.S. Reactive Oxygen Species Signaling and Oxidative Stress: Transcriptional Regulation and Evolution. *Antioxidants* **2024**, *13*, 312. [CrossRef]
91. Kiessling, M.K.; Linke, B.; Brechmann, M.; Süss, D.; Krammer, P.H.; Gülow, K. Inhibition of NF-κB induces a switch from CD95L-dependent to CD95L-independent and JNK-mediated apoptosis in T cells. *FEBS Lett.* **2010**, *584*, 4679–4688. [CrossRef] [PubMed]
92. Son, Y.; Cheong, Y.-K.; Kim, N.-H.; Chung, H.-T.; Kang, D.G.; Pae, H.-O. Mitogen-Activated Protein Kinases and Reactive Oxygen Species: How Can ROS Activate MAPK Pathways? *J. Signal Transduct.* **2011**, *2011*, 792639. [CrossRef] [PubMed]
93. Trinh, V.H.; Nguyen Huu, T.; Sah, D.K.; Choi, J.M.; Yoon, H.J.; Park, S.C.; Jung, Y.S.; Lee, S.-R. Redox Regulation of PTEN by Reactive Oxygen Species: Its Role in Physiological Processes. *Antioxidantsa* **2024**, *13*, 199. [CrossRef]
94. Chang, L.; Karin, M. Mammalian MAP kinase signalling cascades. *Nature* **2001**, *410*, 37–40. [CrossRef] [PubMed]
95. Liu, Y.; He, C. A review of redox signaling and the control of MAP kinase pathway in plants. *Redox Biol.* **2017**, *11*, 192–204. [CrossRef]
96. Matsuzawa, A.; Ichijo, H. Redox control of cell fate by MAP kinase: Physiological roles of ASK1-MAP kinase pathway in stress signaling. *Biochim. Biophys. Acta* **2008**, *1780*, 1325–1336. [CrossRef]
97. Shu, P.; Liang, H.; Zhang, J.; Lin, Y.; Chen, W.; Zhang, D. Reactive oxygen species formation and its effect on CD4+ T cell-mediated inflammation. *Front. Immunol.* **2023**, *14*, 1199233. [CrossRef]
98. Shaulian, E.; Karin, M. AP-1 as a regulator of cell life and death. *Nat. Cell Biol.* **2002**, *4*, E131-6. [CrossRef]
99. Kramer-Stickland, K.; Edmonds, A.; Bair, W.B.; Bowden, G.T. Inhibitory effects of deferoxamine on UVB-induced AP-1 transactivation. *Carcinogenesis* **1999**, *20*, 2137–2142. [CrossRef]
100. Lu, J.; Guo, J.-H.; Tu, X.-L.; Zhang, C.; Zhao, M.; Zhang, Q.-W.; Gao, F.-H. Tiron Inhibits UVB-Induced AP-1 Binding Sites Transcriptional Activation on MMP-1 and MMP-3 Promoters by MAPK Signaling Pathway in Human Dermal Fibroblasts. *PLoS ONE* **2016**, *11*, e0159998. [CrossRef]
101. Ding, M.; Li, J.J.; Leonard, S.S.; Ye, J.P.; Shi, X.; Colburn, N.H.; Castranova, V.; Vallyathan, V. Vanadate-induced activation of activator protein-1: Role of reactive oxygen species. *Carcinogenesis* **1999**, *20*, 663–668. [CrossRef] [PubMed]
102. Lian, S.; Li, S.; Zhu, J.; Xia, Y.; Do Jung, Y. Nicotine stimulates IL-8 expression via ROS/NF-κB and ROS/MAPK/AP-1 axis in human gastric cancer cells. *Toxicology* **2022**, *466*, 153062. [CrossRef] [PubMed]
103. Kamata, H.; Honda, S.-I.; Maeda, S.; Chang, L.; Hirata, H.; Karin, M. Reactive oxygen species promote TNFalpha-induced death and sustained JNK activation by inhibiting MAP kinase phosphatases. *Cell* **2005**, *120*, 649–661. [CrossRef] [PubMed]
104. Rothwarf, D.M.; Karin, M. The NF-kappa B activation pathway: A paradigm in information transfer from membrane to nucleus. *Sci. STKE* **1999**, *1999*, RE1. [CrossRef] [PubMed]
105. Huang, Y.; Wange, R.L. T cell receptor signaling: Beyond complex complexes. *J. Biol. Chem.* **2004**, *279*, 28827–28830. [CrossRef] [PubMed]
106. Schreck, R.; Rieber, P.; Baeuerle, P.A. Reactive oxygen intermediates as apparently widely used messengers in the activation of the NF-kappa B transcription factor and HIV-1. *EMBO J.* **1991**, *10*, 2247–2258. [CrossRef]

107. Israël, N.; Gougerot-Pocidalo, M.A.; Aillet, F.; Virelizier, J.L. Redox status of cells influences constitutive or induced NF-kappa B translocation and HIV long terminal repeat activity in human T and monocytic cell lines. *J. Immunol.* **1992**, *149*, 3386–3393. [CrossRef]
108. Schroeder, A.; Warnken, U.; Röth, D.; Klika, K.D.; Vobis, D.; Barnert, A.; Bujupi, F.; Oberacker, T.; Schnölzer, M.; Nicolay, J.P.; et al. Targeting Thioredoxin-1 by dimethyl fumarate induces ripoptosome-mediated cell death. *Sci. Rep.* **2017**, *7*, 43168. [CrossRef] [PubMed]
109. Gloire, G.; Piette, J. Redox regulation of nuclear post-translational modifications during NF-kappaB activation. *Antioxid. Redox Signal.* **2009**, *11*, 2209–2222. [CrossRef]
110. Hogan, P.G.; Chen, L.; Nardone, J.; Rao, A. Transcriptional regulation by calcium, calcineurin, and NFAT. *Genes Dev.* **2003**, *17*, 2205–2232. [CrossRef]
111. Gülow, K.; Tümen, D.; Kunst, C. The Important Role of Protein Kinases in the p53 Sestrin Signaling Pathway. *Cancers* **2023**, *15*, 5390. [CrossRef]
112. Aaen, S.M.; Horsberg, T.E. A screening of multiple classes of pharmaceutical compounds for effect on preadult salmon lice Lepeophtheirus salmonis. *J. Fish Dis.* **2016**, *39*, 1213–1223. [CrossRef] [PubMed]
113. Rajendran, P.; Nandakumar, N.; Rengarajan, T.; Palaniswami, R.; Gnanadhas, E.N.; Lakshminarasaiah, U.; Gopas, J.; Nishigaki, I. Antioxidants and human diseases. *Clin. Chim. Acta* **2014**, *436*, 332–347. [CrossRef] [PubMed]
114. Ngo, V.; Duennwald, M.L. Nrf2 and Oxidative Stress: A General Overview of Mechanisms and Implications in Human Disease. *Antioxidants* **2022**, *11*, 2345. [CrossRef]
115. Bae, S.H.; Sung, S.H.; Oh, S.Y.; Lim, J.M.; Lee, S.K.; Park, Y.N.; Lee, H.E.; Kang, D.; Rhee, S.G. Sestrins activate Nrf2 by promoting p62-dependent autophagic degradation of Keap1 and prevent oxidative liver damage. *Cell Metab.* **2013**, *17*, 73–84. [CrossRef]
116. Glorieux, C.; Enríquez, C.; González, C.; Aguirre-Martínez, G.; Buc Calderon, P. The Multifaceted Roles of NRF2 in Cancer: Friend or Foe? *Antioxidants* **2024**, *13*, 70. [CrossRef] [PubMed]
117. Bukke, V.N.; Moola, A.; Serviddio, G.; Vendemiale, G.; Bellanti, F. Nuclear factor erythroid 2-related factor 2-mediated signaling and metabolic associated fatty liver disease. *World J. Gastroenterol.* **2022**, *28*, 6909–6921. [CrossRef] [PubMed]
118. Tossetta, G.; Fantone, S.; Marzioni, D.; Mazzucchelli, R. Cellular Modulators of the NRF2/KEAP1 Signaling Pathway in Prostate Cancer. *Front. Biosci. (Landmark Ed)* **2023**, *28*, 143. [CrossRef] [PubMed]
119. Shibata, T.; Ohta, T.; Tong, K.I.; Kokubu, A.; Odogawa, R.; Tsuta, K.; Asamura, H.; Yamamoto, M.; Hirohashi, S. Cancer related mutations in NRF2 impair its recognition by Keap1-Cul3 E3 ligase and promote malignancy. *Proc. Natl. Acad. Sci. USA* **2008**, *105*, 13568–13573. [CrossRef]
120. Padmanabhan, B.; Tong, K.I.; Ohta, T.; Nakamura, Y.; Scharlock, M.; Ohtsuji, M.; Kang, M.-I.; Kobayashi, A.; Yokoyama, S.; Yamamoto, M. Structural basis for defects of Keap1 activity provoked by its point mutations in lung cancer. *Mol. Cell* **2006**, *21*, 689–700. [CrossRef]
121. Kim, Y.R.; Oh, J.E.; Kim, M.S.; Kang, M.R.; Park, S.W.; Han, J.Y.; Eom, H.S.; Yoo, N.J.; Lee, S.H. Oncogenic NRF2 mutations in squamous cell carcinomas of oesophagus and skin. *J. Pathol.* **2010**, *220*, 446–451. [CrossRef] [PubMed]
122. Kwon, J.; Devadas, S.; Williams, M.S. T cell receptor-stimulated generation of hydrogen peroxide inhibits MEK-ERK activation and lck serine phosphorylation. *Free Radic. Biol. Med.* **2003**, *35*, 406–417. [CrossRef] [PubMed]
123. Westendorp, M.O.; Shatrov, V.A.; Schulze-Osthoff, K.; Frank, R.; Kraft, M.; Los, M.; Krammer, P.H.; Dröge, W.; Lehmann, V. HIV-1 Tat potentiates TNF-induced NF-kappa B activation and cytotoxicity by altering the cellular redox state. *EMBO J.* **1995**, *14*, 546–554. [CrossRef] [PubMed]
124. Westendorp, M.O.; Frank, R.; Ochsenbauer, C.; Stricker, K.; Dhein, J.; Walczak, H.; Debatin, K.M.; Krammer, P.H. Sensitization of T cells to CD95-mediated apoptosis by HIV-1 Tat and gp120. *Nature* **1995**, *375*, 497–500. [CrossRef] [PubMed]
125. Froehlich, T.C.; Müller-Decker, K.; Braun, J.D.; Albrecht, T.; Schroeder, A.; Gülow, K.; Goerdt, S.; Krammer, P.H.; Nicolay, J.P. Combined inhibition of Bcl-2 and NFκB synergistically induces cell death in cutaneous T-cell lymphoma. *Blood* **2019**, *134*, 445–455. [CrossRef]
126. Nicolay, J.P.; Müller-Decker, K.; Schroeder, A.; Brechmann, M.; Möbs, M.; Géraud, C.; Assaf, C.; Goerdt, S.; Krammer, P.H.; Gülow, K. Dimethyl fumarate restores apoptosis sensitivity and inhibits tumor growth and metastasis in CTCL by targeting NF-κB. *Blood* **2016**, *128*, 805–815. [CrossRef]
127. Nicolay, J.P.; Melchers, S.; Albrecht, J.D.; Assaf, C.; Dippel, E.; Stadler, R.; Wehkamp, U.; Wobser, M.; Zhao, J.; Burghaus, I.; et al. Dimethyl fumarate treatment in relapsed and refractory cutaneous T-cell lymphoma: A multicenter phase 2 study. *Blood* **2023**, *142*, 794–805. [CrossRef]

International Journal of
Molecular Sciences

MDPI

Review

The Role of Glutathione in Age-Related Macular Degeneration (AMD)

Sylwia Brodzka [1,2], Jędrzej Baszyński [1], Katarzyna Rektor [2], Karolina Hołderna-Bona [1], Emilia Stanek [1], Natalia Kurhaluk [3], Halina Tkaczenko [3], Grażyna Malukiewicz [4], Alina Woźniak [5] and Piotr Kamiński [1,2,*]

1 Division of Ecology and Environmental Protection, Department of Medical Biology and Biochemistry, Faculty of Medicine, Collegium Medicum in Bydgoszcz, Nicolaus Copernicus University in Toruń, M. Skłodowska-Curie St. 9, PL 85-094 Bydgoszcz, Poland; sylwia.brodzka@cm.umk.pl (S.B.); jedrzej.baszynski@cm.umk.pl (J.B.); karolina.holderna@cm.umk.pl (K.H.-B.); emilia.stanek@cm.umk.pl (E.S.)

2 Department of Biotechnology, Institute of Biological Sciences, Faculty of Biological Sciences, University of Zielona Góra, Prof. Z. Szafran St. 1, PL 65-516 Zielona Góra, Poland; k.rektor@wnb.uz.zgora.pl

3 Institute of Biology, Pomeranian University in Słupsk, Arciszewski St. 22 B, PL 76-200 Słupsk, Poland; natalia.kurhaluk@upsl.edu.pl (N.K.); halina.tkaczenko@upsl.edu.pl (H.T.)

4 Department of Eye Diseases, University Hospital No. 1, Faculty of Medicine, Collegium Medicum in Bydgoszcz, Nicolaus Copernicus University in Toruń, M. Skłodowska-Curie St. 9, PL 85-092 Bydgoszcz, Poland; g.malukiewicz@cm.umk.pl

5 Department of Medical Biology and Biochemistry, Faculty of Medicine, Collegium Medicum in Bydgoszcz, Nicolaus Copernicus University, M. Karłowicz St. 24, PL 85-092 Bydgoszcz, Poland; al1103@cm.umk.pl

* Correspondence: piotr.kaminski@cm.umk.pl or p.kaminski@wnb.uz.zgora.pl

Abstract: Age-related macular degeneration (AMD) is a chronic disease that usually develops in older people. Pathogenetic changes in this disease include anatomical and functional complexes. Harmful factors damage the retina and macula. These changes may lead to partial or total loss of vision. The disease can occur in two clinical forms: dry (the progression is slow and gentle) and exudative (wet—progression is acute and severe), which usually starts in the dry form; however, the coexistence of both forms is possible. The etiology of AMD is not fully understood, and the precise mechanisms of the development of this illness are still unknown. Extensive genetic studies have shown that AMD is a multi-factorial disease and that genetic determinants, along with external and internal environmental and metabolic-functional factors, are important risk factors. This article reviews the role of glutathione (GSH) enzymes engaged in maintaining the reduced form and polymorphism in *glutathione S-transferase theta-1* (*GSTT1*) and *glutathione S-transferase mu-1* (*GSTM1*) in the development of AMD. We only chose papers that confirmed the influence of the parameters on the development of AMD. Because GSH is the most important antioxidant in the eye, it is important to know the influence of the enzymes and genetic background to ensure an optimal level of glutathione concentration. Numerous studies have been conducted on how the glutathione system works till today. This paper presents the current state of knowledge about the changes in GSH, GST, GR, and GPx in AMD. GST studies clearly show increased activity in ill people, but for GPx, the results relating to activity are not so clear. Depending on the research, the results also suggest higher and lower GPx activity in patients with AMD. The analysis of polymorphisms in *GST* genes confirmed that mutations lead to weaker antioxidant barriers and may contribute to the development of AMD; unfortunately, a meta-analysis and some research did not confirm that connection. Unspecific results of many of the parameters that make up the glutathione system show many unknowns. It is so important to conduct further research to understand the exact mechanism of defense functions of glutathione against oxidative stress in the human eye.

Keywords: age-related macular degeneration; AMD; oxidative stress; glutathione (GSH); glutathione peroxidase (GPx); glutathione reductase (GR); glutathione S-transferase (GST); *GSTT1*; *GSTM1*

1. Introduction

Age-related macular degeneration (AMD) is a chronic eye disease that usually appears in people after the age of 50; however, in highly developed countries, it also occurs in younger people. AMD is an important cause of blindness in the USA and accounts for 54% of blindness in Caucasian patients, 4.4% in black patients, and 14.3% in Spanish patients. Today, among white people aged 40 and older, AMD is the most common cause of visual impairment and blindness in the USA [1]. According to the increasing life expectancy, it is expected that the frequency of AMD will increase [2–5]. The precise development mechanism of this illness is still unknown. Previous studies have shown that, due to pathological changes in the retinal pigment epithelium (RPE), the macula is damaged, which may lead to partial or complete loss of vision [6].

It is known that age-related macular degeneration occurs in two different pathophysiological forms: dry and exudative (known as wet or neovascular). The dry form of the disease is characterized by mild and gradual degeneration, and in the neovascular form, the disease has an acute and violent nature. Usually, the dry form transforms into the wet form (Figure 1) [7]. This disease can develop in only one eye, but with progression, it may affect both ones. It has been reported that both forms of AMD may occur simultaneously [2–5].

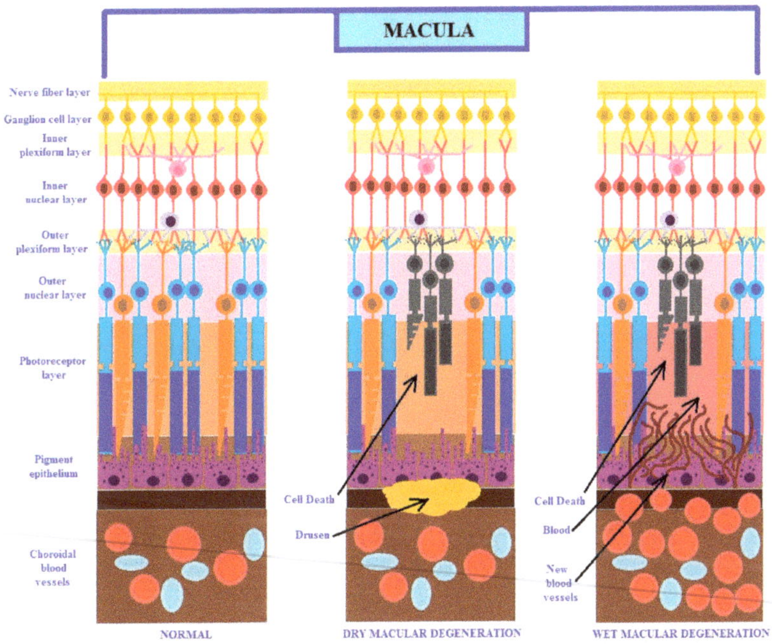

Figure 1. The figure shows the two different clinical forms of macular degeneration (modified from the work of Nayyar et al. (2020) [7]).

AMD is a heterogenic and multi-factorial disease that can also be influenced by numerous environmental and genetic risk factors [3,4,7], which significantly contribute to the development of the disease. Studies have confirmed that environmental factors such as smoking, a high cholesterol diet, carotenoids, vitamins A and E levels, zinc levels, age, sex, and the presence of heavy metal ions and chemicals slightly increase the risk of AMD. Previous studies have shown that there is no clear evidence that oxidative stress is a major factor in AMD development; however, reactive oxygen species (ROS) continue to be an important factor in the course of this disease [2–4,8]. AMD risk factors can be classified as modifiable and non-modifiable. The division of these groups of factors is illustrated in

Figure 2. It is important that modifiable and non-modifiable AMD risk factors are mutually dependent on the epigenetic regulation of gene expression, which is mainly determined by the cellular epigenetic profile [9].

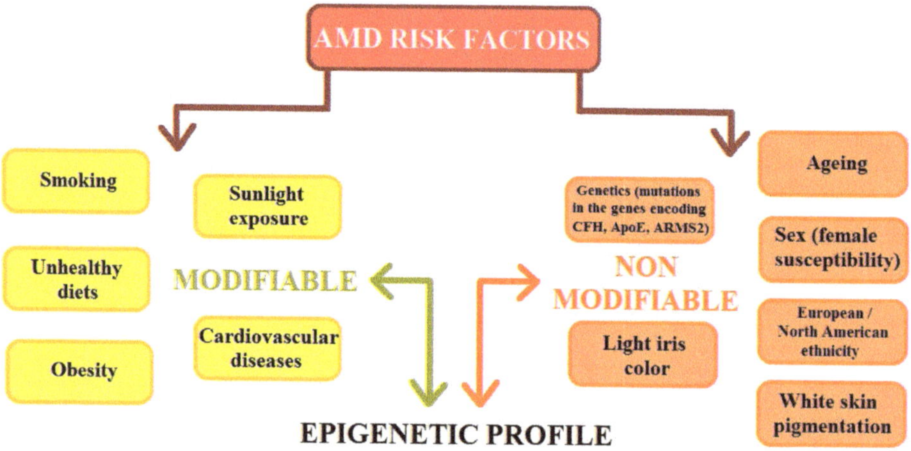

Figure 2. The figure shows the risk factors for age-related macular degeneration (AMD). CFH: complement H activity; ApoE: apolipoprotein E; ARMS2: age-related maculopathy susceptibility 2 (modified from the work of Hyttinen et al. (2023) [9]).

Four processes take part in the development of AMD: lipofuscinogenesis, drusogenesis, inflammation, and neovascularization [10]. As we described earlier, retinal cells are exposed to oxidative stress to a large extent due to the effects of light [2–4]. As a result of aging processes, the cells do not respond to mitogenic factors and lose their ability to proliferate. Studies have confirmed the presence of AGE-exposed genes (advanced glycation products) in RPE cells [11]. They are deposited in lysosomes in the form of lipofuscin (lipofuscinogenesis), which, in turn, is deposited underneath the RPE as a drusen (drusogenesis) [10]. As a result, this disturbs the metabolic exchange between the choroid and retina [12], weakening the activity of photoreceptors [13]. Then, low-intensity inflammatory processes begin, which increase the levels of acute-phase proteins, reducing total antioxidant status (TAS) [4,14]. The next stage is neovascularization, which occurs due to the secretion of proangiogenic factors during the para-inflammatory process. New capillaries come from the choroid and are formed under the retina in the macula area, where they should not occur [4]. As a result of all of these stages, photoreceptors are destroyed in the macula [2–4].

2. The Role of Reactive Oxygen Species (ROS)

The retina is exposed to light and consumes large amounts of oxygen when transforming light into vision. As a result, ROS production takes place [15]. ROS, as a group of compounds (hydrogen peroxide (H_2O_2), superoxide radical anions (O_2^{*-}), and hydroxyl radicals (*OH), react with other molecules and promote lipid peroxidation (LPO) [2–4]. These processes, involving ROS and LPO, are shown in Table 1.

These various particle molecules can form in many ways, e.g., as a product of the respiratory chain in mitochondria, photochemical or enzymatic reactions, as an effect of exposure to UV light, exposure to ionizing radiation, or the influence of metal ions [2–4,16]. ROS play an important role in the regulation of many physiological processes by participating in intracellular signaling [16,20], but they can also cause serious damage to biomolecules via lipid peroxidation (autoxidation). ROS also attack structural and enzymatic proteins via the oxidation of amino acid residues, the formation of transverse bonds and protein aggregates, and proteolysis. The deactivation of key proteins can have serious consequences in important metabolic pathways. ROS can also react with nucleic acids. The inability of the cell to repair the damage incurred may lead to the death of the cell; alternatively, there may be mutations in the DNA, leading to the development of diseases [2–4,16].

Table 1. The table shows the main reactive oxygen species (ROS) and products of lipid peroxidation connected with oxidative stress in the eye (* symbolize free electron in molecule) [15–19].

ROS/Chemical Compounds	Characteristics and Activity
Superoxide radical anions ($O_2{}^{*-}$)	Is a product of the single-electron reduction of oxygen. $O_2{}^{*-}$ reacts relatively fast with compounds that contain thiol groups (like cysteine) or with such reductive compounds as ascorbate or NADH. It is also able to react quickly with metal ions (especially copper) or certain metalloproteins. $O_2{}^{*-}$ reactions are biologically important and lead to the reduction of transition metal ions (Fe^{3+} to Fe^{2+}; Cu^{2+} to Cu^{+}).
Hydrogen peroxide (H_2O_2)	Generated as a consequence of reactions of $O_2{}^{*-}$ molecules that are mediated by superoxide dismutase or non-enzymatically. It is a relatively stable compound but may undergo a disproportionation reaction catalyzed by transition metal ions. H_2O_2 is not excessively reactive. This molecule with superoxide radical anion ($O_2{}^{*-}$) takes part in Fenton and Haber–Weiss reactions and creates a hydroxyl radical (*OH).
Hydroxyl radicals (*OH)	These are some of the most reactive oxidative agents. They are able to react with practically all substances that are present in an organism; however, in lots of cases, the tempo of such reactions is determined by diffusion only (not by an activation barrier). The biologically critical consequence of *OH reactivity is a risk factor for the generation of DNA damages. *OH is produced in Fenton and Haber–Weiss reactions or as a by-product from the process of water radiolysis.
Products of LPO	Characteristics and Activity
Malondialdehyde (MDA)	It is a reactive aldehyde generated as a result of polyunsaturated lipid degradation by ROS. MDA constitutes a highly toxic and mutagenic compound that forms lipoperoxidation end-products. It is capable of reacting with lysine residues on proteins, leading to the formation of secondary oxidation products. It may also participate in provoking DNA damages (reaction with deoxy-guanosine of DNA and the generation of 8-hydroxy-2-deoxy-guanosine) or interact with deoxyadenosine residues.
4-Hydroxynonenal (4-HNE)	This aldehyde compound is highly reactive and able to form adducts with cellular proteins and DNA. 4-HNE is mainly produced by such precursor substances as arachidonic acid and linolenic acid through 15-lipoxygenase. Electrophilic double-bond and nucleophilic amino acid residues on proteins enable 4-HNE to form Michael adducts with such amino acids as cysteine, histidine, and lysine residues. It leads to increases in their molecular masses (by the molecular mass of HNE 156 Da). These adducts are relatively stable and capable of remaining in cells even for several hours, provoking dysfunctions of the targeted biomolecules.

It is known that the retina is the element that is most exposed to oxidative stress. This is due to its continuous exposure to radiation, high concentrations of O_2, high levels of polyunsaturated fatty acids in the external photoreceptors, and the presence of many chromophores (lipofuscin, melanin, rhodopsin, and cytochrome C oxidase in the retina and retinal pigment epithelial cells (RPE) that also generate ROS in phagocytosis photoreceptor disks by RPE [4,21]. Molecular damage caused by ROS results in the formation and activity of advanced glycation products (AGEs). These products may affect cell DNA and increase the expression of genes that promote RPE cell aging [22]. These harmful factors are associated with the progression of AMD [2,4,23,24]. As a result of recognition system disorders, DNA repair genes are not able to efficiently repair DNA damage. In effect, changes accelerate the aging of the organism, leading to the dysfunction of cells and tissues. ROS formation is the result of normal, daily cellular metabolism, which is why aging is unavoidable. To protect molecules from the negative influence of ROS, cells have developed complex defense mechanisms [3,4,16]. According to these studies, photoreceptors and RPE cells are some of the most susceptible to oxidative stress damage. These non-proliferating postmitotic cells do not have any DNA damage detection systems at the cell cycle checkpoints [3,4].

The retinal pigment epithelium (RPE) performs a number of crucial functions, such as in the formation of an external retinal barrier, transport, retinoid retention, phagocytosis, the degradation of segmental photoreceptors, and protection against light and oxidative stress [25]. In this part of the eye, the dominant photoreceptors are cones. They are characterized by higher demand and energy production than rods, leading to higher oxygen requirements. In addition, rod cells and cones differ in their susceptibility to oxidative stress. Unfortunately, cones show greater sensitivity to free radicals [26].

The macula is also permanently exposed to high metabolic activity and oxidative stress due to the high partial pressure of choriocapillaris and polyunsaturated fatty acids (PUFAs), which is related to the external segments of the retina [2–4,27]. This is considered important in inducing drusen formation between RPE cells and Bruch's membrane. Lipofuscin is a chromophore that serves as a primary RPE photo-oxide [28], which, after absorbing high energy photons (especially blue light), undergoes a series of photochemical reactions. These photochemical reactions lead to ROS formation, which, in turn, induces photochemical damage in the retina and RPE cells [2–4,29]. The role of auto-phagocytosis and the RPE homeostasis pathway plays an important role in the aging of cells and the effect of oxidative stress on AMD. Studies on the cellular and molecular pathways of autophagy showed that acute exposure to oxidative stress increased autophagy activity, while chronic exposure to oxidative stress reduced autophagy. Based on animal and human models, elevated levels of autophagy in the early AMD stage have been observed, with activity decreasing in late AMD. Moreover, it has been observed that the reduced effect of auto-phagocytosis makes RPE cells more susceptible to oxidative stress and that the autophagy system is enhanced to protect RPE cells from oxidative damage [25].

3. Antioxidant Defense

To counteract the effects of oxidative stress, a response is triggered in two stages. The first stage is a reaction where the cytochrome P450 monooxygenase system and aldo-keto-reductase (AKR), carboxylesterases (CES), and epoxide hydrolase are involved. This action involves the oxidation and reduction of dangerous compounds. In the next phase, these products are coupled with hydrophilic molecules. The antioxidants involved in the second phase are divided into so-called "direct" and "indirect" categories. The "direct" group contains superoxide dismutase (SOD), glutathione (GSH), and thioredoxin reductase (Trx). The first two enzymes oxidize dangerous compounds and quickly regenerate themselves. The so-called "indirect" enzymes take part in the biosynthesis and regeneration of GSH and Trx. They also participate in the removal of oxidized compounds [11].

An antioxidant barrier cannot function without nuclear factor erythroid-2 related factor 2 (Nrf2). The antioxidants involved in the second phase are regulated by transcriptional master Nrf2/Keap1 (nuclear factor erythroid-2 related factor 2/Kelch Like ECH Associated Protein 1). This factor plays a major role in antioxidant response. It activates the transcription of the leucine zipper by binding to the antioxidant response element (ARE) within the promoter of the target genes. Its role is to maintain redox homeostasis in the cell [2,30,31]. Nrf2 regulates the action of both "direct" and "indirect" enzymes [32,33]. The "direct" group is created by SOD, glutathione peroxidase (GPX), catalase (CAT), and peroxiredoxins (PRDXs). This group of "indirect" enzymes contains transketolase (TKT), glucose-6-phosphate dehydrogenase (G6PD), glyceraldehyde-3-phosphate dehydrogenase (GAPDH), glutathione-redox cycle enzymes (glutathione reductase (GSR), glutathione synthetase (GSS)), selenoproteins, and thioredoxin-related enzymes (thioredoxin (TXN), sulfiredoxin (SRXN)) [34].

4. Glutathione Function in AMD

Glutathione is the most important low-molecular antioxidant from the thiol group [35–37]. It exists in two forms: oxidized glutathione disulfide (GSSG), which is reduced as a tripeptide gamma-glutamyl-cysteinyl-glycine (GSH). The GSH thiol group of cysteine can donate a reducing equivalent ($H^+ + e^-$) to different unstable molecules (e.g., ROS). After donation, glutathione becomes reactive but can easily react with other reactive glutathione and, in effect, may create the glutathione disulfide (GSSG) form [38]. These two molecules (GSH and GSSG) form a redox buffer system [39]. Analyzing the status of this system allows us to determine the status of the oxidative state in the intracellular environment. The GSH:GSSG ratio is specific to intracellular compartments [39,40]. The concentration of GSH in cells depends on their type and ranges from 5 to 10 mM [36,37], while the level of GSH in the serum is much lower (20 μM) because the oxidized form (GSSG) predominates ($2GSH + R_2O_2 \rightarrow GSSG + 2ROH$) [37].

Glutathione occurs in eukaryotic cells (both animals and plants) [35–37]. In the case of physiological pH, GSH has functional groups (one positively charged amino group and two negatively charged carboxyl groups) that allow for reactions with various macromolecules [41,42]. Along with ascorbate, albumin, cysteine, uric acid, creatinine, bilirubin, melanin, beta-carotene, and others, glutathione is one of the most important antioxidants in the human body. To protect cells from oxidative stress, glutathione competes with other antioxidants when reducing ROS. GSH is also involved in different vital functions. This molecule plays a role in xenobiotic metabolism by forming S-Nitroso-glutathione (GSNO), which can be a reservoir for nitric-oxide (NO) [42] and plays a role in the regulation of blood pressure. GSH is necessary for the transformation of prostaglandin H2 into prostaglandin D2, which is also important in leukotriene synthesis. It has been confirmed that GSH inhibits infection caused by influenza virus [43]. On the other hand, a study during the SARS-CoV-2 pandemic showed that cellular glutathione metabolism and redox function are impaired in this disease [44].

It is essential to keep GSH in cells at an appropriate level because when it is significantly lowered, the accumulation of H_2O_2 may occur, which results in cell damage [35,41]. In effect, low GSH concentrations may lead to the development of various diseases, such as neurodegeneration, mitochondrial dysfunction, and even cancer [38,42,45,46].

GSH biosynthesis proceeds across many stages. There are three precursor amino acids (cysteine, glutamate, and glycine) that are combined to form GSH. This molecule is synthesized exclusively in the cytosol. This process requires the sequential action of two ATP-dependent enzymes: glutamate–cysteine ligase (GCL is composed of two subunits: a catalytic GCLC and a modifier GCLM) and glutathione synthetase (GS). GCL catalyzes the first stage of the biosynthetic pathway where glutamate and cysteine are linked, creating γ-glutamyl-cysteine in the presence of ATP and Mg^{2+} or Mn^{2+}. The last stage of GSH synthesis involves the addition of glycine to γ-glutamyl-cysteine, which is catalyzed by GSH synthetase (Figure 3). The hydrolysis of GSH can be conducted by γ-glutamyl cyclo-transferase and cation transport regulator-like protein-1 (CHAC1) to obtain cysteinyl glycine and 5-oxoproline. Then, the breakdown of 5-oxoproline (which is catalyzed by 5-oxoprolinase) creates glutamate, and cysteinyl glycine is split by its respective peptidases to yield cysteine and glycine. These newly liberated amino acids can be reused for the synthesis of GSH [47,48]. The enzymes involved in GSH synthesis and metabolism in different types of retinal cells are shown in Figure 3.

As we said earlier, ROS are important in the development of AMD [2–4,8]. Therefore, research focusing on glutathione shows that this molecule plays a major role in protecting RPE cells against oxidative stress [36,48–50]. In the retina, GSH protects tissue against ROS because it is a thiol reductant and is able to detoxicate photooxidation products [42]. Studies have confirmed the role of GSH supplementation in preventing AMD [50–52]. On the other hand, GSH deficiency leads to the death of RPE cells. Unfortunately, the mechanism of increasing oxidative stress in RPE cells as a result of GSH depletion is still unknown [50].

Research has shown different results in terms of the concentration of GSH in the serum of patients with AMD compared to the controls. Some research has not indicated differences in the GSH levels between patients and healthy people [53,54]. Most studies showed decreased GSH levels in AMD patients compared to people without the disease [47,55–60]. There was even a study in which higher serum GSH concentrations were observed in AMD patients than in the control group [61]. These different results show that GSH is a non-specific parameter when analyzing the development of AMD.

Cellular GSH status in RPE cells is decreased, which has been confirmed in many studies [50,51,62–66]. Only a few studies showed increased cellular GSH levels [65,67]. These novel studies concentrate on mitochondrial GSH (mGSH) because this molecule reaches a 5-10-fold greater concentration in this organelle than in the cytosol [47,68]. GSH is transported to the mitochondria via an efflux pump and takes part in antioxidant defense, the detoxification of xenobiotics, the stabilization of mDNA [47,69], the synthesis of cluster Fe-S (mGSH is a cofactor in this reaction) [47,70], and the electron transport chain (mGSH is a redox regulator) [47,71,72]. Studies on RPE cells showed that mGSH concentration is lower in the development of AMD [47,65,73,74].

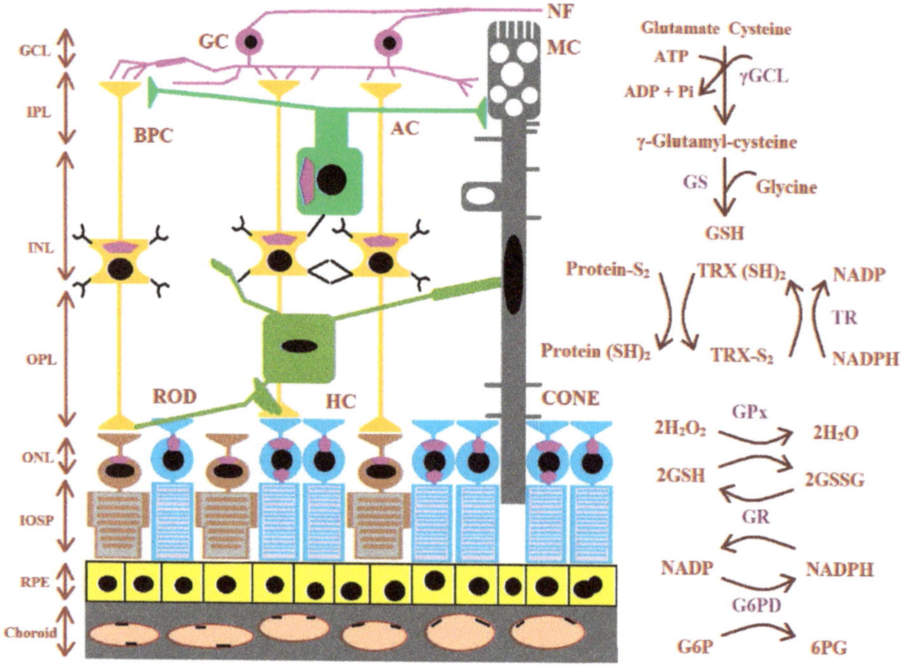

Figure 3. The figure shows the enzymes involved during glutathione synthesis and metabolism in the retina. AC: amacrine cell; BPC: bipolar cell; GC: ganglion cell; GCL: ganglion cell layer; γGCL: γ-glutamate cysteine ligase; G6PD: glucose-6-phosphate dehydrogenase; 6PG: 6-phosphogluconate; GPx: glutathione peroxidase; GR: glutathione reductase; GS: glutathione synthase; GSH: glutathione; ROD: rhodopsin; HC: horizontal cell; INL: inner nuclear layer; IPL: inner plexiform layer; IOSP: inner and outer segments of photoreceptors; NF: nerve fiber; ONL: outer nuclear layer; OPL: outer plexiform layer; protein-S2: protein-S−S; protein (SH)2: protein-SH; RPE: retinal pigment epithelium; TR: thioredoxin reductase; TRX (SH)2: thioredoxin; TRX-S2: oxidized thioredoxin (modified from the work of McBean et al. (2015) [48]).

5. Enzymes Engaged in the Maintenance of Functions of Glutathione

In cells, reduced glutathione (GSH) reacts with oxidants spontaneously or in reactions catalyzed by glutathione peroxidase (GPx) [35,39]. In effect, it creates glutathione disulfide (GSSG), which is composed of two glutathione molecules linked by a disulfide bridge. GSSG can be actively transported out of the cell or react with a protein sulfhydryl group (PSH). As a result of this process, a mixed disulfide (PSSG) is formed, but cellular GSH undergoes depletion [75]. GSSG can be reduced to GSH via glutathione reductase (GR) with the participation of 1 NADPH [24,42]. To protect proteins from harmful oxidative reactions, GSH insult trans-sulphuration reactions catalyzed by gluta-redoxin (GRX) [56,76]. On the other hand, glutathione transferase (GST) can conjugate GSH with many hydrophobic and electrophilic molecules (e.g., carcinogens, therapeutic drugs, and the products of oxidative metabolism), making them less toxic and easier to remove from cells [42,75].

Every fraction of the tetramer GPx contains a selenium atom [35]. This enzyme has eight isoforms distributed across different tissues (GPX1-GPX8) [77–79]. These isoforms are shown in Table 2. In this family, GPx1 is the most abundant member. This isoform is present in every cell and also in the cytosol, mitochondria, and peroxysomes. In some cells, it has been found in the peroxisomal compartment. This isoform plays a crucial role in preventing the harmful reaction of intracellular hydrogen peroxide and is much more effective than catalase [80]. Isoforms GPx1 and GPx4 probably have a protective role in the development of AMD [81,82]. In studies focusing on Polish patients with AMD, mutations in GPx1 were found, and these mutations lead to weaker antioxidant defense. This may have led to the development of AMD in this group of patients [81]. However, studies on mice showed that GPx4 suppresses increasing levels of vascular endothelial growth factor A (VEGF-A), which increase during neovascularization [82]. Studies also showed that lower GPx activity may increase the risk of developing AMD [83]. In several pieces of research, AMD patients were characterized by lower GPx activity [4,84]. Unfortunately, in recent AMD studies, patients had higher GPx activity than that seen in a healthy group [85].

Table 2. The table shows the isoforms of glutathione peroxidase (GPx), their location, and role [79].

Isoform	Location	Role
GPx1	Every cell (cytosol, mitochondria, and peroxysomes)	Plays a major role in protecting against systemic oxidative stress.
GPx2	Gastrointestinal system	In mammals, it is the first barrier that prevents the absorption of hydroperoxides produced by food.
GPx3	Plasma	In the plasma, it is a major ROS scavenger, eliminating all circulating complex hydroperoxides.
GPx4	Cell membrane	It is able to reduce and destroy lipid hydroperoxides.
GPx5	Epididymal tissues	In the epididymis, it regulates oxidative stress damage and protects sperm maturation.
GPx6	Embryonic and olfactory organ epithelial cells	It probably plays a role in the transmission and degradation of odor-related signals.
GPx7	Endoplasmic reticulum	It plays a unique role in maintaining redox homeostasis. It has different responses to oxidative stress than other enzymes from the GPX family.
GPx8	Transmembrane protein	It exerts a role in oxidative stress, takes part in folding endoplasmic reticulum proteins and regulates the concentration of Ca^{2+} in the endoplasmic reticulum.

Most studies on AMD showed that during this disease, GR activity was decreased [83,84,86] or that lower activity may increase the risk of developing AMD [83]. Research suggests that lower GR activity is associated with weaker antioxidant abilities [84]. There are some mechanisms relating to ganglion cell death during macular degeneration, including high intraocular pressure (IOP) and hypoxia, oxidative or nitrative stress, glutamate toxicity, loss of neurotrophic factors, and autoimmune reactions assisting in the degeneration of retinal ganglion cells. The retina is a highly vascularized tissue in the body and is vulnerable to oxidative stress due to its high consumption of oxygen and exposure to light. This tissue has one of the highest oxidative metabolic rates per tissue mass in the human body. In effect, it is important for retinal cells to maintain normal GSH levels and oxidation states. This is due to the fact that a deficiency of this molecule manifests in increased susceptibility to oxidative stress, and the resulting damage is thought to be a key step in the onset and progression of eye diseases [48].

The superfamily of GST enzymes is divided into cytosolic and microsomal groups, and the cytosolic family contains four groups: GSTA (α), GSTM (μ), GSTT (θ), and GSTP (π) [87]. The GSTM isoform is responsible for detoxifying ingredient electrolytes, such as drugs, carcinogens, environmental toxins, or products of oxidative stress [88,89]. GSTT catalyzes the combination of GSH with hydrophobic and electrolyte compounds. The main function of this isoform is to neutralize reactive low-molecular hydrocarbons, such as ethylene oxide [90]. Studies have shown that this isoform of GSTP protects the eye from oxidative stress. This protein can bind zeaxanthin and meso-zeaxanthine in the retina, suggesting that it can modulate antioxidant levels in the human eye [91,92].

6. The Role of Polymorphisms in AMD

It has been confirmed that no single gene is responsible for the development of AMD. As we stated in the beginning, AMD is a multi-factor disease. There are many genetic factors engaged in the development of AMD. However, the progression of this disease is determined in 50–60% of cases by the polymorphisms in three genes that encode the following: CFH (*rs1061170*), ARMS2 (*rs10490924*), and Il-8 (*rs2227306*) [93]. Other polymorphisms are less likely to correlate with the development of AMD, but their influence has been confirmed in other studies [81,93–104]. Genes and their polymorphisms are presented in Table 3.

Table 3. The table presents an overview of the polymorphisms engaged in the development of AMD.

Gene	Polymorphism	Mechanism	Reference
CFH	*rs1061170 (Y402H)*	Hyperactivation of the alternative complement pathway	[93]
ARMS2	*rs10490924*	Changes in the function of retinal mitochondria	[93]
	LOC387715	Reduction in ARMS2 mRNA stability	[93,94]
Il-8	*rs2227306*	Increase in inflammation and angiogenesis	[93]
LIPC	*rs1046817, rs493258, rs10468017*	Changes in HDL levels	[95]
C2	*rs9332739, rs547154*	Lowering the risk of progression to late AMD	[93,95]
CFB	*rs4151667, rs641153*	Lowering the risk of progression to late AMD	[93,95]
TNFRSF10A	*rs13278062*	Enables apoptosis in cells	[95]
HTRA1	*rs11200638*	Elevates the expression of HtrA1	[95]
CETP	*rs3764261*	Changes in the concentration of cholesteryl esters in HDL	[96,97]
GSTM1	*rs1183423000*	Weaker antioxidant protection	[98–103]
GSTT1	*rs1601993659*	Weaker antioxidant protection	[99,102,103]
GSTP1	*rs1695, rs1138272*	Weaker antioxidant protection	[99,103,104]
GPx1	*Pro197Leu*	Weaker ability to scavenge hydrogen peroxide	[81]

The activation or dysregulation of the complement factors of the alternative complements pathway, like complement factor H (CFH), complement component (C2), complement factor I (CFI), and complement factor B (CFB), have been found to be associated with the development of AMD. Their influence is caused by the release of local inflammatory-activating products [93,95]. Complement factors, especially alternative complement pathway genes, have been reported in the pathogenesis of AMD (Figure 4). The data analyzed concerning four SNPs, *rs9332739* and *rs547154* for the *C2* gene and *rs4151667* and *rs641153* for the *CFB* gene, probably suggested that these alleles lower the risk of all AMD pathogenesis in the Caucasian population by 2.0% to 6.0%. It has been observed that the CFB polymorphism (*R32Q*) is greatly correlated with early AMD; on the other hand, it has a protective effect against late AMD in the Caucasian population [95,105]. It has been reported that *CFH* and *ARMS2* SNPs are significantly correlated with early AMD pathogenesis, and it has also been suggested that the polymorphisms of Apolipoprotein E (APOE) are associated with early AMD [106]. Tumor necrosis factor receptor superfamily 10A (TNFRSF10A) signaling plays a crucial role in apoptosis in cells. It has been identified as an important risk factor that is significantly correlated with AMD pathology. Studies focusing on Japanese populations showed a significant association between disease susceptibility and the *TNFRSF10A* gene on chromosome 8p21 (*rs13278062*) [95].

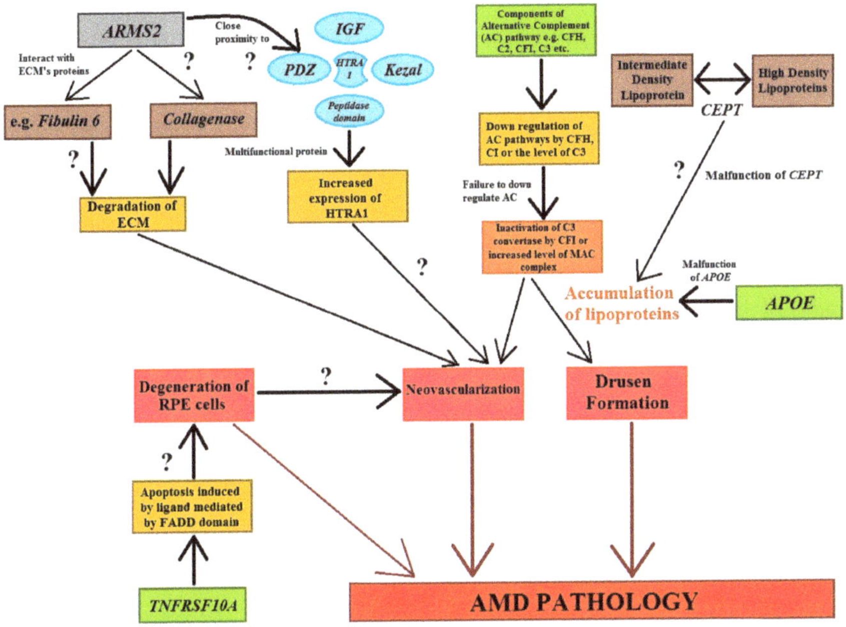

Figure 4. The figure shows the genes and their associations with AMD pathogenesis. Cellular functions, e.g., apoptotic, tumorigenesis, homologous recombination, angiogenesis, and inflammation are regulated by genes that stimulate the cardinal features of AMD abnormalities. Question marks symbolize the expected role of factors in the processes involved in AMD's development (modified from the work of Anand et al. (2016) [95]).

Human hepatic lipase (HL) plays an important role in lipid metabolism. This enzyme converts intermediate-density lipoproteins to low-density lipoproteins and is encoded by the *LIPC* gene. Polymorphism in this gene has shown strong evidence for its relationship with the pathogenesis of AMD [95,96]. Anand et al. (2016) [95] conducted studies and found the protective effect of the *TT* genotype in *LIPC*. Their work further revealed that SNP variation exists in the promoter region (*rs1046817*). This change influences *LIPC* expression, showing a strong association with AMD pathogenesis. The same study showed *LIPC* variance *rs493258* and its connection with the development of AMD [95]. A meta-analysis conducted in 2015 showed the protective role of the *rs10468017* polymorphism in the *LIPC* gene. This mutation elevates the level of HDL, which protects against oxidative stress. Another polymorphism in this gene engaged in a lipid metabolism pathway associated with the development of AMD was confirmed in gene *CETP* [96,97]. This gene encodes cholesteryl ester transfer protein (CETP), which regulates the concentration of cholesteryl esters in HDL molecules [97]. It was confirmed in a meta-analysis that *rs3764261* in *CETP* was associated with an increased risk of developing AMD. In theory, CETP shuttles triglyceride particles from very-low-density lipoproteins (VLDL) and LDL to HDL and creates triglyceride-enriched HDL. In effect, excessive HDL can lead to the accumulation of oxidized lipids in the retina, which is related to the development of AMD [96].

The age-related maculopathy susceptibility 2 (*ARMS2*) gene has been studied widely and replicated in several ethnic groups worldwide (Figure 4). *ARMS2* is predominantly located in high-energy-demanding tissues, like human photoreceptor cells. SNP variation in *ARMS2* genes (*LOC387715*) could reduce the stability of ARMS2 mRNA (Figure 4) [95]. HTRA1 are serine proteases composed of four protein domains for the binding of the insulin-like growth factor binding domain (ILGF), kazal domain, trypsin-like peptidase domain, and PDZ domain to accomplish cellular functions. Moreover, polymorphisms in the *ARMS2/HTRA1* (*rs10490924* and *rs11200638*) locus are significantly associated with early and late AMD; however, instead of this locus, complement factor components (C2, CFH) and complement factor B (CFB) have not shown a positive correlation with AMD abnormalities [95] (Figure 4).

Multi-domain protein DICER1 is an important enzyme in the synthesis of short interfering RNAs (siRNAs∼21–25 nucleotides) from pre-double-stranded RNAs during RNA interference. DICER1 is the miRNA-processing enzyme required for the maturation of miRNAs. If there is any defect in DICER1, then most of these RNAs cannot be generated [107]. About 30% of human genes are regulated by micro-RNA classes of small RNA [95]. However, DICER1 plays an important function in visual activities by degrading toxic RNA. The accumulation of transcripts of Alu RNA in geographic AMD was caused due to dysregulation in DICER1. It has been reported that the accumulation of transcripts of Alu RNA in the geographic AMD was caused by the dysregulation of DICER1. DICER1 was found to be reduced in the advanced form of AMD, i.e., geographic atrophy in AMD patients [105].

The first study on the GST polymorphism in AMD patients was conducted in 2006; the research team analyzed DNA from the blood of patients to determine changes in this gene. The results suggested that the combinations of *GSTM1* (null) and *GSTP1* (mutant), *GSTM1* (null) and *GSTT1* (null) can be risk factors for the development of wet AMD [99]. Further studies on a larger number of individuals have shown that polymorphisms in this gene may correlate with the development of AMD. Studies confirmed that the *GSTM1* (null) polymorphism correlates with both dry and wet AMD stages [100]. In 2012, a Chinese research team analyzed the role of polymorphisms in the *GSTP1* gene (*rs1695* and *rs1138272*) in the development of wet AMD stage. The results showed that only *rs1695* moderately increases the risk of developing this disease [104]. In the same year, a GST polymorphism study was conducted in Iran [100]. The results showed that *GSTM1* and *GSTT1* correlate with a higher risk of the progression of AMD. The *GSTM1* polymorphism is associated with a decline in the expression of the protein, which is a risk factor. If the concentration or activity of this enzyme is lower, ROS cannot be effectively neutralized, affecting the development of AMD [98]. Unfortunately, cohort studies conducted in 2012 did not confirm the role of polymorphisms in genes *GSTM1*, *GSTT1*, and *GSTP1* in the development of AMD [108]. Studies conducted in 2016 showed that the control group was characterized by higher copy numbers of *GSTM1* than AMD patients. Moreover, it was the first report to show a lack of GSTM1 (null) (0/0) or polymorphisms of *GSTM5* (+/0, 0/0) in AMD. The authors suggested that weaker antioxidant functions could intensify oxidative stress and lead to the development of AMD [101].

Unfortunately, a meta-analysis conducted in 2021 showed no significant associations between *GSTM1/GSTT1* polymorphisms and the development of AMD in a Caucasian population [102]. Additionally, a meta-analysis conducted in 2022 showed no significant role of polymorphisms of *GSTM1* (*rs1183423000*), *GSTT1* (*rs1601993659*), and *GSTP1* (*rs1695*) in the development of AMD [103]. Only one study on the polymorphisms in gene *GPx* confirmed the influence of mutations in this gene on the development of AMD. A study on a group of Polish patients showed the polymorphism of *Pro197Leu* in the GPx1 isoform. This mutation is probably related to a weaker ability to scavenge ROS (especially hydrogen peroxide) in the retina, which may lead to the development of AMD [81].

7. Summary and Conclusions

Glutathione is the most important antioxidant in the eye. To maintain GSH function, an organism needs enzymes such as GPx, GST, GR, and GS. The proper functioning of enzymes and maintaining the balance between oxidants and antioxidants help avoid the negative effects of oxidative stress in the eye, which may lead to the development of many eye diseases. Numerous studies have found the exact role of glutathione in the development of AMD. Studies conducted on the activity of enzymes have shown that patients are characterized by lower GR activity. GST studies clearly showed an increase in activity in sick people, but for GPx activity, the results are not so clear. The results also suggest higher and lower GPx activity in patients with AMD. The improper functioning of enzymes may result from polymorphisms in their genes. Numerous studies have examined polymorphisms that can be associated with the development of AMD. The analysis of the polymorphisms in the GST genes confirmed that mutations lead to a weaker antioxidant barrier and may be conducive to the development of AMD. Non-specific results concerning the many parameters that make up the glutathione system reveal many unknowns. Thus, it is important to conduct further research to understand the exact mechanisms of the defensive functions of glutathione against oxidative stress in the human eye.

Author Contributions: Conceptualization: S.B., P.K., N.K., H.T., G.M. and A.W.; Original draft preparation: S.B., K.R., K.H.-B., E.S., J.B. and P.K; Writing—review and editing: S.B., J.B. and P.K. All authors have read and agreed to the published version of the manuscript.

Funding: This research received no funding. The authors declare no competing financial interests. The authors declare that this research did not receive any specific funding.

Acknowledgments: We thank Brendan P. Kavanagh (RCSI, Dublin, Ireland) for his help with improving the quality of the written English. We thank Maria Bogdzińska for her help in the laboratory analysis.

Conflicts of Interest: The authors of "The role of glutathione in age-related macular degeneration (AMD)" have read the manuscript and agreed to submit it in its current form for consideration for publication in the MDPI journal *IJMS*. The undersigned authors, with the consent of all authors, hereby assign to MDPI *IJMS*, the copyright in the above article to be transferred, including supplemental tables, illustrations, or other information submitted in all forms and media throughout the world, in all languages and format, effective when and if the article is accepted for publication. The authors also agree to the following terms: 1. The article submitted is not subject to any prior claim or agreement and is not under consideration for publication elsewhere. 2. The article contains no libelous or other unlawful statements and does not contain any materials that violate the proprietary rights of any other person, company, organization, or nation. 3. The article was prepared jointly with other authors, and the authors agree with the authorship sequence. 4. The participants in the study signed an informed consent form. The authors declare no conflicts of interest. All authors of this MS declare that they have neither personal or financial conflicts of interest to disclose (the authors declare no conflicts of any interest). All authors of the manuscript have read and approved the MS of this paper and declare that it has not been published previously nor is it being considered by any other peer-reviewed journal. The authors hereby assign to the journal the copyright in the above identified article to be transferred, including supplemental tables, illustrations or other information submitted in all forms and media throughout the world, in all languages and formats, effective when and if the article is accepted for publication. We confirm that neither the manuscript nor any parts of its content are currently under consideration or published in another journal. The participants gave written informed consent for the publication of this paper in its current form. The authors have no relevant financial or non-financial or other potential conflict of interest and personal relationships with other people or organizations that could inappropriately influence (bias) this work. All authors have no relationships or affiliations that may be perceived by readers to have influenced, or give the appearance of potentially influencing, what is written in the submitted paper. The authors declare that there are no conflicts of interest that could be perceived as prejudicing the impartiality of the research reported. This paper has not been published elsewhere, and it has not been simultaneously submitted for publication elsewhere in any language. All authors have contributed significantly and are in agreement with the content of the manuscript. All authors have approved the manuscript and agree with its submission to MDPI *IJMS*.

References

1. Congdon, N.; O'Colmain, B.; Klaver, C.C.; Klein, R.; Muñoz, B.; Friedman, D.S.; Kempen, J.; Taylor, H.R.; Mitchell, P. Eye Diseases Prevalence Research Group. Causes and prevalence of visual impairment among adults in the United States. *Arch. Ophthalmol.* **2004**, *122*, 477–485. [PubMed]
2. Tokarz, P.; Kaarniranta, K.; Blasiak, J. Role of antioxidant enzymes and small molecular weight antioxidants in the pathogenesis of age-related macular degeneration (AMD). *Biogerontology* **2013**, *14*, 461–482. [CrossRef] [PubMed]
3. Błasiak, J.; Petrovski, G.; Veréb, Z.; Facskó, A.; Kaarniranta, K. Oxidative Stress, Hypoxia, and Autophagy in the Neovascular Processes of Age-Related Macular Degeneration. *Biomed. Res. Int.* **2014**, *2014*, 768026. [CrossRef]

4. Plestina-Borjan, I.; Katusic, D.; Medvidovic-Grubisic, M.; Supe-Domic, D.; Bucan, K.; Tandara, L.; Rogosic, V. Association of age-related macular degeneration with erythrocyte antioxidant enzymes activity and serum total antioxidant status. *Oxid. Med. Cell Longev.* **2015**, *2015*, 804054. [CrossRef]
5. WHO. *World Report on Vision;* WHO: Geneva, Switzerland, 2019; 180p.
6. Zhang, M.; Jiang, N.; Chu, Y.; Postnikova, O.; Varghese, R.; Horvath, A.; Cheema, A.K.; Golestaneh, N. Dysregulated metabolic pathways in age-related macular degeneration. *Sci. Rep.* **2020**, *10*, 2464. [CrossRef]
7. Nayyar, A.; Gindina, S.; Barron, A.; Hu, Y. Do epigenetic changes caused by commensal microbiota contribute to development of ocular disease? A review of evidence. *Hum. Genom.* **2020**, *14*, 1. [CrossRef]
8. Zarbin, M.A. Current concepts in the pathogenesis of age-related macular degeneration. *Arch. Ophthalmol.* **2004**, *122*, 598–614. [CrossRef] [PubMed]
9. Hyttinen, J.; Błasiak, J.; Kaarniranta, K. Non-Coding RNAs Regulating Mitochondrial Functions and the Oxidative Stress Response as Putative Targets against Age-Related Macular Degeneration (AMD). *Int. J. Mol. Sci.* **2023**, *24*, 2636. [CrossRef] [PubMed]
10. Nowak, J.Z. Age-related macular degeneration (AMD): Pathogenesis and therapy. *Pharmacol. Rep.* **2006**, *58*, 353–363.
11. Handa, J.T.; Verzijl, N.; Matsunaga, H.; Aotaki-Keen, A.; Lutty, G.A.; Te Koppele, J.M.; Miyata, T.; Hjelmeland, L.M. Increase in the advanced glycation end product pentosidine in Bruch's membrane with age. *Investig. Ophthalmol. Vis. Sci.* **1999**, *40*, 775–779.
12. Fisher, R.F. The influence of age on some ocular basement membranes. *Eye* **1987**, *1*, 184–189. [CrossRef] [PubMed]
13. Bok, D. Retinal photoreceptor-pigment epithelium interactions. Friedenwald lecture. *Investig. Ophthalmol. Vis. Sci.* **1985**, *26*, 1659–1694.
14. Fenkci, V.; Fenkci, S.; Yilmazer, M.; Serteser, M. Decreased total antioxidant status and increased oxidative stress in women with polycystic ovary syndrome may contribute to the risk of cardiovascular disease. *Fertil. Steril.* **2003**, *80*, 123–127. [CrossRef]
15. Ruan, Y.; Jiang, S.; Gericke, A. Age-Related Macular Degeneration: Role of Oxidative Stress and Blood Vessels. *Int. J. Mol. Sci.* **2021**, *22*, 1296. [CrossRef] [PubMed]
16. Bartosz, G. *Druga Twarz Tlenu. Wolne Rodniki w Przyrodzie;* Wydawnictwo Naukowe PWN: Warsaw, Poland, 2019; 447p.
17. Daruich, A.; Sauvain, J.-J.; Matet, A.; Eperon, S.; Schweizer, C.; Berthet, A.; Danuser, B.; Behar-Cohen, F. Levels of the oxidative stress biomarker malondialdehyde in tears of patients with central serous chorioretinopathy relate to disease activity. *Mol. Vis.* **2020**, *26*, 722–730. [PubMed]
18. Li, Y.; Zhao, T.; Li, J.; Xia, M.; Li, Y.; Wang, X.; Liu, C.; Zheng, T.; Chen, R.; Kan, D.; et al. Oxidative Stress and 4-hydroxy-2-nonenal (4-HNE): Implications in the Pathogenesis and Treatment of Aging-related Diseases. *J. Immunol. Res.* **2022**, *2022*, 2233906. [CrossRef] [PubMed]
19. Shu, D.Y.; Chaudhary, S.; Cho, K.-S.; Lennikov, A.; Miller, W.P.; Thorn, D.C.; Yang, M.; McKay, T.B. Role of Oxidative Stress in Ocular Diseases: A Balancing Act. *Metabolites* **2023**, *13*, 187. [CrossRef] [PubMed]
20. Finkel, T. Signal transduction by reactive oxygen species. *J. Cell Biol.* **2011**, *194*, 7–15. [CrossRef] [PubMed]
21. Shen, J.K.; Dong, A.; Hackett, S.F.; Bell, W.R.; Green, W.R.; Campochiaro, P.A. Oxidative damage in age-related macular degeneration. *Histol. Histopathol.* **2007**, *22*, 1291–1308.
22. Honda, S.; Farboud, B.; Hjelmeland, L.M.; Handa, J.T. Induction of an aging mRNAretinal pigment epithelial cell phenotype by matrix-containing advanced glycation end products in vitro. *Investig. Ophthalmol. Vis. Sci.* **2001**, *42*, 2419–2425.
23. Hollyfield, J.G.; Bonilha, V.L.; Rayborn, M.E.; Yang, X.; Shadrach, K.G.; Lu, L.; Ufret, R.L.; Salomon, R.G.; Perez, V.L. Oxidative damage-induced inflammation initiates age-related macular degeneration. *Nat. Med.* **2008**, *14*, 194–198. [CrossRef] [PubMed]
24. Thurman, J.M.; Renner, B.; Kunchithapautham, K.; Ferreira, V.P.; Pangburn, M.K.; Ablonczy, Z.; Tomlinson, S.; Holers, V.M.; Rohrer, B. Oxidative stress renders retinal pigment epithelial cells susceptible to complement-mediated injury. *J. Biol. Chem.* **2009**, *284*, 16939–16947. [CrossRef] [PubMed]
25. Mitter, S.K.; Song, C.; Qi, X.; Mao, H.; Rao, H.; Akin, D.; Lewin, A.; Grant, M.; Dunn, W., Jr.; Ding, J.; et al. Dysregulated autophagy in the RPE is associated with increased susceptibility to oxidative stress and AMD. *Autophagy* **2014**, *11*, 1989–2005. [CrossRef] [PubMed]
26. Rogers, B.S.; Symons, R.C.; Komeima, K.; Shen, J.; Xiao, W.; Swaim, M.E.; Gong, Y.Y.; Kachi, S.; Campochiaro, P.A. Differential sensitivity of cones to iron-mediated oxidative damage. *Investig. Ophthalmol. Vis. Sci.* **2007**, *48*, 438–445. [CrossRef] [PubMed]
27. Tate, D.J., Jr.; Miceli, M.V.; Newsome, D.A. Phagocytosis and H_2O_2 induce catalase and metallothionein gene expression in human retinal pigment epithelial cells. *Investig. Ophthalmol. Vis. Sci.* **1995**, *36*, 1271–1279.
28. Kennedy, C.J.; Rakoczy, P.E.; Constable, I.J. Lipofuscin of the retinal pigment epithelium: A review. *Eye* **1995**, *9*, 763–771. [CrossRef] [PubMed]
29. Murdaugh, L.S.; Avalle, L.B.; Mandal, S.; Dill, A.E.; Dillon, J.; Simon, J.D.; Gaillard, E.R. Compositional studies of human RPE lipofuscin. *J. Mass Spectrom.* **2010**, *45*, 1139–1147. [CrossRef] [PubMed]
30. Thimmulappa, R.K.; Mai, K.H.; Srisuma, S.; Kensler, T.W.; Yamamoto, M.; Biswal, S. Identification of Nrf2-regulated genes induced by the chemopreventive agent sulforaphane by oligonucleotide microarray. *Cancer Res.* **2002**, *62*, 5196–5203.
31. Nguyen, T.; Sherratt, P.J.; Pickett, C.B. Regulatory mechanisms controlling gene expression mediated by the antioxidant response element. *Ann. Rev. Pharmacol. Toxicol.* **2003**, *43*, 233–260. [CrossRef]
32. Cano, M.; Thimmalappula, R.; Fujihara, M.; Nagai, N.; Sporn, M.; Wang, A.L.; Neufeld, A.H.; Biswal, S.; Handa, J.T. Cigarette smoking, oxidative stress, the anti-oxidant response through Nrf2 signaling, and Age-related Macular Degeneration. *Vis. Res.* **2010**, *50*, 652–664. [CrossRef]

33. Yates, J.R.; Sepp, T.; Matharu, B.K.; Khan, J.C.; Thurlby, D.A.; Shahid, H.; Clayton, D.G.; Hayward, C.; Morgan, J.; Wright, A.F.; et al. Complement C3 variant and the risk of age-related macular degeneration. *N. Engl. J. Med.* **2007**, *357*, 553–561. [CrossRef] [PubMed]
34. Álvarez-Barrios, A.; Álvarez, L.; García, M.; Artime, E.; Pereiro, R.; González-Iglesias, H. Antioxidant Defenses in the Human Eye: A Focus on Metallothioneins. *Antioxidants* **2021**, *10*, 89. [CrossRef] [PubMed]
35. Łukaszewicz-Hussain, A. Rola glutationu i enzymów z nim związanych w procesach antyoksydacyjnych organizmu. *Med. Pr.* **2003**, *54*, 473–479. [PubMed]
36. Bukowska, B. Funkcje glutationu oraz czynniki zmniejszające jego stężenie. *Med. Pr.* **2005**, *56*, 69–80. [PubMed]
37. Bilska, A.; Kryczyk, A.; Włodek, L. Różne oblicza biologicznej roli glutationu. *Postępy Hig. Med. Doświadczalnej* **2007**, *61*, 438–453.
38. Pizzorno, J. Glutathione! *Integr. Med.* **2014**, *13*, 8–12.
39. Lu, Z.G.; May, A.; Dinh, B.; Lin, V.; Su, F.; Tran, C.; Adivikolanu, H.; Ehlen, R.; Che, B.; Wang, Z.H.; et al. The interplay of oxidative stress and ARMS2-HTRA1 genetic risk in neovascular AMD. *Vessel Plus* **2021**, *5*, 4. [CrossRef] [PubMed]
40. Kaludercic, N.; Deshwal, S.; Di Lisa, F. Reactive oxygen species and redox compartmentalization. *Front. Physiol.* **2014**, *5*, 285. [CrossRef] [PubMed]
41. Lushchak, V.I. Glutathione homeostasis and functions: Potential targets for medical interventions. *J. Amino-Acids* **2012**, *2012*, 736837. [CrossRef]
42. Giustarini, D.; Milzani, A.; Dalle-Donne, I.; Rossi, R. How to Increase Cellular Glutathione. *Antioxidants* **2023**, *12*, 1094. [CrossRef]
43. Cai, J.; Chen, Y.; Seth, S.; Furukawa, S.; Compans, R.W.; Jones, D.P. Inhibition of influenza infection by glutathione. *Free. Radic. Biol. Med.* **2003**, *34*, 928–936. [CrossRef] [PubMed]
44. Bartolini, D.; Stabile, A.M.; Bastianelli, S.; Giustarini, D.; Pierucci, S.; Busti, C.; Vacca, C.; Gidari, A.; Francisci, D.; Castronari, R.; et al. SARS-CoV2 infection impairs the metabolism and redox function of cellular glutathione. *Redox Biol.* **2021**, *45*, 102041. [CrossRef]
45. Forman, H.J.; Zhang, H.; Rinna, A. Glutathione: Overview of its protective roles, measurement, and biosynthesis. *Mol. Asp. Med.* **2009**, *30*, 1–12. [CrossRef]
46. Lee, D.H.; Jacobs, D.R., Jr. Hormesis and public health: Can glutathione depletion and mitochondrial dysfunction due to very low-dose chronic exposure to persistent organic pollutants be mitigated? *J. Epidemiol. Commun. Health* **2015**, *69*, 294–300. [CrossRef]
47. Sreekumar, P.G.; Ferrington, D.A.; Kannan, R. Glutathione Metabolism and the Novel Role of Mitochondrial GSH in Retinal Degeneration. *Antioxidants* **2021**, *10*, 661. [CrossRef] [PubMed]
48. McBean, G.J.; Aslan, M.; Griffiths, H.R.; Torrão, R.C. Thiol redox homeostasis in neurodegenerative disease. *Redox Biol.* **2015**, *5*, 186–194. [CrossRef]
49. Marc, R.E.; Jones, B.; Watt, C.; Vazquez-Chona, F.; Vaughan, D.; Organisciak, D. Extreme retinal remodeling triggered by light damage: Implications for age related macular degeneration. *Mol. Vis.* **2008**, *14*, 782–805. [PubMed]
50. Sun, Y.; Zheng, Y.; Wang, C.; Liu, Y. Glutathione depletion induces ferroptosis, autophagy, and premature cell senescence in retinal pigment epithelial cells. *Cell Death Dis.* **2018**, *9*, 753. [CrossRef]
51. Terluk, M.R.; Ebeling, M.C.; Fisher, C.R.; Kapphahn, R.J.; Yuan, C.; Kartha, R.V.; Montezuma, S.R.; Ferrington, D.A. N-Acetyl-L-cysteine Protects Human Retinal Pigment Epithelial Cells from Oxidative Damage: Implications for Age-Related Macular Degeneration. *Oxid. Med. Cell. Longev.* **2019**, *2019*, 5174957. [CrossRef]
52. Biswas, P.; Dellanoce, C.; Vezzoli, A.; Mrakic-Sposta, S.; Malnati, M.; Beretta, A.; Accinni, R. Antioxidant Activity with Increased Endogenous Levels of Vitamin C, E and A Following Dietary Supplementation with a Combination of Glutathione and Resveratrol Precursors. *Nutrients* **2020**, *12*, 3224. [CrossRef]
53. Yildirim, Z.; Ucgun, N.I.; Yildirim, F. The role of oxidative stress and antioxidants in the pathogenesis of age-related macular degeneration. *Clinics* **2011**, *66*, 743–746. [PubMed]
54. Brantley, M.A.; Osborn, M.P.; Sanders, B.J.; Rezaei, K.A.; Lu, P.; Li, C.; Milne, G.L.; Cai, J.; Sternberg, P., Jr. Plasma biomarkers of oxidative stress and genetic variants in age-related macular degeneration. *Am. J. Ophthalmol.* **2012**, *153*, 460–467. [CrossRef] [PubMed]
55. Samiec, P.S.; Drews-Botsch, C.; Flagg, E.W.; Kurtz, J.C.; Sternberg, P., Jr.; Reed, R.L.; Jones, D.P. Glutathione in human plasma: Decline in association with aging, age-related macular degeneration, and diabetes. *Free. Radic. Biol. Med.* **1998**, *24*, 699–704. [CrossRef] [PubMed]
56. Ayalasomayajula, S.P.; Kompella, U.B. Induction of vascular endothelial growth factor by 4-hydroxynonenal and its prevention by glutathione precursors in retinal pigment epithelial cells. *Eur. J. Pharmacol.* **2002**, *449*, 213–220. [CrossRef] [PubMed]
57. Coral, K.; Raman, R.; Rathi, S.; Rajesh, M.; Sulochana, K.N.; Angayarkanni, N.; Paul, P.G.; Ramakrishnan, S. Plasma homocysteine and total thiol content in patients with exudative age-related macular degeneration. *Eye* **2006**, *20*, 203–207. [CrossRef] [PubMed]
58. Ates, O.; Alp, H.H.; Mumcu, U.; Azizi, S.; Cinici, E.; Kiziltunc, A.; Baykal, O. The effect of L-carnitine treatment on levels of malondialdehyde and glutathione in patients with age related macular degeneration. *Eur. J. Med.* **2008**, *40*, 1–5.
59. Javadzadeh, A.; Ghorbanihaghjo, A.; Bahreini, E.; Rashtchizadeh, N.; Argani, H.; Alizadeh, S. Plasma oxidized LDL and thiol-containing molecules in patients with exudative age-related macular degeneration. *Mol. Vis.* **2010**, *16*, 2578–2584. [PubMed]
60. Bharathselvi, M.; Biswas, S.; Raman, R.; Selvi, R.; Coral, K.; Narayanansamy, A.; Ramakrishnan, S.; Sulochana, K.N. Homocysteine & its metabolite homocysteine-thiolactone & deficiency of copper in patients with age related macular degeneration–A pilot study. *Indian J. Med. Res.* **2016**, *143*, 756.

61. Nowak, M.; Swietochowska, E.; Wielkoszyński, T.; Marek, B.; Karpe, J.; Górski, J.; Głogowska-Szelag, J.; Kos-Kudła, B.; Ostrowska, Z. Changes in blood antioxidants and several lipid peroxidation products in women with age-related macular degeneration. *Eur. J. Ophthalmol.* **2003**, *13*, 281–286. [CrossRef]
62. Bertram, K.M.; Baglole, C.J.; Phipps, R.P.; Libby, R.T. Molecular regulation of cigarette smoke induced-oxidative stress in human retinal pigment epithelial cells: Implications for age-related macular degeneration. *Am. J. Physiol.-Cell Physiol.* **2009**, *297*, 1200–1210. [CrossRef]
63. Ferrington, D.A.; Ebeling, M.C.; Kapphahn, R.J.; Terluk, M.R.; Fisher, C.R.; Polanco, J.R.; Roehrich, H.; Leary, M.M.; Geng, Z.; Dutton, J.R.; et al. Altered bioenergetics and enhanced resistance to oxidative stress in human retinal pigment epithelial cells from donors with age-related macular degeneration. *Redox Biol.* **2017**, *13*, 255–265. [CrossRef] [PubMed]
64. Alhasani, R.H.; Almarhoun, M.; Zhou, X.; Reilly, J.; Patterson, S.; Zeng, Z.; Shu, X. Tauroursodeoxycholic Acid Protects Retinal Pigment Epithelial Cells from Oxidative Injury and Endoplasmic Reticulum Stress In Vitro. *Biomedicines* **2020**, *8*, 367. [CrossRef] [PubMed]
65. Kularatne, R.N.; Bulumulla, C.; Catchpole, T.; Takacs, A.; Christie, A.; Stefan, M.C.; Csaky, K.G. Protection of human retinal pigment epithelial cells from oxidative damage using cysteine prodrugs. *Free Radic. Biol. Med.* **2020**, *152*, 386–394. [CrossRef] [PubMed]
66. Zhou, J.; Chen, F.; Yan, A.; Xia, X. Madecassoside protects retinal pigment epithelial cells against hydrogen peroxide-induced oxidative stress and apoptosis through the activation of Nrf2/HO-1 pathway. *Biosci. Rep.* **2020**, *40*, 40. [CrossRef] [PubMed]
67. Tosi, G.M.; Giustarini, D.; Franci, L.; Minetti, A.; Imperatore, F.; Caldi, E.; Fiorenzani, P.; Aloisi, A.M.; Sparatore, A.; Rossi, R.; et al. Superior Properties of N-acetylcysteine Ethyl Ester over N-Acetyl Cysteine to Prevent Retinal Pigment Epithelial Cells Oxidative Damage. *Int. J. Mol. Sci.* **2021**, *22*, 600. [CrossRef] [PubMed]
68. Ribas, V.; García-Ruiz, C.; Fernández-Checa, J.C. Glutathione and mitochondria. *Front. Pharmacol.* **2014**, *5*, 151. [CrossRef] [PubMed]
69. Raimundo, N. Mitochondrial pathology: Stress signals from the energy factory. *Trends Mol. Med.* **2014**, *20*, 282–292. [CrossRef] [PubMed]
70. Lu, S.C. Glutathione synthesis. *Biochim. Biophys. Acta (BBA)-Gen. Subj.* **2013**, *1830*, 3143–3153. [CrossRef]
71. Sun, F.; Zhou, Q.; Pang, X.; Xu, Y.; Rao, Z. Revealing various coupling of electron transfer and proton pumping in mitochondrial respiratory chain. *Curr. Opin. Struct. Biol.* **2013**, *23*, 526–538. [CrossRef]
72. Venditti, P.; Di Stefano, L.; Di Meo, S. Mitochondrial metabolism of reactive oxygen species. *Mitochondrion* **2013**, *13*, 71–82. [CrossRef]
73. Dou, G.; Sreekumar, P.G.; Spee, C.; He, S.; Ryan, S.J.; Kannan, R.; Hinton, D.R. Deficiency of αB crystallin augments ER stress-induced apoptosis by enhancing mitochondrial dysfunction. *Free Radic. Biol. Med.* **2012**, *53*, 1111–1122. [CrossRef] [PubMed]
74. Matsunaga, D.; Sreekumar, P.G.; Ishikawa, K.; Terasaki, H.; Barron, E.; Cohen, P.; Kannan, R.; Hinton, D.R. Humanin Protects RPE Cells from Endoplasmic Reticulum Stress-Induced Apoptosis by Upregulation of Mitochondrial Glutathione. *PLoS ONE* **2016**, *11*, e0165150. [CrossRef]
75. Lu, S.C. Regulation of glutathione synthesis. *Mol. Aspects. Med.* **2009**, *30*, 42–59. [CrossRef]
76. Fujii, J.; Osaki, T.; Soma, Y.; Matsuda, Y. Critical Roles of the Cysteine-Glutathione Axis in the Production of γ-Glutamyl Peptides in the Nervous System. *Int. J. Mol. Sci.* **2023**, *24*, 8044. [CrossRef] [PubMed]
77. Zhang, L.; Wang, X.; Cueto, R.; Effi, C.; Zhang, Y.; Tan, H.; Qin, X.; Ji, Y.; Yang, X.; Wang, H. Biochemical basis and metabolic interplay of redox regulation. *Redox Biol.* **2019**, *26*, 101284. [CrossRef]
78. Egea, G.; Jiménez-Altayó, F.; Campuzano, V. Reactive Oxygen Species and Oxidative Stress in the Pathogenesis and Progression of Genetic Diseases of the Connective Tissue. *Antioxidants* **2020**, *9*, 1013. [CrossRef] [PubMed]
79. Pei, J.; Pan, X.; Wei, G.; Hua, Y. Research progress of glutathione peroxidase family (GPX) in redoxidation. *Front. Pharmacol.* **2023**, *14*, 1147414. [CrossRef]
80. Lubos, E.; Loscalzo, J.; Handy, D.E. Glutathione peroxidase-1 in health and disease: From molecular mechanisms to therapeutic opportunities. *Antiox. Redox Signal.* **2011**, *15*, 1957–1997. [CrossRef]
81. Mrowicka, M.; Mrowicki, J.; Szaflik, J.P.; Szaflik, M.; Ulinska, M.; Szaflik, J.; Majsterek, I. Analysis of antioxidative factors related to AMD risk development in the polish patients. *Acta Ophthalmol.* **2017**, *95*, 530–536. [CrossRef]
82. Roggia, M.F.; Imai, H.; Shiraya, T.; Noda, Y.; Ueta, T. Protective role of glutathione peroxidase 4 in laser-induced choroidal neovascularization in mice. *PLoS ONE* **2014**, *9*, e98864. [CrossRef]
83. Cohen, S.M.; Olin, K.L.; Feuer, W.J.; Hjelmeland, L.; Keen, C.L.; Morse, L.S. Low glutathione reductase and peroxidase activity in age-related macular degeneration. *Br. J. Ophthalmol.* **1994**, *78*, 791–794. [CrossRef] [PubMed]
84. Zafrilla, P.; Losada, M.; Perez, A.; Caravaca, G.; Mulero, J. Biomarkers of oxidative stress in patients with wet age related macular degeneration. *J. Nutr. Health Aging* **2013**, *17*, 219–222. [CrossRef]
85. Ulańczyk, Z.; Grabowicz, A.; Cecerska-Heryć, E.; Śleboda-Taront, D.; Krytkowska, E.; Mozolewska-Piotrowska, K.; Safranow, K.; Kawa, M.P.; Dołęgowska, B.; Machalińska, A. Dietary and Lifestyle Factors Modulate the Activity of the Endogenous Antioxidant System in Patients with Age-Related Macular Degeneration: Correlations with Disease Severity. *Antioxidants* **2020**, *10*, 954. [CrossRef] [PubMed]

86. Michalska-Małecka, K.; Kabiesz, A.; Nowak, M.; Śpiewak, D. Age related macular degeneration—Challenge for future: Pathogenesis and new perspectives for the treatment. *Eur. Geriatr. Med.* **2015**, *6*, 69–75. [CrossRef]
87. Nasr, A.S.; Sami, R.M.; Ibrahim, N.Y.; Darwish, D.O. Glutathione S transferase (GSTP 1, GSTM 1, and GSTT 1) gene polymorphisms in Egyptian patients with acute myeloid leukemia. *Indian J. Cancer* **2015**, *52*, 490–495. [CrossRef] [PubMed]
88. Hashibe, M.; Brennan, P.; Strange, R.C.; Bhisey, R.; Cascorbi, I.; Lazarus, P.; Boffetta, P. Metaand pooled analyses of GSTM1, GSTT1, GSTP1, and CYP1A1 genotypes and risk of head and neck cancer. *Cancer Epidemiol. Biomark. Prev.* **2003**, *12*, 1509–1517.
89. Kwon, D.D.; Lee, J.W.; Han, D.J. Relationship between the Glutathione-S Transferase P1, M1, and T1 Genotypes and Prostate Cancer Risk in Korean Subjects. *Korean J. Urol.* **2011**, *52*, 247–252. [CrossRef] [PubMed]
90. Arockiaraj, J.; Gnanam, A.J.; Palanisamy, R.; Bhatt, P.; Kumaresan, V.; Chaurasia, M.K.; Pasupuleti, M.; Ramaswamy, H.; Arasu, A.; Sathyamoorthi, A. A cytosolic glutathione s-transferase, GST-theta from freshwater prawn Macrobrachium rosenbergii: Molecular and biochemical properties. *Gene* **2014**, *546*, 437–442. [CrossRef] [PubMed]
91. Li, B.X.; Vachali, P.; Bernstein, P.S. Human ocular carotenoid-binding proteins. *Photochem. Photobiol. Sci.* **2010**, *9*, 1418–1425. [CrossRef]
92. Mrowicka, M.; Mrowicki, J.; Kucharska, E.; Majsterek, I. Lutein and Zeaxanthin and Their Roles in Age-Related Macular Degeneration-Neurodegenerative Disease. *Nutrients* **2022**, *14*, 827. [CrossRef]
93. Cascella, R.; Ragazzo, M.; Strafella, C.; Missiroli, F.; Borgiani, P.; Angelucci, F.; Marsella, L.T.; Cusumano, A.; Novelli, G.; Ricci, F.; et al. Age-related macular degeneration: Insights into inflammatory genes. *J. Ophthalmol.* **2014**, *2014*, 582842. [CrossRef] [PubMed]
94. Kato, Y.; Oguchi, Y.; Omori, T.; Kasai, A.; Ogasawara, M.; Sugano, Y.; Itagaki, K.; Ojima, A.; Ishida, Y.; Machida, T.; et al. Age-Related Maculopathy Susceptibility 2 and Complement Factor H Polymorphism and Intraocular Complement Activation in Neovascular Age-Related Macular Degeneration. *Ophthalmol. Sci.* **2022**, *2*, 100167. [CrossRef] [PubMed]
95. Anand, A.; Sharma, K.; Sharma, S.K.; Singh, R.; Sharma, N.k.; Prasad, K. AMD genetics in India: The missing links. *Front. Aging Neurosci.* **2016**, *8*, 115. [CrossRef] [PubMed]
96. Wang, Y.F.; Han, Y.; Zhang, R.; Qin, L.; Wang, M.X.; Ma, L. CETP/LPL/LIPC gene polymorphisms and susceptibility to age-related macular degeneration. *Sci. Rep.* **2015**, *27*, 15711. [CrossRef]
97. Liutkeviciene, R.; Vilkeviciute, A.; Streleckiene, G.; Kriauciuniene, L.; Chaleckis, R.; Deltuva, V.P. Associations of cholesteryl ester transfer protein (CETP) gene variants with predisposition to age-related macular degeneration. *Gene* **2017**, *636*, 30–35. [CrossRef] [PubMed]
98. Othman, H.; Gholampour, A.R.; Saadat, I.; Farvardin-Jahromoi, M.; Saadat, M. Age-related macular degeneration and genetic polymorphisms of glutathione S-transferases M1 (GSTM1) and T1 (GSTT1). *Mol. Biol. Rep.* **2012**, *39*, 3299–3303. [CrossRef]
99. Oz, O.; Aras Ates, N.; Tamer, L.; Yildirim, O.; Adigüzel, U. Glutathione S-transferase M1, T1, and P1 gene polymorphism in exudative age-related macular degeneration: A preliminary report. *Eur. J. Ophthalmol.* **2006**, *16*, 105–110. [CrossRef] [PubMed]
100. Güven, M.; Görgün, E.; Ünal, M.; Yenerel, M.; Batar, B.; Küçümen, B.; Dinç, U.A.; Güven, G.S.; Ulus, T.; Yüksel, A. Glutathione S-transferase M1, GSTT1 and GSTP1 genetic polymorphisms and the risk of age-related macular degeneration. *Ophthalmic Res.* **2011**, *46*, 31–37. [CrossRef]
101. Hunter, A.A.; Smit-McBride, Z.; Anderson, R.; Bordbari, M.H.; Ying, G.S.; Kim, E.S.; Park, S.S.; Telander, D.G.; Dunaief, J.L.; Hjelmeland, L.M.; et al. GSTM1 and GSTM5 Genetic polymorphisms and Expression in Age-Related Macular Degeneration. *Curr. Eye Res.* **2016**, *41*, 410–416.
102. Saadat, M. Evaluation of associations of GSTM1/GSTT1 null genotypes with the susceptibility to age-related macular degeneration: A meta-analysis. *Scr. Medica* **2021**, *52*, 38–41. [CrossRef]
103. Chen, X.; Luo, Y. Association of GSTM1, GSTT1, and GSTP1 Ile105Val polymorphisms with risk of age-related macular degeneration: A meta-analysis. *Ophthalmic Genet.* **2022**, *43*, 615–621. [CrossRef] [PubMed]
104. Gu, H.; Sun, E.; Cui, L.; Yang, X.; Lim, A.; Xu, J.; Snellingen, T.; Liu, X.; Wang, N.; Liu, N. Association of glutathione S-transferase pi isoform single-nucleotide polymorphisms with exudative age-related macular degeneration in a Chinese population. *Retina* **2012**, *32*, 1967–1972. [CrossRef] [PubMed]
105. Deng, Y.; Qiao, L.; Du, M.; Qu, C.; Wan, L.; Li, J.; Huang, L. Age-related macular degeneration: Epidemiology, genetics, pathophysiology, diagnosis, and targeted therapy. *Genes Dis.* **2021**, *9*, 62–79. [CrossRef] [PubMed]
106. Adams, M.K.; Simpson, J.A.; Richardson, A.J.; English, D.R.; Aung, K.Z.; Makeyeva, G.A.; Guymer, R.H.; Giles, G.G.; Hopper, J.; Robman, L.D.; et al. Apolipoprotein E gene associations in age-related macular degeneration: The Melbourne Collaborative Cohort Study. *Am. J. Epidemiol.* **2012**, *175*, 511–518. [CrossRef] [PubMed]
107. Kaarniranta, K.; Pawlowska, E.; Szczepanska, J.; Blasiak, J. DICER1 in the Pathogenesis of Age-related Macular Degeneration (AMD)—Alu RNA Accumulation versus miRNA Dysregulation. *Aging Dis.* **2020**, *11*, 851–862. [CrossRef]
108. Lechanteur, Y.T.; Peters, W.H.; te Morsche, R.H.; Fauser, S.; Den Hollander, A.I.; Hoyng, C.C. No Association between Genotypes in Glutathione S-transferase and Age-related Macular Degeneration. *Investig. Ophthalmol. Vis. Sci.* **2012**, *53*, 3307.

MDPI AG
Grosspeteranlage 5
4052 Basel
Switzerland
Tel.: +41 61 683 77 34

International Journal of Molecular Sciences Editorial Office
E-mail: ijms@mdpi.com
www.mdpi.com/journal/ijms

www.ingramcontent.com/pod-product-compliance
Lightning Source LLC
Chambersburg PA
CBHW061811210926
51850CB00006B/670
* 9 7 8 3 7 2 5 8 4 9 2 1 5 *